Communications and Control Engineering

Communications and Control Engineering is a high-level academic monograph series publishing research in control and systems theory, control engineering and communications. It has worldwide distribution to engineers, researchers, educators (several of the titles in this series find use as advanced textbooks although that is not their primary purpose), and libraries.

The series reflects the major technological and mathematical advances that have a great impact in the fields of communication and control. The range of areas to which control and systems theory is applied is broadening rapidly with particular growth being noticeable in the fields of finance and biologically inspired control. Books in this series generally pull together many related research threads in more mature areas of the subject than the highly specialised volumes of *Lecture Notes in Control and Information Sciences*. This series's mathematical and control-theoretic emphasis is complemented by *Advances in Industrial Control* which provides a much more applied, engineering-oriented outlook.

Indexed by SCOPUS and Engineering Index.

Publishing Ethics: Researchers should conduct their research from research proposal to publication in line with best practices and codes of conduct of relevant professional bodies and/or national and international regulatory bodies. For more details on individual ethics matters please see:
https://www.springer.com/gp/authors-editors/journal-author/journal-author-helpdesk/publishing-ethics/14214

Andrii Mironchenko

Input-to-State Stability

Theory and Applications

Springer

Andrii Mironchenko
Faculty of Computer Science
and Mathematics
University of Passau
Passau, Germany

ISSN 0178-5354 ISSN 2197-7119 (electronic)
Communications and Control Engineering
ISBN 978-3-031-14673-2 ISBN 978-3-031-14674-9 (eBook)
https://doi.org/10.1007/978-3-031-14674-9

Mathematics Subject Classification: 93C15, 34H05, 93A15, 93B52, 93B53, 93C10, 93D05, 93D09, 93D30, 93D15

This Springer imprint is published by the registered company Springer Nature Switzerland AG
The registered company address is: Gewerbestrasse 11, 6330 Cham, Switzerland

This book provides a comprehensive introduction to the input-to-state stability theory, encompassing the foundational results in the ISS theory of ODE systems and some of its central applications.

Chapter 1 contains preliminary results needed for the development of ISS theory in subsequent chapters. We consider the existence and uniqueness theory for ODEs with measurable right-hand sides (Carathéodory theory). We derive important results concerning reachability sets of nonlinear control systems and regularity of the flow of nonlinear systems with Lipschitz continuous right-hand side.

Chapter 2 is central to this book. Here we introduce the notion of input-to-state stability and prove powerful characterizations of this property, which show a fundamental role of ISS in the stability theory of nonlinear control systems and provide basic tools for the investigation of ISS. We develop a Lyapunov theory for the ISS property, including Lyapunov characterization of global and local ISS. We characterize ISS in terms of uniform attraction times (uniform asymptotic gain property) and in terms of robust stability as well as prove an ISS superposition theorem, which states that ISS is equal to the combination of global asymptotic stability and so-called asymptotic gain property. We prove two such results: one for systems with mild regularity assumptions and another for systems whose right-hand side is Lipschitz continuous w.r.t. the state. We show that the existence of an ISS Lyapunov function for an ISS system is equivalent to the existence of an exponential ISS Lyapunov function. Using this fact and the fundamental results in the topology of $\mathbb{R}^n$, we show that for every ISS system, there is a change of coordinates that makes the system exponentially ISS.

In Chap. 3 we investigate large-scale interconnections of control systems. The main question in this respect is under which conditions the system, consisting of ISS components, is itself ISS. The powerful small-gain theorems answer this question: if all subsystems are ISS with given ISS gains (characterizing the influence of the systems on one another), then the whole interconnection will be ISS if a certain small-gain condition is satisfied. Moreover, the small-gain methods help to construct an ISS Lyapunov function for the whole interconnection, provided that ISS Lyapunov functions for its subsystems are known.

Among various generalizations and refinements of the ISS notion, the concept of integral input-to-state stability (iISS) plays a significant role. In Chap. 4 we develop the basics of the iISS theory, including the iISS superposition theorems and Lyapunov characterizations of 0-GAS and iISS properties. We compare the properties of ISS and integral ISS systems and analyze the concept of strong integral ISS, which offers a half-way between the strength of input-to-state stability and the generality of integral input-to-state stability. We prove that bilinear systems, which are almost never ISS, are always strongly iISS, if they are globally asymptotically stable for a zero input. We show that cascades of strongly iISS systems are again strongly iISS, and develop small-gain theorems for feedback interconnections of 2 strongly iISS systems.

Having developed the core of ISS theory, we proceed in Chap. 5 to the problem of input-to-state stabilization of nonlinear systems. We start with a basic result on ISS feedback redesign, which makes it possible to "robustify" stabilizing controllers that are not robust w.r.t. actuator disturbances. Next, we introduce ISS control Lyapunov functions and develop the universal formula for input-to-state stabilizing controllers for systems, affine w.r.t. controls and disturbances. We present the backstepping approach, which is one of the most powerful methods for designing nonlinear controllers. Practical implementation of the stabilizing controllers imposes numerous challenging problems: in many cases, it is impossible to measure the whole state of the system; it is not possible to update the controller at each moment of time; the magnitude of all physical inputs (like force, torque, thrust, stroke, etc.) is a priori limited, etc. The ISS framework helps to overcome these challenges systematically. We present how the ISS property of a system helps to develop efficient event-based controllers for nonlinear systems and present methods for design of robust observers for nonlinear systems. At the end of the chapter, we give a short overview of further applications of ISS theory to nonlinear control problems.

Even though we concentrate on the ISS theory of finite-dimensional systems, we emphasize the importance of a more general view of infinite-dimensional ISS theory, which not only allows us to analyze more general system classes, but also provides new perspectives and better understanding of the classical ISS theory for ODEs. To make the transition from the finite-dimensional to the infinite-dimensional case as easy as possible, we use continuous ISS Lyapunov functions from the beginning of the book and extensively use Dini derivatives. Our proof of Lyapunov-based small-gain theorems is based on Dini derivatives as well, which makes it extendable to finite networks of infinite-dimensional systems. We also present ISS superposition theorems for systems with very mild regularity assumptions, which can be transferred to the case of infinite ODE networks and more general infinite-dimensional systems.

Chapter 6 is devoted to infinite ODE systems, which is a subclass of infinite-dimensional systems, which has recently attracted significant attention in part due to the desire to obtain scalable control strategies for finite networks of unknown size. Besides its importance in applications, infinite ODE systems illustrate very nicely many challenges arising in the stability theory of infinite-dimensional systems (nonuniform convergence rates, dependence of the stability properties on the choice of the norm in the state space, impossibility of a naive extension of superposition theorems for regular ODE systems to the infinite-dimensional case, distinction between forward completeness and boundedness of reachability sets, etc.).

In Chap. 7 we provide references to other literature, which a reader may consult to deepen her/his knowledge in particular subjects of ISS theory. We give a brief overview of the ISS theory for other classes of systems, such as systems of partial differential equations (PDEs), hybrid, impulsive, and time-delay systems, with particular stress on the ISS theory of PDEs.

In addition to the main body of the paper, we include 3 Appendices presenting auxiliary results for the development of the ISS theory, which are, however, interesting in their own right.

In Appendix A, we introduce a comparison function formalism, which significantly simplifies definitions of stability notions and their applications. We prove basic facts concerning marginal functions and Dini derivatives of continuous functions. Furthermore, we show various comparison principles that are instrumental for

proving direct ISS and integral ISS Lyapunov theorems but can also be effectively used outside the ISS context.

In Appendix B, we develop stability theory for systems with disturbances. The choice of the topics is highly selective, and its primary objective is to ensure a firm basis for the development of the input-to-state stability theory. We present a characterization of the global asymptotic stability in terms of comparison functions and in terms of uniform attraction times, as well as give a proof of a converse Lyapunov theorem based on the Sontag's lemma on $\mathcal{KL}$-functions. For the important special case of systems with uniformly bounded disturbances, we prove additional results, in particular, the equivalence between global asymptotic stability and uniform global asymptotic stability.

Finally, Appendix C is devoted to the stability theory of nonlinear monotone discrete-time systems. We characterize the uniform global asymptotic stability of such systems and relate this property to various small-gain conditions. If monotone discrete-time systems are induced by operators with additional structure (gain operators, max-preserving operators, homogeneous subadditive operators, linear operators), we can prove stronger and more constructive results. We also study so-called paths of strict decay of continuous monotone operators. We rely on these results in Chap. 3 for the small-gain analysis of interconnected systems.

We assume that a reader has basic knowledge of analysis, Lebesgue integration theory, linear algebra, and the theory of ordinary differential equations. No prior knowledge of dynamical systems, stability theory and/or control theory is required, and all necessary notions will be introduced in the subsequent chapters. This book is intended for an active reader and contains numerous exercises of varying difficulty, which complement and widen the material developed in the monograph. They are an integral part of the book. Dismiss them at your peril.

The author thanks all his coauthors for illuminating discussions and for the exchange of ideas. Special thanks go to Sergey Dashkovskiy for introducing me to the input-to-state stability framework and to Fabian Wirth for illuminating discussions and the long-lasting cooperation in Passau.

I appreciated the thoughtful comments of the reviewers and the support and professionalism of Oliver Jackson from Springer's *Communications and Control Engineering* series. I thank Gökhan Göksu, Iasson Karafyllis, Alexander Kilian, and Fabian Wirth, who have read parts of the manuscript and given me valuable comments. I thank Elena Petukhova for her help with the preparation of the figures.

Last but not least, I thank my daughter Darina, whose desire to study the nature around us always inspired me.

Please send any corrections and comments, no matter how small, to the email: andersmir@gmail.com.

Passau, Germany Andrii Mironchenko

Preface

The concept of input-to-state stability (ISS), introduced by E. Sontag, empowers us to study the stability of control systems

$$\dot{x} = f(x, u)$$

with respect to initial conditions and external inputs. ISS unified the Lyapunov and input–output stability theories and revolutionized our view on stabilization of nonlinear systems, design of robust nonlinear observers, stability of nonlinear interconnected control systems, nonlinear detectability theory, and supervisory adaptive control. This made ISS the dominating stability paradigm in nonlinear control theory, with such diverse applications as robotics, mechatronics, electrical and aerospace engineering, and systems biology, to name a few.

Such an overwhelming success has become possible due to several key properties of ISS systems. The first of them is the equivalence between ISS and the existence of a smooth ISS Lyapunov function, which naturally generalizes the theorems of Massera and Kurzweil for undisturbed systems and relates input-to-state stability to the theory of dissipative systems. The second result is an ISS superposition theorem that characterizes ISS as a combination of the global asymptotic stability in absence of disturbances together with a kind of a global attractivity property for the system with inputs. Finally, the nonlinear small-gain theorem provides a powerful criterion for ISS of a network of systems consisting of ISS components. These results fostered the development of new techniques for the design of nonlinear stabilizing controllers that make the system ISS with respect to disturbances appearing due to actuator and observation errors, hidden dynamics of a system, and external disturbances.

The importance of the ISS concept for nonlinear control theory has led to the development of related notions, refining and/or generalizing ISS in some sense: integral input-to-state stability (iISS), input-to-output stability (IOS), input/output-to-state stability (IOSS), input-to-state dynamical stability (ISDS), incremental ISS, etc. On the other hand, significant research efforts have been devoted to the extension of ISS theory to further classes of systems, such as partial differential equations, time-delay, impulsive, hybrid, and discrete-time systems.

Contents

Abbreviations and Symbols

Abbreviations

0-GAS	Global asymptotic stability at zero (Definition 2.3)
0-UAS	Uniform asymptotic stability at zero (Definition 2.33)
AG	Asymptotic gain (Definition 2.44)
AS	Asymptotic stability (Definition B.14)
ATT	Attractivity (Definition B.14)
BEBS	Bounded energy-bounded state (Definition 4.2)
BECS	Bounded energy-convergent state (Definition 4.2)
BEFBS	Bounded energy-frequently bounded state (property) (Definition 4.19)
BEWCS	Bounded energy-weakly convergent state (Definition 4.19)
BIC	Boundedness-implies-continuation (Definition 1.19)
BRS	Boundedness of reachability sets/Bounded reachability sets (Definition 1.30)
bUAG	Uniform asymptotic gain on bounded sets (Definition 2.44)
bULIM	Uniform limit on bounded sets (Definition 2.47)
CEP	Continuity at the equilibrium point (Definition 6.19)
CIUCS	Convergent input-uniformly convergent state (Definition 2.6)
CLF	Control Lyapunov function (Definition 5.6)
eISS	Exponential input-to-state stability (Definition 2.68)
GAS	Global asymptotic stability (Definition B.14)
GATT	Global attractivity (Definition B.14)
i-IOSS	Incremental input/output-to-state stability (Definition 5.17)
iISS	Integral input-to-state stability (Definition 4.1)
IOS	Input-to-output stability (Definition 2.88)
ISS	Input-to-state stability/Input-to-state stable (Definition 2.2)
i-UIOSS	Incremental uniform input/output-to-state stability (Definition 5.17)
LIM	Limit (property) (Definition 2.47)
LISS	Local input-to-state stability (Definition 2.29)
MAF	Monotone aggregation function (Definition 3.23)

MBI	Monotone bounded invertibility (Definition C.14)
ODE, ODEs	Ordinary differential equation(s)
PDE, PDEs	Partial differential equation(s)
REP	Robust equilibrium point (Definition B.2)
RFC	Robust forward completeness (Definition B.3)
sAG	Strong asymptotic gain (Definition 2.44)
sLIM	Strong limit (property) (Definition 2.47)
UAG	Uniform asymptotic gain (Definition 2.44)
UBEBS	Uniform bounded energy-bounded state (property) (Definition 4.19)
UGAS	Uniform global asymptotic stability (Definition B.20)
UGATT	Uniform global attractivity (Definition B.21)
UGS	Uniform global stability (Definition 2.40)
ULIM	Uniform limit (property) (Definition 2.47)
ULS	Uniform local stability (Definition 2.40)
WRS	Weak robust stability (Definition 2.61)

Symbols

Sets and Numbers

$\mathbb{N}$	Set of natural numbers: $\mathbb{N} := \{1, 2, \ldots\}$
$\mathbb{Z}$	Set of integer numbers
$\mathbb{R}$	Set of real numbers
$\mathbb{Z}_+$	Set of nonnegative integer numbers $= \mathbb{N} \cup \{0\}$
$\mathbb{R}_+$	Set of nonnegative real numbers
$\mathbb{C}$	Set of complex numbers
$\mathrm{Re} z$	Real part of a complex number z
$\mathrm{Im} z$	Imaginary part of a complex number z
S^n	$:= \underbrace{S \times \ldots \times S}_{n \text{ times}}$
x^T	Transposition of a vector $x \in \mathbb{R}^n$
$\mathbb{1}$	$:= (1, 1, \ldots, 1) \in \mathbb{R}^n$
I	Identity matrix
$\lvert\cdot\rvert$	Euclidean norm in the space $\mathbb{R}^s$, $s \in \mathbb{N}$
$\lVert A \rVert$	For a matrix $A \in \mathbb{R}^{n \times n}$ we denote $\lVert A \rVert := \sup_{x \neq 0} \frac{\lvert Ax \rvert}{\lvert x \rvert}$
$\lambda_{\min}(P)$	For a positive definite matrix $P \in \mathbb{R}^{n \times n}$ the minimal eigenvalue of P
$\lambda_{\max}(P)$	For a positive definite matrix $P \in \mathbb{R}^{n \times n}$ the maximal eigenvalue of P

Various Notation

dist (z, Z)	$:= \inf\{\lvert y-z\rvert : y \in Z\}$. The distance between $z \in \mathbb{R}^n$ and $Z \subset \mathbb{R}^n$.
B_r	Open ball of radius r around $0 \in \mathbb{R}^n$, i.e., $\{x \in \mathbb{R}^n : \lvert x\rvert < r\}$
$B_r(Z)$	Open ball of radius r around $Z \subset \mathbb{R}^n$, i.e., $\{x \in \mathbb{R}^n : \text{dist}(x, Z) < r\}$
$B_{r,\,\mathcal{U}}$	Open ball of radius r around 0 in a normed vector space $\mathcal{U}$, i.e., $\{u \in \mathcal{U} : \lVert u\rVert_{\mathcal{U}} < r\}$
int A	(Topological) interior of a set A (in a given topology)
$\bar{A}$	Closure of a set A (in a given topology)
∇f	Gradient of a function $f : \mathbb{R}^n \to \mathbb{R}$
$f \circ g$	Composition of maps f and g: $f \circ g(s) = f(g(s))$
∂G	Boundary of a domain G
μ	Lebesgue measure on $\mathbb{R}$
id	Identity operator
$\overline{\lim}$	$:=$ lim sup (limit superior)
$\underline{\lim}$	$:=$ lim inf (limit inferior)
$t \to a-0$	t converges to a from the left
$t \to a+0$	t converges to a from the right
$t \to -0$	t converges to 0 from the left
$t \to +0$	t converges to 0 from the left
$x \vee y$	The supremum (aka join) of vectors $x,\ y \in \mathbb{R}^n_+$

Sequence Spaces

(x_k)	Shorthand notation for a sequence $(x_k)_{k\in\mathbb{N}}$
e_k	k-th standard basis vector of ℓ^p, for some $p \in [1, +\infty)$
$\lVert x\rVert_{\ell^p}$	$:= \left(\sum_{k=1}^{\infty} \lvert x_k\rvert^p\right)^{1/p}$, for $p \in [1, +\infty)$ and $x = (x_k)$
$\lVert x\rVert_{\ell^\infty}$	$:= \sup_{k\in\mathbb{N}} \lvert x_k\rvert$, for $x = (x_k)$
ℓ^p	Space of the sequences $x = (x_k)$ so that $\lVert x\rVert_{\ell^p} < \infty$

Function Spaces

$C(X, U)$	Linear space of continuous functions from X to U. We associate with this space the map $u \mapsto \lVert u\rVert_{C(X,U)} := \sup_{x\in X} \lVert u(x)\rVert_U$, That may attain $+\infty$ as a value
$PC_b(\mathbb{R}_+, U)$	Space of piecewise continuous and right-continuous functions from $\mathbb{R}_+$ to U with a bounded norm $\lVert u\rVert_{PC_b(\mathbb{R}_+,U)} = \lVert u\rVert_{C(\mathbb{R}_+,U)} < \infty$

$AC(\mathbb{R}_+, U)$ Space of absolutely continuous functions from $\mathbb{R}_+$ to U with $\|u\|_{C(\mathbb{R}_+,U)} < \infty$

$C(X)$ $:= C(X, X)$

$C_0(\mathbb{R})$ $:= \{f \in C(\mathbb{R}) : \forall \varepsilon > 0$ there is a compact set $K_\varepsilon \subset \mathbb{R} : |f(s)| < \varepsilon \ \forall s \in \mathbb{R}\backslash K_\varepsilon\}$

$C^k(\mathbb{R}^n, \mathbb{R}_+)$ (Where $k \in \mathbb{N} \cup \{\infty\}$, $n \in \mathbb{N}$) linear space of k times continuously differentiable functions from $\mathbb{R}^n$ to $\mathbb{R}_+$.

$L(X, U)$ Space of bounded linear operators from X to U

$L(X)$ $:= L(X, X)$

$\|f\|_\infty$ $:= \text{ess sup}_{x \geq 0}|f(x)| = \inf_{D \subset \mathbb{R}_+,\ \mu(D)=0} \sup_{x \in \mathbb{R}_+ \backslash D}|f(x)|$

$L^\infty(\mathbb{R}_+, \mathbb{R}^m)$ Set of Lebesgue measurable functions with $\|f\|_\infty < \infty$

$L^\infty_{loc}(\mathbb{R}_+, \mathbb{R}^m)$ $:= \{u : \mathbb{R}_+ \to \mathbb{R}^m : u_t \in L^\infty(\mathbb{R}_+, \mathbb{R}^m)\ \forall t \in \mathbb{R}_+\}$, where $u_t(s) := \begin{cases} u(s) & \text{, if } s \in [0, t], \\ 0 & \text{, if } s > t. \end{cases}$ The set of Lebesgue measurable and locally essentially bounded functions

$\mathcal{U}$ Space of input functions. For the most part of the book $\mathcal{U} = L^\infty(\mathbb{R}_+, \mathbb{R}^m)$

$\|f\|_{L^p(0,d)}$ $:= \left(\int_0^d |f(x)|^p dx\right)^{\frac{1}{p}}$, $p \in [1, +\infty)$

$L^p(0, d)$ Space of p-th power integrable functions $f : (0, d) \to \mathbb{R}$ with $\|f\|_{L^p(0,d)} < \infty$, $p \in [1, +\infty)$

Comparison Functions

$\mathcal{P}$ $:= \{\gamma \in C(\mathbb{R}_+) : \gamma(0) = 0 \text{ and } \gamma(r) > 0 \text{ for } r > 0\}$

$\mathcal{K}$ $:= \{\gamma \in \mathcal{P} : \gamma \text{ is strictly increasing}\}$

$\mathcal{K}_\infty$ $:= \{\gamma \in \mathcal{K} : \gamma \text{ is unbounded}\}$

$\mathcal{L}$ $:= \left\{\gamma \in C(\mathbb{R}_+) : \gamma \text{ is strictly decreasing with } \lim_{t\to\infty} \gamma(t) = 0\right\}$

$\mathcal{KL}$ $:= \{\beta \in C(\mathbb{R}_+ \times \mathbb{R}_+, \mathbb{R}_+) : \beta(\cdot, t) \in \mathcal{K},\ \forall t \geq 0,\ \beta(r, \cdot) \in \mathcal{L},\ \forall r > 0\}$

Chapter 1
Ordinary Differential Equations with Measurable Inputs

In this chapter, we recall basic concepts and results of the theory of ordinary differential equations (ODEs) with measurable inputs ("Carathéodory theory"). In Sect. 1.2, we study existence, uniqueness, and continuous dependence on initial states of solutions for initial value problems. Afterward, we study the properties of reachability sets and the regularity of the flow map for forward complete systems.

1.1 Ordinary Differential Equations with Inputs

In this book, we deal with systems of ordinary differential equations (ODEs) with external inputs:

$$\dot{x} = f(x, u), \tag{1.1a}$$

$$x(0) = x_0. \tag{1.1b}$$

Here $x(t) \in \mathbb{R}^n$, $u(t) \in \mathbb{R}^m$ and $f : \mathbb{R}^n \times \mathbb{R}^m \to \mathbb{R}^n$ is continuous on $\mathbb{R}^n \times \mathbb{R}^m$, and $x_0 \in \mathbb{R}^n$ is a given initial condition. Furthermore, we assume that an input u belongs to the space $\mathcal{U} := L^\infty(\mathbb{R}_+, \mathbb{R}^m)$ of Lebesgue measurable globally essentially bounded functions $u : \mathbb{R}_+ \to \mathbb{R}^m$ endowed with the essential supremum norm

$$\|u\|_\infty := \operatorname{ess\,sup}_{t\geq 0}|u(t)| = \inf_{D\subset\mathbb{R}_+,\ \mu(D)=0}\ \sup_{t\in\mathbb{R}_+\setminus D} |u(t)|, \tag{1.2}$$

where by $|x| := \sqrt{x_1^2 + \cdots + x_k^2}$ we denote the Euclidean norm of $x \in \mathbb{R}^k$ for any $k \in \mathbb{N}$. We refer to Sect. 1.7 for a precise definition of the L^∞ space.

Since the space of admissible inputs $\mathcal{U}$ contains functions, which are not continuous, the right-hand side $(x, t) \mapsto f(x, u(t))$ of (1.1) becomes after substitution of such inputs a merely Lebesgue measurable with respect to time function, and hence

A. Mironchenko, *Input-to-State Stability*, Communications and Control Engineering,
https://doi.org/10.1007/978-3-031-14674-9_1

the system (1.1) does not possess in general classical continuously differentiable solutions. Therefore we look for solutions with a lesser degree of smoothness.

Let I be an interval in $\mathbb{R}$.

Definition 1.1 A function $g : I \to \mathbb{R}^n$ is called *absolutely continuous*, if for any $\varepsilon > 0$ there exists $\delta > 0$ such that for any finite $r \in \mathbb{N}$ and any pairwise disjoint intervals $(a_k, b_k) \subset I$, $k = 1, \dots, r$ with $\sum_{k=1}^{r} |b_k - a_k| \leq \delta$, it follows that $\sum_{k=1}^{r} |g(b_k) - g(a_k)| \leq \varepsilon$.

The space of all absolutely continuous functions $g : I \to \mathbb{R}^n$ we denote by $AC(I, \mathbb{R}^n)$. By $L^1(I, \mathbb{R}^n)$ we denote the space of Lebesgue integrable functions $v : I \to \mathbb{R}^n$ endowed with the (finite) L^1-norm

$$\|v\|_{L^1(I,\mathbb{R}^n)} := \int_a^b |v(t)|dt. \tag{1.3}$$

Fundamental theorem of calculus can be extended to absolutely continuous functions (see, e.g., [4, Theorem 33.2.6, p. 340]):

Theorem 1.2 (Newton–Leibniz theorem) *Every $g \in AC([a, b], \mathbb{R}^n)$ is differentiable almost everywhere, its derivative g' belongs to $L^1([a, b], \mathbb{R}^n)$ and for any $t \in [a, b]$ it holds that*

$$\int_a^t \frac{dg}{ds}(s)ds = g(t) - g(a). \tag{1.4}$$

Conversely, the indefinite integral of a Lebesgue integrable function is absolutely continuous (see, e.g., [4, Theorem 33.2.5, p. 338]):

Theorem 1.3 *For $f \in L^1([a, b], \mathbb{R}^n)$ the indefinite integral $F : [a, b] \to \mathbb{R}^n$: $F(x) = \int_a^x f(t)dt$ is absolutely continuous on [a,b].*

The differentiability of absolutely continuous functions in a.e. sense motivates us to seek solutions of (1.1) is this class of functions.

Definition 1.4 Let $\tau > 0$ be given. A map $\zeta : [0, \tau) \to \mathbb{R}^n$ is called a *solution* of (1.1) over $[0, \tau)$ corresponding to a given initial condition $x_0 \in \mathbb{R}^n$ and to an input $u \in L^\infty(\mathbb{R}_+, \mathbb{R}^m)$ if ζ is absolutely continuous, $\zeta(0) = x_0$ and the equality (1.1) after substitution of an input u and of ζ instead of x holds for almost every $t \in [0, \tau)$.

A map $\zeta : \mathbb{R}_+ \to \mathbb{R}^n$ is called a solution of (1.1) over $\mathbb{R}_+$ corresponding to a given initial condition $x_0 \in \mathbb{R}^n$ and to an input $u \in L^\infty(\mathbb{R}_+, \mathbb{R}^m)$ if ζ is a solution of (1.1) (for the given data x_0, u) over $[0, \tau)$ for all $\tau > 0$.

Integrating (1.1) from 0 to t and exploiting (1.4), we have

Proposition 1.5 *Any solution of (1.1) satisfies the integral equation*

$$x(t) = x_0 + \int_0^t f\big(x(s), u(s)\big)ds \tag{1.5}$$

on its domain of definition. Conversely, absolutely continuous solutions of (1.5) *solve (1.1) in a.e. sense.*

1.2 Existence and Uniqueness Theory

Having in mind Eq. (1.5), we develop in this section basic solution theory for systems (1.1) with measurable inputs (so-called "*Carathéodory theory*") along the lines of the classical theory of continuously differentiable solutions.

Definition 1.6 We call $f : \mathbb{R}^n \times \mathbb{R}^m \to \mathbb{R}^n$.

(i) *Lipschitz continuous (with respect to the first argument) on bounded subsets* if for all $C > 0$ there is $L(C) > 0$, such that

$$|x| \le C, \;\; |y| \le C, \;\; |v| \le C \;\; \Rightarrow \;\; |f(y, v) - f(x, v)| \le L(C)|y - x|. \tag{1.6}$$

(ii) *Lipschitz continuous (with respect to the first argument) on bounded subsets, uniformly with respect to the second argument* if for all $C > 0$ there is $L(C) > 0$, such that

$$|x| \le C, \;\; |y| \le C, \;\; v \in \mathbb{R}^m \;\; \Rightarrow \;\; |f(y, v) - f(x, v)| \le L(C)|y - x|. \tag{1.7}$$

(iii) *Uniformly globally Lipschitz continuous (with respect to the first argument)* if there is $L > 0$ so that

$$x, y \in \mathbb{R}^n, \;\; v \in \mathbb{R}^m \;\; \Rightarrow \;\; |f(y, v) - f(x, v)| \le L|y - x|. \tag{1.8}$$

We will omit the indication "with respect to the first argument" wherever this is clear from the context. The following example illustrates the differences between various concepts of Lipschitz continuity.

Example 1.7 Let $x(t) \in \mathbb{R}$, $u \in L^\infty(\mathbb{R}_+, \mathbb{R})$, and $t \ge 0$.

(i) $f : (x, u) \mapsto xu$ is Lipschitz continuous on bounded subsets, but not uniformly with respect to the second argument.
(ii) $f : (x, u) \mapsto x^2 + u$ is Lipschitz continuous on bounded subsets, uniformly with respect to the second argument, but not uniformly globally Lipschitz continuous.

(iii) $f : (x, u) \mapsto \arctan x + u$ is uniformly globally Lipschitz continuous.

In the rest of this section, we rely on the following assumption on the nonlinearity f in (1.1), which allows for development of a rich existence and uniqueness theory for the system (1.1).

Assumption 1.8 We suppose throughout this chapter that:

(i) f is Lipschitz continuous on bounded subsets of $\mathbb{R}^n$.
(ii) f is continuous on $\mathbb{R}^n \times \mathbb{R}^m$.

Definition 1.9 Let X be a metric space with a metric ρ. A *fixed point* of a mapping $F : Z \subseteq X \to Z$ is an element $x \in Z$ such that $F(x) = x$.

F is called a *contraction* if there is a *contraction constant* $\theta \in [0, 1)$ such that

$$\rho(F(x), F(y)) \leq \theta\rho(x, y), \quad x, y \in Z. \tag{1.9}$$

We also use the notation $F^n(x) = F(F^{n-1}(x))$, $F^0(x) = x$.

The following result is well known (see, e.g., [4, Theorem 8.8.1, p. 66]):

Theorem 1.10 (Banach fixed point theorem) *Let Z be a nonempty closed subset of a complete metric space X with a metric ρ and let $F : Z \to Z$ be a contraction. Then F has a unique fixed point $\bar{x} \in Z$ and it holds that*

$$\rho(F^n(x), \bar{x}) \leq \frac{\theta^n}{1-\theta}\rho(F(x), x), \quad x \in Z. \tag{1.10}$$

Define the *distance* from $z \in \mathbb{R}^n$ to the set $Z \subset \mathbb{R}^n$ by

$$\text{dist}\,(z, Z) := \inf\{|y - z| : y \in Z\}.$$

Further, define an open ball of radius r around Z by

$$B_r(Z) := \{y \in \mathbb{R}^n : \text{dist}\,(y, Z) < r\}.$$

Slightly abusing the notation, we denote also $B_r(x) := B_r(\{x\})$ and $B_r := B_r(0)$. Finally, for a set $\mathcal{S} \subset \mathbb{R}^m$ define the set of inputs with the essential image in $\mathcal{S}$ as $\mathcal{U}_\mathcal{S}$:

$$\mathcal{U}_\mathcal{S} := \{u \in \mathcal{U} : u(t) \in \mathcal{S}, \text{ for a.e. } t \in \mathbb{R}_+\}.$$

In the underlying pages we will be interested in systems (1.1) possessing unique solutions for every $x_0 \in \mathbb{R}^n$ and every $u \in \mathcal{U}$. A sufficient criterion for that is given by the following theorem.

Theorem 1.11 (Picard–Lindelöf theorem) *Let Assumption 1.8 hold. For any compact set $W \subset \mathbb{R}^n$, any compact set $\mathcal{S} \subset \mathbb{R}^m$ and any $\delta > 0$, there is a time*

$\tau = \tau(W, \mathcal{S}, \delta) > 0$, such that for any $x_0 \in W$, and $u \in \mathcal{U}_S$ there is a unique solution ζ of (1.1) on $[0, \tau]$, and $\zeta([0, \tau]) \subset B_\delta(W)$.

Proof Fix $C > 0$ such that $W \subset B_C$ and $\mathcal{U}_S \subset B_{C,\mathcal{U}}$. Take any $\delta > 0$ and define

$$K = K(C, \delta) := C + \delta + \sup_{v \in \mathbb{R}^m : |v| \le C} |f(0, v)|.$$

As f is continuous by assumption, K is well defined and finite.

Consider for any $t^* > 0$ the set

$$Y_{t^*} := \left\{x \in C([0, t^*], \mathbb{R}^n), \operatorname{dist}(x(t), W) \le \delta \ \ \forall t \in [0, t^*]\right\}, \tag{1.11}$$

endowed with the metric $\rho_{t^*}(x, y) := \sup_{s \in [0,t^*]} |x(s) - y(s)|$, which makes Y_{t^*} a complete metric space, see Exercise 1.2.

Fix arbitrary $x_0 \in W$ and arbitrary $u \in \mathcal{U}_S$. We are going to prove that for small enough t^* the space Y_{t^*} is invariant under the operator Φ_{t^*}, defined for all $x \in Y_{t^*}$ by

$$\Phi_{t^*}(x)(t) = x_0 + \int_0^t f(x(s), u(s))ds, \quad t \in [0, t^*].$$

Indeed, let $x \in Y_{t^*}$. Then $\Phi_{t^*}(x) \in C([0, t^*], \mathbb{R}^n)$, and for any $t \in [0, t^*]$

$$\begin{aligned}\operatorname{dist}(\Phi_{t^*}(x)(t), W) &\le |\Phi_{t^*}(x)(t) - x_0| \le \int_0^t |f(x(s), u(s))|ds \\ &\le \int_0^t |f(0, u(s))| + |f(x(s), u(s)) - f(0, u(s))|ds.\end{aligned}$$

Since $\|u\|_\infty \le C \le K$, and $\sup_{s \in [0,t^*]} |x(s)| \le C + \delta \le K$, and as f is Lipschitz continuous on bounded subsets of $\mathbb{R}^n$ with a certain Lipschitz constant $L(K)$, for almost all $s \in [0, t^*]$ it holds that

$$|f(x(s), u(s)) - f(0, u(s))| \le L(K)|x(s)|.$$

We continue the estimates for any $t \in [0, t^*]$ as

$$\begin{aligned}\operatorname{dist}(\Phi_{t^*}(x)(t), W) &\le \int_0^{t^*} K + L(K)|x(s)|ds \\ &\le t^*\big(K + L(K)K\big).\end{aligned}$$

Choosing $0 < t_1^* < \frac{\delta}{K+L(K)K}$, we see that for all $t^* \in (0, t_1^*]$ and all $x \in Y_{t^*}$ it holds that

$$\text{dist}\,(\Phi_{t^*}(x)(t), W) \leq \delta, \quad t \in [0, t^*].$$

This means that Y_{t^*} is invariant with respect to Φ_{t^*} for all $t^* \in (0, t_1^*]$.

Choose $t_2^* \in (0, \min\{t_1^*, \frac{1}{2L(K)}\})$. Then for any $x, y \in Y_{t^*}$ and any $t^* \leq t_2^*$

$$\begin{aligned}
\rho_{t^*}(\Phi_{t^*}(x), \Phi_{t^*}(y)) &= \sup_{t\in[0,t^*]} |\Phi_{t^*}(x)(t) - \Phi_{t^*}(y)(t)| \\
&= \sup_{t\in[0,t^*]} \left| \int_0^t f(x(s), u(s))ds - \int_0^t f(y(s), u(s))ds \right| \\
&\leq \int_0^{t^*} |f(x(s), u(s)) - f(y(s), u(s))| ds \\
&\leq L(K) \int_0^{t^*} |x(s) - y(s)| ds \\
&\leq L(K) t^* \rho_{t^*}(x, y) \\
&\leq \frac{1}{2} \rho_{t^*}(x, y).
\end{aligned}$$

Overall,

$$\rho_{t_2^*}(\Phi_{t_2^*}(x), \Phi_{t_2^*}(y)) \leq \frac{1}{2} \rho_{t_2^*}(x, y), \quad x, y \in Y_{t_2^*},$$

and according to Theorem 1.10 there exists a unique continuous solution of $x(t) = \Phi_{t_2^*}(x)(t)$ on $[0, t_2^*]$. Absolute continuity of x follows by Theorem 1.3. Hence, x is a solution of (1.1) on $[0, t_2^*]$. Note that t_2^* depends only on W, $\mathcal{S}$, and δ, as required. □

Our next aim is to study the prolongations of solutions and their asymptotic properties.

Definition 1.12 Let $x_1(\cdot)$, $x_2(\cdot)$ be solutions of (1.1) defined on the intervals $[0, \tau_1)$ and $[0, \tau_2)$, respectively, $\tau_1, \tau_2 > 0$. We call x_2 an *extension* of x_1 if $\tau_2 > \tau_1$, and $x_2(t) = x_1(t)$ for all $t \in [0, \tau_1)$.

Lemma 1.13 *Let Assumption 1.8 hold. Take any $x_0 \in \mathbb{R}^n$ and $u \in \mathcal{U}$. Any two solutions of (1.1) coincide on their common domain of existence.*

Proof Let ϕ_1, ϕ_2 be two solutions of (1.1) corresponding to $x_0 \in \mathbb{R}^n$ and $u \in \mathcal{U}$, defined over $[0, t_1)$ and $[0, t_2)$, respectively. Assume that $t_1 \leq t_2$ (the other case is analogous).

By local existence and uniqueness, there is some positive $t_3 \leq t_1$ (we take t_3 to be the maximal of such times), such that ϕ_1 and ϕ_2 coincide on $[0, t_3)$. If $t_3 = t_1$, then the claim is shown. If $t_3 < t_1$, then by continuity $\phi_1(t_3) = \phi_2(t_3)$. Now $\psi_1 : t \mapsto \phi_1(t_3 + t)$ and $\psi_2 : t \mapsto \phi_2(t_3 + t)$ are two solutions for the problem

$$\dot{x}(t) = f(x(t), u(t + t_3)), \quad x(0) = \phi_1(t_3),$$

By local existence and uniqueness, ϕ_1 and ϕ_2 coincide on $[0, t_3 + \varepsilon)$ for some $\varepsilon > 0$, which contradicts to the maximality of t_3. Hence ϕ_1 and ϕ_2 coincide on $[0, t_1)$. □

Definition 1.14 A solution $x(\cdot)$ of (1.1) is called:

(i) *maximal* if there is no solution of (1.1) that extends $x(\cdot)$,
(ii) *global* if $x(\cdot)$ is defined on $\mathbb{R}_+$.

A central property of the system (1.1) is

Definition 1.15 We say that the system (1.1) is *well posed* if for every initial value $x_0 \in \mathbb{R}^n$ and every external input $u \in \mathcal{U}$, a unique maximal solution of (1.1) that we denote by $\phi(\cdot, x_0, u) : [0, t_m(x_0, u)) \to \mathbb{R}^n$ exists, where $0 < t_m(x_0, u) \leq \infty$. We call $t_m(x_0, u)$ the *maximal existence time* of a solution corresponding to (x_0, u).

The map ϕ, defined in Definition 1.15, and describing the evolution of the system (1.1), is called the *flow map*, or just *flow*. The domain of definition of the flow ϕ is

$$D_\phi := \cup_{x_0 \in \mathbb{R}^n,\ u \in \mathcal{U}} [0, t_m(x_0, u)) \times \{(x_0, u)\}.$$

In the following pages we will always deal with the maximal solutions, and an initial condition we usually denote just $x \in \mathbb{R}^n$.

We proceed with:

Theorem 1.16 (Well-posedness) *Let Assumption 1.8 hold. Then (1.1) is well posed.*

Proof Pick any $x_0 \in \mathbb{R}^n$ and $u \in \mathcal{U}$. Let $\mathcal{S}$ be the set of all solutions ξ of (1.1), defined on a corresponding domain of definition $I_\xi = [0, t_\xi)$.

Define $I := \cup_{\xi \in \mathcal{S}} I_\xi = [0, t^*)$ for some $t^* > 0$ that can be either finite or infinite. For any $t \in I$ there is $\xi \in \mathcal{S}$, such that $t \in I_\xi$. Define $\xi_{\max}(t) := \xi(t)$. In view of Lemma 1.13, the map $\xi_{\max}$ does not depend on the choice of ξ above and thus is well defined on I.

Furthermore, for each $t \in (0, t^*)$ there is $\xi \in \mathcal{S}$, such that $\xi_{\max} = \xi$ on $[0, t + \varepsilon) \subset I$, for a certain $\varepsilon > 0$, and thus $\phi(\cdot, x_0, u) := \xi_{\max}$ is a solution of (1.1a) corresponding to x_0, u. Finally, it is a maximal solution by construction. □

The following proposition establishes an important *cocycle property* (aka *semigroup property*) of the flow:

Proposition 1.17 (Cocycle property) *For all $x \in \mathbb{R}^n$, $u \in \mathcal{U}$ and all $t, \tau \geq 0$ with $t + \tau < t_m(x, u)$ we have*

$$\phi(t + \tau, x, u) = \phi(t, \phi(\tau, x, u), u(\tau + \cdot)). \tag{1.12}$$

Proof Pick any $x \in \mathbb{R}^n$, any $u \in \mathcal{U}$, and the corresponding unique maximal solution $\phi(\cdot, x, u)$ of (1.1) on $[0, t_m(x, u))$. Pick any $\tau \in [0, t_m(x, u))$ and let y be the maximal solution of the initial value problem

$$\begin{aligned}\dot{y}(t) &= f(y(t), u(t)), \quad t > \tau,\\ y(\tau) &= \phi(\tau, x, u),\end{aligned}$$

for the same x, u as above. Since the solution of (1.1) subject to given x, u is unique on $[0, t_m(x, u))$, and $y(\cdot)$ intersects with this solution at $t = \tau$, it holds that $y(t) = \phi(t, x, u)$ for $t \in [\tau, t_m(x, u))$.

Denote $z(t) := y(t + \tau)$ for $t \in [0, t_m(x, u) - \tau)$. Then z solves the differential equation

$$\frac{d}{dt}z(t) = \frac{d}{dt}y(t+\tau) = \frac{d}{d(t+\tau)}y(t+\tau) = f(y(t+\tau), u(t+\tau)) = f(z(t), u(t+\tau))$$

subject to initial condition $z(0) = \phi(\tau, x, u)$. Due to the uniqueness of the solution of this problem and by the definition of the flow ϕ we have that $z(t) = \phi(t, \phi(\tau, x, u), u(\tau + \cdot))$ for $t \in [0, t_m(x, u) - \tau)$. □

We proceed with a general asymptotic property of the flow.

Proposition 1.18 *Let Assumption 1.8 hold. Pick any $x \in \mathbb{R}^n$ and any $u \in \mathcal{U}$. If $t_m(x, u)$ is finite, then $|\phi(t, x, u)| \to \infty$ as $t \to t_m(x, u) - 0$ (i.e., t converges to $t_m(x, u)$ from the left).*

Proof Pick any $x \in \mathbb{R}^n$, any $u \in \mathcal{U}$, and consider the corresponding maximal solution $\phi(\cdot, x, u)$, defined on $[0, t_m(x, u))$. Assume that $t_m(x, u) < +\infty$, but at the same time $\underline{\lim}_{t \to t_m(x,u)-0} |\phi(t, x, u)| < \infty$. Then there is a sequence (t_k), such that $t_k \to t_m(x, u)$ as $k \to \infty$ and $\lim_{k\to\infty} |\phi(t_k, x, u)| < \infty$. Then also $\sup_{k\in\mathbb{N}} |\phi(t_k, x, u)| =: C < \infty$.

Let $\tau(C) > 0$ be a uniform existence time for the solutions starting in the ball $\overline{B_C}$ subject to inputs of a magnitude not exceeding $\|u\|_\infty$ that exists and is positive in view of Theorem 1.11. Then the solution of (1.1) starting in $\phi(t_k, x, u)$, corresponding to the input $u(\cdot + t_k)$, exists and is unique on $[0, \tau(C)]$ by Theorem 1.11, and by the cocycle property, we have that $\phi(\cdot, x, u)$ can be prolonged to $[0, t_k + \tau(C))$. Since $t_k \to t_m(x, u)$ as $k \to \infty$, this contradicts to the maximality of the solution corresponding to (x, u).

Hence $\underline{\lim}_{t\to t_m(x,u)-0} |\phi(t, x, u)| = \infty$, which implies the claim. □

Definition 1.19 Consider a well-posed system (1.1). We say that (1.1) satisfies the *boundedness-implies-continuation (BIC) property*, if all uniformly bounded solutions are global, that is, for any $x \in \mathbb{R}^n$ and any $u \in \mathcal{U}$

$$\sup_{t\in[0,t_m(x,u))} |\phi(t, x, u)| < \infty \quad \Rightarrow \quad t_m(x, u) = +\infty.$$

As a corollary of Proposition 1.18, we obtain that any uniformly bounded maximal solution is a global solution.

Proposition 1.20 (Boundedness-implies-continuation (BIC) property) *If Assumption 1.8 holds, then (1.1) satisfies the BIC property.*

Definition 1.21 If $t_m(x, u)$ is finite, it is called *(finite) escape time* for a pair (x, u).

Example 1.22 A simple example of a system possessing a finite escape time is

$$\dot{x} = x^3, \tag{1.13}$$

where $x(t) \in \mathbb{R}$. Straightforward computation shows that the flow of this system is given by $\phi(t, x_0) = \frac{x_0}{\sqrt{1-2tx_0^2}}$, where $\phi(\cdot, 0)$ is defined on $\mathbb{R}_+$, and for each $x_0 \in \mathbb{R}\backslash\{0\}$ $\phi(\cdot, x_0)$ is defined on $[0, \frac{1}{2x_0^2})$. In other words, for each $x_0 \neq 0$ the solution of (1.13) escapes to infinity in time $\frac{1}{2x_0^2}$.

The systems, whose trajectories exist globally, deserve a special name.

Definition 1.23 A system (1.1) is called *forward complete* if for all $x \in \mathbb{R}^n$ and all $u \in \mathcal{U}$ the solution, corresponding to initial condition x and input u exists and is unique on $[0, +\infty)$.

By definition, forward complete systems are necessarily well posed and satisfy BIC property.

Systems that fail to have unique backward solutions may have quite specific behavior, as the next example shows.

Example 1.24 Consider the equation

$$\dot{y}(t) = -y^3(t) - y^{1/3}(t), \tag{1.14}$$

and denote its solution at time $t \in [0, +\infty)$ subject to initial condition $y \in \mathbb{R}$ by $\phi(t, y)$ (which is unique, see, e.g., item (ii) in Exercise 1.3).

We are going to show that there is a finite time t_c such that $\phi(t_c, y) = 0$ for all $y \in \mathbb{R}$ (*"settling time"*).

The solution $\phi(\cdot, y)$ subject to positive/negative y can be bounded from above/below by the solution $\phi_1(\cdot, y)$ of the equation

$$\dot{y}(t) = -y^3(t), \quad t > 0, \tag{1.15}$$

with the same initial condition $y \in \mathbb{R}$ (show this).

The solution ϕ_1 of (1.15) subject to initial condition $y \in \mathbb{R}$ reads as

$$\phi_1(t, y) = \operatorname{sgn}(y)\sqrt{\frac{1}{2t + y^{-2}}}, \quad \operatorname{sgn} x = \begin{cases} 1, & x > 0, \\ 0, & x = 0, \\ -1, & x < 0. \end{cases}$$

In particular, for $t \geq 1$ and any $y \in \mathbb{R}$ we have

$$|\phi(t, y)| \leq |\phi_1(t, y)| \leq \sqrt{\frac{1}{2t + y^{-2}}} \leq \sqrt{\frac{1}{2t}} \leq \frac{1}{\sqrt{2}}.$$

On the other hand, the solution of (1.14) subject to positive/negative y can be bounded from above/below by the solution ϕ_2 of the equation

$$\dot{y}(t) = -y^{1/3}(t), \tag{1.16}$$

with the same initial condition y.

The solution of (1.16) is given by

$$\phi_2(t, y) = \begin{cases} \operatorname{sgn}(y)\Big(y^{2/3} - \frac{2}{3}t\Big)^{3/2}, & \text{if } t \in [0, \frac{3}{2}y^{2/3}], \\ 0, & \text{if } t \geq \frac{3}{2}y^{2/3}. \end{cases}$$

Using the cocycle property and the fact that $|\phi(t, y)| \leq |\phi_2(t, y)|$, we obtain for $t \geq t_c := 1 + \frac{3}{2}\big(\frac{1}{\sqrt{2}}\big)^{2/3} = 1 + \frac{3}{2^{4/3}}$ that

$$\phi(t, y) = \phi(t - 1, \phi(1, y)) = 0.$$

Thus, $\phi(t, y) = 0$ for all $y \in \mathbb{R}$ and all $t \geq t_c$.

Remark 1.25 Example 1.24 shows that the flow map of a forward complete system with forward unique solutions is not a homeomorphism and may map open sets into closed sets (in this case $\phi(t_c, \cdot)$ maps $\mathbb{R}$ into $\{0\}$).

We will need the well-known Grönwall's inequality ([7, Lemma 2.7]):

Proposition 1.26 (Grönwall's inequality) *Let $a > 0$ and let $w_1, w_2, y \in C([0, a], \mathbb{R})$. Let also w_2 be nonnegative and w_1 be nondecreasing. If*

$$y(t) \leq w_1(t) + \int_0^t w_2(s)y(s)ds, \quad t \in [0, a], \tag{1.17}$$

then

$$y(t) \leq w_1(t)e^{\int_0^t w_2(s)ds}, \quad t \in [0, a]. \tag{1.18}$$

For a subset A of a certain normed linear space, by int A we denote the *(topological) interior of* A and by $\overline{A}$ its *closure*.

Let us prove a basic result on the continuous dependence on initial state.

Theorem 1.27 (Continuous dependence) *Let $f : \mathbb{R}^n \times \mathbb{R}^m \to \mathbb{R}^n$ satisfy Assumption 1.8. Take any $x_0 \in \mathbb{R}^n$, $u \in \mathcal{U}$ and any $\tau < t_m(x_0, u)$. Then there are $R > 0$ and $L = L(x_0, R, u, \tau) > 0$, such that $t_m(x, u) \geq \tau$ for all $x \in B_R(x_0)$, and*

$$|\phi(t,x,u)-\phi(t,x_0,u)| \leq |x-x_0|e^{Lt}, \quad x\in B_R(x_0), \quad t\in[0,\tau]. \tag{1.19}$$

Proof Take a compact set $W\subset\mathbb{R}^n$ large enough such that $S:=\phi([0,\tau],x_0,u)\subset \operatorname{int}(W)$.

By Picard–Lindelöf theorem, there is $R_1>0$, and time $t_1\in(0,\tau]$, such that all solutions, starting in $B_{R_1}(S)$, exist on $[0,t_1]$, and $\phi([0,t_1],B_{R_1}(S),u)\subset W$.

As f is Lipschitz continuous in the first argument, take a Lipschitz constant $L=L(W)>0$ of f on the set W. We obtain that for all $x\in B_{R_1}(x_0)$ and all $t\in[0,t_1]$ it holds that

$$\begin{aligned}
&|\phi(t,x,u)-\phi(t,x_0,u)|\\
&\quad=\Big|x+\int_0^t f(\phi(s,x,u),u(s))\mathrm{d}s-x_0-\int_0^t f(\phi(s,x_0,u),u(s))\mathrm{d}s\Big|\\
&\quad\leq|x-x_0|+\int_0^t \big|f(\phi(s,x,u),u(s))-f(\phi(s,x_0,u),u(s))\big|\mathrm{d}s\\
&\quad\leq|x-x_0|+L\int_0^t \big|\phi(s,x,u)-\phi(s,x_0,u)\big|\mathrm{d}s.
\end{aligned}$$

By Grönwall's inequality (Proposition 1.26) we obtain that for all $x\in B_{R_1}(x_0)$ and all $t\in[0,t_1]$ the following holds:

$$|\phi(t,x,u)-\phi(t,x_0,u)| \leq |x-x_0|e^{Lt}. \tag{1.20}$$

Now take $R_2:=R_1e^{-Lt_1}$, such that all solutions, starting in $B_{R_2}(x_0)$, exist on $[0,t_1]$, and using (1.20), we obtain that $\phi([0,t_1],B_{R_2}(x_0),u)\subset B_{R_1}(S)$. Hence, using (1.20), we obtain that for $x\in B_{R_2}(x_0)$ the estimate (1.20) is valid for $t\in[0,\min\{2t_1,\tau\}]$. Recurrently, defining $R_k:=R_{k-1}e^{-Lt_1}$, we obtain the claim of the lemma. □

Remark 1.28 **(Solutions backward in time)** Theorems 1.11, 1.27 provide sufficient conditions for local existence and uniqueness of solutions of (1.1), and for the continuous dependence of solutions on initial data forward in time, i.e., on an interval $[0,\tau)$ for a certain τ.

Now let $x(\cdot)$ be the solution of (1.1). Define $y(t):=x(-t)$. Then $y(0)=x(0)$ and y satisfies the equation

$$\frac{\mathrm{d}y(t)}{\mathrm{d}t}=\frac{\mathrm{d}x(-t)}{\mathrm{d}t}=\frac{\mathrm{d}x(-t)}{d(-t)}\frac{d(-t)}{\mathrm{d}t}=-f(x(-t),u(-t))=-f(y(t),u(-t)).$$

As f is Lipschitz if $-f$ is Lipschitz, the same conditions on f as in Theorems 1.11, 1.27 ensure the existence, uniqueness, and continuous dependence of solutions of (1.1) also backward in time.

On the other hand, for most of this book existence and uniqueness of solutions will be required for the future time only, and backward uniqueness can be sacrificed. Asymptotically stable systems are often forward complete systems and at the same time possess nonunique solutions backwards in time, see, e.g., Exercise 1.3. Consequently, many results in this book will hold if we assume merely the existence and uniqueness of solutions in positive times without further requirements imposed on the nonlinearity f.

1.3 Boundedness of Reachability Sets

Let (1.1) be a forward complete system.

Definition 1.29 A state y is called *reachable from* x at time t, if there exists $u \in \mathcal{U}$ so that $\phi(t, x, u) = y$. We say that y is reachable from the set $C \subset \mathbb{R}^n$ at time t, if there exist $x \in C$ and $u \in \mathcal{U}$ so that $\phi(t, x, u) = y$.

For each subset $\mathcal{S} \subset \mathbb{R}^m$, each $T \geq 0$, and each subset $C \subseteq \mathbb{R}^n$, we denote the sets of the states that can be reached from C by inputs from $\mathcal{U}_\mathcal{S} = \{u \in \mathcal{U} : u(t) \in \mathcal{S},\ \text{for a.e. } t \in \mathbb{R}_+\}$ at time T

$$\mathcal{R}^T_\mathcal{S}(C) := \{\phi(T, x, u) : u \in \mathcal{U}_\mathcal{S},\ x \in C\},$$

at a time not exceeding T:

$$\mathcal{R}^{\leq T}_\mathcal{S}(C) := \{\phi(t, x, u) : 0 \leq t \leq T,\ u \in \mathcal{U}_\mathcal{S},\ x \in C\},$$

and at arbitrary time

$$\mathcal{R}_\mathcal{S}(C) := \bigcup_{T \geq 0} \mathcal{R}^T_\mathcal{S}(C) = \{\phi(t, x, u) : t \geq 0,\ u \in \mathcal{U}_\mathcal{S},\ x \in C\}.$$

For short, we denote $\mathcal{R}^T(C) := \mathcal{R}^T_{\mathbb{R}^m}(C)$, $\mathcal{R}^{\leq T}(C) := \mathcal{R}^{\leq T}_{\mathbb{R}^m}(C)$ and $\mathcal{R}(C) := \mathcal{R}_{\mathbb{R}^m}(C)$.

Definition 1.30 We say that (1.1) has *bounded (finite-time) reachability sets (BRS)*, if (1.1) is forward complete and for each bounded $\mathcal{S} \subset \mathbb{R}^m$, each bounded subset $C \subseteq \mathbb{R}^n$ and each $T \geq 0$ the reachability set $\mathcal{R}^{\leq T}_\mathcal{S}(C)$ is bounded.

We will also call the above property of (1.1) as *boundedness of reachability sets* and abbreviate it again by BRS.

Under Assumption 1.8, Theorem 1.11 ensures that for each bounded $\mathcal{S} \subset \mathbb{R}^m$ and each bounded $C \subseteq \mathbb{R}^n$ there is a *small enough* $T > 0$ so that the reachability set $\mathcal{R}^{\leq T}_\mathcal{S}(C)$ is bounded.

The following important result shows boundedness of $\mathcal{R}^{\leq T}_\mathcal{S}(C)$ for any $T > 0$, provided that (1.1) is forward complete.

Theorem 1.31 *Let Assumption 1.8 hold. If (1.1) is forward complete, then (1.1) has BRS.*

For the proof of Theorem 1.31 several technical lemmas will be needed.

Lemma 1.32 *Let Assumption 1.8 hold, $\mathcal{S} \subset \mathbb{R}^m$ be compact, and let $T > 0$ be any real number. If $\overline{\mathcal{R}_{\mathcal{S}}^{\leq T}(x)}$ is compact for some $x \in \mathbb{R}^n$, then for any $\varepsilon > 0$ there is $r > 0$ such that $\overline{\mathcal{R}_{\mathcal{S}}^{\leq T}(B_r(x))}$ is compact and*

$$\overline{\mathcal{R}_{\mathcal{S}}^{\leq T}(B_r(x))} \subset M_\varepsilon := B_\varepsilon\Big(\mathcal{R}_{\mathcal{S}}^{\leq T}(x)\Big).$$

Proof Let $\mathcal{S} \subset \mathbb{R}^m$ be compact, $x \in \mathbb{R}^n$, and let $T > 0$ be so that $\overline{\mathcal{R}_{\mathcal{S}}^{\leq T}(x)}$ is compact. Take any $\varepsilon > 0$. Then the set M_ε is precompact. Denote

$$C := \max\Big\{\sup\{|y| : y \in M_\varepsilon\},\ \sup\{|u| : u \in \mathcal{S}\}\Big\}.$$

Since f is Lipschitz continuous on bounded balls, there is $L > 0$: for all $y_1, y_2 \in B_C$ and for all $u \in B_{C,\mathbb{R}^m}$ it holds that

$$|f(y_1, u) - f(y_2, u)| \leq L|y_1 - y_2|.$$

Set $r := \varepsilon e^{-LT}$ and pick any $y \in B_r(x) \subset M_\varepsilon$ and any $u \in \mathcal{S}$ (for an illustration see Fig. 1.1).

Define the escape time for y from M_ε by

$$t_e = t_e(y, u) := \inf\{t \geq 0 : \phi(t, y, u) \notin M_\varepsilon\}.$$

Assume that $t_e < T$. Since $\phi([0, t_e], x, u) \subset \overline{M}_\varepsilon$, for any $t \in [0, t_e]$ it holds that

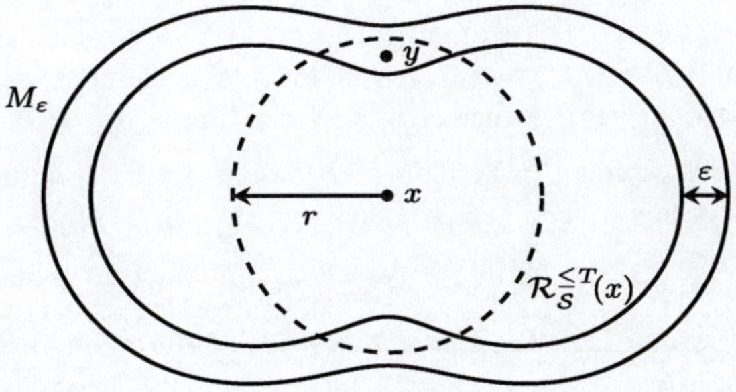

Fig. 1.1 Illustration for the proof of Lemma 1.32.

$$|\phi(t,x,u)-\phi(t,y,u)| = \left|x-y+\int_0^t f(\phi(s,x,u),u)-f(\phi(s,y,u),u)\mathrm{d}s\right|$$
$$\leq |x-y|+\int_0^t \left|f(\phi(s,x,u),u)-f(\phi(s,y,u),u)\right|\mathrm{d}s$$
$$\leq |x-y|+L\int_0^t \left|\phi(s,x,u)-\phi(s,y,u)\right|\mathrm{d}s.$$

By Grönwall's inequality (Proposition 1.26) we proceed for $t \in [0, t_e)$ to

$$|\phi(t,x,u)-\phi(t,y,u)| \leq |x-y|e^{Lt} \leq re^{Lt_e} = \varepsilon e^{L(t_e-T)} < \varepsilon.$$

Since ϕ is continuous with respect to the first argument,

$$|\phi(t_e,x,u)-\phi(t_e,y,u)| \leq \varepsilon e^{L(t_e-T)} < \varepsilon.$$

holds as well. Thus, $\phi(t_e, y, u)$ is in the interior of M_ε, which contradicts to the definition of t_e. Hence $t_e \geq T$ and thus $\phi(t, y, u)$ stays within M_ε for all $t \in [0, T]$.□

Lemma 1.33 *Let two sets $K \subset \mathbb{R}^n$ and $\mathcal{S} \subset \mathbb{R}^m$ be compact and let $T > 0$ be any real number. Then $\overline{\mathcal{R}_\mathcal{S}^{\leq T}(K)}$ is compact if and only if $\overline{\mathcal{R}_\mathcal{S}^{\leq T}(x)}$ is compact for any $x \in K$.*

Proof Let both $K \subset \mathbb{R}^n$ and $\mathcal{S} \subset \mathbb{R}^m$ be compact and let $T > 0$ be any real number.

"⇒". Evidently, compactness of $\overline{\mathcal{R}_\mathcal{S}^{\leq T}(K)}$ implies compactness of $\overline{\mathcal{R}_\mathcal{S}^{\leq T}(x)}$ for any $x \in K$.

"⇐". Assume that $\overline{\mathcal{R}_\mathcal{S}^{\leq T}(x)}$ is compact for any $x \in K$ and at the same time $\overline{\mathcal{R}_\mathcal{S}^{\leq T}(K)}$ is not compact (and hence not bounded). Then there exists a sequence $(z_k) \subset \mathcal{R}_\mathcal{S}^{\leq T}(K)$ so that $|z_k| \to \infty$ as $k \to \infty$. Pick $t_k \in [0, T]$, $x_k \in K$, $u_k \in \mathcal{S}$ so that $z_k = \phi(t_k, x_k, u_k)$, $k \in \mathbb{N}$. Since K is compact, a sequence (x_k) contains a converging subsequence $(x_{k_j})_{j\in\mathbb{N}}$ so that $\lim_{j\to\infty} x_{k_j} =: x^* \in K$.

According to Lemma 1.32 there is an $\varepsilon > 0$ so that $\overline{\mathcal{R}_\mathcal{S}^{\leq T}(B_\varepsilon(x^*))}$ is compact. For this ε there is $N > 0$ such that $x_{k_j} \in B_\varepsilon(x^*)$ for all $j \geq N$. This means that $z_{k_j} \in \overline{\mathcal{R}_\mathcal{S}^{\leq T}(B_\varepsilon(x^*))}$ for all $j \geq N$, which contradicts to unboundedness of $(z_{k_j})_{j\in\mathbb{N}}$. □

Lemma 1.34 *Let Assumption 1.8 hold. Furthermore, let $\mathcal{S} \subset \mathbb{R}^m$ be compact, $W \subset \mathbb{R}^n$ and let $T > 0$ be any real number so that $\overline{\mathcal{R}_\mathcal{S}^{\leq T}(W)}$ is compact. Then there is $\delta > 0$ so that $\overline{\mathcal{R}_\mathcal{S}^{\leq(T+\delta)}(W)}$ is compact.*

Proof As $\overline{\mathcal{R}_\mathcal{S}^{\leq T}(W)}$ and $\mathcal{S}$ are compact, there is $C > 0$ so that $\overline{\mathcal{R}_\mathcal{S}^{\leq T}(W)} \subset B_C$ and $\mathcal{S} \subset B_{C,\mathbb{R}^m}$.

As Assumption 1.8 holds, we can use Theorem 1.11 with the above C to obtain $\tau = \tau(C)$ so that for each $x \in \mathcal{R}_{\mathcal{S}}^{\leq T}(W)$ and each $v \in L^\infty(\mathbb{R}_+, \mathcal{S})$ there is a unique solution $\phi(\cdot, x, v)$ of (1.1), which is defined at the interval $[0, \tau(C)]$, and lies in a ball B_K for a finite K.

As $x = \phi(t, y, u)$ for some $y \in W$, $t \in [0, T]$ and $u \in L^\infty(\mathbb{R}_+, \mathcal{S})$, we apply above considerations with $v(\cdot) := u(\cdot + t)$ and obtain by using the cocycle property that

$$\phi(r, x, v) = \phi(r, \phi(t, y, u), u(\cdot + t)) = \phi(t + r, y, u),$$

for $r \in [0, \tau(C)]$.

Thus, $\phi(s, y, u) \in B_K$ for any $y \in W$, any $u \in L^\infty(\mathbb{R}_+, \mathcal{S})$, and any $s \in [0, T + \tau(C)]$, which shows the claim with $\delta := \tau(C)$. □

Lemma 1.35 *For any compact $\mathcal{S} \subset \mathbb{R}^m$, any $M \subset \mathbb{R}^n$, and any $T > 0$ the following properties hold:*

(i) $\mathcal{R}_{\mathcal{S}}^{T}(\overline{M}) \subset \overline{\mathcal{R}_{\mathcal{S}}^{T}(M)}$.
(ii) $\mathcal{R}_{\mathcal{S}}^{\leq T}(\overline{M}) \subset \overline{\mathcal{R}_{\mathcal{S}}^{\leq T}(M)}$.
(iii) $\overline{\mathcal{R}_{\mathcal{S}}^{\leq T}(\overline{M})} = \overline{\mathcal{R}_{\mathcal{S}}^{\leq T}(M)}$.

Proof (i) Pick any $x \in \overline{M}$, any $u \in \mathcal{S}$ and any $T > 0$. Due to continuity of (1.1) on initial states (Theorem 1.27), for any $\varepsilon > 0$ there is $y \in M$ so that

$$\sup_{t \in [0,T]} |\phi(t, x, u) - \phi(t, y, u)| \leq \varepsilon.$$

Since $\varepsilon > 0$ can be chosen arbitrarily small and T is again arbitrary, this shows (i).

(ii) Follows from (i).

(iii) As $M \subset \overline{M}$, it holds that $\overline{\mathcal{R}_{\mathcal{S}}^{\leq T}(\overline{M})} \supset \overline{\mathcal{R}_{\mathcal{S}}^{\leq T}(M)}$. By (ii) and since $\overline{\mathcal{R}_{\mathcal{S}}^{\leq T}(M)}$ is closed, we have that $\overline{\mathcal{R}_{\mathcal{S}}^{\leq T}(\overline{M})} \subset \overline{\mathcal{R}_{\mathcal{S}}^{\leq T}(M)}$. □

Now we proceed to the proof of Theorem 1.31 (for an illustration see Fig. 1.2).

Proof By Lemma 1.33, it is enough to show that $\overline{\mathcal{R}_{\mathcal{S}}^{\leq T}(x)}$ is compact for any $x \in \mathbb{R}^n$, any compact $\mathcal{S}$ and any finite $T > 0$.

We proceed in 6 steps.

Step 1. Pick any $\xi_0 \in \mathbb{R}^n$, any compact $\mathcal{S}$ and define

$$\tau := \sup\left\{T \geq 0 : \overline{\mathcal{R}_{\mathcal{S}}^{\leq T}(\xi_0)} \text{ is compact}\right\}. \tag{1.21}$$

By Lemma 1.34 applied with $W := \{\xi_0\}$, it holds that $\tau > 0$.

Assume that $\tau < \infty$. If $\overline{\mathcal{R}_{\mathcal{S}}^{\leq \tau}(\xi_0)}$ is compact, then by Lemma 1.34 there is a $\delta > 0$ so that $\overline{\mathcal{R}_{\mathcal{S}}^{\leq(\tau+\delta)}(\xi_0)}$ is compact, which contradicts to the definition of τ. Thus, $\overline{\mathcal{R}_{\mathcal{S}}^{\leq \tau}(\xi_0)}$ is not compact.

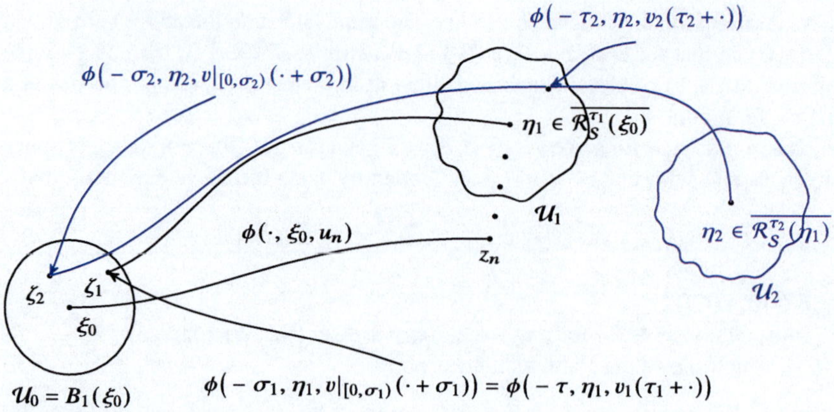

Fig. 1.2 Illustration for the proof of Theorem 1.31.

Define $\tau_1 := \frac{\tau}{2}$. Then $\overline{\mathcal{R}_S^{\tau_1}(\xi_0)}$ is a compact set. Assume that for all $\eta \in \overline{\mathcal{R}_S^{\tau_1}(\xi_0)}$ the set $\overline{\mathcal{R}_S^{\leq(\tau-\tau_1)}(\eta)}$ is compact. By Lemma 1.33 also the set $\overline{\mathcal{R}_S^{\leq(\tau-\tau_1)}(\overline{\mathcal{R}_S^{\tau_1}(\xi_0)})}$ is compact. On the other hand, by cocycle property

$$\mathcal{R}_S^{\leq\tau}(\xi_0) \subset \mathcal{R}_S^{\leq\tau_1}(\xi_0) \cup \mathcal{R}_S^{\leq\tau-\tau_1}(\mathcal{R}_S^{\tau_1}(\xi_0)) \subset \overline{\mathcal{R}_S^{\leq\tau_1}(\xi_0)} \cup \overline{\mathcal{R}_S^{\leq\tau-\tau_1}(\overline{\mathcal{R}_S^{\tau_1}(\xi_0)})}.$$

As the sets on the right-hand side are compact, we obtain that $\mathcal{R}_S^{\leq\tau}(\xi_0)$ is precompact, a contradiction.

Hence, there is $\eta_1 \in \overline{\mathcal{R}_S^{\tau_1}(\xi_0)}$ *such that the set* $\overline{\mathcal{R}_S^{\leq(\tau-\tau_1)}(\eta_1)}$ *is not compact.*

At the same time, $\overline{\mathcal{R}_S^{\leq t}(\eta_1)}$ is compact for all $t \in [0, \tau - \tau_1)$.

As $\eta_1 \in \overline{\mathcal{R}_S^{\tau_1}(\xi_0)}$, there is a sequence $(z_n) \subset \mathcal{R}_S^{\tau_1}(\xi_0)$, such that $z_n \to \eta_1$ as $n \to \infty$. Take $(u_n) \subset L^\infty(\mathbb{R}_+, S)$ such that $z_n = \phi(\tau_1, \xi_0, u_n)$ for all $n \in \mathbb{N}$. Define

$$K_1 := \overline{\mathcal{R}_S^{\leq\tau_1}(\xi_0)}.$$

By backward uniqueness (see Remark 1.28), for each $n \in \mathbb{N}$ it holds that $\phi(s, z_n, u_n(\tau_1 + \cdot)) \in K_1$, for all $s \in [-\tau_1, 0]$. Thus, there is a bounded ball of radius R_1, such that $\phi(s, z_n, u_n(\tau_1 + \cdot)) \in B_{R_1}$ for all $s \in [-\tau_1, 0]$.

Step 2. *Let us show the following claim:*

$$\begin{aligned}&|\phi(-\tau_1, \eta_1, u_n(\tau_1 + \cdot)) - \xi_0|\\ =&|\phi(-\tau_1, \eta_1, u_n(\tau_1 + \cdot)) - \phi(-\tau_1, z_n, u_n(\tau_1 + \cdot))| \to 0, \quad n \to \infty. \qquad (1.22)\end{aligned}$$

First of all, let us modify the right-hand side f outside of $K_1 \times \mathcal{U}_S$ in a way that the modified f satisfies Assumption 1.8, has compact support, and is hence globally bounded. Hence, the solutions of the modified system are complete. This ensures that the solution $\phi_m(\cdot, \eta_1, u)$ of the modified exists and is unique for any $u \in \mathcal{U}_S$ on $[-\tau_1, 0]$.

Using the continuous dependence on initial conditions in the backward direction (Theorem 1.27, Remark 1.28), we obtain that for each $u \in \mathcal{U}_S$, each $\varepsilon > 0$ there is $R > 0$, such that for all $x \in B_R(\eta_1)$ the solution $\phi_m(\cdot, x, u)$ exists on $[-\tau_1, 0]$, and

$$\sup_{s\in[-\tau_1,0]} |\phi_m(s, \eta_1, u) - \phi_m(s, x, u)| \leq \varepsilon.$$

Now for $x \in \mathcal{R}_S^{\tau_1}(\xi_0) \cap B_R(\eta_1)$, we have $\phi_m(s, x, u) = \phi(s, x, u)$, $s \in [-\tau_1, 0]$ as we have modified the right-hand side f only outside of $K_1 \times \mathcal{U}_S$. As for each $\varepsilon > 0$ we can choose suitable $x \in \mathcal{R}_S^{\tau_1}(\xi_0)$, we obtain that $\phi_m([-\tau_1, 0], \eta_1, u) \subset K_1$ for any $u \in \mathcal{U}_S$. In particular, $\phi_m([-\tau_1, 0], \eta_1, u) = \phi([-\tau_1, 0], \eta_1, u)$.

We obtain that

$$\cup_{n\in\mathbb{N}}\Big(\phi([-\tau_1, 0], \eta_1, u_n(\tau_1 + \cdot)) \cup \phi([-\tau_1, 0], z_n, u_n(\tau_1 + \cdot))\Big) \subset K_1,$$

and by the compactness of K_1 and Grönwall's inequality, we obtain (1.22).

Step 3. Hence there is $n_0 \in \mathbb{N}$, such that

$$|\phi(-\tau_1, \eta_1, u_{n_0}(\tau_1 + \cdot)) - \xi_0| < \frac{1}{2}. \tag{1.23}$$

Define

$$v_1 := u_{n_0}, \quad \eta_0 := \phi(-\tau_1, \eta_1, v_1(\tau_1 + \cdot)). \tag{1.24}$$

With this new notation we have that $|\eta_0 - \xi_0| < \frac{1}{2}$.

By continuity on initial conditions for the reversed system, there is a neighborhood $\mathcal{U}_1$ of η_1, such that

$$|\phi(-\tau_1, \xi, v_1(\tau_1 + \cdot)) - \eta_0| < \frac{1}{2} \quad \forall \xi \in \mathcal{U}_1. \tag{1.25}$$

Combining (1.23) and (1.25), and using the notation $\mathcal{U}_0 := B_1(\xi_0)$, we obtain that

$$\phi(-\tau_1, \xi, v_1(\tau_1 + \cdot)) \in \mathcal{U}_0, \quad \xi \in \mathcal{U}_1. \tag{1.26}$$

Step 4. Define

$$\tau_k := 2^{-k}\tau, \quad \sigma_k := \tau_1 + \cdots + \tau_k = \tau(1 - 2^{-k}), \quad k \in \mathbb{N}. \tag{1.27}$$

Let further $\sigma_0 := 0$. Using above argumentation, we obtain recursively for each $k \in \mathbb{N}$ a point $\eta_k \in \overline{\mathcal{R}_S^{\tau_k}}(\eta_{k-1})$, a control v_k, and a neighborhood $\mathcal{U}_k$ contained in $B_1(\eta_k)$, such that

- $\overline{\mathcal{R}_S^{\leq \tau - \sigma_k}(\eta_k)}$ is not compact,
- $\phi(-\tau_k, \xi, v_k(\cdot + \tau_k)) \in \mathcal{U}_{k-1}$, for any $\xi \in \mathcal{U}_k$.

Define v as a concatenation of v_k:

$$v(t) := v_k(t), \quad t \in [\sigma_{k-1}, \sigma_k), \quad k \in \mathbb{N}.$$

Clearly, $v \in \mathcal{U}_S$. For each $k \in \mathbb{N}$ define

$$\zeta_k := \phi(-\sigma_k, \eta_k, v|_{[0,\sigma_k)}(\cdot + \sigma_k)).$$

Inductively, we obtain that

$$\phi(-(\sigma_k - \sigma_i), \eta_k, v|_{[0,\sigma_k)}(\cdot + \sigma_k)) \in \mathcal{U}_{k-i}, \quad i = 0, \ldots, k.$$

This implies that $\zeta_k \in \mathcal{U}_0$, for all $k \in \mathbb{N}$. As $\overline{\mathcal{U}_0}$ is compact, there is a convergent subsequence of (ζ_k), which we denote again by (ζ_k), which converges to certain $\zeta_0 \in \overline{\mathcal{U}_0}$.

Step 5. *Let us show the following claim: for each compact set S, there is k* ≥ 1, *such that* $\eta_k \notin S$.

Assume that this is not the case, and $(\eta_k) \subset S$. By compactness of S, there is time $\delta > 0$, such that

$$\mathcal{R}^{\leq \delta}(\eta) \subset S_1 = \{\xi : d(\xi, S) \leq 1\} \quad \forall \eta \in S.$$

In particular, $\mathcal{R}_S^{\leq \tau - \sigma_k}(\eta_k) \subset S_1$ for all $k \geq 1$ large enough such that $\tau - \sigma_k < \delta$. This contradicts however to noncompactness of $\overline{\mathcal{R}_S^{\leq \tau - \sigma_k}(\eta_k)}$ for all $k \geq 1$.

Step 6. Now assume that $\phi(\tau, \zeta_0, v)$ is well defined. By continuity on initial conditions (Theorem 1.27), this means that also $\phi(\tau, \zeta_k, v)$ is well defined and converges uniformly to $\phi(\tau, \zeta_0, v)$, for k large enough. Hence, there is a compact set W, such that $\phi(t, \zeta_k, v) \in W$ for all $t \in [0, \tau]$ and all $k \in \mathbb{N}$. However,

$$\phi(\sigma_k, \zeta_k, v) = \phi(\sigma_k, \zeta_k, v|_{[0,\sigma_k)}) = \eta_k,$$

contradicting the above claim. Hence, $\phi(\tau, \zeta_0, v)$ is not defined, which contradicts to forward completeness of Σ. □

We call a function $h : \mathbb{R}_+^3 \to \mathbb{R}_+$ *increasing*, if $(r_1, r_2, r_3) \leq (R_1, R_2, R_3)$ implies that $h(r_1, r_2, r_3) \leq h(R_1, R_2, R_3)$, where we use the componentwise partial order on $\mathbb{R}_+^3$. We call h *strictly increasing* if $(r_1, r_2, r_3) \leq (R_1, R_2, R_3)$ and $(r_1, r_2, r_3) \neq (R_1, R_2, R_3)$ imply $h(r_1, r_2, r_3) < h(R_1, R_2, R_3)$.

BRS has a simple restatement in terms of increasing functions, which will be useful in the sequel.

Lemma 1.36 *Consider a forward complete system (1.1) satisfying Assumption 1.8. Then there exists a continuous, increasing function $\mu : \mathbb{R}^3_+ \to \mathbb{R}_+$, such that for all $x \in \mathbb{R}^n$, $u \in \mathcal{U}$ and all $t \geq 0$ we have*

$$|\phi(t, x, u)| \leq \mu(|x|, \|u\|_\infty, t). \tag{1.28}$$

Proof Define $\tilde{\mu} : \mathbb{R}^3_+ \to \mathbb{R}_+$ by

$$\tilde{\mu}(C_1, C_2, \tau) := \sup_{|x| \leq C_1,\ \|u\|_\infty \leq C_2,\ t \in [0,\tau]} |\phi(t, x, u)|, \tag{1.29}$$

which is well defined since reachability sets of (1.1) are bounded, see Theorem 1.31. Clearly, $\tilde{\mu}$ is increasing by definition. In particular, it is locally integrable.

Define $\hat{\mu} : (0, +\infty)^3 \to \mathbb{R}_+$ by setting for $C_1, C_2, \tau > 0$

$$\hat{\mu}(C_1, C_2, \tau) := \frac{1}{C_1 C_2 \tau} \int_{C_1}^{2C_1} \int_{C_2}^{2C_2} \int_{\tau}^{2\tau} \tilde{\mu}(r_1, r_2, s) dr_1 dr_2 ds + C_1 C_2 \tau.$$

By construction, $\hat{\mu}$ is strictly increasing and continuous on $(0, +\infty)^3$.

We can enlarge the domain of definition of $\hat{\mu}$ to all of $\mathbb{R}^3_+$ using monotonicity. To this end we define for $C_2, \tau > 0$: $\hat{\mu}(0, C_2, \tau) := \lim_{C_1 \to +0} \hat{\mu}(C_1, C_2, \tau)$, for $C_1 \geq 0$, $\tau > 0$: $\hat{\mu}(C_1, 0, \tau) := \lim_{C_2 \to +0} \hat{\mu}(C_1, C_2, \tau)$ and for $C_1, C_2 \geq 0$ we define $\hat{\mu}(C_1, C_2, 0) := \lim_{\tau \to +0} \hat{\mu}(C_1, C_2, \tau)$. All these limits are well defined as $\hat{\mu}$ is increasing on $(0, +\infty)^3$ and we obtain that the resulting function is increasing on $\mathbb{R}^3_+$. Note that the construction does not guarantee that $\hat{\mu}$ is continuous. In order to obtain continuity choose a continuous strictly increasing function $\nu : \mathbb{R}_+ \to \mathbb{R}_+$ with $\nu(r) > \max\{\hat{\mu}(0, 0, r), \hat{\mu}(0, r, 0), \hat{\mu}(r, 0, 0)\}$, $r \geq 0$ and define for $(C_1, C_2, \tau) \geq (0, 0, 0)$

$$\mu(C_1, C_2, \tau) := \max\left\{\nu\big(\max\{C_1, C_2, \tau\}\big), \hat{\mu}(C_1, C_2, \tau)\right\} + C_1 C_2 \tau. \tag{1.30}$$

The function μ is continuous as $\mu(C_1, C_2, \tau) = \nu\big(\max\{C_1, C_2, \tau\}\big) + C_1 C_2 \tau$ whenever C_1, C_2 , or τ is small enough. At the same time we have for $C_1 > 0$, $C_2 > 0$, and $\tau > 0$ that

$$\begin{aligned} \mu(C_1, C_2, \tau) \geq \hat{\mu}(C_1, C_2, \tau) &\geq \frac{1}{C_1 C_2 \tau} \int_{C_1}^{2C_1} \int_{C_2}^{2C_2} \int_{\tau}^{2\tau} dr_1 dr_2 ds\, \tilde{\mu}(C_1, C_2, \tau) + C_1 C_2 \tau \\ &\geq \tilde{\mu}(C_1, C_2, \tau). \end{aligned}$$

Consequently, the claim holds with this μ. □

1.4 Regularity of the Flow

Using that forward completeness is equivalent to BRS for ODEs with right-hand sides satisfying Assumption 1.8, we analyze in this section the continuous dependence of the flow map at $(x, u) = (0, 0)$ and on $\mathbb{R}^n \times \mathbb{R}^m$, as well as Lipschitz regularity with respect to the state.

Definition 1.37 We say that (1.1) *depends continuously on inputs and on initial states*, if for all $x \in \mathbb{R}^n$, $u \in \mathcal{U}$, $\tau > 0$, and $\varepsilon > 0$ there exists $\delta = \delta(x, u, \tau, \varepsilon) > 0$, such that for all $x' \in \mathbb{R}^n$ and for all $u' \in \mathcal{U}$ it holds that

$$|x - x'| < \delta, \quad \|u - u'\|_\infty < \delta, \quad t \in [0, \tau] \;\Rightarrow\; |\phi(t, x, u) - \phi(t, x', u')| < \varepsilon.$$

If above property holds just for $x = 0$ and $u \equiv 0$, then we say that (1.1) *depends continuously on inputs and on initial states at* $x = 0$ *and* $u \equiv 0$.

Our assumptions ensure the following result:

Proposition 1.38 *Let Assumption 1.8 hold and* $f(0, 0) = 0$. *Then (1.1) depends continuously on inputs and on initial states at* $x = 0$ *and* $u \equiv 0$.

Proof Consider the following auxiliary system

$$\dot{x}(t) = \tilde{f}(x(t), u(t)) := f(\mathrm{sat}(x(t)), \mathrm{sat}(u(t))), \tag{1.31a}$$

$$x(0) = x_0, \tag{1.31b}$$

where the saturation function is given for the vectors z in $\mathbb{R}^n$ or $\mathbb{R}^m$ by

$$\mathrm{sat}(z) := \begin{cases} z, & |z| \le 1, \\ \frac{z}{|z|}, & \text{otherwise.} \end{cases}$$

As f satisfies Assumption 1.8, one can show that $\tilde{f}$ is uniformly globally Lipschitz continuous. Hence (1.31) is forward complete by Exercise 1.9.

The flow of (1.31) we denote by $\tilde{\phi}(t, x, u)$. As $f(x, u) = \tilde{f}(x, u)$ whenever $|x| \le 1$ and $|u| \le 1$, it holds also

$$\phi(t, x, u) = \tilde{\phi}(t, x, u),$$

if $\|u\|_\infty \le 1$, $\phi(\cdot, x, u)$ exists on $[0, t]$, and $|\phi(s, x, u)| \le 1$ for all $s \in [0, t]$.

Pick any $\tau > 0$, any $\varepsilon \in (0, 1)$, any $\delta \in (0, \varepsilon)$, as well as arbitrary $x \in B_\delta$ and $u \in B_{\delta, \mathcal{U}}$. Then for any $t > 0$

$$
\begin{aligned}
|\tilde{\phi}(t, x, u)| &= \Big|x + \int_0^t \tilde{f}(\tilde{\phi}(s, x, u), u(s))ds\Big| \\
&\leq |x| + \int_0^t \big|\tilde{f}(\tilde{\phi}(s, x, u), u(s))\big|ds \\
&\leq |x| + \int_0^t \big|\tilde{f}(\tilde{\phi}(s, x, u), u(s)) - \tilde{f}(0, u(s))\big|ds + \int_0^t \big|\tilde{f}(0, u(s))\big|ds.
\end{aligned}
$$

Since $\tilde{f}(0, \cdot)$ coincides with $f(0, \cdot)$ on the unit ball around the origin, and in view of uniform global Lipschitz continuity of $\tilde{f}$, there is $L > 0$, such that for all $t \in [0, \tau]$

$$
\begin{aligned}
|\tilde{\phi}(t, x, u)| &\leq |x| + \int_0^t L|\tilde{\phi}(s, x, u)|ds + t \sup_{|v| \leq \delta} |f(0, v)| \\
&\leq |x| + \tau \sup_{|v| \leq \delta} |f(0, v)| + \int_0^t L|\tilde{\phi}(s, x, u)|ds.
\end{aligned}
$$

According to Grönwall's inequality (1.18), we obtain

$$
\begin{aligned}
|\tilde{\phi}(t, x, u)| &\leq \big(|x| + \tau \sup_{|v| \leq \delta} |f(0, v)|\big)e^{Lt} \\
&\leq \big(\delta + \tau \sup_{|v| \leq \delta} |f(0, v)|\big)e^{L\tau}.
\end{aligned}
$$

In view of Assumption 1.8, f is a continuous function of its arguments. As $f(0, 0) = 0$, we can choose δ small enough to ensure that

$$
\big(\delta + \tau \sup_{|v| \leq \delta} |f(0, v)|\big)e^{L\tau} \leq \varepsilon.
$$

As $\varepsilon < 1$, we see that $|\phi(s, x, u)| \leq 1$ on $[0, \tau] \cap [0, t_m(x, u))$. But if $t_m(x, u) < \tau$, then ϕ is uniformly bounded on $[0, t_m(x, u))$, which contradicts to Proposition 1.18 (BIC property). Thus, $\phi(s, x, u) = \tilde{\phi}(s, x, u)$ and $|\phi(s, x, u)| \leq \varepsilon$ for $s \in [0, \tau]$.

This shows that (1.1) depends continuously on inputs and on initial states at $x = 0$ and $u \equiv 0$ with this choice of δ. □

Definition 1.39 We say that a forward complete system (1.1) *depends continuously on initial states*, if for all $x \in \mathbb{R}^n$, $u \in \mathcal{U}$, $\tau > 0$, and $\varepsilon > 0$ there exists $\delta = \delta(x, u, \tau, \varepsilon) > 0$, such that

$$|x - x'| < \delta,\ t \in [0, \tau] \ \Rightarrow\ |\phi(t, x, u) - \phi(t, x', u)| < \varepsilon.$$

The next result shows that the flow of (1.1) is Lipschitz continuous on bounded balls with respect to initial states.

Theorem 1.40 (Lipschitz continuity of the flow) *Let Assumption 1.8 hold, and let (1.1) be forward complete. Then for any $C > 0$ and any $\tau > 0$ there exists a constant $L > 0$, such that*

$$\begin{aligned} &x_1^0, x_2^0 \in B_C,\ u \in B_{C,\mathcal{U}},\ t \in [0, \tau] \\ &\qquad \Rightarrow\ |\phi(t, x_1^0, u) - \phi(t, x_2^0, u)| \le |x_1^0 - x_2^0| e^{Lt}. \end{aligned} \tag{1.32}$$

In particular, (1.1) depends continuously on initial states.

Proof Take any $C > 0$ and pick any $x_1^0, x_2^0 \in B_C$, and any $u \in B_{C,\mathcal{U}}$. Let $x_i(t) = \phi(t, x_i^0, u)$, $i = 1, 2$ be the corresponding solutions.

Due to (1.5), we have for all $t \ge 0$:

$$|x_1(t) - x_2(t)| \le |x_1^0 - x_2^0| + \int_0^t |f(x_1(r), u(r)) - f(x_2(r), u(r))| dr.$$

Using boundedness of reachability sets for the system (1.1) (see Theorem 1.31), we have for any fixed $\tau > 0$

$$K := \sup_{|z| \le C,\ t \in [0,\tau],\ v \in B_{C,\mathcal{U}}} |\phi(t, z, v)| < \infty.$$

Looking at $t = 0$, we see that $K \ge C$, and using Lipschitz continuity of f on bounded balls, i.e., (1.6) with $L(K)$ instead of $L(C)$, we obtain

$$|x_1(t) - x_2(t)| \le |x_1^0 - x_2^0| + \int_0^t L(K) |x_1(r) - x_2(r)| dr.$$

Using Grönwall's inequality we obtain (1.32).

For any $\tau > 0$, any $\varepsilon > 0$ and any $u \in B_{C,\mathcal{U}}$ define $\tilde{\delta}(u, \tau, \varepsilon) := \varepsilon e^{-L(K)\tau}$. Now for any $x \in B_C$, define $\delta(x, u, \tau, \varepsilon) := \min\{C - |x|, \tilde{\delta}\}$. It is easy to see that (1.1) depends continuously on initial states with this choice of δ. □

In order to prove continuity of (1.1) with respect to initial states *and* inputs globally on $\mathbb{R}^n \times \mathcal{U}$, stronger conditions should be imposed on the nonlinearity f.

Theorem 1.41 *Let there exist $q \in \mathcal{K}_\infty$ such that for all $C > 0$ there is $L(C) > 0$: for all $x_1, x_2 \in B_C$ and all $u_1, u_2 \in B_{C,\mathbb{R}^m}$ it holds that*

$$|f(x_1, u_1) - f(x_2, u_2)| \le L(C)\big(|x_1 - x_2| + q(|u_1 - u_2|)\big). \tag{1.33}$$

Then for each $x, y \in \mathbb{R}^n$, each $v, u \in \mathcal{U}$, and each $t^* > 0$ in the common interval of existence of $\phi(\cdot, x, u)$ and $\phi(\cdot, y, v)$ there is a constant $L = L(t^*, x, y, u, v)$ so that

$$t \in [0, t^*] \;\Rightarrow\; |\phi(t, x, u) - \phi(t, y, v)| \le \big(|x - y| + tLq(\|u - v\|_{[0,t]})\big)e^{Lt}. \tag{1.34}$$

Proof The assumptions on f ensure that f satisfies Assumption 1.8. By Theorem 1.11, for each $x, y \in \mathbb{R}^n$ and each $u, v \in \mathcal{U}$ there is $t^* > 0$, such that the corresponding solutions $\phi(\cdot, x, u)$ and $\phi(\cdot, y, v)$ both exist on $[0, t^*]$. Pick $C > 0$ so that $\|u\|_\infty \le C$, $\|v\|_\infty \le C$, $\phi([0, t^*], x, u) \subset B_C$ and $\phi([0, t^*], y, v) \subset B_C$. Exploiting (1.33), we obtain for any $t \in [0, t^*]$ that

$$\begin{aligned}
&|\phi(t, x, u) - \phi(t, y, v)| \\
&\quad= \Big|x + \int_0^t f(\phi(s, x, u), u(s))ds - y - \int_0^t f(\phi(s, x, v), v(s))ds\Big| \\
&\quad\le |x - y| + \int_0^t \big|f(\phi(s, x, u), u(s)) - f(\phi(s, y, v), v(s))\big|ds \\
&\quad\le |x - y| + \int_0^t L(C)\big(|\phi(s, x, u) - \phi(s, y, v)| + q(|u(s) - v(s)|)\big)ds \\
&\quad\le |x - y| + tL(C)q(\|u - v\|_{[0,t]}) + \int_0^t L(C)|\phi(s, x, u) - \phi(s, y, v)|ds.
\end{aligned}$$

By Grönwall's inequality (Proposition 1.26) we obtain (1.34). □

1.5 Uniform Crossing Times

For a given forward complete system (1.1), a point $x \in \mathbb{R}^n$, a subset $\Omega \subseteq \mathbb{R}^n$, and a control u, denote by

$$\tau(x, \Omega, u) := \inf\{t \ge 0 : \ \phi(t, x, u) \in \Omega\}, \tag{1.35}$$

the *first crossing time* of the set Ω by the trajectory $\phi(\cdot, x, u)$, with the convention that $\tau(x, \Omega, u) = +\infty$ if $\phi(t, x, u) \notin \Omega$ for all $t \geq 0$.

Recall the well-known Arzelá–Ascoli precompactness criterion for a family of continuous functions (see, e.g., [4, Theorem 11.4.4, p. 102]):

Theorem 1.42 (Arzelá–Ascoli) *Consider a compact interval* $[a, b] \subset \mathbb{R}$ *and a sequence of continuous functions* (f_k)*:*

$$f_k : [a, b] \to \mathbb{R}^n$$

that is:

(i) Uniformly bounded*: there is a bound S, such that*

$$|f_k(x)| \leq S \quad \forall k \in \mathbb{N},\ x \in [a, b].$$

(ii) Uniformly equicontinuous*: for all* $\varepsilon > 0$ *there is* $\delta > 0$*, such that for all* $x, y \in [a, b]$ *it holds that*

$$|x - y| < \delta \quad \Rightarrow \quad |f_k(x) - f_k(y)| < \varepsilon \quad \forall k \in \mathbb{N}.$$

Then there is a uniformly convergent on $[a, b]$ *subsequence of* (f_k)*.*

The following result shows that if the trajectories of a certain control system starting in $a \in \mathbb{R}^n$ do not cross an open subset $\Omega \subset \mathbb{R}^n$ in a uniform time for all inputs with a given bounded "essential image", then for any compact subset $K \subset \Omega$ there are points arbitrarily close to a that do not cross K.

Theorem 1.43 *Let (1.1) be a forward complete system and let Assumption 1.8 hold. Assume the following are given:*

- *an open subset* Ω *of the state space* $\mathbb{R}^n$*;*
- *a bounded subset* $S \subset \mathbb{R}^m$*;*
- *a point* $a \in \mathbb{R}^n$*.*

such that

$$\sup\{\tau(a, \Omega, u) : u \in \mathcal{U}_S\} = +\infty.$$

Then for any compact $K \subset \Omega$ *and any neighborhood W of the point a there is* $q \in W$ *and* $v \in \mathcal{U}_S$ *such that*

$$\tau(q, K, v) = +\infty.$$

Proof We proceed in four steps (for an illustration see Fig. 1.3).

Step 1. Let $p_0 := a$ be as in the assumptions. Then for each $k \in \mathbb{N}$ we may pick some $u_k \in \mathcal{U}_S$ such that $\tau(p_0, \Omega, u_k) \geq k$, and thus $\phi(t, p_0, u_k) \notin \Omega$ for $t \in [0, k]$.

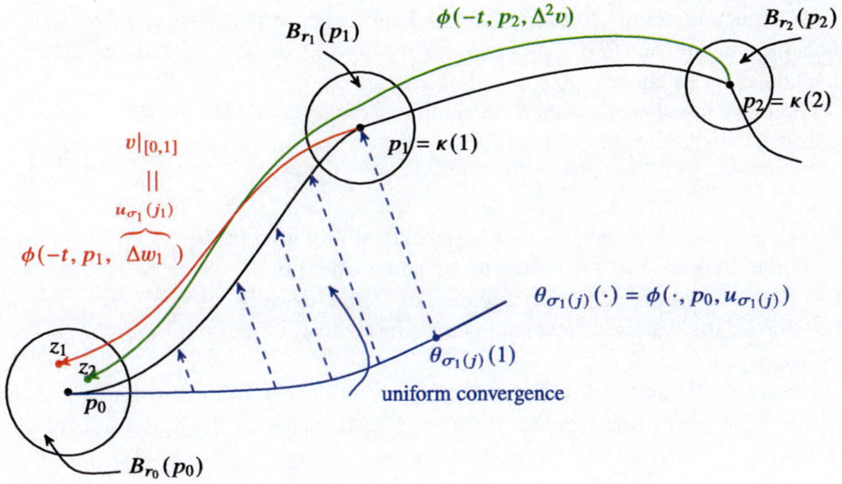

Fig. 1.3 Illustration for the proof of Theorem 1.43.

Define for each $k \geq 1$ the functions θ_k by

$$\theta_k(t) := \phi(t, p_0, u_k), \quad t \geq 0.$$

By Theorem 1.31, the reachability set $S_1 := \overline{\mathcal{R}_{\mathcal{S}}^{\leq 1}(\{p_0\})}$ is compact, which shows that the sequence (θ_k) is uniformly bounded on $[0, 1]$. Furthermore, by continuity of f it holds that

$$M := \sup\{|f(\xi, \lambda)| : \xi \in S_1, \lambda \in \mathcal{S}\} < +\infty.$$

Hence for all $j \in \mathbb{N}$ and any $t, s \in [0, 1]$ we have

$$\theta_j(t) - \theta_j(s) = \int_s^t f(\phi(r, p_0, u_j), u_j(r))\mathrm{d}r,$$

and thus

$$|\theta_j(t) - \theta_j(s)| \leq M|t - s|, \quad t \in [0, 1], \quad j \in \mathbb{N}.$$

This implies that (θ_k) is uniformly equicontinuous on $[0, 1]$, and invoking Arzelá–Ascoli theorem, we can find a subsequence $(\sigma_1(j))_{j\geq 1}$ of $(j)_{j\geq 1}$ such that $(\theta_{\sigma_1(j)})_{j\geq 1}$ converges uniformly on $[0, 1]$ to a continuous function κ_1.

Next consider $(\theta_{\sigma_1(j)})_{j\geq 1}$ as a sequence of functions defined on $[1, 2]$. Arguing as above, one shows existence of a subsequence $(\sigma_2(j))_{j\geq 1}$ of $(\sigma_1(j))_{j\geq 1}$ such that $(\theta_{\sigma_2(j)})_{j\geq 1}$ converges uniformly on $[1, 2]$ to a function κ_2. Since $(\sigma_2(j))_{j\geq 1}$ is a subsequence of $(\sigma_1(j))_{j\geq 1}$, it holds that $\kappa_2(1) = \kappa_1(1)$.

By induction, for each $k \geq 1$ there is a subsequence $(\sigma_{k+1}(j))_{j\geq 1}$ of $(\sigma_k(j))_{j\geq 1}$ such that the sequence $(\theta_{\sigma_{k+1}(j)})_{j\geq 1}$ converges uniformly on $[k, k+1]$ to a continuous function κ_{k+1}. As above, $\kappa_k(k) = \kappa_{k+1}(k)$ for all $k \geq 1$.

Gluing κ_k together, we obtain the continuous function $\kappa : \mathbb{R}_+ \to \mathbb{R}^n$:

$$\kappa(t) := \kappa_k(t), \quad t \in [k-1, k), \quad k \geq 1. \tag{1.36}$$

Note that on each interval $[k-1, k]$, κ is the uniform limit of $(\theta_{\sigma_k(j)})_{j\geq 1}$.

As the images of θ_j, $j \in \mathbb{N}$ live in the complement of Ω, which is a closed set, the image of κ is also in the complement of Ω and hence is outside K. If κ were a trajectory of the system corresponding to some control v, the result would be proved (with $q = p_0$).

However, κ is not necessary a trajectory of the system. In the remaining part of the proof, we show however that it can be approximated by the trajectories of the system.

Step 2. For each control u with values in $\mathcal{S}$, we denote by Δu the control given by $\Delta u(t) = u(t+1)$ for each t in the domain of u (so, for instance, the domain of Δu is $[-1, +\infty)$ if the domain of u was $\mathbb{R}_+$). We will also consider iterates of the Δ operator, $\Delta^k u$, corresponding to a shift by k.

Since K is compact and Ω is open, we may pick an $r > 0$ such that $B_r(K) \subset \Omega$. Let W be any neighborhood of a. We pick an $r_0 < r/2$ so that $\overline{B_{r_0}(p_0)} \subset W$. Finally, let

$$p_k = \kappa(k), \qquad \text{for each } k \geq 1.$$

By construction, p_0 and p_1 are in S_1.

Next, for each $j \geq 1$, we wish to study the trajectory $\phi\big(-t, p_1, \Delta u_{\sigma_1(j)}\big)$ for $t \in [0, 1]$. This may be *a priori* undefined for all such t. However, since S_1 is compact, we may pick another compact set $\tilde{S}_1$ containing $B_r(S_1)$ in its interior, and we may also pick a function $\tilde{f} : \mathbb{R}^n \times \mathbb{R}^m \to \mathbb{R}^n$, which is equal to f for all $(x, u) \in \tilde{S}_1 \times \mathcal{S}$, is Lipschitz continuous in x, and has compact support. Then the modified system

$$\dot{x} = \tilde{f}(x, u)$$

is complete, meaning that solutions exist and are unique for all $t \in (-\infty, \infty)$. We use $\tilde{\phi}(t, \xi, u)$ to denote solutions of this new system. Observe that for each trajectory $\tilde{\phi}(t, \xi, u)$ that remains in $\tilde{S}_1$, $\phi(t, \xi, u)$ is also defined and coincides with $\tilde{\phi}(t, \xi, u)$. In particular,

$$\tilde{\phi}\big(-t, \theta_{\sigma_1(j)}(1), \Delta u_{\sigma_1(j)}\big) = \phi\big(-t, \theta_{\sigma_1(j)}(1), \Delta u_{\sigma_1(j)}\big) \tag{1.37}$$

for each j, since by backward uniqueness these both equal $\phi\big(1-t, p_0, u_{\sigma_1(j)}\big)$ for each $t \in [0, 1]$.

By Theorem 1.31, the set of states that can be reached from S_1, using the complete modified system in negative times $t \in [-1, 0]$, is included in some compact set.

In view of (1.37), using Lipschitz continuity in x of $\tilde{f}$ and using Grönwall's inequality, there is some $L \geq 0$ so that for all $j \geq 1$ and all $t \in [0, 1]$

$$
\begin{aligned}
&\left|\tilde{\phi}\big(-t, p_1, \Delta u_{\sigma_1(j)}\big) - \tilde{\phi}\big(-t, \theta_{\sigma_1(j)}(1), \Delta u_{\sigma_1(j)}\big)\right| \\
&\qquad = \left|\tilde{\phi}\big(-t, p_1, \Delta u_{\sigma_1(j)}\big) - \phi\big(-t, \theta_{\sigma_1(j)}(1), \Delta u_{\sigma_1(j)}\big)\right| \leq L|p_1 - \theta_{\sigma_1(j)}(1)|.
\end{aligned}
$$

Since $\theta_{\sigma_1(j)}(1) \to p_1$, as $j \to \infty$, it follows that there exists some j_1 such that for all $j \geq j_1$

$$
\left|\tilde{\phi}\big(-t, p_1, \Delta u_{\sigma_1(j)}\big) - \phi\big(-t, \theta_{\sigma_1(j)}(1), \Delta u_{\sigma_1(j)}\big)\right| < \frac{r_0}{2}, \quad t \in [0, 1]. \tag{1.38}
$$

As $r_0 < \frac{r}{2}$, this implies in particular that $\tilde{\phi}\big(-t, p_1, \Delta u_{\sigma_1(j)}\big) \in B_{r/4}(S_1) \subset \tilde{S}_1$ for all $t \in [0, 1]$ and all $j \geq j_1$, and hence "$\sim$" can be dropped in (1.38) for all $j \geq j_1$. By continuity of the $\tilde{\phi}$ with respect to initial condition, take $0 < r_1 < r_0$ such that

$$
\left|\tilde{\phi}\big(-t, p, \Delta u_{\sigma_1(j_1)}\big) - \phi\big(-t, p_1, \Delta u_{\sigma_1(j_1)}\big)\right| < \frac{r_0}{2} \tag{1.39}
$$

for all $t \in [0, 1]$, and all $p \in B_{r_1}(p_1)$. As this implies in particular that $\tilde{\phi}\big(-t, p, \Delta u_{\sigma_1(j_1)}\big) \in B_{r/2}(S_1) \subset \tilde{S}_1$, again tildes can be dropped.

Combining (1.38) and (1.39), it follows that for each $p \in B_{r_1}(p_1)$, the trajectory $\phi\big(-t, p, \Delta u_{\sigma_1(j_1)}\big)$ is defined for all $t \in [0, 1]$ and

$$
\left|\phi\big(-t, p, \Delta u_{\sigma_1(j_1)}\big) - \phi\big(-t, \theta_{\sigma_1(j_1)}, \Delta u_{\sigma_1(j_1)}\big)\right| < r_0 \tag{1.40}
$$

for all $t \in [0, 1]$. Let

$$
w_1(t) = u_{\sigma_1(j_1)}(t).
$$

As $\phi\big(-1, \theta_{\sigma_1(j_1)}, \Delta u_{\sigma_1(j_1)}\big) = p_0$, the estimate (1.40) implies that

$$
p \in B_{r_1}(p_1) \quad \Rightarrow \quad \phi(-1, p, \Delta w_1) \in B_{r_0}(p_0).
$$

Furthermore, since $\phi\big(-t, \theta_{\sigma_1(j_1)}, \Delta u_{\sigma_1(j_1)}\big) \notin \Omega$ for all $t \in [0, 1]$, and as $r_0 < \frac{r}{2}$, the estimate (1.40) also implies

$$
p \in B_{r_1}(p_1) \quad \Rightarrow \quad \phi(-t, p, \Delta w_1) \notin B_{r/2}(K) \quad \forall t \in [0, 1].
$$

Step 3. *In what follows, we will prove by induction that for each $i \geq 1$, there exists $0 < r_i < r_{i-1}$ and w_i of the form $w_i = \Delta^{i-1} u_{\sigma_i(j)}$ for some j so that for all $p \in B_{r_i}(p_i)$, the trajectory $\phi(-t, p, \Delta w_i)$ is defined for all $t \in [0, 1]$, and it holds that*

$$
\phi(-t, p, \Delta w_i) \notin B_{\frac{r}{2}}(K) \quad \forall t \in [0, 1],
$$

and

$$\phi(-1, p, \Delta w_i) \in B_{r_{i-1}}(p_{i-1}).$$

The case $i = 1$ has already been shown in the above argument. Assume now that the above conclusion is true for $i = 1, 2, \ldots, k$.

Consider, for each j, the trajectory $\phi\big(-t, p_{k+1}, \Delta^{k+1} u_{\sigma_{k+1}(j)}\big)$. Again, this may be *a priori* undefined for all such t. But, again, by modifying the system in a compact set $\tilde{S}_{k+1}$ containing a neighborhood of $\mathcal{R}_S^{\leq k+1}(p_0)$, one can show that there exists some $\bar{j}_{k+1} \geq k+1$ so that for all $j \geq \bar{j}_{k+1}$, $\phi\big(-t, p_{k+1}, \Delta^{k+1} u_{\sigma_{k+1}(j)}\big)$ is defined for all $t \in [0, 1]$, and

$$\left|\phi\big(-t, p_{k+1}, \Delta^{k+1} u_{\sigma_{k+1}(j)}\big) - \phi\big(-t, \theta_{\sigma_{k+1}(j)}(k+1), \Delta^{k+1} u_{\sigma_{k+1}(j)}\big)\right| < \frac{r_k}{4}, \quad (1.41)$$

for all $0 \leq t \leq 1$. We note here that for any j, the expression $\phi\big(-t, \theta_{\sigma_{k+1}(j)}(k+1), \Delta^{k+1} u_{\sigma_{k+1}(j)}\big)$ is defined for all $t \in [0, k+1]$ and

$$\phi\big(-1, \theta_{\sigma_{k+1}(j)}(k+1), \Delta^{k+1} u_{\sigma_{k+1}(j)}\big) = \theta_{\sigma_{k+1}(j)}(k).$$

Since $\theta_{\sigma_k(j)}(k) \to p_k$ as $j \to \infty$, and $(\sigma_{k+1}(j))_{j\geq 1}$ is a subsequence of $(\sigma_k(j))_{j\geq 1}$, it follows that there exists some $\tilde{j}_{k+1} \geq 0$ such that

$$\left|\phi\big(t, \theta_{\sigma_{k+1}(j)}(k), \Delta^k u_{\sigma_k(j)}\big) - \phi\big(t, p_k, \Delta^k u_{\sigma_k(j)}\big)\right| < \frac{r_k}{4} \quad (1.42)$$

for all $t \in [0, 1]$, for all $j \geq \tilde{j}_{k+1}$. Let $j_{k+1} = \max\{\bar{j}_{k+1}, \tilde{j}_{k+1}\}$, and let

$$w_{k+1}(t) := \Delta^k u_{\sigma_{k+1}(j_{k+1})}(t), \quad t \in [0, 1].$$

Then (1.41), applied with $j = j_{k+1}$, says that

$$\left|\phi(-t, p_{k+1}, \Delta w_{k+1}) - \phi\big(-t, \theta_{\sigma_{k+1}(j_{k+1})}(k+1), \Delta w_{k+1}\big)\right| < \frac{r_k}{4} \quad (1.43)$$

for all $t \in [0, 1]$. Using the case $t = 1$ of this equation, as well as (1.42) applied with $j = j_{k+1}$ and $t = 0$ and the equality $\phi\big(-t, \theta_{\sigma_{k+1}(j_{k+1})}(k+1), \Delta w_{k+1}\big) = \theta_{\sigma_{k+1}(j_{k+1})}(k)$, we conclude that

$$\phi(-1, p_{k+1}, \Delta w_{k+1}) \in B_{\frac{r_k}{2}}(p_k). \quad (1.44)$$

Pick any r_{k+1} so that $0 < r_{k+1} < r_k$ and for every $p \in B_{r_{k+1}}(p_{k+1})$, the expression $\phi(-t, p, \Delta w_{k+1})$ is defined for all $t \in [0, 1]$, and

$$\left|\phi(-t, p, \Delta w_{k+1}) - \phi(-t, p_{k+1}, \Delta w_{k+1})\right| < \frac{r_k}{2} \quad (1.45)$$

for all $t \in [0, 1]$. Then for such a choice of r_{k+1}, it holds for each $p \in B_{r_{k+1}}(p_{k+1})$, by (1.43) that

$$\left|\phi(-t, p, \Delta w_{k+1}) - \phi\big(-t, \theta_{\sigma_{k+1}(j_{k+1})}(k+1), \Delta w_{k+1}\big)\right| < \frac{3r_k}{4} < \frac{r}{2}$$

for all $t \in [0, 1]$, which implies that for each $p \in B_{r_{k+1}}(p_{k+1})$

$$\phi(-t, p, \Delta w_{k+1}) \notin B_{\frac{r}{2}}(K) \quad \forall\, t \in [0, 1].$$

Moreover, it follows from (1.44) and (1.45) that

$$p \in B_{r_{k+1}}(p_{k+1}) \quad \Rightarrow \quad \phi(-1, p, \Delta w_{k+1}) \in B_{r_k}(p_k).$$

This completes the induction step.

Step 4. Finally, we define a control v on $\mathbb{R}_+$ as follows:

$$v(t) = w_k(t - (k-1)), \quad \text{if} \quad t \in [k-1, k),$$

for each integer $k \geq 1$. Then $v \in \mathcal{U}_S$ and

$$\Delta^k v(t) = \Delta\Delta^{k-1} v(t) = \Delta v(t + (k-1)) = \Delta w_k(t), \quad t \in [0, 1].$$

For each k, we consider the trajectory $\phi(-t, p_k, \Delta^k v)$ for $t \in [0, k]$. Inductively,

$$\phi(-i, p_k, \Delta^k v) \in B_{r_{k-i}}(p_{k-i}) \quad \forall i \leq k, \tag{1.46}$$

thus this trajectory is well defined. We now let

$$z_k := \phi(-k, p_k, \Delta^k v), \quad k \in \mathbb{N}. \tag{1.47}$$

Note that also from (1.46) we can conclude that

$$\phi(-t, p_k, \Delta^k v) \notin B_{\frac{r}{2}}(K) \quad \forall t \in [0, k], \tag{1.48}$$

which is equivalent to

$$\phi(t, z_k, v) \notin B_{\frac{r}{2}}(K) \quad \forall t \in [0, k]. \tag{1.49}$$

As $z_k \in B_{r_0}(p_0)$ for all $k \in \mathbb{N}$, there exists a subsequence of (z_k) that converges to some point $q \in W$. To simplify notation, we still use (z_k) to denote one such convergent subsequence. We will finish our proof by showing that $\phi(t, q, v) \notin K$ for all $t \geq 0$.

Fix any integer $N > 0$. Using uniform Lipschitz continuity of solutions as a function of initial states in $B_{r_0}(p_0)$ (Theorem 1.40) and using the control v on the interval $[0, N]$, we know that there exists some $L_1 > 0$ such that

$$\left|\phi(t, q, v) - \phi(t, z_k, v)\right| \le L_1|q - z_k| \quad \forall\, t \in [0, N], \quad \forall\, k.$$

Hence, there exists some $k_0 > 0$ such that

$$\left|\phi(t, q, v) - \phi(t, z_{k_0}, v)\right| < \frac{r}{4} \quad \forall\, t \in [0, N]. \tag{1.50}$$

Without loss of generality, we assume that $k_0 \ge N$. Combining (1.50) with (1.49), it follows that $\phi(t, q, v) \notin B_{r/4}(K)$ for all $t \in [0, N]$. As N was arbitrary, it follows that $\phi(t, q, v) \notin K$ for all $t \ge 0$. □

Now we use Theorem 1.43 to show that if a compact set K can be reached from any state by means of any bounded input in a finite time, then any open set containing K can be reached in a *uniform* finite time from any point of any compact set C. More precisely:

Theorem 1.44 *Let (1.1) be a forward complete system and let Assumption 1.8 hold. Assume the following are given:*

- *a compact subset C of the state space $\mathbb{R}^n$;*
- *an open subset Ω of the state space $\mathbb{R}^n$;*
- *a compact subset $K \subset \Omega$;*
- *a bounded subset $S \subset \mathbb{R}^m$,*

so that for all $x \in \mathbb{R}^n$, all $u \in \mathcal{U}_S$, there is $t \ge 0$ such that $\phi(t, x, u) \in K$. Then:

$$\sup\{\tau(x, \Omega, u) : x \in C, u \in \mathcal{U}_S\} < +\infty.$$

Proof Pick any open set Ω_0 and $\varepsilon > 0$, such that

$$K \subset \Omega_0 \subset B_\varepsilon(\Omega_0) \subset \Omega.$$

Assume that there is $a \in \mathbb{R}^n$ such that $\sup_{u \in \mathcal{U}_S} \tau(a, \Omega_0, u) = +\infty$.

Applying Theorem 1.43 with Ω_0 instead of Ω and with $W := \mathbb{R}^n$, we obtain that there is $q \in W = \mathbb{R}^n$ and $v \in \mathcal{U}_S$, such that $\tau(q, K, v) = +\infty$, which contradicts to the assumptions of the theorem.

Thus, for each $a \in C$ there is $T_a < \infty$:

$$u \in \mathcal{U}_S \quad \Rightarrow \quad \exists t \in [0, T_a] : \ \phi(t, a, u) \in \Omega_0.$$

As C is compact, Theorem 1.40 ensures that for each $a \in C$ there is $L_a > 0$, such that

$$t \in [0, T_a], \quad \xi \in C, \quad u \in \mathcal{U}_S \quad \Rightarrow \quad \big|\phi(t, \xi, u) - \phi(t, a, u)\big| \le L_a |\xi - a|.$$

Choose $\delta_a := \frac{\varepsilon}{L_a}$. Above implication shows that:

$$t \in [0, T_a], \quad \xi \in B_{\delta_a}(a), \quad u \in \mathcal{U}_S \Rightarrow \big|\phi(t, \xi, u) - \phi(t, a, u)\big| \le \varepsilon. \tag{1.51}$$

As C is compact, by Heine–Borel lemma, we can find from the open cover $\cup_{a\in C} B_{\delta_a}(a)$ of C a finite subcover $\cup_{i=1}^{s} B_{\delta_{a_i}}(a_i)$ of C, for some $s \in \mathbb{N}$. Define $T := \max_{i=1}^{s} T_{a_i}$.

Now for each $\xi \in C$ one can find $a_i \in C$ such that $|\xi - a_i| \le \delta_{a_i}$. Furthermore, for any $u \in \mathcal{U}_S$ there is $t \le T_{a_i} \le T$ such that $\phi(t, a_i, u) \in \Omega_0$. Applying (1.51) with this t, we see that

$$\phi(t, \xi, u) \in B_\varepsilon(\Omega_0) \subset \Omega.$$

This finishes the proof. □

Theorem 1.44 is essential for the derivation of the precise characterizations of ISS for nonlinear systems (1.1) with f satisfying Assumption 1.8. The following important corollary will be useful for deriving characterizations of uniform global asymptotic stability and input-to-state stability in Appendix B and Chap. 2, respectively.

Corollary 1.45 *In addition to the assumptions of Theorem 1.44, let $\Omega \subset C$. Then there exists $T \ge 0$ so that*

$$\mathcal{R}_S(C) = \mathcal{R}_S^{\le T}(C),$$

and in particular $\mathcal{R}_S(C)$ is bounded.

Proof Set $T := \sup\{\tau(x, \Omega, u) : \ x \in C, u \in \mathcal{U}_S\} < \infty$, which is finite by Theorem 1.44. Assume that there exists $y \in \mathcal{R}_S(C) \backslash \mathcal{R}_S^{\le T}(C)$. Let $t^* > T, x \in C, u \in \mathcal{U}_S$ be so that $y = \phi(t^*, x, u)$. Let also t' be so that $t' := \max\{t \in [0, t^*] : \phi(t, x, u) \in C\}$. As C is compact, $\phi(t', x, u) \in C$.

Since y cannot be reached from $\phi(t', x, u) \in C$ in time less than T, it holds that $t^* - t' > T$. However, $\tau\big(\phi(t', x, u), \Omega, u(t' + \cdot)\big) \in (0, T]$, and hence there exists $t_1 \in (t', t^*)$ so that $\phi(t_1, x, u) \in \Omega \subset C$, which contradicts to the definition of t'.

Thus, $\mathcal{R}_S(C) = \mathcal{R}_S^{\le T}(C)$, which is a bounded set due to Theorem 1.31. □

1.6 Lipschitz and Absolutely Continuous Functions

As we have already seen, Lipschitz continuous and absolutely continuous functions play a fundamental role in the solution theory of ordinary differential equations. In this section, we study some properties of these function classes and relate them to other important spaces of functions.

For $k \in \mathbb{N} \cup \{\infty\}$, we denote by $C^k([a, b], \mathbb{R}^n)$ the linear space of k times continuously differentiable on $[a, b]$ functions with values in $\mathbb{R}^n$. We start with a

Proposition 1.46 *Every $g \in C^1([a, b], \mathbb{R}^n)$ is Lipschitz continuous on $[a, b]$.*

Proof By mean value theorem for each $x, y \in [a, b]$ there is $\xi \in (x, y)$ so that

$$g(x) - g(y) = g'(\xi)(x - y).$$

As g' is continuous, we obtain that $|g(x) - g(y)| \le L|x - y|$, with $L := \sup_{r \in [a,b]} |g'(r)|$. □

Lipschitz continuity is far weaker than continuous differentiability. Consider, e.g., $x \mapsto |x|$, which is Lipschitz continuous on $\mathbb{R}$ but is not continuously differentiable at 0. However:

Proposition 1.47 *Every Lipschitz continuous on $[a, b]$ function $g : [a, b] \to \mathbb{R}^n$ is absolutely continuous.*

Proof Let $g : [a, b] \to \mathbb{R}^n$ be Lipschitz continuous on $[a, b]$ and let L be a Lipschitz constant of g on $[a, b]$. Pick any $\varepsilon > 0$ and define $\delta := \frac{1}{L}\varepsilon$. For any finite r and any pairwise disjoint intervals $(a_k, b_k) \subset [a, b]$, with $\sum_{k=1}^{r} |b_k - a_k| \le \delta$ it follows that

$$\sum_{k=1}^{r} |g(b_k) - g(a_k)| \le \sum_{k=1}^{r} L|b_k - a_k| \le L\delta = \varepsilon.$$

This shows the absolute continuity of g. □

The converse statement to Proposition 1.47 does not hold, as the next example shows.

Example 1.48 Consider the function $g : [0, 1] \to \mathbb{R}$: $g(t) = \sqrt{t}$.

For all $t, s > 0$ it holds that $\sqrt{t} - \sqrt{s} = \frac{1}{\sqrt{t}+\sqrt{s}}(t - s)$, which shows that g is not Lipschitz continuous, as $\frac{1}{\sqrt{t}+\sqrt{s}} \to \infty$ as $t, s \to 0$.

On the other hand, for any $x \in [0, 1]$ it holds that $\sqrt{t} = \int_0^t \frac{1}{2\sqrt{z}} dz$. As $z \mapsto \frac{1}{2\sqrt{z}}$ is a Lebesgue integrable function on $[0, 1]$, g is absolutely continuous by Theorem 1.3.

The next result shows that the compositions of Lipschitz continuous and absolutely continuous functions are absolutely continuous.

Proposition 1.49 *Let $h : \mathbb{R}^n \to \mathbb{R}$ be Lipschitz continuous on bounded balls and $g : [a, b] \to \mathbb{R}^n$ be absolutely continuous. Then $h \circ g : [a, b] \to \mathbb{R}$ is absolutely continuous.*

Proof As g is continuous, $\operatorname{Im}(g) := \{g(s) : s \in [a, b]\}$ is compact. Let L be a Lipschitz constant for h on this set.

As $g : [a, b] \to \mathbb{R}^n$ is absolutely continuous, for any $\varepsilon > 0$ there exists $\delta > 0$ so that for any finite r and any pairwise disjoint intervals $(a_k, b_k) \subset [a, b]$, $k = 1, \ldots, r$ with $\sum_{k=1}^{r} |b_k - a_k| \leq \delta$ it follows that $\sum_{k=1}^{r} |g(b_k) - g(a_k)| \leq \frac{\varepsilon}{L}$. By Lipschitz continuity of h we have that

$$\sum_{k=1}^{r} |h \circ g(b_k) - h \circ g(a_k)| \leq L \sum_{k=1}^{r} |g(b_k) - g(a_k)| \leq \varepsilon.$$

□

The assumption of Lipschitz continuity of h cannot be weakened in the above proposition to merely absolute continuity, see Exercise 1.14.

Definition 1.50 We call $f : \mathbb{R}^n \times \mathbb{R}^m \to \mathbb{R}^n$ *locally Lipschitz continuous with respect to the first argument*, if for every $(x, u) \in \mathbb{R}^n \times \mathbb{R}^m$ there is $\delta > 0$ and $L > 0$ such that for every $(y, v) \in \mathbb{R}^n \times \mathbb{R}^m$: $|x - y| \leq \delta$ and $|u - v| \leq \delta$ it holds that

$$|x - y| \leq \delta,\ |v - u| \leq \delta \quad \Rightarrow \quad |f(y, v) - f(x, v)| \leq L|y - x|. \tag{1.52}$$

Using the compactness argument, we can show the following:

Proposition 1.51 *Let $f : \mathbb{R}^n \times \mathbb{R}^m \to \mathbb{R}^n$ be continuous on $\mathbb{R}^n \times \mathbb{R}^m$. Then f is Lipschitz continuous on bounded balls with respect to the first argument if and only if f is locally Lipschitz with respect to the first argument.*

Proof "$\Rightarrow$". Clear.

"$\Leftarrow$". We argue by contradiction. Assume that f is not Lipschitz continuous on bounded balls. Then there is some $r > 0$ and sequences $(x_k) \subset B_r$, $(y_k) \subset B_r$, $(u_k) \subset B_r$ such that for each $k > 0$ it holds that

$$|f(x_k, u_k) - f(y_k, u_k)| > k|x_k - y_k|. \tag{1.53}$$

As B_r is precompact, there are convergent subsequences $(x_{k_s})_{s\in\mathbb{N}}$ of (x_k) and $(u_{k_s})_{s\in\mathbb{N}}$ of (u_k), which are convergent to some $x^* \in \overline{B_r}$ and $u^* \in \overline{B_r}$. As f is continuous on $\mathbb{R}^n \times \mathbb{R}^m$, $\sup_{z\in\overline{B_r},\ v\in\overline{B_r}} |f(z, v)| = M < \infty$. Hence from (1.53) we obtain that

$$|x_k - y_k| \leq \frac{1}{k}|f(x_k, u_k) - f(y_k, u_k)| \leq \frac{2M}{k} \to 0, \quad k \to \infty. \tag{1.54}$$

Hence $(y_{k_s})_{s\in\mathbb{N}}$ is also convergent to x^*.

By local Lipschitz continuity of f, there is some $\delta > 0$ and $L > 0$ such that for all x, y: $|x - x^*| \leq \delta$ and $|y - x^*| \leq \delta$ and for all v: $|v - u^*| \leq \delta$ it holds that

$$|f(x, v) - f(y, v)| \leq L|x - y|. \tag{1.55}$$

Pick $N > 0$ such that $|x_{k_s} - x^*| \leq \delta$, $|y_{k_s} - x^*| \leq \delta$ and $|u_{k_s} - u^*| \leq \delta$ for all $s \geq N$. Then it holds that

$$|f(x_{k_s}, u_{k_s}) - f(y_{k_s}, u_{k_s})| \leq L|x_{k_s} - y_{k_s}|, \quad s \geq N. \tag{1.56}$$

This however contradicts to (1.53). □

1.7 Spaces of Measurable and Integrable Functions

In this section, we recall the definition of L^p-spaces.

Definition 1.52 Consider a pair $(X, \mathbb{F})$, where X is a nonempty set, and $\mathbb{F}$ is a scalar field, endowed with a map: $(x_1, x_2) \to x_1 + x_2$ from $X \times X$ to X, which we call *addition*, and a map: $(\alpha, x) \to \alpha x$ from $\mathbb{F} \times X$ to X, called *scalar multiplication*.

X is called a *linear vector space*, provided that these maps satisfy the following axioms for all $x, y, z \in X$ and all $\alpha, \beta \in \mathbb{F}$:

(i) $x + y = y + x$ (*commutativity*).
(ii) $(x + y) + z = x + (y + z)$ (*associativity*).
(iii) There exists a unique element 0 in X such that $x + 0 = x$ (the existence of the *zero element*).
(iv) For each $x \in X$, there exists a unique element $-x \in X$ such that $x + (-x) = 0$ (*invertibility*).
(v) $\alpha(\beta x) = (\alpha\beta)x$.
(vi) $(\alpha + \beta)x = \alpha x + \beta x$.
(vii) $\alpha(x + y) = \alpha x + \alpha y$.
(viii) $1x = x$, where 1 is the unit element of the scalar field $\mathbb{F}$.

A vector space over the scalar field $\mathbb{F} := \mathbb{R}$ is called a *real vector space*, and vector space over $\mathbb{F} := \mathbb{C}$ is called a *complex vector space*.

In addition to the linear structure, we would like to endow the spaces that we consider with the center of coordinate and with the possibility to measure the distance between objects.

Definition 1.53 Let X be a linear vector space. A *norm* in X is a function $\|\cdot\|_X : X \to \mathbb{R}_+$, such that:

(i) $\|x\|_X = 0$ if and only if $x = 0$;
(ii) $\|\alpha x\|_X = |\alpha|\|x\|_X$ for all $x \in X$ and $\alpha \in \mathbb{F}$;
(iii) $\|x + y\|_X \leqslant \|x\|_X + \|y\|_X$ for all $x, y \in X$ (the *triangle inequality*).

If only conditions (ii)–(iii) hold, then $\|\cdot\|_X$ is called *seminorm*.

Definition 1.54 A *normed vector space* is a linear vector space X with a norm $\|\cdot\|_X$ on it, and it is denoted by $(X, \|\cdot\|_X)$. We write simply X provided that the meaning is clear from the context.

Definition 1.55 A sequence (x_n) of elements in a normed vector space $(X, \|\cdot\|_X)$ is called a *Cauchy sequence* if for all $\varepsilon > 0$ there is $N \in \mathbb{N}$ so that for all $m, n \geq N$ it holds that $\|x_n - x_m\|_X \leq \varepsilon$.

Clearly, every convergent sequence is a Cauchy sequence. Normed vector spaces, in which the converse implication holds, play an important role in analysis.

Definition 1.56 A normed vector space X is called *complete* if every Cauchy sequence has a limit in X.

Completeness is a very important property, as it makes it possible to determine whether the sequence is convergent without knowing the limit of the sequence. This motivates the following definition

Definition 1.57 A *Banach space* is a complete, normed vector space.

Important examples of Banach spaces are L^p-spaces.

L^p-space. Let $p \geq 1$ be a fixed real number, $n \in \mathbb{N}$ and let $-\infty \leq a < b \leq \infty$. Consider the set of measurable functions $x : [a, b] \to \mathbb{R}^n$ satisfying $\int_a^b |x(t)|^p \mathrm{d}t < \infty$, endowed with the map

$$\|x\| := \Big(\int_a^b |x(t)|^p \mathrm{d}t \Big)^{1/p}.$$

This is a linear vector space with addition and scalar multiplication defined by:

$$(x + y)(t) = x(t) + y(t), \qquad (\alpha x)(t) = \alpha x(t).$$

The norm axiom (ii) holds trivially, the axiom (iii) follows by Minkowski inequality, which shows that $\|\cdot\|$ is a seminorm. However, $\|\cdot\|$ is not a norm on this linear space, as $\|x\| = 0$ only implies that $x(t) = 0$ almost everywhere.

Now consider the space of equivalence classes $[x]$ of functions that equal x almost everywhere. Clearly, these equivalence classes form a linear space and

$$\|[x]\|_{L^p([a,b],\mathbb{R}^n)} := \|x_1\|, \quad \text{for any } x_1 \in [x]$$

defines a norm. We call this normed vector space $L^p([a, b], \mathbb{R}^n)$.

Usually, we write x_1 instead of $[x]$, where x_1 is any element of the equivalence class $[x]$.

Theorem 1.58 *If $p \in [1, \infty)$, and $-\infty \leq a < b \leq \infty$, then $L^p([a, b], \mathbb{R}^n)$ is a Banach space.*

Proof See [3, p. 11, Theorem 1.21]. □

L^∞-space. Let $-\infty \le a < b \le \infty$ and consider all measurable functions $x : [a, b] \to \mathbb{R}^n$ satisfying the condition $\operatorname{ess\,sup}_{t\in(a,b)} |x(t)| < \infty$. Following the construction of L^p-spaces, we construct equivalence classes $[x]$ containing functions that equal x almost everywhere on (a, b). With the norm

$$\|[x]\|_\infty := \operatorname{ess\,sup}_{t\in(a,b)} |x_1(t)|, \qquad \text{for any } x_1 \in [x],$$

this space is a normed vector space, which we denote by $L^\infty([a, b], \mathbb{R}^n)$. Following the standard conventions, we usually write x_1 instead of $[x]$, where x_1 is any element of $[x]$.

Proposition 1.59 *Let $-\infty \le a < b \le \infty$. The space $L^\infty([a, b], \mathbb{R}^n)$ is a Banach space.*

Proof For finite a, b see [3, p. 12, Proposition 1.24]. □

1.8 Concluding Remarks

For Theorem 1.2 see, e.g., [4, Theorem 6, p. 340], and Theorem 1.3 can be found in [4, Theorem 5, p. 338].

Existence and uniqueness theory for the ODEs with a right-hand side, which is measurable in t, has been developed by Carathéodory. For much more on this theory you can consult, e.g., [1]. In particular, for a more general version of Theorem 1.11 see [1, Chap. 2, Theorem 2.1].

The results in Sect. 1.3 are due to [5, Section 5], and the results in Sect. 1.5 are due to [6, Sect. III]. We follow [5, 6] rather closely in these sections.

Proposition 1.51 can be extended without many difficulties also to the functions f defined over the compact metric spaces. This result belongs to the folklore, although the author is not aware of a particular place where it was shown. If f does not depend on u, one can find the result in [8], which is probably not the first reference.

In our presentation of the material in Sect. 1.7 we are close to [2, Sect. A.2.1].

1.9 Exercises

Properties of ODE systems

Exercise 1.1 We say that $f : \mathbb{R}^n \times \mathbb{R}^m \to \mathbb{R}^n$ is *one-sided Lipschitz continuous on bounded subsets* of $\mathbb{R}^n$ if for all $C > 0$ there is $L = L(C) > 0$, such that

$$|x| \le C, \ |y| \le C, \ |v| \le C \ \Rightarrow \ \big(f(y, v) - f(x, v)\big)^T (x - y) \le L|y - x|^2. \tag{1.57}$$

$$\mathrm{H\ddot{o}l}([a,b]) \qquad\qquad\qquad\qquad \mathrm{BV}([a,b])$$

$$\cup \qquad\qquad\qquad\qquad \cup$$

$$C^1([a,b]) \subset \mathrm{Lip}([a,b]) \subset AC([a,b]) \subset \mathrm{BV}([a,b]) \cap C([a,b]) \subset C([a,b])$$

Fig. 1.4 Relationships between basic function classes on compact interval of the real line. Here we denote by Lip($[a, b]$) the space of Lipschitz continuous functions on $[a, b]$, Höl ($[a, b]$) the space of Hölder continuous functions on $[a, b]$ (see Definition 1.61), and by BV($[a, b]$) the space of functions of bounded variation defined on $[a, b]$ (see Exercise 1.14).

Let f be continuous on $\mathbb{R}^n \times \mathbb{R}^m$, and one-sided Lipschitz continuous on bounded subsets. Show that for any $u \in \mathcal{U}$ and any $x_0 \in \mathbb{R}^n$ there is at most one (maximal) solution of (1.1).

Exercise 1.2 Show that the space (1.11) with the metric ρ_{t^*} is a complete metric space.

Exercise 1.3 (i) Consider an initial value problem

$$\dot{x}(t) = x^{1/3}(t), \quad x(0) = 0. \tag{1.58}$$

Clearly, $x \equiv 0$ is a solution of (1.58). Find by means of a separation of variables another solution of (1.58). Can you find more solutions?

(ii) Show that initial value problem

$$\dot{x}(t) = -x^{1/3}(t), \quad x(0) = x_0 \tag{1.59}$$

has a unique solution on $[0, +\infty)$ for any $x_0 \in \mathbb{R}$.

(iii) Is there some relation between solutions of equations in (i) and (ii)?

Exercise 1.4 Investigate the existence and uniqueness of solutions to the initial value problem

$$\dot{x} = f(x), \quad x(0) = x_0, \quad \text{where } f(x) = \begin{cases} 1, & \text{if } x < 0, \\ -1, & \text{if } x \geq 0, \end{cases}$$

for all $x_0 \in \mathbb{R}$ in positive and negative time directions. Find the maximal interval of the existence of solutions.

Exercise 1.5 Let $f : \mathbb{R}^2 \to \mathbb{R}^2$ be Lipschitz continuous and let

$$f\left(\begin{bmatrix} x_1 \\ x_2 \end{bmatrix}\right) = \begin{bmatrix} -x_2 \\ x_1 \end{bmatrix}.$$

for all $(x_1, x_2) \in \mathbb{R}^2$ such that $x_1^2 + x_2^2 = 1$. Find the maximal existence interval of a solution of the initial value problem

$$\dot{x} = f(x), \quad x(0) = x_0, \quad \text{where } |x_0| < 1.$$

Exercise 1.6 (Group property) Let (1.1) be complete. That is, the solution of (1.1) exists and is unique on $(-\infty, \infty)$ for any $x \in \mathbb{R}^n$ and any $u \in \mathcal{U} = L^\infty(\mathbb{R}, \mathbb{R}^m)$. Show that for all $x \in \mathbb{R}^n$, $u \in \mathcal{U}$ and all $t, \tau \in \mathbb{R}$ the equality (1.12) holds.

Exercise 1.7 Using the integral representation (1.5) for the solutions of (1.1), give an alternative proof for the cocycle property (Proposition 1.17).

Exercise 1.8 Consider a system

$$\dot{x}(t) = f(x(t), u(t), x_0), \quad x(0) = x_0.$$

Assume that this system is well posed in the sense that for all inputs u and all initial conditions x_0 there is a unique maximal solution, existing on a certain interval of a positive length. We can again define the solution map $\phi(\cdot, x_0, u)$. Does the cocycle property hold for this solution map?

Exercise 1.9 Let f in (1.1) be continuous on $\mathbb{R}^n \times \mathbb{R}^m$ and uniformly globally Lipschitz continuous. Show that (1.1) is a forward complete system with BRS.

Exercise 1.10 (Invariance of reachability sets) Show that for any $C \subset \mathbb{R}^n$ and any $S \subset \mathbb{R}^m$ the reachability set $\mathcal{R}_S(C)$ is invariant with respect to the flow, that is, the following holds:

$$y \in \mathcal{R}_S(C),\ t \geq 0,\ u \in \mathcal{U}_S \quad \Rightarrow \quad \phi(t, y, u) \in \mathcal{R}_S(C).$$

Function spaces

Exercise 1.11 Let $f, g \in AC([a, b], \mathbb{R})$. Show that $f + g \in AC([a, b], \mathbb{R})$ and $f \cdot g \in AC([a, b], \mathbb{R})$.

Exercise 1.12 Let $f, g : \mathbb{R}^n \to \mathbb{R}^n$ be Lipschitz continuous on bounded balls. Show that $s \mapsto \max\{f(s), g(s)\}$ is also Lipschitz continuous on bounded balls.

Hint: show first the *Birkhoff's inequality*, which states that for all $x_1, x_2, y_1, y_2 \in \mathbb{R}$:

$$\big|\max\{x_1, x_2\} - \max\{y_1, y_2\}\big| \leq \max\big\{|x_1 - x_2|, |y_1 - y_2|\big\}. \tag{1.60}$$

Exercise 1.13 (Banach space of Lipschitz continuous functions) Let M denote a metric space and let $\mathrm{Lip}_b(M)$ denote the space of all bounded Lipschitz continuous real-valued functions on M, endowed with the norm, defined for any $f \in \mathrm{Lip}_b(M)$ as the maximum of the supremum (infinity) norm and the Lipschitz constant of f.

Show that $\mathrm{Lip}_b(M)$ with this norm is a Banach space.

Definition 1.60 We say that a *function* $g : [a, b] \to \mathbb{R}^n$ *is of bounded variation on* $[a, b]$, if there is a constant $C > 0$ such that for every $p \in \mathbb{N}$, and for every partition of $[a, b]$ of the following form

$$a = x_0 < x_1 < \cdots < x_p = b,$$

it holds that

$$\sum_{k=1}^{p-1} |f(x_k) - f(x_{k+1})| \leq C. \tag{1.61}$$

The *variation* of g on $[a, b]$ is defined by

$$V_a^b(g) := \sup_{P=(x_k)_{k=1}^{p(P)} \subset [a,b]} \sum_{k=1}^{p(P)-1} |f(x_k) - f(x_{k+1})|,$$

where the supremum is taken over all partitions of $[a, b]$.

Exercise 1.14 (Absolute continuity and functions of bounded variation)

(i) Show that every absolutely continuous function $g : [a, b] \to \mathbb{R}^n$ is of bounded variation on $[a, b]$.
(ii) Find a function on a bounded interval $[a, b]$, which is uniformly continuous, having a bounded variation but which is not absolutely continuous.
Hint: consider, e.g., Cantor's stair function.
(iii) Give an example of a uniformly continuous function on a bounded interval $[a, b]$, which does not have a bounded variation (and hence is not absolutely continuous).
Hint: consider $g(t) := \begin{cases} 0, & \text{if } t = 0 \\ t \sin \frac{1}{t}, & \text{otherwise,} \end{cases}$ defined on $[0, \frac{2}{\pi}]$.
(iv) Let $h, g \in AC([a, b], \mathbb{R})$. Show that $h \circ g$ is not necessarily absolutely continuous.
Hint: consider, e.g., $h = f \circ g$ with $f(t) = \sqrt{t}$ and $g(t) = t^2 |\sin \frac{1}{t}|$ on $[0, 1]$.

Definition 1.61 We call $f : \mathbb{R}^n \to \mathbb{R}^m$ *Hölder continuous on bounded subsets* of $\mathbb{R}^n$ if there is an $\alpha > 0$ so that for all $C > 0$ there is $L(C) > 0$, such that

$$|x| \leq C, \ |y| \leq C \quad \Rightarrow \quad |f(y) - f(x)| \leq L(C)|y - x|^\alpha. \tag{1.62}$$

Exercise 1.15 (Hölder continuity) Show the following statements:

(i) If $f : \mathbb{R} \to \mathbb{R}$ is Hölder continuous with $\alpha > 1$, then f is a constant function.
(ii) If $f : \mathbb{R}^n \to \mathbb{R}^m$ is Hölder continuous with a certain $\alpha > 0$, then f is Hölder continuous with any $\beta \in (0, \alpha)$.

References

1. Coddington EA, Levinson N (1955) Theory of ordinary differential equations. Tata McGraw-Hill Education
2. Curtain R, Zwart H (2020) Introduction to infinite-dimensional systems theory: a state-space approach. Springer
3. Fabian M, Habala P, Hájek P, Montesinos V, Zizler V (2011) Banach space theory: the basis for linear and nonlinear analysis. Springer
4. Kolmogorov AN, Fomin SV (1975) Introductory real analysis. Dover
5. Lin Y, Sontag ED, Wang Y (1996) A smooth converse Lyapunov theorem for robust stability. SIAM J Control Optim 34(1):124–160
6. Sontag ED, Wang Y (1996) New characterizations of input-to-state stability. IEEE Trans Autom Control 41(9):1283–1294
7. Teschl G (2012) Ordinary differential equations and dynamical systems. Am Math Soc
8. Xu X, Liu L, Feng G (2020) On Lipschitz conditions of infinite dimensional systems. Automatica 117(108947)

Chapter 2
Input-to-State Stability

In the previous chapter, we have analyzed well-posedness and properties of reachability sets for nonlinear control systems of the form

$$\dot{x} = f(x, u), \tag{2.1a}$$

$$x(0) = x_0, \tag{2.1b}$$

where $x(t) \in \mathbb{R}^n$, $u \in \mathcal{U} = L^\infty(\mathbb{R}_+, \mathbb{R}^m)$, and $x_0 \in \mathbb{R}^n$ is a given initial condition.

In this chapter, which is foundational for the rest of the book, we study the stability of control systems (2.1). We define the notion of input-to-state stability that unifies the asymptotic stability in the sense of Lyapunov with input–output stability. In Sect. 2.2, we introduce ISS Lyapunov functions that will be our primary tool for the analysis of input-to-state stability. In Sect. 2.3, we show that local input-to-state stability under mild regularity requirements on the right-hand side is equivalent to the local asymptotic stability of the undisturbed system (2.1). Finally, we prove several fundamental results, which throw light on the essence of the ISS concept. We establish ISS superposition theorems (Theorems 2.51 and 2.52), showing that ISS is equivalent to a combination of uniform local stability and asymptotic gain property. We prove a converse ISS Lyapunov theorem (Theorem 2.63), as well as a criterion for ISS in terms of stability properties involving integrals of trajectories. Finally, we show that every ISS system becomes exponentially ISS under a suitable nonlinear change of coordinates.

We generally assume in this chapter:

Assumption 2.1 Suppose that the following holds:

(i) f is continuous on $\mathbb{R}^n \times \mathbb{R}^m$.
(ii) (2.1) is well-posed: for any initial condition $x \in \mathbb{R}^n$ and any input $u \in \mathcal{U}$, there is a unique maximal solution of (2.1), which we denote by $\phi(\cdot, x, u)$.
(iii) (2.1) satisfies the BIC property.

A. Mironchenko, *Input-to-State Stability*, Communications and Control Engineering,
https://doi.org/10.1007/978-3-031-14674-9_2

Our Assumption 2.1 on the nonlinearity f is rather mild. In particular, in view of Theorem 1.16 and Proposition 1.20, it is less restrictive than Assumption 1.8.

To motivate the definition of the input-to-state stability for nonlinear systems, let us look first at linear systems

$$\dot{x} = Ax + Bu \tag{2.2}$$

with a *Hurwitz matrix* $A \in \mathbb{R}^{n\times n}$ (i.e., the spectrum of A lies in $\{z \in \mathbb{C} : \mathrm{Re} z < 0\}$) and some $B \in \mathbb{R}^{n\times m}$. The solution of (2.2) subject to initial condition $x \in \mathbb{R}^n$ and input $u \in \mathcal{U}$ is given by the variation of constants formula

$$\phi(t, x, u) = e^{At}x + \int_0^t e^{A(t-s)}Bu(s)ds. \tag{2.3}$$

Since A is a Hurwitz matrix, it holds that $\|e^{At}\| \le Me^{-\lambda t}$ for some $M > 0$, $\lambda > 0$, and all $t \ge 0$[1]. Consequently

$$\begin{aligned}|\phi(t, x, u)| &\le |e^{At}x| + \int_0^t \|e^{A(t-s)}\||Bu(s)|ds \\ &\le Me^{-\lambda t}|x| + M\int_0^t e^{-\lambda(t-s)}ds\|B\|\|u\|_\infty.\end{aligned}$$

We can easily compute the integral on the right-hand side for any $t > 0$:

$$\int_0^t e^{-\lambda(t-s)}ds = \frac{1}{|\lambda|}(1 - e^{-\lambda t}) \le \frac{1}{|\lambda|}.$$

Substitution of this inequality into the previous calculations shows that the trajectories of (2.2) satisfy the estimate

$$|\phi(t, x, u)| \le Me^{-\lambda t}|x| + \frac{M}{|\lambda|}\|B\|\|u\|_\infty \quad \forall t, x, u.$$

Here the term $Me^{-\lambda t}|x|$ describes the transient behavior of the system (2.2) and is a $\mathcal{KL}$ function of $(|x|, t)$. On the other hand, the term $\frac{M}{|\lambda|}\|B\|\|u\|_\infty$ describes the asymptotic deviation of the trajectory from the equilibrium, or, as the control theorists

[1] For $A \in \mathbb{R}^{n\times n}$ we denote $\|A\| := \sup_{|x|=1} |Ax|$.

say, the *asymptotic gain* of the system. In particular, inputs of a bounded magnitude induce only bounded asymptotic deviations of the system from the origin.

Proper extension of this idea to a nonlinear setting leads to the notion of input-to-state stability.

2.1 Basic Definitions and Results

The next notion will be central in this book:

Definition 2.2 System (2.1) is called *input-to-state stable (ISS)*, if (2.1) is forward complete and there exist $\beta \in \mathcal{KL}$ and $\gamma \in \mathcal{K}$ such that for all $x \in \mathbb{R}^n$, all $u \in \mathcal{U}$, and all $t \geq 0$ the following holds

$$|\phi(t, x, u)| \leq \beta(|x|, t) + \gamma(\|u\|_\infty). \tag{2.4}$$

The function γ is called *gain* and describes the influence of input on the system. The map β describes the transient behavior of the system.

Definition 2.3 We call (2.1) *globally asymptotically stable at zero (0-GAS)*, if (2.1) is GAS (see Definition B.14) for $u \equiv 0$.

Remark 2.4 In view of Proposition B.37, if $f(\cdot, 0)$ is Lipschitz continuous, 0-GAS is equivalent to the existence of $\beta \in \mathcal{KL}$ such that for all $x \in \mathbb{R}^n$ and all $t \geq 0$ it holds that

$$|\phi(t, x, 0)| \leq \beta(|x|, t).$$

This shows that ISS is a natural way to extend the global asymptotic stability notion to systems with inputs.

The derivations in the beginning of the chapter show that a linear system (2.2) is ISS $\Leftrightarrow$ (2.2) is 0-GAS. For nonlinear systems, ISS is a much stronger notion than 0-GAS, which can be easily seen from the following example:

$$\dot{x} = -x + (1 + x^2)u, \quad x(t),\ u(t) \in \mathbb{R}.$$

This system is 0-GAS, but for $u \equiv 1$, the trajectory of this system exhibits blow-up for any initial condition (show this!).

In Fig. 2.1b a trajectory of a typical ISS system is depicted. Substituting $u \equiv 0$ into the definition of ISS, we see immediately that any ISS system is 0-GAS. On the other hand, taking the limit superior $t \to \infty$, we see that the trajectory of any ISS system satisfies for certain $\gamma \in \mathcal{K}$, any $x \in \mathbb{R}^n$, and any $u \in \mathcal{U}$ the inequality

$$\overline{\lim_{t\to\infty}} |\phi(t, x, u)| \leq \gamma(\|u\|_\infty), \tag{2.5}$$

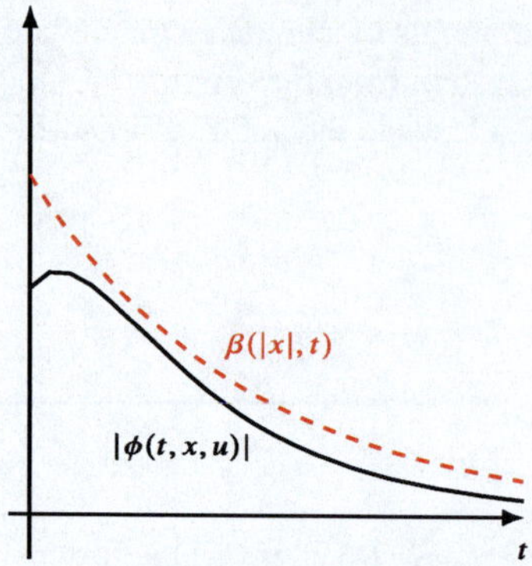

(a) Typical solution of an ISS system with $u \equiv 0$

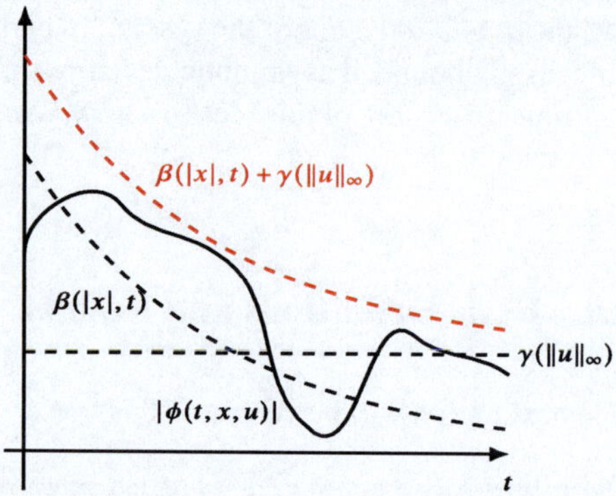

(b) Typical solution of an ISS system with $u \not\equiv 0$

Fig. 2.1 Typical behavior of trajectories of an ISS system

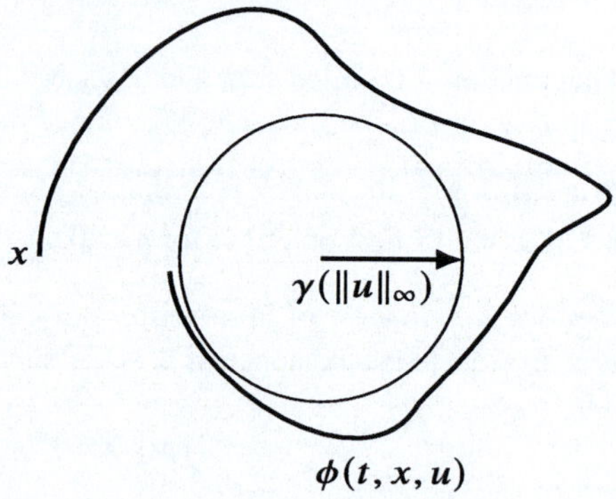

Fig. 2.2 Typical trajectory of an AG system in the phase space

called the *asymptotic gain (AG) property*. For short, the system satisfying AG property will be called an AG system. A trajectory of a typical AG system is described in Fig. 2.2. Every trajectory of an AG system converges to the neighborhood of an origin with a radius $\gamma(\|u\|_\infty)$.

Remark 2.5 We have defined ISS property in the "summation form". Alternatively, ISS can be defined in the *"maximum form"*: (2.1) is ISS if and only if (2.1) is forward complete and there exist $\beta \in \mathcal{KL}$ and $\gamma \in \mathcal{K}$ such that for all $x \in \mathbb{R}^n$, all $u \in \mathcal{U}$, and all $t \geq 0$ the following holds

$$|\phi(t, x, u)| \leq \max\left\{\beta(|x|, t), \gamma(\|u\|_\infty)\right\}. \tag{2.6}$$

Sum and max formulations are qualitatively equivalent due to the inequality $\max\{r, s\} \leq r + s \leq 2\max\{r, s\}$, which holds for any $r, s \in \mathbb{R}_+$. Certainly β and γ may be different in max and sum formulations. Many other equivalent formulations for ISS are possible. A choice of a "right" formulation for ISS depends on the application.

ISS systems have many interesting properties. We start by analyzing the asymptotic behavior of ISS systems for asymptotically vanishing inputs.

Definition 2.6 We say that a forward complete system has a *convergent input-uniformly convergent state (CIUCS) property*, if for each $u \in \mathcal{U}$ such that $\lim_{t\to\infty} \|u(t+\cdot)\|_\infty = 0$, and for any $r > 0$, it holds that

$$\lim_{t\to\infty} \sup_{|x|\le r} |\phi(t,x,u)| = 0.$$

Proposition 2.7 *If (2.1) is ISS, then (2.1) has CIUCS property.*

Proof Let (2.1) be ISS and assume without loss of generality that the corresponding gain γ is a $\mathcal{K}_\infty$-function. Pick any $r > 0$, any $\varepsilon > 0$, and any $u \in \mathcal{U}$ so that $\lim_{t\to\infty} \|u(t+\cdot)\|_\infty = 0$.

Let us show that there is a time $t_\varepsilon = t_\varepsilon(r,u) > 0$ so that

$$|x| \le r, \quad t \ge t_\varepsilon \;\Rightarrow\; |\phi(t,x,u)| \le \varepsilon.$$

Choose t_1 so that $\|u(\cdot + t_1)\|_\infty \le \gamma^{-1}(\frac{\varepsilon}{2})$. Due to the cocycle property and ISS of (2.1) we have for any $x \in B_r$

$$\begin{aligned}
|\phi(t+t_1,x,u)| &= \big|\phi\big(t,\phi(t_1,x,u),u(\cdot+t_1)\big)\big| \\
&\le \beta\big(|\phi(t_1,x,u)|,t\big) + \gamma\big(\|u(\cdot+t_1)\|_\infty\big) \\
&\le \beta\big(|\phi(t_1,x,u)|,t\big) + \frac{\varepsilon}{2} \\
&\le \beta\big(\beta(|x|,t_1)+\gamma(\|u\|_\infty),t\big) + \frac{\varepsilon}{2} \\
&\le \beta\big(\beta(r,0)+\gamma(\|u\|_\infty),t\big) + \frac{\varepsilon}{2}.
\end{aligned}$$

Pick any t_2 in a way that $\beta\big(\beta(r,0)+\gamma(\|u\|_\infty),t_2\big) \le \frac{\varepsilon}{2}$. This ensures that for all $t \ge 0$

$$\begin{aligned}
|\phi(t+t_2+t_1,x,u)| &\le \beta\big(\beta(r,0)+\gamma(\|u\|_\infty),t+t_2\big) \\
&\quad + \frac{\varepsilon}{2} \le \beta\big(\beta(r,0)+\gamma(\|u\|_\infty),t_2\big) + \frac{\varepsilon}{2} = \varepsilon.
\end{aligned}$$

Since $\varepsilon > 0$ can be chosen arbitrary small, the claim of the proposition follows. □

The definition of ISS can be refined in various directions. For example, in the definition of ISS, we have assumed in advance that the system is forward complete, and thus the expression (2.4) makes sense. However, forward completeness of (2.1) may not be known in advance. In this case it makes sense to require the ISS estimate (2.4) only for $t \in [0, t_m(x,u))$, where $t_m(x,u)$ is the maximal existence time of a solution starting in x and subject to input u. On the other hand, since $\phi(t,x,u)$ does not depend on the values of u that are larger than t, it would be more precise to have

$\operatorname{ess\,sup}_{0\le s\le t}|u(s)|$ instead of $\operatorname{ess\,sup}_{s\ge 0}|u(s)|$ in the ISS estimate (2.4). Finally, we have studied ISS w.r.t. inputs from the class $\mathcal{U} = L^\infty(\mathbb{R}_+, \mathbb{R}^m)$. For some applications, this class of inputs may be too restrictive, and it is of interest to study the response of (2.1) w.r.t. inputs that are only locally bounded but may have infinite L^∞-norm on the whole $\mathbb{R}_+$. In the following proposition, we show that an alternative notion of ISS, including all these refinements, is equivalent to ISS as defined in Definition 2.2.

Define for any $u : \mathbb{R}_+ \to \mathbb{R}^m$ and any $t > 0$ the *restriction of u to the segment* $[0, t]$ as

$$u_t(s) := \begin{cases} u(s), & \text{if } s \in [0, t], \\ 0, & \text{if } s > t. \end{cases} \tag{2.7}$$

and denote

$$L^\infty_{\text{loc}}(\mathbb{R}_+, \mathbb{R}^m) := \{u : \mathbb{R}_+ \to \mathbb{R}^m : u_t \in L^\infty(\mathbb{R}_+, \mathbb{R}^m)\ \forall t \in \mathbb{R}_+\}. \tag{2.8}$$

Proposition 2.8 *Let Assumption 2.1 hold. (2.1) is ISS $\Leftrightarrow$ there exist $\beta \in \mathcal{KL}$ and $\gamma \in \mathcal{K}$ such that for all $x \in \mathbb{R}^n$, all $u \in L^\infty_{\text{loc}}(\mathbb{R}_+, \mathbb{R}^m)$, and all $t \in [0, t_m(x, u))$ the following holds*

$$|\phi(t, x, u)| \le \beta(|x|, t) + \gamma\big(\operatorname{ess\,sup}_{0\le s\le t}|u(s)|\big). \tag{2.9}$$

Proof "$\Leftarrow$". Let the estimate (2.9) holds for every $t \in [0, t_m(x, u))$. Assume that $t_m(x, u) < \infty$ for certain $x \in \mathbb{R}^n$ and $u \in L^\infty_{\text{loc}}(\mathbb{R}_+, \mathbb{R}^m)$. For these x, u we have $\operatorname{ess\,sup}_{0\le s\le t_m(x,u)}|u(s)| < \infty$ and hence $\overline{\lim}_{t\to t_m(x,u)-0} |\phi(t, x, u)|$ is bounded. By BIC property (as a part of Assumption 2.1), the solution $\phi(\cdot, x, u)$ can be prolonged to the interval $[0, t_m(x, u) + \varepsilon)$ for some $\varepsilon > 0$. This contradicts to the fact that $t_m(x, u)$ is a maximal existence time of a corresponding solution. Thus, $t_m(x, u) = +\infty$ for all x, u, which means that (2.1) is forward complete and (2.9) holds for all $t \ge 0$. Finally, since $\operatorname{ess\,sup}_{0\le s\le t}|u(s)| \le \|u\|_\infty$ for all $u \in L^\infty(\mathbb{R}_+, \mathbb{R}^m)$, (2.9) implies that (2.1) is ISS according to Definition 2.2.

"$\Rightarrow$". Let (2.1) be forward complete and ISS w.r.t. inputs from the class $L^\infty(\mathbb{R}_+, \mathbb{R}^m)$ with certain $\beta \in \mathcal{KL}$, $\gamma \in \mathcal{K}$. Pick any $u \in L^\infty_{\text{loc}}(\mathbb{R}_+, \mathbb{R}^m)$ and any $x \in \mathbb{R}^n$. Note that for any $t \ge 0$ and for any input u_t, defined by (2.7), the *causality property* $\phi(t, x, u) = \phi(t, x, u_t)$ holds. As $u_t \in L^\infty(\mathbb{R}_+, \mathbb{R}^m)$, ISS of (2.1) w.r.t. inputs from class $L^\infty(\mathbb{R}_+, \mathbb{R}^m)$ implies for all $t \ge 0$

$$|\phi(t, x, u)| = |\phi(t, x, u_t)| \le \beta(|x|, t) + \gamma\big(\operatorname{ess\,sup}_{0\le s\le t}|u_t(s)|\big).$$

This proves the claim with $t_m(x, u) = +\infty$. $\square$

2.2 ISS Lyapunov Functions

As a rule, to show a certain stability property of a nonlinear system, one constructs a corresponding Lyapunov function. In this section, we introduce ISS Lyapunov functions that will be our main tool for ISS analysis of nonlinear control systems. In fact, ISS Lyapunov functions are much more than just certificates for the ISS property. As we will see in Sect. 3.5, having ISS Lyapunov functions for all subsystems of the network, it is possible to show the stability of the interconnection provided that the so-called small-gain condition holds. Furthermore, in Chap. 5, we will see how one can use ISS Lyapunov functions to design various types of stabilizing controllers.

2.2.1 *Direct ISS Lyapunov Theorems*

There are two common ways to introduce the concept of a Lyapunov function.

Definition 2.9 A continuous function $V : \mathbb{R}^n \to \mathbb{R}_+$ is called an *ISS Lyapunov function in a dissipative form* for (2.1), if there exist $\psi_1, \psi_2 \in \mathcal{K}_\infty$, $\alpha \in \mathcal{K}_\infty$, and $\xi \in \mathcal{K}$ such that

$$\psi_1(|x|) \le V(x) \le \psi_2(|x|) \quad \forall x \in \mathbb{R}^n, \tag{2.10}$$

and for any $x \in \mathbb{R}^n$, $u \in \mathcal{U}$ the following inequality holds:

$$\dot{V}_u(x) \le -\alpha(V(x)) + \xi(\|u\|_\infty), \tag{2.11}$$

where *Lie derivative* $\dot{V}_u(x)$ corresponding to the pair (x, u) is the upper right-hand Dini derivative at zero of the function $t \mapsto V(\phi(t, x, u))$, that is:

$$\dot{V}_u(x) := \overline{\lim_{t \to +0}} \frac{1}{t}\big(V(\phi(t, x, u)) - V(x)\big). \tag{2.12}$$

The dissipative form of a Lyapunov function resembles the storage function in a theory of dissipative systems, see Sect. 2.11. This explains the name "dissipative".

Another common definition of an ISS Lyapunov function is:

Definition 2.10 A continuous function $V : \mathbb{R}^n \to \mathbb{R}_+$ is called an *ISS Lyapunov function in implication form* for (2.1), if there exist $\psi_1, \psi_2 \in \mathcal{K}_\infty$, $\alpha \in \mathcal{P}$, and $\chi \in \mathcal{K}$ such that (2.10) is satisfied and for any $x \in \mathbb{R}^n$, $u \in \mathcal{U}$ the following implication holds:

$$V(x) \ge \chi(\|u\|_\infty) \quad \Rightarrow \quad \dot{V}_u(x) \le -\alpha(V(x)). \tag{2.13}$$

The function χ will be called a *Lyapunov gain*, and α will be called *decay rate* of a Lyapunov function V.

We start with

Proposition 2.11 *If V is a continuous ISS Lyapunov function in a dissipative form, then V is an ISS Lyapunov function in implication form.*

Proof Let V be a continuous ISS Lyapunov function in a dissipative form so that (2.11) is satisfied. Whenever $\|u\|_\infty \le \xi^{-1}(\frac{1}{2}\alpha(V(x)))$ holds, which is the same as $V(x) \ge \alpha^{-1}(2\xi(\|u\|_\infty))$, the inequality $\dot{V}_u(x) \le -\frac{1}{2}\alpha(V(x))$ is satisfied. □

Specifying Definitions 2.9, 2.10 to systems without inputs (by setting $u \equiv 0$), we see that an ISS Lyapunov function is always a strict Lyapunov function (as introduced in Definition B.24). For $u \not\equiv 0$, the existence of an ISS Lyapunov function does not guarantee convergence of all trajectories to zero. However, according to the implication (2.13), if $V(x) \ge \chi(\|u\|_\infty)$ holds, the Lie derivative of V is negative and thus the trajectory converges to the region $\{x \in \mathbb{R}^n : V(x) \le \chi(\|u\|_\infty)\}$, which after some extra effort shows ISS property of (2.1). We render this informal argument more precisely in Theorem 2.12, which is a basis for a great part of applications of ISS theory.

Theorem 2.12 (Direct ISS Lyapunov theorem) *Let Assumption 2.1 hold. Then existence of an ISS Lyapunov function (in dissipative or implication form) for (2.1) implies ISS of (2.1).*

Proof As dissipative ISS Lyapunov functions are also ISS Lyapunov functions in implication form by Proposition 2.11, it suffices to show the claim only for ISS Lyapunov functions in implication form.

Pick any input $u \in \mathcal{U}$ and consider the set

$$S := \left\{x \in \mathbb{R}^n : V(x) \le \chi(\|u\|_\infty)\right\}.$$

Step 1. We are going to prove that S is forward invariant in the following sense:

$$x \in S \ \wedge \ v \in \mathcal{U} : \ \|v\|_\infty \le \|u\|_\infty \quad \Rightarrow \quad \phi(t, x, v) \in S \quad \forall t \ge 0. \tag{2.14}$$

Assume that this implication does not hold for certain $x \in S$ and $v \in \mathcal{U} : \ \|v\|_\infty \le \|u\|_\infty$. There are two alternatives.

First of all, it may happen that the maximal solution, corresponding to initial condition x and input v, exists on a finite time interval $[0, t_m(x, v))$ and $\phi(t, x, v) \in S$ for $t \in [0, t_m(x, v))$. However, this cannot be true in view of BIC property of (2.1).

The second alternative is that there exists $t_1 \in (0, t_m(x, v))$ and $\varepsilon > 0$ so that $V(\phi(t_1, x, v)) > \chi(\|u\|_\infty) + \varepsilon$. Define

$$t^* := \inf\left\{t \in [0, t_m(x, u)) : V(\phi(t, x, v)) > \chi(\|u\|_\infty) + \varepsilon\right\}.$$

As $x \in S$, and $\phi(\cdot, x, v)$ is continuous,

$$V\left(\phi(t^*, x, v)\right) = \chi(\|u\|_\infty) + \varepsilon \ge \chi(\|v\|_\infty).$$

Due to (2.13), it holds that

$$D^+V\big(\phi(t^*,x,v)\big) \le -\alpha\big(V(\phi(t^*,x,v))\big).$$

Denote $y(\cdot) := V(\phi(\cdot,x,v))$. Assume that there is $(t_k) \subset (t^*,+\infty)$, such that $t_k \to t^*$, $k \to \infty$ and $y(t_k) \ge y(t^*)$. Then

$$0 \le \overline{\lim_{k\to\infty}} \frac{y(t_k) - y(t^*)}{t_k - t^*} \le D^+y(t^*) \le -\alpha(y(t^*)) < 0,$$

a contradiction. Thus, there is some $\delta > 0$ such that $y(t) < y(t^*) = \chi(\|u\|_\infty) + \varepsilon$ for all $t \in (t^*, t^* + \delta)$. This contradicts to the definition of t^*. The statement (2.14) is shown.

Overall, for any $x \in S$ we have

$$\psi_1(|\phi(t,x,u)|) \le V(\phi(t,x,u)) \le \chi(\|u\|_\infty) \quad \forall t \ge 0,$$

which implies that for all $t \ge 0$

$$|\phi(t,x,u)| \le \psi_1^{-1} \circ \chi(\|u\|_\infty). \tag{2.15}$$

Step 2. Now let $x \notin S$, i.e., $V(x) > \chi(\|u\|_\infty)$. Define

$$\tau := \inf\{t \in [0, t_m(x,u)) : V\big(\phi(t,x,u)\big) \le \chi(\|u\|_\infty)\},$$

if this infimum is taken over a nonempty set. Otherwise, let $\tau := t_m(x,u)$. As $\phi(\cdot,x,u)$ is continuous, $\tau \in (0, t_m(x,u)]$. For $t \in [0,\tau)$, we have that

$$V\big(\phi(t,x,u)\big) \ge \chi(\|u\|_\infty) \ge \chi(\|u(t+\cdot)\|_\infty),$$

and thus for all $t \in [0,\tau)$ by the cocycle property and in view of the implication (2.13)

$$\begin{aligned}
\dot V_{u(t+\cdot)}(\phi(t,x,u)) &= \overline{\lim_{h\to+0}} \frac{1}{h}\Big(V\big(\phi\big(h,\phi(t,x,u),u(t+\cdot)\big)\big) - V\big(\phi(t,x,u)\big)\Big)\\
&= \overline{\lim_{h\to+0}} \frac{1}{h}\Big(V\big(\phi(t+h,x,u)\big) - V\big(\phi(t,x,u)\big)\Big)\\
&= D^+V(\phi(t,x,u))\\
&\le -\alpha\big(V(\phi(t,x,u))\big).
\end{aligned}$$

By comparison principle (Proposition A.35), there exists $\beta \in \mathcal{KL}$ (that depends on α solely) so that

$$V(\phi(t,x,u)) \le \beta(V(x),t), \quad t \in [0,\tau).$$

Thus,

$$\psi_1(|\phi(t,x,u)|) \leq V(\phi(t,x,u)) \leq \beta\left(\psi_2(|x|),t\right), \quad t \leq \tau. \tag{2.16}$$

In view of the BIC property of (2.1), it follows that $\tau < t_m(x,u)$, and thus

$$V\big(\phi(\tau,x,u)\big) = \chi(\|u\|_\infty).$$

Hence, $\phi(\tau,x,u) \in S$. For $t \geq \tau$, we have by the cocycle property that

$$\phi(t,x,u) = \phi\big(t-\tau, \phi(\tau,x,u), u(\tau+\cdot)\big).$$

As $\|u(\tau+\cdot)\|_\infty \leq \|u\|_\infty$, and $\phi(\tau,x,u) \in S$, the property (2.14) ensures that $\phi(t,x,u) \in S$ for all $t \geq \tau$, and thus for $t \geq \tau$ the estimate (2.15) is valid. We conclude from (2.15) and (2.16) that

$$|\phi(t,x,u)| \leq \tilde{\beta}\left(|x|,t\right) + \gamma(\|u\|_\infty) \quad \forall t > 0,$$

where $\tilde{\beta}(r,t) := \psi_1^{-1}\big(\beta(\psi_2(r),t)\big)$ for any $r,t \geq 0$. This proves that system (2.1) is ISS. □

2.2.2 *Scalings of ISS Lyapunov Functions*

Having one ISS Lyapunov function, one can scale it to construct a family of other ISS Lyapunov functions, which may possess better properties.

Definition 2.13 A *nonlinear scaling* is a function $\mu \in \mathcal{K}_\infty$, continuously differentiable on $(0,+\infty)$, satisfying $\mu'(s) > 0$ for all $s > 0$ and such that $\lim_{s\to 0}\mu'(s)$ exists, and is finite (i.e., belongs to $\mathbb{R}_+$).

We will need the following lemma on convex functions

Lemma 2.14 *Let $f : \mathbb{R}_+ \to \mathbb{R}$ be convex and differentiable. Then f' is nondecreasing on $(0,+\infty)$.*

Proof By definition, the convexity of f means that for any $x_1, x_2 \in \mathbb{R}_+$ and any $x \in (x_1,x_2)$ it holds that

$$f(\lambda_1 x_1 + \lambda_2 x_2) \leq \lambda_1 f(x_1) + \lambda_2 f(x_2), \tag{2.17}$$

for all $\lambda_1, \lambda_2 \in (0,1)$ with $\lambda_1 + \lambda_2 = 1$.

For any $x \in (x_1,x_2)$ take $\lambda_1 := \frac{x_2-x}{x_2-x_1}$ and $\lambda_2 := \frac{x-x_1}{x_2-x_1}$. Then $\lambda_1+\lambda_2 = 1$ and $x = \lambda_1 x_1 + \lambda_2 x_2$.

From (2.17) we obtain that

$$
\begin{aligned}
(x_2 - x)f(x) + (x - x_1)f(x) &= (x_2 - x_1)f(x) \\
&\le (x_2 - x)f(x_1) + (x - x_1)f(x_2),
\end{aligned}
$$

which implies

$$
\frac{f(x_2) - f(x)}{x_2 - x} \ge \frac{f(x) - f(x_1)}{x - x_1}.
$$

Taking the limit for $x \to x_2 - 0$, and the limit for $x \to x_1 + 0$, and recalling that f is differentiable, and thus the right and left derivatives coincide at each point, we obtain that

$$
f'(x_2) \ge \frac{f(x_2) - f(x_1)}{x_2 - x_1} \ge f'(x_1).
$$

This implies the claim. □

Proposition 2.15 *Let V be an ISS Lyapunov function* in a dissipative form *for (2.1). Then $\mu \circ V$ is an ISS Lyapunov function in a dissipative form for any convex nonlinear scaling μ.*

Proof Let V be as in Definition 2.9, μ be any convex nonlinear scaling and let $W := \mu \circ V$. Clearly, W satisfies the sandwich inequality

$$
\mu \circ \psi_1(|x|) \le W(x) \le \mu \circ \psi_2(|x|) \quad \forall x \in \mathbb{R}^n. \tag{2.18}
$$

By Lemma A.30, we have for any $x \in \mathbb{R}^n$ and any $u \in \mathcal{U}$:

$$
\begin{aligned}
\dot{W}_u(x) &= \mu'(V(x))\dot{V}_u(x) \le \mu'(V(x))\big(-\alpha(V(x)) + \xi(\|u\|_\infty)\big) \\
&= -\frac{1}{2}\mu'(V(x))\alpha(V(x)) + \mu'(V(x))\big(-\frac{1}{2}\alpha(V(x)) + \xi(\|u\|_\infty)\big).
\end{aligned}
$$

If $-\frac{1}{2}\alpha(V(x)) + \xi(\|u\|_\infty) \ge 0$, then $V(x) \le \alpha^{-1}\big(2\xi(\|u\|_\infty)\big)$. As μ is convex and continuously differentiable, Lemma 2.14 ensures that μ' is nondecreasing. Thus

$$
\dot{W}_u(x) \le -\frac{1}{2}\mu'(V(x))\alpha(V(x)) + \mu' \circ \alpha^{-1}\big(2\xi(\|u\|_\infty)\big)\xi(\|u\|_\infty). \tag{2.19}
$$

If $-\frac{1}{2}\alpha(V(x)) + \xi(\|u\|_\infty) \le 0$, then $\dot{W}_u(x) \le -\frac{1}{2}\mu'(V(x))\alpha(V(x))$, and (2.19) trivially holds. Overall, for all x, u we have

$$
\begin{aligned}
\dot{W}_u(x) \le &-\frac{1}{2}\mu'(\mu^{-1} \circ W(x))\alpha(\mu^{-1} \circ W(x)) \\
&+ \mu' \circ \alpha^{-1}\big(2\xi(\|u\|_\infty)\big)\xi(\|u\|_\infty),
\end{aligned} \tag{2.20}
$$

and thus W is an ISS Lyapunov function in a dissipative form. □

The assumption of convexity is essential and cannot be removed, see Exercise 2.9. At the same time, convexity is not needed for the corresponding result for ISS Lyapunov function in implication form.

Proposition 2.16 *Let V be an ISS Lyapunov function* in implication form *for (2.1). Then $\mu \circ V$ is an ISS Lyapunov functions in implication form for any nonlinear scaling μ.*

Proof We leave the proof as Exercise 2.9. □

In the dissipative form of the ISS Lyapunov function, we require the decay rate α to be a $\mathcal{K}_\infty$-function (see, however, Exercise 2.8), whereas in the implication form of the ISS Lyapunov function concept, α can be merely a $\mathcal{P}$-function. As the following proposition shows, by suitable scaling of the ISS Lyapunov function (in implication form), we can always make the decay rate a $\mathcal{K}_\infty$-function. Later, in Theorem 2.70, we will show a stronger result that the decay rate can always be chosen to be even a linear function.

Proposition 2.17 *Let $V \in C(\mathbb{R}^n, \mathbb{R}_+)$ be an ISS Lyapunov function in implication form with a decay rate $\alpha \in \mathcal{P}$. Then there is a nonlinear scaling $\xi \in \mathcal{K}_\infty$, which belongs to $C^1(\mathbb{R}_+, \mathbb{R}_+)$, is infinitely differentiable on $(0, +\infty)$, and such that $\xi \circ V(\cdot)$ is an ISS Lyapunov function in implication form with a decay rate belonging to $\mathcal{K}_\infty$.*

Proof Let $V \in C(\mathbb{R}^n, \mathbb{R}_+)$ be an ISS Lyapunov function in implication form with the bounds $\psi_1, \psi_2 \in \mathcal{K}_\infty$, a decay rate $\alpha \in \mathcal{P}$, and a Lyapunov gain $\chi \in \mathcal{K}_\infty$. By Proposition A.13, there are $\omega \in \mathcal{K}_\infty$ and $\sigma \in \mathcal{L}$ such that

$$\alpha(r) \geq \omega(r)\sigma(r), \quad r \geq 0. \tag{2.21}$$

Without loss of generality, we can assume that σ is infinitely differentiable on $(0, +\infty)$ (otherwise take $\frac{1}{2}\sigma$ and smooth it).

From (2.13) it follows that for all $x \in \mathbb{R}^n$ and $u \in \mathcal{U}$

$$V(x) \geq \chi(\|u\|_\infty) \quad \Rightarrow \quad \dot{V}_u(x) \leq -\omega(V(x))\sigma(V(x)), \tag{2.22}$$

and thus for all $x \in \mathbb{R}^n$ we have

$$V(x) \geq \chi(\|u\|_\infty) \quad \Rightarrow \quad \frac{1}{\sigma(V(x))}\dot{V}_u(x) \leq -\omega(V(x)). \tag{2.23}$$

Define $\xi \in C^1(\mathbb{R}_+, \mathbb{R}_+)$ by

$$\xi(r) := \int_0^r \frac{1}{\sigma(s)} ds, \quad r \geq 0. \tag{2.24}$$

As σ is infinitely differentiable and never zero, ξ is well-defined on $\mathbb{R}_+$ and infinitely differentiable. It is easy to see that $\xi \in \mathcal{K}_\infty$ is a nonlinear scaling. Define $W := \xi \circ V$ and observe by exploiting Proposition A.30 that $\dot{W}_u(x) = \frac{1}{\sigma(V(x))}\dot{V}_u(x)$ and thus (2.23) can be transformed into

$$W(x) \geq \xi \circ \chi(\|u\|_\infty) \quad \Rightarrow \quad \dot{W}_u(x) \leq -\omega \circ \xi^{-1}(W(x)), \tag{2.25}$$

and $\omega \circ \xi^{-1} \in \mathcal{K}_\infty$ as a composition of $\mathcal{K}_\infty$-functions. As

$$\xi \circ \psi_1(|x|) \leq W(x) \leq \xi \circ \psi_2(|x|) \quad \forall x \in \mathbb{R}^n, \tag{2.26}$$

we verify that W is an ISS Lyapunov function in implication form with $\mathcal{K}_\infty$-decay rate. □

2.2.3 *Continuously Differentiable ISS Lyapunov Functions*

Theorem 2.12 is quite general, as it allows for arbitrary continuous ISS Lyapunov functions, which are not necessarily differentiable. In this form, Theorem 2.12 can be extended to a very broad class of infinite-dimensional systems, see Chap. 6. We pay for the generality with the need to compute the Lie derivative of V in the inequality 2.13. In this section, we give a computationally efficient reformulation of the *continuously differentiable* ISS Lyapunov function concept.

For systems satisfying Assumption 1.8 with continuous inputs, Lie derivative of a continuously differentiable V can be easily computed.

Lemma 2.18 *Let Assumption 1.8 hold and let $V \in C^1(\mathbb{R}^n, \mathbb{R})$. Then for all $x \in \mathbb{R}^n$ and all continuous inputs $u \in \mathcal{U}$ it holds that*

$$\dot{V}_u(x) = \nabla V(x) \cdot f(x, u(0)). \tag{2.27}$$

Proof Pick any continuous input $u \in \mathcal{U}$ and any $x \in \mathbb{R}^n$. As f is assumed to be continuous w.r.t. the second argument, and u is continuous, the solution $\phi(\cdot, x, u)$ to (2.1) with an initial condition x is continuously differentiable, and thus $V(\phi(\cdot, x, u))$ is a composition of differentiable functions and hence

$$\begin{aligned}\dot{V}_u(x) = \overline{\lim_{t \to +0}} \frac{V(\phi(t, x, u)) - V(x)}{t} &= \frac{d}{dt} V(\phi(t, x, u))|_{t=0} \\ &= \nabla V(x) \cdot \dot{\phi}(t, x, u)|_{t=0} \\ &= \nabla V(x) \cdot f(x, u(0)).\end{aligned}$$

□

Next we restate the definition of a *continuously differentiable* ISS Lyapunov function in a computationally more efficient manner.

Proposition 2.19 *Let Assumption 1.8 hold. A function $V \in C^1(\mathbb{R}^n, \mathbb{R}_+)$ is an ISS Lyapunov function in implication form for (2.1) if and only if there exist $\psi_1, \psi_2 \in \mathcal{K}_\infty$, $\alpha \in \mathcal{P}$, and $\chi \in \mathcal{K}$ such that (2.10) holds and for all $x \in \mathbb{R}^n$ and for any input value $v \in \mathbb{R}^m$ the following implication holds:*

$$V(x) \geq \chi(|v|) \quad \Rightarrow \quad \nabla V(x) \cdot f(x, v) \leq -\alpha(V(x)). \tag{2.28}$$

Proof "⇒". Let $V \in C^1(\mathbb{R}^n, \mathbb{R}_+)$ be an ISS Lyapunov function as in Definition 2.10. Pick any $v \in \mathbb{R}^m$ and consider a constant input $u(\cdot) \equiv v$. Then (2.13) shows for all $x \in \mathbb{R}^n$ that

$$V(x) \geq \chi(|v|) \quad \Rightarrow \quad \dot{V}_u(x) \leq -\alpha(V(x)). \tag{2.29}$$

Since $u(\cdot) \equiv v$ is constant and $f(x, v)$ (for a fixed v) is Lipschitz continuous, we see as in Lemma 2.18 that $\dot{V}_u(x) = \nabla V(x) \cdot f(x, v)$. Substituting this expression into (2.29), we obtain the claim.

"⇐". Pick any $\varepsilon > 0$, $u \in \mathcal{U}$, any $x \in \mathbb{R}^n$ so that

$$V(x) \geq (1 + \varepsilon)\chi(\|u\|_\infty). \tag{2.30}$$

As both V and the trajectory $\phi(\cdot, x, u)$ are continuous, there is a time $t > 0$ so that for almost all $s \in [0, t]$ it holds that $V(\phi(s, x, u)) \geq \chi(|u(s)|)$.

As $V \in C^1(\mathbb{R}^n, \mathbb{R}_+)$ and ϕ is absolutely continuous w.r.t. t as a solution of an ODE, the map $s \mapsto V(\phi(s, x, u))$ is absolutely continuous by Propositions 1.49, 1.46. Hence, this map is differentiable on $[0, t]$ almost everywhere and thus for a.e. $s \in [0, t]$ it holds that

$$\frac{d}{ds} V(\phi(s, x, u)) = \nabla V(\phi(s, x, u)) \cdot f(\phi(s, x, u), u(s)) \leq -\alpha\big(V(\phi(s, x, u))\big).$$

As $s \mapsto V(\phi(s, x, u))$ is absolutely continuous, we can apply the fundamental theorem of calculus (Theorem 1.2):

$$\frac{1}{t}\big(V(\phi(t, x, u)) - V(x)\big) = \frac{1}{t}\int_0^t \frac{d}{ds} V(\phi(s, x, u))ds \leq \frac{1}{t}\int_0^t -\alpha\big(V(\phi(s, x, u))\big)ds,$$

and computing the limit superior $t \to +0$ from both sides, we obtain for all $x \in \mathbb{R}^n$ and $u \in \mathcal{U}$ satisfying (2.30) that:

$$\dot{V}_u(x) \leq -\alpha(|x|),$$

and thus V is an ISS Lyapunov function in implication form according to Definition 2.10. □

Similarly, for dissipative ISS Lyapunov functions the following criterion can be shown:

Proposition 2.20 *Let Assumption 1.8 hold. A function $V \in C^1(\mathbb{R}^n, \mathbb{R}_+)$ is an ISS Lyapunov function in the dissipative form for (2.1) if and only if there exist $\psi_1, \psi_2 \in \mathcal{K}_\infty$, $\alpha \in \mathcal{K}_\infty$, and $\xi \in \mathcal{K}$ such that (2.10) holds and for all $x \in \mathbb{R}^n$ and for any input value $v \in \mathbb{R}^m$ the following dissipation inequality holds:*

$$\nabla V(x) \cdot f(x, v) \leq -\alpha(V(x)) + \xi(|v|). \tag{2.31}$$

Proof "⇒". Same as in Proposition 2.19.

"⇐". Pick any $x \in X$ and $u \in \mathcal{U}$. As $V \in C^1(\mathbb{R}^n, \mathbb{R}_+)$ and ϕ is absolutely continuous w.r.t. t as a solution of an ODE, the map $s \mapsto V(\phi(s, x, u))$ is differentiable on $[0, t]$ almost everywhere and at all such points it holds that

$$\begin{aligned}\frac{d}{ds}V(\phi(s, x, u)) &= \nabla V(\phi(s, x, u)) \cdot f(\phi(s, x, u), u(s)) \\ &\leq -\alpha\big(V(\phi(s, x, u))\big) + \xi(|u(s)|).\end{aligned}$$

Applying the fundamental theorem of calculus (Theorem 1.2), we obtain

$$\begin{aligned}\frac{1}{t}\big(V(\phi(t, x, u)) - V(x)\big) &= \frac{1}{t}\int_0^t \frac{d}{ds}V(\phi(s, x, u))ds \\ &\leq \frac{1}{t}\int_0^t -\alpha\big(V(\phi(s, x, u))\big) + \xi(|u(s)|)ds \\ &\leq \frac{1}{t}\int_0^t -\alpha\big(V(\phi(s, x, u))\big)ds + \xi(\|u\|_\infty),\end{aligned}$$

and computing the limit superior $t \to +0$ from both sides, we obtain for all $x \in \mathbb{R}^n$ and $u \in \mathcal{U}$ that:

$$\dot{V}_u(x) \leq -\alpha(|x|) + \xi(\|u\|_\infty),$$

and thus V is an ISS Lyapunov function in the dissipative form according to Definition 2.9. □

As we have seen in Proposition 2.11, any ISS Lyapunov function in a dissipative form is also an ISS Lyapunov function in implication form. Whether the converse holds is less clear, as in the case of ISS Lyapunov functions in implication form, we do not know anything about the behavior of $\dot{V}_u(x)$ in the region $V(x) \leq \chi(\|u\|_\infty)$.

However, for continuously differentiable ISS Lyapunov functions and sufficiently regular nonlinearities f also the converse holds.

Proposition 2.21 *Let Assumption 1.8 hold and $V \in C^1(\mathbb{R}^n, \mathbb{R}_+)$. Then V is an ISS Lyapunov function in implication form if and only if V is an ISS Lyapunov function in a dissipative form.*

Proof "$\Leftarrow$". Was shown for continuous V in Proposition 2.11.

"$\Rightarrow$". Let V be an ISS Lyapunov function in implication form. By Proposition 2.19, for V the implication (2.28) holds for all $x \in \mathbb{R}^m$ and all $u \in \mathbb{R}^m$. Define $\bar{\xi}(r) := \max\{0, \xi^*(r)\}$, where

$$\xi^*(r) := \max\{\nabla V(x) \cdot f(x, v) + \alpha(V(x)) : |v| \le r,\ V(x) \le \chi(r)\}.$$

The maximum in the definition of ξ^* exists since we are maximizing a continuous function over a compact set. Furthermore, V is a Lyapunov function for a system (2.1) with $u \equiv 0$, which shows that $x \equiv 0$ is an equilibrium for the undisturbed system (2.1), and thus $f(0, 0) = 0$. Hence

$$\xi^*(0) = \nabla V(0) \cdot f(0, 0) + \alpha(V(0)) = 0,$$

and thus $\bar{\xi}(0) = 0$. Similarly, $\bar{\xi}$ is continuous at zero (since $\xi^*(r) \to 0$ as $r \to +0$).

By construction, it holds that $\bar{\xi}$ is nondecreasing and $\bar{\xi}(0) = 0$. Thus, by Proposition A.16, the function $\bar{\xi}$ can be upper-bounded by a $\mathcal{K}_\infty$-function ξ in a way that $\xi(r) \ge \bar{\xi}(r)$ for all $r \in \mathbb{R}_+$. Now

$$V(x) \le \chi(|u|) \quad \Rightarrow \quad \nabla V(x) \cdot f(x, u) + \alpha(V(x)) \le \xi(|u|). \tag{2.32}$$

with α from (2.28) and with the constructed ξ.

Combining this inequality with (2.28) we see that (2.31) holds for all x, u and by Proposition 2.20, V is an ISS Lyapunov function in a dissipative form. □

Let us illustrate the usefulness of the ISS Lyapunov function concept on an academic example.

Example 2.22 Consider a scalar system

$$\dot{x} = -x + u \ln(|x| + 1), \quad x(t), u(t) \in \mathbb{R}. \tag{2.33}$$

We show ISS of this system by verifying that a continuously differentiable function $V(x) = x^2$, $x \in \mathbb{R}$ is an ISS Lyapunov function for (2.33) in implication form.

For all $x \in \mathbb{R}$ and $u \in \mathbb{R}$ we have:

$$\begin{aligned}\dot{V}(x) = \frac{\mathrm{d}V}{\mathrm{d}x}(x) \cdot f(x, u) &= 2x(-x + u\log(|x| + 1)) \\ &\le -2x^2 + 2|x||u|\log(|x| + 1).\end{aligned}$$

Define

$$\chi^{-1}(r) := \begin{cases} \frac{r}{2\log(r+1)} - \frac{1}{2} & \text{, if } r > 0, \\ 0 & \text{, if } r = 0. \end{cases}$$

It is an easy exercise to show that χ^{-1} is continuous and strictly increasing to infinity. In other words, $\chi^{-1} \in \mathcal{K}_\infty$ and thus also $\chi \in \mathcal{K}_\infty$.

Then $\chi(|u|) \le |x|$ (which is the same as $|u| \le \chi^{-1}(|x|) = \frac{|x|}{2\log(|x|+1)} - \frac{1}{2}$) implies

$$\nabla V(x) \cdot f(x, u) \le -|x|^2 - |x| \log(|x| + 1) =: -\alpha(|x|).$$

This shows that V is an ISS Lyapunov function in implication form for (2.33) and hence (2.33) is ISS (we are using here Exercise 2.6).

2.2.4 *Example: Robust Stabilization of a Nonlinear Oscillator*

Consider the equation of motion of the *nonlinear elastic oscillator*

$$\ddot{y}(t) = -k_1(y(t)) + u(t), \tag{2.34}$$

where the *elastic law* $k_1 : \mathbb{R} \to \mathbb{R}$ is a Lipschitz continuous function satisfying $k_1(0) = 0$, $sk_1(s) > 0$ for all $s \in \mathbb{R}\backslash\{0\}$, and there is $\eta_1 \in \mathcal{K}_\infty$ such that $|k_1(s)| \ge \eta_1(|s|)$ for all $s \in \mathbb{R}$. An input u represents an external force applied to the oscillator.

In the absence of external forces, i.e., if $u \equiv 0$, this system is globally stable, but it is not globally asymptotically stable, which can be shown by means of a suitable Lyapunov function (see below). A natural question is how to choose the force u to make this system stable in a certain sense. We divide our analysis into several parts.

Step 1. Let us use a so-called linear *proportional-derivative (PD) control law*:

$$u(t) := -ay(t) - \mu\dot{y}(t) + d(t), \tag{2.35}$$

where $a, \mu > 0$ and $d \in L^\infty(\mathbb{R}_+, \mathbb{R})$ is an unknown disturbance input, which appears since any realistic control law is never exact and is subject to disturbances of various kinds (e.g., other forces acting on the system).

Substituting this control law into (2.34), and defining

$$x_1 := y, \; x_2 := \dot{y}, \; k(x_1) := k_1(x_1) + ax_1,$$

and denoting $x := (x_1, x_2) \in \mathbb{R}^2$, we obtain a closed-loop system

$$\dot{x}_1 = x_2, \tag{2.36a}$$

$$\dot{x}_2 = -k(x_1) - \mu x_2 + d. \tag{2.36b}$$

We are going to prove that (2.36) is input-to-state stable, and thus the control law u stabilizes the system (2.34), robustly (in ISS sense) with respect to additive disturbances. Consider the following ISS Lyapunov function candidate

$$V(x) := \frac{\mu^2}{2}x_1^2 + \mu x_1 x_2 + x_2^2 + 2\int_0^{x_1} k(r)dr, \quad x = (x_1, x_2) \in \mathbb{R}^2. \tag{2.37}$$

As k is continuous, V is continuously differentiable. Let us show that V is an ISS Lyapunov function for the system (2.36) using Proposition 2.20.

Clearly, $V(0) = 0$ and for a suitable $C > 0$ and any $x \in \mathbb{R}^2$ it holds that

$$V(x) \geq \frac{\mu^2}{2}x_1^2 + \mu x_1 x_2 + x_2^2 \geq C|x|^2.$$

This implies that V is positive definite and radially unbounded, and Proposition A.14 and Corollary A.11 show that (2.10) holds for certain $\psi_1, \psi_2 \in \mathcal{K}_\infty$.

By assumptions on k_1, it holds that

$$|k(r)| = a|r| + |k_1(r)| \geq a|r| + \eta_1(|r|) =: \eta(|r|), \quad r \in \mathbb{R}. \tag{2.38}$$

Now let us compute

$$\begin{aligned}
\nabla V(x) \cdot f(x, d) &= \big(\mu^2 x_1 + \mu x_2 + 2k(x_1)\big)x_2 + \big(\mu x_1 + 2x_2\big)\big(-k(x_1) - \mu x_2 + d\big)\\
&= -\mu x_2^2 - \mu x_1 k(x_1) + \big(\mu x_1 + 2x_2\big)d\\
&\leq -\mu x_2^2 - \mu|x_1 k(x_1)| + \mu|x_1||d| + 2|x_2||d|\\
&\leq -\mu x_2^2 - \mu|x_1|\eta(|x_1|) + \mu|x_1||d| + 2|x_2||d|,
\end{aligned}$$

where in the last estimate we have used (2.38).

Employing $\mathcal{K}_\infty$-inequality A.2 with $r = \frac{1}{2}$ and $\alpha := \eta$, we have that

$$|x_1||d| \leq \frac{1}{2}|x_1|\eta(|x_1|) + |d|\eta^{-1}(2|d|). \tag{2.39}$$

Using this estimate and utilizing Young's inequality

$$2|x_2||d| \leq \frac{\mu}{2}|x_2|^2 + \frac{2}{\mu}|d|^2,$$

we obtain

$$\begin{aligned}\nabla V(x)\cdot f(x,d) &\leq -\mu x_2^2 - \mu|x_1|\eta(|x_1|) + \frac{\mu}{2}|x_1|\eta(|x_1|)\\ &\quad +\mu|d|\eta^{-1}(2|d|) + \frac{\mu}{2}|x_2|^2 + \frac{2}{\mu}|d|^2\\ &= -\frac{1}{2}\mu\big(x_2^2 + |x_1|\eta(|x_1|)\big) + \mu|d|\eta^{-1}(2|d|) + \frac{2}{\mu}|d|^2.\end{aligned}$$

As $(x_1, x_2) \mapsto \frac{1}{2}\mu(x_2^2 + |x_1|\eta(|x_1|))$ is positive definite and radially unbounded, by Proposition A.14 there is some $\alpha \in \mathcal{K}_\infty$, such that $\frac{1}{2}\mu(x_2^2 + |x_1|\eta(|x_1|)) \geq \alpha(|x|)$, and thus

$$\nabla V(x)\cdot f(x,d) \leq -\alpha(|x|) + \mu|d|\eta^{-1}(2|d|) + \frac{2}{\mu}|d|^2,$$

which shows that $V \in C^1(\mathbb{R}^2, \mathbb{R}_+)$ is an ISS Lyapunov function in a dissipative form, and by Proposition 2.21 the system (2.36) is ISS.

In other words, the nonlinear oscillator (2.34) can be globally asymptotically stabilized (robustly with respect to disturbances d) by means of PD controller (2.35).

Step 2. At the same time, a *proportional controller* (obtained by setting $\mu := 0$ into (2.35))

$$u(t) := -ay(t) + d(t) \tag{2.40}$$

where $a > 0$ and $d \in L^\infty(\mathbb{R}_+, \mathbb{R})$ is a disturbance input, makes the system (2.34) merely globally stable (for any $a > 0$) if $d \equiv 0$, and furthermore for any initial condition there is a disturbance of an arbitrarily small magnitude for which the corresponding trajectory of (2.34) with the feedback law (2.40) converges to infinity as $t \to \infty$. To see this, we again write the equations of motion as a planar system of first-order ODEs

$$\dot{x}_1 = x_2, \tag{2.41a}$$
$$\dot{x}_2 = -k(x_1) + d. \tag{2.41b}$$

We consider the function $W : \mathbb{R}^2 \to \mathbb{R}_+$, obtained from V by setting $\mu := 0$:

$$W(x) := x_2^2 + 2\int_0^{x_1} k(r)\mathrm{d}r, \quad x = (x_1, x_2) \in \mathbb{R}^2. \tag{2.42}$$

It holds that

$$\dot{W}(x) = 2x_2\big(-k(x_1) + d\big) + 2k(x_1)x_2 = 2x_2 d. \tag{2.43}$$

If $d \equiv 0$, then the "energy" W is conserved, that is,

$$W(x(t)) = W(x(0)), \quad t \ge 0, \tag{2.44}$$

for any solution x of (2.34) with $d \equiv 0$.

As W is positive definite and radially unbounded, by Proposition A.14 and Corollary A.11 there are $\omega_1, \omega_2 \in \mathcal{K}_\infty$ such that

$$\omega_1(|x|) \le W(x) \le \omega_2(|x|), \quad x \in \mathbb{R}^2.$$

Consequently,

$$\omega_1(|x(t)|) \le W(x(t)) = W(x(0)) \le \omega_2(|x(0)|),$$

and thus

$$|x(t)| \le \omega_1^{-1} \circ \omega_2(|x(0)|), \quad t \ge 0.$$

At the same time, if $x(t) \to 0$ as $t \to \infty$, then also $W(x(t)) \to 0$ as $t \to \infty$ since W is a continuous function. However, this contradicts to (2.44).

As the energy of the undisturbed system is conserved, and in view of (2.43), one can easily design for any initial condition x_0 a disturbance d or arbitrarily small magnitude, such that $W(\phi(t, x_0, d)) \to \infty$ as $t \to \infty$.

2.2.5 *Lyapunov Criterion for ISS of Linear Systems*

Unfortunately, there is no general method for constructing ISS Lyapunov functions for nonlinear control systems. In contrast, for linear ISS systems (2.2), it is always possible to find a quadratic ISS Lyapunov function, and in this section, we explain how.

Definition 2.23 A symmetric matrix $M \in \mathbb{R}^{n\times n}$ is called

(i) *positive definite* (we write $M > 0$) if $x^T Mx > 0$ for all $x \neq 0$;
(ii) *negative definite* (we write $M < 0$) if $x^T Mx < 0$ for all $x \neq 0$.

We will need the following technical result, see [46, Lemma 5.7.18].

Proposition 2.24 *Let $M, N \in \mathbb{R}^{p\times p}$ be two Hurwitz matrices. Then for any $Q \in \mathbb{R}^{p\times p}$ there exists the unique solution of the* Sylvester equation

$$MX + XN = Q, \tag{2.45}$$

given by $P = \int_0^\infty e^{Mt} Q e^{Nt} dt$.

Next theorem provides a criterion for ISS of a linear system (2.2) and a construction of ISS Lyapunov functions for general ISS linear systems.

Theorem 2.25 *The following statements are equivalent:*

(i) (2.2) is ISS.
(ii) (2.2) is 0-GAS.
(iii) A in (2.2) is Hurwitz.
(iv) For any $Q > 0$ there exists a unique $P > 0$ that solves the Lyapunov equation

$$A^T P + PA = -Q. \tag{2.46}$$

(v) There are $Q > 0$ and $P > 0$ that satisfy (2.46).

Moreover, the function

$$V(x) = x^T Px \tag{2.47}$$

with P as in (iv) is an ISS Lyapunov function for (2.2).

Proof (i) ⇒ (ii). Clear.

(ii) ⇔ (iii). Well-known, see, e.g., [53, Corollary 3.6].

(iii) ⇒ (iv). According to Proposition 2.24, for any $Q \in \mathbb{R}^{n\times n}$ there exists a unique solution of (2.46) given by $P = \int_0^\infty (e^{At})^T Q e^{At} dt$. It is easy to see that if $Q > 0$ then also $P > 0$.

(iv) ⇒ (v). Clear.

(v) ⇒ (i). Pick any $Q > 0$ and let P be the corresponding solution of (2.46). We show next that V defined by (2.47) is an ISS Lyapunov function for (2.2).

Since $P > 0$, it holds that

$$\lambda_{\min}(P)|x|^2 \le V(x) \le \lambda_{\max}(P)|x|^2,$$

where $\lambda_{\min}(P)$ and $\lambda_{\max}(P)$ are the smallest and the largest eigenvalues of P (both positive real numbers). Thus, the condition (2.10) is satisfied. Moreover,

$$\begin{aligned}
\dot V(x) &= \dot x^T Px + x^T P\dot x \\
&= x^T A^T Px + x^T PAx + 2x^T PBu \\
&= x^T (A^T P + PA)x + 2x^T PBu \\
&= -x^T Qx + 2x^T PBu \\
&\le -\lambda_{\min}(Q)|x|^2 + 2|x|\|P\|\|B\||u| \\
&\le -\lambda_{\min}(Q)|x|^2 + \varepsilon|x|^2 + \frac{1}{\varepsilon}\|P\|^2\|B\|^2|u|^2 \\
&\le \big(\varepsilon - \lambda_{\min}(Q)\big)\frac{1}{\lambda_{\max}(P)} V(x) + \frac{1}{\varepsilon}\|P\|^2\|B\|^2|u|^2,
\end{aligned}$$

for all $x \neq 0$ and all $\varepsilon \in (0, \lambda_{\min}(Q)) > 0$ and all $u \in \mathbb{R}^m$. Choosing $\varepsilon > 0$ small enough, this shows that V is an (exponential) ISS Lyapunov function for (2.2), and hence (2.2) is ISS. □

2.2.6 Example: Stability Analysis of Neural Networks

Quadratic Lyapunov functions, constructed in Theorem 2.25, can also be used for the stability analysis of nonlinear systems.

Consider a system of the form

$$\dot{x} = Ax + B\sigma(x) + Cw(x)G(u), \tag{2.48}$$

where $x \in \mathbb{R}^n$, $\sigma : \mathbb{R}^n \to \mathbb{R}^n$, $w : \mathbb{R}^n \to \mathbb{R}^{n\times n}$, $u \in \mathbb{R}^m$, $G : \mathbb{R}^m \to \mathbb{R}^n$.

Such systems can model *dynamic neural networks*. In this case, the usual choice for the entries of σ and w is *sigmoid* functions, e.g.,

$$\sigma(x) := (\sigma_i(x_i))_{i=1}^n, \quad \sigma_i(r) = \frac{a_i}{1+e^{-b_i r}} + k_i, \quad x = (x_i)_{i=1}^n, \quad r > 0,$$

where $a_i, b_i, k_i > 0$.

Assumption 2.26 Suppose that the following holds:

(i) σ, w are globally Lipschitz continuous and uniformly bounded on $\mathbb{R}^n$, i.e.,

$$\sup_{x\in\mathbb{R}^n} |\sigma(x)| < \infty, \quad \sup_{x\in\mathbb{R}^n} |w(x)| < \infty.$$

(ii) A is Hurwitz.
(iii) G is continuous.
(iv) $\exists \xi \in \mathcal{K}_\infty$: $|G(u)| \le \xi(|u|)$ for all $u \in \mathbb{R}^m$.

Assumption 2.26 implies that (2.48) is well-posed. Furthermore, the undisturbed system (2.48) (with $u \equiv 0$) has always a nontrivial equilibrium:

Proposition 2.27 *Let Assumption 2.26 hold. For $u \equiv 0$, the equilibrium points of the system (2.48) are given by the solutions of*

$$x^* = -A^{-1}B\sigma(x^*). \tag{2.49}$$

Furthermore, (2.48) has at least one equilibrium point.

Proof As A is Hurwitz, it is invertible, and we easily see that the fixed points of (2.48) satisfy (2.49).

To show the existence of a fixed point, consider a map

$$p : x \mapsto -A^{-1}B\sigma(x), \quad x \in \mathbb{R}^n.$$

As σ is globally Lipschitz, p is well-defined and continuous. Define $M := \sup_{x\in\mathbb{R}^n} |\sigma(x)| < \infty$ and consider the closed ball

$$S := \{x \in \mathbb{R}^n : |x| \le \|A^{-1}\|\|B\|M\}.$$

For any $x \in S$ we have that

$$|p(x)| \leq \|A^{-1}\|\|B\||\sigma(x)| \leq \|A^{-1}\|\|B\|M,$$

and thus S is invariant w.r.t. p. Brouwer's Theorem A.39 implies that p has a fixed point in S. □

To apply the ISS Lyapunov technique, we perform a change of coordinates:

$$y := x - x^*.$$

In the new coordinates, the system (2.48) has the form

$$\begin{aligned} \dot{y} &= A(y + x^*) + B\sigma(y + x^*) + Cw(y + x^*)G(u) \\ &= Ay + B\big(\sigma(y + x^*) - \sigma(x^*)\big) + Cw(y + x^*)G(u) + Ax^* + B\sigma(x^*) \\ &= Ay + B\big(\sigma(y + x^*) - \sigma(x^*)\big) + Cw(y + x^*)G(u), \end{aligned} \tag{2.50}$$

where in the last computation we have used (2.49).

Using Lyapunov technique, we can obtain the following sufficient condition for ISS of (2.50).

Proposition 2.28 *Let Assumption 2.26 hold. Let $P = P^T > 0$ satisfy Lyapunov inequality*

$$A^T P + PA \leq -I.$$

Let L_σ be the global Lipschitz constant for σ. If

$$2\|P\|\|B\|L_\sigma < 1, \tag{2.51}$$

then the system (2.50) is ISS.

Proof Consider an ISS Lyapunov function candidate

$$V(y) := y^T P y, \quad y \in \mathbb{R}^n.$$

Further define $M := \sup_{x \in \mathbb{R}^n} |w(x)| < \infty$. Differentiating, we obtain that

$$\begin{aligned} \dot{V}_u(y) &\leq -|y|^2 + 2y^T P\big(B\big(\sigma(y + x^*) - \sigma(x^*)\big) + Cw(y + x^*)G(u)\big) \\ &\leq -|y|^2 + 2|y|\|P\|\|B\||\sigma(y + x^*) - \sigma(x^*)| \\ &\quad + 2|y|\|P\|\|C\||w(y + x^*)||G(u)| \\ &\leq -|y|^2 + 2|y|^2\|P\|\|B\|L_\sigma + 2M|y|\|P\|\|C\|\xi(|u|). \end{aligned}$$

Using Cauchy's inequality, we obtain that for any $\varepsilon > 0$

$$\begin{aligned}\dot{V}_u(y) &\le -|y|^2 + 2|y|^2\|P\|\|B\|L_\sigma + \varepsilon|y|^2 + \frac{1}{\varepsilon}M^2\|P\|^2\|C\|^2\xi^2(|u|)\\ &= -\big(1 - 2\|P\|\|B\|L_\sigma - \varepsilon\big)|y|^2 + \frac{1}{\varepsilon}M^2\|P\|^2\|C\|^2\xi^2(|u|).\end{aligned}$$

Assuming (2.51) and choosing $\varepsilon > 0$ small enough, we see that V is an ISS Lyapunov function for (2.50), and thus (2.50) is ISS. □

2.3 Local Input-to-State Stability

Consider a local version of the ISS property:

Definition 2.29 System (2.1) is called *locally input-to-state stable (locally ISS, LISS)*, if there exist $\beta \in \mathcal{KL}$, $\gamma \in \mathcal{K}$, and $r > 0$ such that

$$t \ge 0,\ |x| \le r,\ \|u\|_\infty \le r \quad \Rightarrow \quad |\phi(t, x, u)| \le \beta(|x|, t) + \gamma(\|u\|_\infty). \tag{2.52}$$

In this section, we show that local ISS is equivalent to the local asymptotic stability of the undisturbed system (2.1), provided that the nonlinearity f satisfies natural regularity requirements. This implies that the *local* asymptotic stability property for undisturbed systems (2.1) is intrinsically robust against small disturbances.

Definition 2.30 A continuous function $V : D \to \mathbb{R}_+$, $D \subset \mathbb{R}^n$ is called a *LISS Lyapunov function in a dissipative form*, if there exist $r > 0$, $\psi_1, \psi_2 \in \mathcal{K}_\infty$, $\alpha \in \mathcal{P}$, and $\xi \in \mathcal{K}$ such that $B_r \subset D$,

$$\psi_1(|x|) \le V(x) \le \psi_2(|x|), \quad x \in B_r, \tag{2.53}$$

and

$$|x| \le r,\ \|u\|_\infty \le r \quad \Rightarrow \quad \dot{V}_u(x) \le -\alpha(V(x)) + \xi(\|u\|_\infty). \tag{2.54}$$

In view of Remark 2.31, without loss of generality, one can assume that $\alpha \in \mathcal{K}_\infty$.

Remark 2.31 Assume that V satisfies all the properties of Definition 2.30, with $\alpha \in \mathcal{P}$. Note that whenever $|x| \le r$, we have that $V(x) \le \psi_2(r)$. Define $\tilde{\alpha} \in \mathcal{P}$ by $\tilde{\alpha}(s) = \alpha(s)$, $s \in [0, \psi_2(r)]$ and $\tilde{\alpha}$ is continuous, and monotonically increasing to infinity on $[\psi_2(r), +\infty)$. Then clearly (2.54) holds with $\tilde{\alpha}$ instead of α. Finally, since $\lim_{r\to\infty}\tilde{\alpha}(r) = +\infty$, by Lemma A.28 there is $\rho \in \mathcal{K}_\infty$ such that $\tilde{\alpha} \ge \rho$ on $\mathbb{R}_+$. Hence (2.54) holds also with ρ instead of α.

Analogously to Theorem 2.12 one can obtain a direct LISS Lyapunov theorem:

Theorem 2.32 *If there is a LISS Lyapunov function for (2.1), then (2.1) is LISS.*

Proof One of the strategies to prove this result is first to transform a LISS Lyapunov function into implication form as in the proof of Proposition 2.11, and then follow the lines of the proof of Theorem 2.12. The details are left to the reader. □

The counterpart of LISS for undisturbed systems with $u = 0$ is

Definition 2.33 System (2.1) is called *uniformly asymptotically stable at zero (0-UAS)*, if there exist $\beta \in \mathcal{KL}$ and $r > 0$ such that

$$t \geq 0, \quad |x| \leq r \quad \Rightarrow \quad |\phi(t, x, 0)| \leq \beta(|x|, t). \tag{2.55}$$

The corresponding Lyapunov function concept can be introduced as

Definition 2.34 A continuous function $V : D \to \mathbb{R}_+$, $D \subset \mathbb{R}^n$ is called a *0-UAS Lyapunov function*, if there exist $r > 0$, $\psi_1, \psi_2 \in \mathcal{K}_\infty$, and $\alpha \in \mathcal{P}$, such that $B_r \subset D$, (2.53) holds, and

$$|x| \leq r \quad \Rightarrow \quad \dot{V}_0(x) \leq -\alpha(V(x)). \tag{2.56}$$

In view of Remark 2.31, without loss of generality one can assume that $\alpha \in \mathcal{K}_\infty$.

Similarly to Theorem B.37, one can show the following criterion of the 0-UAS property:

Proposition 2.35 *Consider a forward complete system (2.1) with a locally Lipschitz $f(\cdot, 0)$. The following statements are equivalent:*

(i) (2.1) is 0-UAS.
(ii) 0 is an asymptotically stable equilibrium, that is, (2.1) is Lyapunov stable and locally attractive for the input $u = 0$, see Definitions B.9, B.14.
(iii) (2.1) possesses an infinitely differentiable 0-UAS Lyapunov function.

We have seen that nonlinear 0-GAS systems are seldom ISS. In this section, we prove that 0-UAS is equivalent to LISS, provided that f satisfies Assumption 1.8.

First, relying upon Proposition 2.35, we show that LISS implies the existence of a LISS Lyapunov function. We start with a technical result:

Lemma 2.36 *Assumption 1.8 implies that there exist $\sigma \in \mathcal{K}$ and $\rho > 0$ so that for all $v \in B_{\rho, \mathbb{R}^m}$ and all $x \in B_\rho$ we have*

$$|f(x, v) - f(x, 0)| \leq \sigma(|v|). \tag{2.57}$$

Proof According to Assumption 1.8, f is continuous. Due to the compactness of closed balls in finite-dimensional spaces, the following function is well-defined on $\mathbb{R}_+$:

$$\sigma(r) := \sup_{v \in B_{r, \mathbb{R}^m}} \sup_{x \in B_r} |f(x, v) - f(x, 0)| + r.$$

Clearly, $\sigma(0) = 0$ and σ is strictly increasing and unbounded. By a variation of Lemma A.27, σ is also continuous. Hence $\sigma \in \mathcal{K}_\infty$. □

The next result is closely related to the known fact about robustness of the 0-UAS property [15, Corollary 4.2.3]. It shows that a Lipschitz continuous 0-UAS Lyapunov function for (2.1) is also a LISS Lyapunov function for (2.1).

Proposition 2.37 *Let Assumption 1.8 hold, $f(0,0) = 0$, and let V be a Lipschitz continuous 0-UAS Lyapunov function for (2.1). Then V is also a LISS Lyapunov function for (2.1).*

Proof Let $V : D \to \mathbb{R}_+$, $D \subset \mathbb{R}^n$, with $0 \in \text{int}(D)$ be a Lipschitz continuous 0-UAS Lyapunov function for (2.1), which satisfies (2.56) for a certain $r > 0$.

Since Assumption 1.8 holds, (2.1) depends continuously on states and inputs at $(x, u) = (0, 0)$ due to Proposition 1.38. Hence there exist $r_2 > 0$ and $t^* > 0$ so that for all $x \in B_r$ and all $u \in B_{r_2,\mathcal{U}}$ the solution $\phi(\cdot, x, u)$ exists on $[0, t^*]$ and $|\phi(s, x, u)| \le 2r$ for all $s \in [0, t^*]$.

Pick $\sigma \in \mathcal{K}$ as in Lemma 2.36 and assume without loss of generality that $r_2 < \rho$, $2r < \rho$, and that ρ is small enough such that $\overline{B_\rho} \subset D$.

We are going to prove that V is a LISS Lyapunov function for (2.1). To this end, we derive a dissipative estimate for $\dot{V}_u(x)$ for all $x \in B_r$ and all $u \in B_{r_2,\mathcal{U}}$. We have:

$$\begin{aligned}
\dot{V}_u(x) &= \overline{\lim_{t\to+0}} \frac{1}{t}\Big(V\big(\phi(t,x,u)\big) - V(x)\Big)\\
&= \overline{\lim_{t\to+0}} \frac{1}{t}\Big(V\big(\phi(t,x,0)\big) - V(x) + V\big(\phi(t,x,u)\big) - V\big(\phi(t,x,0)\big)\Big)\\
&= \dot{V}_0(x) + \overline{\lim_{t\to+0}} \frac{1}{t}\Big(V\big(\phi(t,x,u)\big) - V\big(\phi(t,x,0)\big)\Big)
\end{aligned}$$

Since V is a 0-UAS Lyapunov function for (2.1), due to (2.56) it holds for a certain $\alpha \in \mathcal{K}_\infty$ that

$$\dot{V}_u(x) \le -\alpha(|x|) + \overline{\lim_{t\to+0}} \frac{1}{t}\Big|V\big(\phi(t,x,u)\big) - V\big(\phi(t,x,0)\big)\Big|.$$

Since $\phi(t, x, u) \in \overline{B_{2r}} \subset D$ for all $x \in B_r$, all $u \in B_{r_2,\mathcal{U}}$ and all $t \in [0, t^*]$, and since V is Lipschitz continuous on D, there exists $L > 0$ so that for all such t, x, u

$$\dot{V}_u(x) \le -\alpha(|x|) + L\, \overline{\lim_{t\to+0}} \frac{1}{t}\big|\phi(t,x,u) - \phi(t,x,0)\big|. \tag{2.58}$$

Let us estimate $|\phi(t, x, u) - \phi(t, x, 0)|$ for $t \in [0, t^*]$.

$$
\begin{aligned}
\big|\phi(t,x,u)-\phi(t,x,0)\big| &= \Big|\int_0^t \Big(f(\phi(s,x,u),u(s)) - f(\phi(s,x,0),0)\Big)\mathrm{d}s\Big| \\
&\leq \int_0^t \Big|f(\phi(s,x,u),u(s)) - f(\phi(s,x,0),0)\Big|\mathrm{d}s \\
&\leq \int_0^t \Big|f(\phi(s,x,u),u(s)) - f(\phi(s,x,0),u(s))\Big|\mathrm{d}s \\
&\quad + \int_0^t \Big|f(\phi(s,x,0),u(s)) - f(\phi(s,x,0),0)\Big|\mathrm{d}s.
\end{aligned}
$$

Recalling that $|\phi(t,x,u)| \leq 2r < \rho$ for all $t \in [0,t^*]$ and that $|u(s)| \leq r_2$ for almost all s, due to Lipschitz continuity of f w.r.t. the first argument, there exists $L_2 > 0$:

$$
\begin{aligned}
\big|\phi(t,x,u)-\phi(t,x,0)\big| &\leq L_2 \int_0^t \big|\phi(s,x,u)-\phi(s,x,0)\big|\mathrm{d}s + \int_0^t \sigma(|u(s)|)\mathrm{d}s \\
&\leq L_2 t \sup_{0\leq s\leq t} \big|\phi(s,x,u)-\phi(s,x,0)\big| + t\sigma\big(\operatorname{ess\,sup}_{0\leq s\leq t}|u(s)|\big).
\end{aligned}
$$

The right-hand side of the above inequality is nondecreasing in t and consequently it holds that

$$
\begin{aligned}
&\sup_{0\leq s\leq t} \big|\phi(s,x,u)-\phi(s,x,0)\big| \\
&\qquad \leq ML_2 t \sup_{0\leq s\leq t} \big|\phi(s,x,u)-\phi(s,x,0)\big| + Mt\sigma\big(\operatorname{ess\,sup}_{0\leq s\leq t}|u(s)|\big).
\end{aligned}
$$

Pick t small enough so that $1 - ML_2 t > 0$. Then

$$
\sup_{0\leq s\leq t} \big|\phi(s,x,u)-\phi(s,x,0)\big| \leq \frac{Mt}{1-ML_2 t}\sigma\big(\operatorname{ess\,sup}_{0\leq s\leq t}|u(s)|\big).
$$

Using this estimate in (2.58), and taking the limit $t \to +0$, we obtain for all $x \in B_r$ and all $u \in B_{r_2,\,\mathcal{U}}$ the LISS estimate

$$
\dot{V}_u(x) \leq -\alpha(|x|) + LM\sigma\big(\|u\|_\infty\big). \tag{2.59}
$$

This shows that V is a LISS Lyapunov function for (2.1). □

An immediate consequence of Proposition 2.37 is the following characterization of the LISS property:

Theorem 2.38 (Characterization of LISS) *Let Assumption 1.8 hold and* $f(0,0) = 0$. *The following properties are equivalent:*

(i) *(2.1) is 0-UAS.*
(ii) *There is a Lipschitz continuous 0-UAS Lyapunov function for (2.1).*
(iii) *There is a Lipschitz continuous LISS Lyapunov function for (2.1).*
(iv) *(2.1) is LISS.*

Proof The proposition is a consequence of Theorem B.31 and Propositions 2.37, 2.32 and of an obvious fact that LISS implies 0-UAS. □

Let $f(\cdot, 0)$ be continuously differentiable in a neighborhood of 0, and let $f(x, 0) = Ax + o(|x|)$, for $|x| \to 0$, where $A \in \mathbb{R}^{n\times n}$. Here we use small-$o$ notation.

Then the system

$$\dot{x} = Ax \tag{2.60}$$

is called a linearization of (2.1) for $u \equiv 0$ at $x = 0$.

As a consequence of Theorem 2.38 we obtain an easy-to-check sufficient condition for local ISS.

Corollary 2.39 (Linearization method) *Let Assumption 1.8 hold and* $f(0,0) = 0$. *Let further* $f(\cdot, 0)$ *be continuously differentiable in a neighborhood of* 0 *and let (2.60) be the linearization of (2.1) for* $u \equiv 0$ *at* $x = 0$. *If (2.60) is asymptotically stable (i.e.,* A *is a Hurwitz matrix), then (2.1) is LISS.*

Proof According to the linearization theorem, see, e.g., [45, Theorem 19], asymptotic stability of (2.60) implies uniform asymptotic stability of (2.1) for $u \equiv 0$. Now Theorem 2.38 shows that (2.1) is LISS. □

2.4 Asymptotic Properties for Control Systems

In Theorem 2.38, we have obtained characterizations of LISS in terms of Lyapunov functions and in terms of 0-UAS. As we will see, the development of a corresponding theory for global ISS property will require much more effort. In this section, we introduce counterparts of stability and attractivity for systems with inputs and study some relations between such properties.

2.4.1 Stability

Let us define the following stability notions for systems with inputs.

Definition 2.40 A forward complete system (2.1) is called:

(i) *Uniformly locally stable (ULS)*, if $\exists\sigma, \gamma \in \mathcal{K}_\infty$ and $r > 0$ such that

$$|x| \le r,\ \|u\|_\infty \le r,\ t \ge 0 \ \Rightarrow\ |\phi(t, x, u)| \le \sigma(|x|) + \gamma(\|u\|_\infty). \quad (2.61)$$

(ii) *Uniformly globally stable (UGS)*, if $\exists\sigma, \gamma \in \mathcal{K}_\infty$ such that

$$x \in \mathbb{R}^n,\ u \in \mathcal{U},\ t \ge 0 \ \Rightarrow\ |\phi(t, x, u)| \le \sigma(|x|) + \gamma(\|u\|_\infty). \quad (2.62)$$

(iii) *Uniformly globally bounded (UGB)*, if $\exists\sigma, \gamma \in \mathcal{K}_\infty$ and $c > 0$ such that

$$x \in \mathbb{R}^n,\ u \in \mathcal{U},\ t \ge 0 \ \Rightarrow\ |\phi(t, x, u)| \le \sigma(|x|) + \gamma(\|u\|_\infty) + c. \quad (2.63)$$

(iv) *Uniformly locally stable at zero (0-ULS)*, if (2.1) is ULS for $u \equiv 0$.

Uniform local stability can be restated in a more classical $\varepsilon - \delta$ language:

Lemma 2.41 (Characterization of uniform local stability) *Consider a forward complete system (2.1). The following statements are equivalent:*

(i) (2.1) is ULS.
(ii) The following holds:

$$\forall\varepsilon > 0\ \exists\delta > 0 :\ |x| \le \delta,\ \|u\|_\infty \le \delta \ \Rightarrow\ \sup_{t\ge0} |\phi(t, x, u)| \le \varepsilon. \quad (2.64)$$

(iii) (2.1) has CEP property and is ultimately locally stable, *that is: for any $\varepsilon > 0$ there exist $T = T(\varepsilon)$ and $\delta = \delta(\varepsilon)$ so that*

$$t \ge T,\ |x| \le \delta,\ \|u\|_\infty \le \delta \ \Rightarrow\ |\phi(t, x, u)| \le \varepsilon. \quad (2.65)$$

Proof (i) $\Rightarrow$ (ii). Let (2.1) be ULS. Then there exist $\sigma, \gamma \in \mathcal{K}_\infty$ and $r > 0$ such that (2.61) holds.

Take any $\varepsilon > 0$ and choose $\delta = \delta(\varepsilon) := \min\{\sigma^{-1}(\frac{\varepsilon}{2}), \gamma^{-1}(\frac{\varepsilon}{2})\}$. Clearly, with this δ, the property (2.64) follows.

(ii) $\Rightarrow$ (i). Let (2.64) hold. Define

$$\delta(\varepsilon) := \max\{s : |x| \le s,\ \|u\|_\infty \le s \ \Rightarrow\ \sup_{t\ge0} |\phi(t, x, u)| \le \varepsilon\}.$$

The estimate (2.64) and the definition of $\delta(\cdot)$ show that δ is a nondecreasing function, $\delta(\varepsilon) > 0$ for all $\varepsilon > 0$, and $\delta(\varepsilon) \le \varepsilon$ for all $\varepsilon > 0$, and $\delta(0) = 0$. These properties imply that δ is continuous at 0. Now by Proposition A.16 there is $\delta_2 \in \mathcal{K}$ such that $\delta \ge \delta_2$ pointwise on $\mathbb{R}_+$.

Define $\gamma := \delta_2^{-1}$. Then

$$|\phi(t, x, u)| \le \gamma\big(\max\{|x|, \|u\|_\infty\}\big) \le \gamma(|x|) + \gamma(\|u\|_\infty),$$

which shows ULS.

(ii) ⇔ (iii). CEP property of (2.1) means precisely that for any $\varepsilon > 0$ and any $T > 0$ there is a $\delta > 0$ so that

$$t \le T,\ |x| \le \delta,\ \|u\|_\infty \le \delta \quad \Rightarrow \quad |\phi(t, x, u)| \le \varepsilon. \tag{2.66}$$

Combining this estimate with (2.65), we immediately see that (ii) ⇔ (iii). □

In particular, item (iii) of Lemma 2.41 shows that continuity of ϕ w.r.t. initial states and inputs at $(x, u) = (0, 0)$ is a "local in time" component of ULS property.

Next, we show (among other things) that the boundedness of finite-time reachability sets is a "local in time" component of uniform global boundedness.

Lemma 2.42 (Characterization of UGB) *Consider a forward complete system (2.1). The following statements are equivalent:*

(i) (2.1) is UGB.

(ii) There is a continuous function $\mu : \mathbb{R}_+ \times \mathbb{R}_+ \to \mathbb{R}_+$, strictly increasing w.r.t. both arguments so that

$$x \in \mathbb{R}^n,\ u \in \mathcal{U},\ t \ge 0 \ \Rightarrow\ |\phi(t, x, u)| \le \mu(|x|, \|u\|_\infty). \tag{2.67}$$

(iii) There are $\sigma_1, \gamma \in \mathcal{K}_\infty$ and $c > 0$ so that

$$x \in \mathbb{R}^n,\ u \in \mathcal{U},\ t \ge 0 \ \Rightarrow\ |\phi(t, x, u)| \le \sigma_1(|x| + c) + \gamma(\|u\|_\infty). \tag{2.68}$$

(iv) (2.1) has bounded input bounded state (BIBS) property*: For any $r > 0$ it holds that*

$$\sup_{|x|\le r,\ \|u\|_\infty \le r,\ t\ge 0} |\phi(t, x, u)| < \infty. \tag{2.69}$$

(v) (2.1) has BRS and for any $r > 0$ there is $T(r) > 0$ such that

$$\sup_{|x|\le r,\ \|u\|_\infty \le r,\ t\ge T(r)} |\phi(t, x, u)| < \infty. \tag{2.70}$$

Proof (i) ⇔ (ii). If (2.1) is UGB, then set $\mu(|x|, \|u\|_\infty) := \sigma(|x|) + \gamma(\|u\|_\infty) + c$. Clearly, μ is continuous and increasing w.r.t. both arguments.

Conversely, let μ be as in the formulation of the lemma. Then

$$|\phi(t, x, u)| \le \mu(|x|, |x|) + \mu(\|u\|_\infty, \|u\|_\infty).$$

Set $\sigma(s) := \mu(s, s) - \mu(0, 0)$. Clearly, $\sigma \in \mathcal{K}$. We obtain

$$|\phi(t, x, u)| \le \sigma(|x|) + \sigma(\|u\|_\infty) + 2\mu(0, 0),$$

which shows that (2.1) is UGB.

(i) $\Leftrightarrow$ (iii). Let (2.1) be UGB. Then there are $\sigma, \gamma \in \mathcal{K}_\infty$ and $c > 0$ so that for any $x \in \mathbb{R}^n$, any $u \in \mathcal{U}$, and any $t \geq 0$ it holds that

$$\begin{aligned}|\phi(t, x, u)| &\leq \sigma(|x|) + \gamma(\|u\|_\infty) + c \\ &\leq \sigma(|x| + c) + |x| + c + \gamma(\|u\|_\infty) \\ &=: \sigma_1(|x| + c) + \gamma(\|u\|_\infty)\end{aligned}$$

for $\sigma_1(r) := \sigma(r) + r$, $r \geq 0$. Clearly, $\sigma_1 \in \mathcal{K}_\infty$ and hence (2.68) holds.

Conversely, let (2.68) hold. Then there are $\sigma_1, \gamma \in \mathcal{K}_\infty$ and $c > 0$ so that for any $x \in \mathbb{R}^n$, any $u \in \mathcal{U}$, and any $t \geq 0$ it holds that

$$\begin{aligned}|\phi(t, x, u)| &\leq \sigma_1(|x| + c) + \gamma(\|u\|_\infty) \\ &\leq \sigma_1(2|x|) + \gamma(\|u\|_\infty) + \sigma_1(2c),\end{aligned}$$

and thus (2.1) is UGB.

(i) $\Rightarrow$ (iv). Clear.

The implications (i) $\Leftrightarrow$ (i) $\Leftrightarrow$ (v) are left as exercises for the reader. □

These characterizations make boundedness of reachability sets and continuity of (2.1) *at (0,0) a "bridge" between basic existence and uniqueness theory and stability theory.*

Our next result is a decomposition of a global stability (compare it with Definition B.10 and Proposition B.13).

Proposition 2.43 (Characterization of UGS) *Consider a forward complete system (2.1). (2.1) is UGS $\Leftrightarrow$ (2.1) is UGB $\wedge$ ULS.*

Proof Clearly, UGS implies UGB $\wedge$ ULS. Let us prove that UGB $\wedge$ ULS $\Rightarrow$ UGS. Assume that (2.1) is ULS and UGB. This means that there exist $\sigma_1, \gamma_1, \sigma_2, \gamma_2 \in \mathcal{K}_\infty$, and $r, c > 0$ so that for all $x \in B_r$ and all $u \in B_{r,\mathcal{U}}$ it holds that

$$|\phi(t, x, u)| \leq \sigma_1(|x|) + \gamma_1(\|u\|_\infty),$$

and for all $x \in \mathbb{R}^n$ and all $u \in \mathcal{U}$ the following estimate holds:

$$|\phi(t, x, u)| \leq \sigma_2(|x|) + \gamma_2(\|u\|_\infty) + c.$$

Assume without loss of generality that $\sigma_2(s) \geq \sigma_1(s)$ and $\gamma_2(s) \geq \gamma_1(s)$ for all $s \geq 0$. Pick $k_1, k_2 > 0$ so that $c = k_1\sigma_2(r)$ and $c = k_2\gamma_2(r)$. Then

$$c \leq k_1\sigma_2(|x|) + k_2\gamma_2(\|u\|_\infty),$$

for any $x \in \mathbb{R}^n$ and any $u \in \mathcal{U}$ so that either $|x| \geq r$ or $\|u\|_\infty \geq r$. Thus, for all $x \in \mathbb{R}^n$ and all $u \in \mathcal{U}$ it holds that

$$|\phi(t, x, u)| \le (1 + k_1)\sigma_2(|x|) + (1 + k_2)\gamma_2(\|u\|_\infty).$$

This shows uniform global stability of (2.1). □

2.4.2 Asymptotic Gains

Next, we introduce a generalization of global attractivity property for systems with inputs.

Definition 2.44 We say that a forward complete system (2.1) has

(i) *asymptotic gain (AG) property*, if $\exists \gamma \in \mathcal{K}_\infty \cup \{0\}$ such that for all $\varepsilon > 0$, for all $x \in \mathbb{R}^n$ and for all $u \in \mathcal{U}$ there exists $\tau_a = \tau_a(\varepsilon, x, u) < \infty$:

$$t \ge \tau_a \quad \Rightarrow \quad |\phi(t, x, u)| \le \varepsilon + \gamma(\|u\|_\infty). \tag{2.71}$$

(ii) *strong asymptotic gain (sAG) property*, if $\exists \gamma \in \mathcal{K}_\infty \cup \{0\}$ such that for all $x \in \mathbb{R}^n$ and for all $\varepsilon > 0$ there exists $\tau_a = \tau_a(\varepsilon, x) < \infty$:

$$t \ge \tau_a,\ u \in \mathcal{U} \quad \Rightarrow \quad |\phi(t, x, u)| \le \varepsilon + \gamma(\|u\|_\infty). \tag{2.72}$$

(iii) *uniform asymptotic gain on bounded sets (bUAG) property*, if $\exists \gamma \in \mathcal{K}_\infty \cup \{0\}$ such that for all $\varepsilon, \delta > 0$ there is $\tau_a = \tau_a(\varepsilon, \delta) < \infty$:

$$t \ge \tau_a,\ u \in B_{\delta,\mathcal{U}},\ x \in B_\delta \quad \Rightarrow \quad |\phi(t, x, u)| \le \varepsilon + \gamma(\|u\|_\infty). \tag{2.73}$$

(iv) *uniform asymptotic gain (UAG) property*, if $\exists \gamma \in \mathcal{K}_\infty \cup \{0\}$ such that for all $\varepsilon, \delta > 0$ there is $\tau_a = \tau_a(\varepsilon, \delta) < \infty$:

$$t \ge \tau_a,\ u \in \mathcal{U},\ x \in B_\delta \quad \Rightarrow \quad |\phi(t, x, u)| \le \varepsilon + \gamma(\|u\|_\infty). \tag{2.74}$$

Remark 2.45 Note that (2.1) satisfies AG property iff there exists $\gamma \in \mathcal{K}_\infty$ so that

$$x \in \mathbb{R}^n,\ u \in \mathcal{U} \quad \Rightarrow \quad \overline{\lim_{t\to+\infty}} |\phi(t, x, u)| \le \gamma(\|u\|_\infty). \tag{2.75}$$

Recall that a system having the AG property we also call AG system, and similarly with sAG, bUAG, and UAG properties.

All properties AG, sAG, bUAG, and UAG imply that all trajectories converge to the ball of radius $\gamma(\|u\|_\infty)$ around the origin as $t \to \infty$. The difference between them is in the kind of dependence of τ_a on states and inputs. In bUAG and UAG systems, this time depends (besides ε) only on the norm of the state. In sAG systems, it depends on the state x (and may vary for the states with the same norm) but does not depend on u. In AG systems, τ_a depends both on x and on u. For systems without inputs, AG and sAG properties are reduced to global attractivity (see Definition B.14), whereas UAG property is resolved to uniform global attractivity (see Definition B.21).

The next example shows that one should in general "pay" in order to obtain additional uniformity of AG. However, in this example this cost can be made arbitrarily small (but nonzero).

Example 2.46 Consider the system

$$\dot{x}(t) = -\frac{1}{1+|u(t)|}x(t), \tag{2.76}$$

where $u \in \mathcal{U} := L^\infty(\mathbb{R}_+, \mathbb{R})$ is a globally bounded input and $x(t) \in \mathbb{R}$ is a state.

Pick any $u \in \mathcal{U}$ and $x \in \mathbb{R}$. The corresponding solution of (2.76) is given by

$$\begin{aligned} |\phi(t,x,u)| &= e^{-\int_0^t \frac{1}{1+|u(s)|}ds}|x| \\ &\le e^{-\int_0^t \frac{1}{1+\|u\|_\infty}ds}|x| \\ &= e^{-\frac{1}{1+\|u\|_\infty}t}|x|. \end{aligned} \tag{2.77}$$

For all $x \in B_r$ and all $u \in B_{r,\mathcal{U}}$ it holds that

$$|\phi(t,x,u)| \le e^{-\frac{1}{1+r}t}r.$$

For any $\varepsilon > 0$ take τ to be any time such that $e^{-\frac{1}{1+r}\tau}r \le \varepsilon$. Then

$$x \in B_r, \quad u \in B_{r,\mathcal{U}}, \quad t \ge \tau \quad \Rightarrow \quad |\phi(t,x,u)| \le \varepsilon,$$

which shows bUAG property with zero gain (and thus, also AG property with zero gain).

Next we show that (2.76) is UAG. To this end, we continue computations (2.77) to obtain for any $r > 0$

$$|\phi(t,x,u)| \le e^{-\frac{1}{1+\max\{\|u\|_\infty, \frac{1}{r}|x|\}}t} \max\left\{r\|u\|_\infty, |x|\right\}. \tag{2.78}$$

If $r\|u\|_\infty \le |x|$, then (2.78) implies

$$|\phi(t,x,u)| \le e^{-\frac{1}{1+\frac{1}{r}|x|}t}|x|. \tag{2.79}$$

Otherwise, if $r\|u\|_\infty \ge |x|$, then (2.78) leads to

$$|\phi(t,x,u)| \le e^{-\frac{1}{1+\|u\|_\infty}t}r\|u\|_\infty \le r\|u\|_\infty. \tag{2.80}$$

Combining (2.79) and (2.80) we obtain for all $x \in \mathbb{R}^n$, $u \in \mathcal{U}$, $t \geq 0$ and for any $r > 0$ that

$$|\phi(t, x, u)| \leq e^{-\frac{1}{1+\frac{1}{r}|x|}t}|x| + r\|u\|_\infty. \tag{2.81}$$

This shows that (2.76) is ISS with arbitrarily small linear gain. It is easy to obtain from the last estimate that (2.76) is also UAG with arbitrarily small linear gain. On the other hand, (2.76) is not sAG (and, thus, UAG), with the gain $\gamma \equiv 0$. To see this, pick $\varepsilon := \frac{1}{2}$, $x = 1$ and consider constant inputs $u(\cdot) \equiv c$. Then

$$\phi(t, 1, u) = e^{-\frac{1}{1+c}t}.$$

If (2.76) were sAG with the zero gain, then it would exist a time τ_a, which does not depend on u, so that for all c and all $t \geq \tau_a$ it holds that

$$e^{-\frac{1}{1+c}t} \leq \frac{1}{2}.$$

However, this is false since $e^{-\frac{1}{1+c}\tau_a}$ monotonically increases to 1 as $c \to \infty$. Thus, (2.76) is not sAG with zero gain (and hence also not UAG with a zero gain).

2.4.3 Limit Property

Now we define several concepts that are weaker than corresponding asymptotic gain properties. They will be ultimately useful for the characterization of ISS.

Definition 2.47 We say that a forward complete system (2.1) has:

(i) *Limit (LIM) property*, if there exists $\gamma \in \mathcal{K} \cup \{0\}$ such that

$$x \in \mathbb{R}^n, \quad u \in \mathcal{U} \quad \Rightarrow \quad \inf_{t \geq 0} |\phi(t, x, u)| \leq \gamma(\|u\|_\infty). \tag{2.82}$$

(ii) *Strong limit (sLIM) property*, if there exists $\gamma \in \mathcal{K} \cup \{0\}$ so that for every $\varepsilon > 0$ and for every $x \in \mathbb{R}^n$ there exists $\tau = \tau(\varepsilon, x)$ such that for all $u \in \mathcal{U}$ there is a $t \leq \tau$:

$$|\phi(t, x, u)| \leq \varepsilon + \gamma(\|u\|_\infty). \tag{2.83}$$

(iii) *Uniform limit on bounded sets (bULIM) property*, if there exists $\gamma \in \mathcal{K} \cup \{0\}$ so that for every $\varepsilon > 0$ and for every $r > 0$ there exists a $\tau = \tau(\varepsilon, r)$ such that for all $x \in B_r$ and all $u \in B_{r,\mathcal{U}}$ there is a $t \leq \tau$ such that (2.83) holds.

(iv) *Uniform limit (ULIM) property*, if there exists $\gamma \in \mathcal{K} \cup \{0\}$ so that for every $\varepsilon > 0$ and for every $r > 0$ there exists a $\tau = \tau(\varepsilon, r)$ such that for all $x \in B_r$ and all $u \in \mathcal{U}$ there is a $t \le \tau$ such that (2.83) holds.

Whereas AG, sAG, and (b)UAG properties characterize the convergence of trajectories to the ball of radius $\gamma(\|u\|_\infty)$ around the origin as $t \to \infty$, the limit properties characterize the approachability of trajectories to the same ball.

For systems without inputs, LIM and sLIM properties are reduced to a so-called *weak attractivity* of the undisturbed system, whereas (b)ULIM property is resolved to *uniform weak attractivity* of the undisturbed system.

Next, we show that all variations of the limit property coincide for forward complete systems (2.1). First, we state an auxiliary lemma based on Corollary 1.45.

Lemma 2.48 *Let (2.1) be a forward complete system, satisfying LIM property, and let Assumption 1.8 hold. Then for every* $\varepsilon > 0$, $r > 0$, *and* $R > 0$ *there is* $T(\varepsilon, r, R)$ *such that*

$$x \in \overline{B_r}, \quad u \in B_{R,\mathcal{U}} \quad \Rightarrow \quad \exists t \le T(\varepsilon, r, R) : |\phi(t, x, u)| \le \varepsilon + \gamma(R), \tag{2.84}$$

where γ *is the gain from the LIM property.*

Proof Assume (2.1) has LIM property, and let $\gamma \in \mathcal{K}_\infty$ be the corresponding gain. Fix $\varepsilon > 0$, $r > 0$, and $R > 0$. By the LIM property,

$$x \in \mathbb{R}^n, \quad u \in B_{R,\mathcal{U}} \quad \Rightarrow \quad \exists t \ge 0 : |\phi(t, x, u)| \le \frac{1}{2}\varepsilon + \gamma(R).$$

Define

$$S := B_{R,\mathbb{R}^m}, \qquad K := \{z \in \mathbb{R}^n : |z| \le \frac{1}{2}\varepsilon + \gamma(R)\},$$
$$C := \{z \in \mathbb{R}^n : |z| \le r\}, \qquad \Omega := \{z \in \mathbb{R}^n : |z| < \varepsilon + \gamma(R)\}.$$

The assumptions of Theorem 1.44 are fulfilled with the above sets, and applying it, we obtain that there is $T = T(\varepsilon, r, R)$ such that (2.84) holds. □

Proposition 2.49 *Let (2.1) be a forward complete system and let Assumption 1.8 hold. Then LIM, sLIM, bULIM, and ULIM are equivalent properties for (2.1).*

Proof It is clear that ULIM ⇒ bULIM ⇒ LIM and ULIM ⇒ sLIM ⇒ LIM.

Let us show that LIM implies ULIM. Fix $\varepsilon > 0$ and $r > 0$ and define $R_1 := \gamma^{-1}(\max\{r - \varepsilon, 0\})$. Then for each $u \in \mathcal{U}$: $\|u\|_\infty \ge R_1$ and each $x \in \overline{B_r}$ it holds that

$$|\phi(0, x, u)| = |x| \le r \le \varepsilon + \gamma(R_1) \le \varepsilon + \gamma(\|u\|_\infty),$$

and the time $t(\varepsilon, r, u)$ in the definition of ULIM can be chosen for such u as $t := 0$.

Invoking Lemma 2.48, we obtain $T_1 := T(\frac{\varepsilon}{2}, r, R_1)$ such that for all $x \in \overline{B_r}$ and $u \in \mathcal{U} : \|u\|_\infty \le R_1$ there is a time $t \le T_1$ such that

$$|\phi(t, x, u)| \le \varepsilon + \gamma(R_1) - \frac{\varepsilon}{2}. \tag{2.85}$$

Define

$$R_2 := \gamma^{-1}\Big(\max\big\{\gamma(R_1) - \frac{\varepsilon}{2}, 0\big\}\Big) = \gamma^{-1}\Big(\max\big\{r - \frac{3\varepsilon}{2}, 0\big\}\Big).$$

From (2.85) we obtain for all u with $R_2 \le \|u\|_\infty \le R_1$ that for the above t

$$|\phi(t, x, u)| \le \varepsilon + \gamma(\|u\|_\infty).$$

For $k \in \mathbb{N}$ define the times $T_k := T(\frac{\varepsilon}{2}, r, R_k)$ and

$$R_k := \gamma^{-1}\Big(\max\big\{\gamma(R_{k-1}) - \frac{\varepsilon}{2}, 0\big\}\Big) = \gamma^{-1}\Big(\max\big\{r - \frac{(k+1)\varepsilon}{2}, 0\big\}\Big).$$

Repeating the previous argument we see that for all $x \in \overline{B_r}$ and all $u \in \mathcal{U}$ with $R_{k+1} \le \|u\|_\infty \le R_k$ there is a time $t \le T_k$ such that $|\phi(t, x, u)| \le \varepsilon + \gamma(\|u\|_\infty)$.

The procedure ends after finitely many steps because eventually $r - \frac{(k+1)\varepsilon}{2}$ becomes negative. Take $\tau(\varepsilon, r) := \max\big\{T_k : k = 1, \ldots, \lfloor \frac{2r}{\varepsilon} \rfloor + 1\big\}$, where $\lfloor \cdot \rfloor$ denotes the integer part of a real number. Then ULIM property holds with this τ. □

Remark 2.50 Proposition 2.49 is a key component in the proof of the fundamental ISS superposition theorem for sufficiently regular systems (Theorem 2.52). It is based upon a deep Corollary 1.45 on the topology of the reachability sets of nonlinear finite-dimensional spaces and cannot be generalized to general infinite-dimensional systems (even linear ones, see [35] for a discussion).

2.5 ISS Superposition Theorems

ISS of (2.1) trivially implies that (2.1) is a globally stable, 0-UGAS (the property of a system without inputs) and AG system, which shows the influence of input over the system. In this section, we show a profound converse result, called *ISS superposition theorem*, showing that combinations of stability and asymptotic gain/limit properties imply ISS. ISS superposition theorems are a crucial tool to derive small-gain theorems in a trajectory formulation as well as to relate ISS to integral ISS-like stability notions, semiglobal ISS, etc.

Our first superposition theorem is

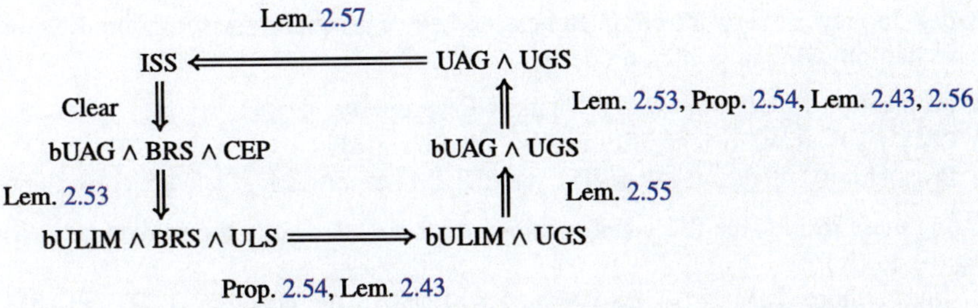

Fig. 2.3 Proof of ISS superposition theorem with mild regularity assumptions (Theorem 2.51)

Theorem 2.51 (ISS superposition theorem with mild regularity assumptions) *Let (2.1) be forward complete. The following statements are equivalent:*

(i) (2.1) is ISS.
(ii) (2.1) is bUAG $\wedge$ BRS $\wedge$ CEP.
(iii) (2.1) is bULIM $\wedge$ BRS $\wedge$ ULS.
(iv) (2.1) is bULIM $\wedge$ UGS.

Proof Follows from Lemmas/Propositions (2.53)–(2.57). A detailed scheme of the proof is presented in Fig. 2.3. □

Besides the fact that the ISS characterizations in Theorem 2.51 are valid under mild regularity assumptions, this result can be extended without significant differences in the statement and the proof to a very general class of control systems, encompassing not only ODEs but also infinite networks of ODEs, partial differential equations, time-delay systems, etc., see Chap. 6. Even in such a general context, the ISS superposition Theorem 2.51 can be used to develop the small-gain theory for finite and infinite networks, the noncoercive ISS Lyapunov theory, and other significant results.

At the same time, the general setting of Theorem 2.51 does not allow using important results on boundedness of reachability sets, uniformity of crossing times, as well as an equivalence between LIM and ULIM properties, which we have developed. To use such results, we need to assume Lipschitz regularity of the right-hand side on bounded balls of $\mathbb{R}^n$. With this at hand, we can characterize the (uniform) ISS property in terms of *nonuniform* limit and asymptotic gain properties.

Theorem 2.52 (ISS superposition theorem) *Let (2.1) be forward complete, $f(0, 0) = 0$, and let Assumption 1.8 hold. The following statements are equivalent:*

(i) (2.1) is ISS.
(ii) (2.1) is bUAG.
(iii) (2.1) is AG and ULS.
(iv) (2.1) is LIM and ULS.
(v) (2.1) is LIM and UGS.
(vi) (2.1) is AG and UGS.
(vii) (2.1) is LIM and 0-ULS.
(viii) (2.1) is LIM and 0-UGAS.
(ix) (2.1) is AG and 0-UGAS.

Proof In view of forward completeness and since f satisfies Assumption 1.8, we have the following important properties:

- Proposition 1.38 implies that (2.1) has CEP property.
- (2.1) has bounded reachability sets by Theorem 1.31.
- Proposition 2.49 ensures that LIM and bULIM properties for (2.1) are equivalent.

Using these results, the ISS superposition Theorem 2.51 shows the equivalences (i) $\Leftrightarrow$ (ii) $\Leftrightarrow$ (iv).

Furthermore, LIM $\wedge$ 0-ULS $\Rightarrow$ LIM $\wedge$ 0-UGAS by Theorem B.37, and LIM $\wedge$ 0-UGAS $\Rightarrow$ LIM $\wedge$ LISS via Theorem 2.38, which implies LIM $\wedge$ ULS. Thus, LIM $\wedge$ 0-ULS $\Leftrightarrow$ LIM $\wedge$ ULS, and we obtain the rest of the equivalences from this criterion. □

In the remaining part of this section, we prove (as in Fig. 2.3 counterclockwise) the auxiliary results needed to establish Theorem 2.51.

Lemma 2.53 *Let (2.1) be forward complete. If (2.1) possesses bUAG and CEP properties, then (2.1) is ULS.*

Proof Take any $\varepsilon > 0$. Let $\tau_a = \tau_a(\varepsilon/2, 1)$. Pick any $\delta_1 \in (0, 1]$ so that $\gamma(\delta_1) < \varepsilon/2$. Then for all $x \in B_1$ and all $u \in B_{\delta_1, \mathcal{U}}$ we have

$$\sup_{t \geq \tau_a} |\phi(t, x, u)| \leq \frac{\varepsilon}{2} + \gamma(\|u\|_\infty) < \varepsilon. \tag{2.86}$$

This shows the property (2.65). Lemma 2.41 justifies that (2.1) is ULS. □

We proceed with

Proposition 2.54 *If (2.1) has BRS and satisfies bULIM property, then (2.1) is UGB.*

Proof Pick any $r > 0$ and set $\varepsilon := \frac{r}{2}$. Since (2.1) has the bULIM property, there exists a $\tau = \tau(r)$ (more precisely, $\tau = \tau\left(\frac{r}{2}, \max\{r, \gamma^{-1}(\frac{r}{4})\}\right)$ from bULIM property), such that

$$x \in \overline{B_r},\ \ \|u\|_\infty \leq \gamma^{-1}(\tfrac{r}{4}) \ \Rightarrow\ \exists t \leq \tau :\ |\phi(t, x, u)| \leq \frac{r}{2} + \gamma(\|u\|_\infty) \leq \frac{3r}{4}. \tag{2.87}$$

Without loss of generality, we can assume that τ is increasing in r. In particular, it is locally integrable. Defining $\bar{\tau}(r) := \frac{1}{r}\int_r^{2r} \tau(s)ds$ we see that $\bar{\tau}(r) \geq \tau(r)$ and $\bar{\tau}$ is continuous. For any $r_2 > r_1 > 0$ via the change of variables $s = \frac{r_2}{r_1}w$, and since $\tau(\frac{r_2}{r_1}w) > \tau(w)$ for all w, we have also that

$$\begin{aligned}
\bar{\tau}(r_2) &= \frac{1}{r_2}\int_{r_2}^{2r_2} \tau(s)ds = \frac{1}{r_2}\int_{r_1}^{2r_1} \tau\left(\frac{r_2}{r_1}w\right)\frac{r_2}{r_1}dw \\
&> \frac{1}{r_1}\int_{r_1}^{2r_1} \tau(w)dw = \bar{\tau}(r_1),
\end{aligned}$$

and thus $\bar{\tau}$ is increasing. We define further $\bar{\tau}(0) := \lim_{r\to+0} \bar{\tau}(r)$ (the limit exists as $\bar{\tau}$ is increasing).

Since (2.1) has BRS, Lemma 1.36 implies that there exists a continuous, increasing function $\mu : \mathbb{R}^3_+ \to \mathbb{R}_+$, such that for all $x \in \mathbb{R}^n$, $u \in \mathcal{U}$, and all $t \geq 0$ the estimate 1.28 holds. This implies that

$$x \in \overline{B_r},\ \|u\|_\infty \leq \gamma^{-1}(\tfrac{r}{4}),\ t \leq \tau(r) \quad \Rightarrow \quad |\phi(t, x, u)| \leq \tilde{\sigma}(r), \tag{2.88}$$

where $\tilde{\sigma} : r \mapsto \mu\big(r, \gamma^{-1}(\tfrac{r}{4}), \tau(r)\big)$ is continuous and increasing, since μ, γ, τ are continuous increasing functions. By setting $t := 0$ into (2.88) it is clear that $\tilde{\sigma}(r) \geq r$ for any $r > 0$.

Assume that there exist $x \in \overline{B_r}$, $u \in \mathcal{U}$ with $\|u\|_\infty \leq \gamma^{-1}(\tfrac{r}{4})$ and $t \geq 0$ so that $|\phi(t, x, u)| > \tilde{\sigma}(r)$. Define

$$t_m := \sup\{s \in [0, t] : |\phi(s, x, u)| \leq r\} \geq 0.$$

The quantity t_m is well-defined, since $|\phi(0, x, u)| = |x| \leq r$.

In view of the cocycle property and since $u(\cdot + t_m) \in \mathcal{U}$, it holds that

$$\phi(t, x, u) = \phi\big(t - t_m, \phi(t_m, x, u), u(\cdot + t_m)\big).$$

Assume that $t - t_m \leq \tau(r)$. Since $|\phi(t_m, x, u)| \leq r$, (2.88) implies that $|\phi(t, x, u)| \leq \tilde{\sigma}(r)$ for all $t \in [t_m, t]$, a contradiction. Otherwise, if $t - t_m > \tau(r)$, then due to (2.87) there exists $t^* < \tau(r)$, so that

$$\big|\phi\big(t^*, \phi(t_m, x, u), u(\cdot + t_m)\big)\big| = |\phi(t^* + t_m, x, u)| \leq \frac{3r}{4},$$

which contradicts to the definition of t_m, since $t_m + t^* < t$. Hence

$$x \in \overline{B_r},\ \ \|u\|_\infty \leq \gamma^{-1}(\tfrac{r}{4}),\ \ t \geq 0 \quad \Rightarrow \quad |\phi(t, x, u)| \leq \tilde{\sigma}(r). \tag{2.89}$$

Denote $\sigma(r) := \tilde{\sigma}(r) - \tilde{\sigma}(0)$, for any $r \geq 0$. Clearly, $\sigma \in \mathcal{K}_\infty$.

For each $x \in \mathbb{R}^n$ and $u \in \mathcal{U}$ we can pick $r := \max\{|x|, 4\gamma(\|u\|_\infty)\}$ and (2.89) immediately shows for all $x \in \mathbb{R}^n$, $u \in \mathcal{U}$, $t \geq 0$ that

$$\begin{aligned}|\phi(t, x, u)| &\leq \sigma\big(\max\{|x|, 4\gamma(\|u\|_\infty)\}\big) + \tilde{\sigma}(0)\\ &\leq \sigma(|x|) + \sigma\big(4\gamma(\|u\|_\infty)\big) + \tilde{\sigma}(0),\end{aligned}$$

which shows UGB of (2.1). □

Our next result is:

Lemma 2.55 *If (2.1) is forward complete, bULIM and UGS, then (2.1) is bUAG.*

Proof Without loss of generality, assume that γ in the definitions of bULIM and UGS is the same (otherwise, take the maximum of both).

Pick any $\varepsilon > 0$ and any $r > 0$. By the bULIM property, there exists $\gamma \in \mathcal{K}_\infty$, same for all ε and r, and $\tau = \tau(\varepsilon, r)$ so that for any $x \in \overline{B_r}$, $u \in \overline{B_{r,\mathcal{U}}}$ there exists a $t \leq \tau$ satisfying $|\phi(t, x, u)| \leq \varepsilon + \gamma(\|u\|_\infty)$.

In view of the cocycle property, we have from the UGS property that for the above x, u, t and any $s \geq 0$

$$\begin{aligned} |\phi(s+t, x, u)| &= \big|\phi\big(s, \phi(t, x, u), u(s+\cdot)\big)\big| \\ &\leq \sigma(|\phi(t, x, u)|) + \gamma(\|u\|_\infty) \\ &\leq \sigma(\varepsilon + \gamma(\|u\|_\infty)) + \gamma(\|u\|_\infty). \end{aligned}$$

Now let $\tilde{\varepsilon} := \sigma(2\varepsilon) > 0$. Using the evident inequality $\sigma(a+b) \leq \sigma(2a) + \sigma(2b)$, which holds for any $a, b \geq 0$, we proceed to

$$|\phi(s+t, x, u)| \leq \tilde{\varepsilon} + \tilde{\gamma}(\|u\|_\infty) \quad \forall s \geq 0,$$

where $\tilde{\gamma}(r) = \sigma(2\gamma(r)) + \gamma(r)$.

Overall, for any $\tilde{\varepsilon} > 0$ and any $r > 0$ there exists $\tilde{\tau} = \tilde{\tau}(\tilde{\varepsilon}, r) = \tau(\frac{1}{2}\sigma^{-1}(\tilde{\varepsilon}), r) = \tau(\varepsilon, r)$, so that

$$|x| \leq r, \quad \|u\|_\infty \leq r, \quad t \geq \tau \quad \Rightarrow \quad |\phi(t, x, u)| \leq \tilde{\varepsilon} + \tilde{\gamma}(\|u\|_\infty).$$

This shows that (2.1) is bUAG. □

The following lemma shows how bUAG property can be "upgraded" to the UAG property.

Lemma 2.56 *Let (2.1) be forward complete. If (2.1) is UGS and bUAG, then (2.1) is UAG.*

Proof Pick arbitrary $\varepsilon > 0$, $r > 0$ and let τ and γ be as in the formulation of the bUAG property. Let $x \in B_r$ and let $u \in \mathcal{U}$ be arbitrary. If $\|u\|_\infty \leq r$, then (2.73) is the desired estimate.

Let $\|u\|_\infty > r$. Hence it holds that $\|u\|_\infty > |x|$. Due to uniform global stability of (2.1), it holds for all t, x, u that

$$|\phi(t, x, u)| \leq \sigma(|x|) + \gamma(\|u\|_\infty),$$

where we assume that γ is the same as in the definition of a bUAG property (otherwise, pick the maximum of both). For $\|u\|_\infty > |x|$ we obtain that

$$|\phi(t, x, u)| \leq \sigma(\|u\|_\infty) + \gamma(\|u\|_\infty),$$

and thus for all $x \in \mathbb{R}^n$, $u \in \mathcal{U}$ it holds that

$$t \geq \tau \quad \Rightarrow \quad |\phi(t, x, u)| \leq \varepsilon + \gamma(\|u\|_\infty) + \sigma(\|u\|_\infty),$$

which shows UAG property with the asymptotic gain $\gamma + \sigma$. □

The final technical lemma of this section is:

Lemma 2.57 *Let (2.1) be forward complete. If (2.1) is UAG and UGS, then (2.1) is ISS.*

Proof Fix arbitrary $r \in \mathbb{R}_+$. We are going to construct a function $\beta \in \mathcal{KL}$ so that (2.4) holds.

From global stability it follows that there exist $\gamma, \sigma \in \mathcal{K}_\infty$ such that for all $t \geq 0$, all $x \in \overline{B_r}$, and all $u \in \mathcal{U}$ we have

$$|\phi(t, x, u)| \leq \sigma(r) + \gamma(\|u\|_\infty). \tag{2.90}$$

Define $\varepsilon_n := 2^{-n}\sigma(r)$, for $n \in \mathbb{Z}_+$. The UAG property implies that there exists a sequence of times $\tau_n := \tau(\varepsilon_n, r)$, which we may without loss of generality assume to be strictly increasing, such that for all $x \in \overline{B_r}$ and all $u \in \mathcal{U}$

$$|\phi(t, x, u)| \leq \varepsilon_n + \gamma(\|u\|_\infty) \quad \forall t \geq \tau_n.$$

From (2.90) we see that we may set $\tau_0 := 0$. Define $\omega(r, \tau_n) := \varepsilon_{n-1}$, for $n \in \mathbb{N}$, and $\omega(r, 0) := 2\varepsilon_0 = 2\sigma(r)$.

Now extend the definition of ω to a function $\omega(r, \cdot) \in \mathcal{L}$. We obtain for $t \in (\tau_n, \tau_{n+1})$, $n = 0, 1, \ldots$ and $x \in B_r$ that $|\phi(t, x, u)| \leq \varepsilon_n + \gamma(\|u\|_\infty) < \omega(r, t) + \gamma(\|u\|_\infty)$. Doing this for all $r \in \mathbb{R}_+$ we obtain the definition of the function ω.

Now define $\hat{\beta}(r, t) := \sup_{0 \leq s \leq r} \omega(s, t) \geq \omega(r, t)$ for $(r, t) \in \mathbb{R}^2_+$. According to Proposition A.17, $\hat{\beta}$ can be upper-bounded by certain $\beta \in \mathcal{KL}$ and the estimate (2.4) is satisfied with such a β (See the proof of Theorem B.22 for some more details). □

2.6 Converse ISS Lyapunov Theorem

In this section, we show that ISS is equivalent to the existence of an ISS Lyapunov function, provided the nonlinearity f is Lipschitz continuous on bounded balls of $\mathbb{R}^n \times \mathbb{R}^m$. We follow [48] rather closely in this section.

First, assume that u is not an input but a control that we can use to stabilize the control system (2.1) globally. We design u as a nonlinear feedback law that is subject to multiplicative disturbances with a magnitude, bounded by 1:

$$u(t) := d(t)\varphi(x(t)), \tag{2.91}$$

where

$$d \in \mathcal{D} := \{d : \mathbb{R}_+ \to D, \text{ measurable}\}, \quad D = \{d \in \mathbb{R}^m : |d| \le 1\}, \tag{2.92}$$

and $\varphi : \mathbb{R}^n \to \mathbb{R}_+$ is Lipschitz continuous on bounded balls.

Applying this feedback law to (2.1), we obtain the closed-loop system

$$\begin{aligned} \dot{x}(t) &= f(x(t), d(t)\varphi(x(t))) \\ &=: g(x(t), d(t)). \end{aligned} \tag{2.93}$$

Denote the maximal solution of (2.93) at time t, starting at $x \in \mathbb{R}^n$ and with disturbance $d \in \mathcal{D}$ by $\phi_\varphi(t, x, d)$. On the interval of its existence $\phi_\varphi(t, x, d)$ coincides with the solution of (2.1) for the input $u(t) = d(t)\varphi(x(t))$.

Remark 2.58 Forward completeness of (2.1) does not imply forward completeness of (2.93). For example, consider $\dot{x} = -x + u$, $u(t) = d \cdot x^2(t)$ for $d > 0$.

For the construction of an ISS Lyapunov function for the system (2.1), we would like to use the converse Lyapunov theorem for UGAS systems with disturbances (Theorem B.31) for the system (2.93). One of the requirements of Theorem B.31 is Lipschitz continuity of g with respect to x, which ensures the existence and uniqueness of solutions of (2.93) corresponding to any initial conditions.

In general, Assumption 1.8 does not guarantee Lipschitz continuity of g, and to ensure this, we assume that f is Lipschitz continuous also with respect to u.

Definition 2.59 We call $f : \mathbb{R}^n \times \mathbb{R}^m \to \mathbb{R}^n$ Lipschitz continuous on bounded balls of $\mathbb{R}^n \times \mathbb{R}^m$, if for each $R > 0$ there is $L > 0$ such that

$$\begin{aligned} &x_1, x_2 \in \overline{B_R}, \quad u_1, u_2 \in \overline{B_{R,U}} \\ &\qquad \Rightarrow \quad |f(x_1, u_1) - f(x_2, u_2)| \le L(|x_1 - x_2| + |u_1 - u_2|). \end{aligned} \tag{2.94}$$

Lemma 2.60 *Let f be Lipschitz continuous on bounded balls of $\mathbb{R}^n \times \mathbb{R}^m$. Then g is Lipschitz continuous on bounded balls, uniformly with respect to the second argument, i.e., $\forall R > 0\ \exists L_g(R) > 0$, such that $\forall x, y \in \overline{B_R}$, $\forall v \in D$, it holds that*

$$|g(x, v) - g(y, v)| \le L_g(R)|x - y|.$$

In particular, the system (2.93) is well-posed in the sense of Definition 1.15 (i.e., the solution of (2.93) exists locally and is unique for any initial condition and any disturbance d).

Proof Pick arbitrary $R > 0$ and consider any $x, y \in \overline{B_R}$ and any $v \in D$. It holds that

$$|g(x, v) - g(y, v)| = |f(x, v\varphi(x)) - f(y, v\varphi(y))|. \tag{2.95}$$

Since φ is Lipschitz continuous with the corresponding Lipschitz constant $L_\varphi(R)$, which we assume to be greater than 1, we have

$$|v\varphi(x)| \leq |\varphi(x)| \leq |\varphi(x) - \varphi(0)| + |\varphi(0)| \leq L_\varphi(R)|x| + |\varphi(0)| \leq L_\varphi(R)R + |\varphi(0)|,$$

and analogously $|v\varphi(y)| \leq L_\varphi(R)R + |\varphi(0)|$.

Continuing the estimate (2.95), we obtain

$$\begin{aligned}|g(x,v)-g(y,v)| &\leq L_f(L_\varphi(R)R + |\varphi(0)|)\Big(|x-y| + |v\varphi(x) - v\varphi(y)|\Big)\\ &\leq L_f(L_\varphi(R)R + |\varphi(0)|)|x-y| + L_f(L_\varphi(R)R + |\varphi(0)|)L_\varphi(R)|x-y|.\end{aligned}$$

This shows the claim of the lemma. □

We will need the following property, which formalizes stabilizability of (2.1) w.r.t. feedbacks with multiplicative disturbances.

Definition 2.61 System (2.1) is called *weakly robustly stable (WRS)*, if there exist a Lipschitz continuous on bounded balls function $\varphi : \mathbb{R}^n \to \mathbb{R}_+$ and $\psi \in \mathcal{K}_\infty$ such that $\varphi(x) \geq \psi(|x|)$ for all $x \in \mathbb{R}^n$, and (2.93) is well-posed and uniformly globally asymptotically stable over $\mathcal{D}$ (as defined in Definition B.20).

Remark 2.62 In our terminology, we follow [48]. In particular, the term "robust stability" is already reserved for a stronger concept, see Exercise 2.30.

Our main result of this section is:

Theorem 2.63 (Smooth converse ISS Lyapunov theorem) *Let f be Lipschitz continuous on bounded balls of $\mathbb{R}^n \times \mathbb{R}^m$, and $f(0,0) = 0$. Then the following statements are equivalent:*

(i) (2.1) is ISS.
(ii) (2.1) is WRS.
(iii) There exists an infinitely differentiable ("smooth") ISS Lyapunov function in implication form for (2.1).
(iv) There exists an infinitely differentiable ("smooth") ISS Lyapunov function in a dissipative form for (2.1).

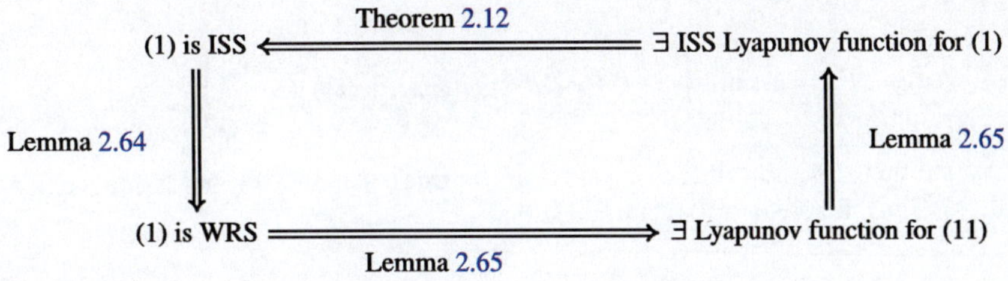

Fig. 2.4 Scheme of the proof of the ISS converse Lyapunov theorem

Proof A scheme of the proof of Theorem 2.63 can be found in Fig. 2.4.

(iii) $\Rightarrow$ (iv). Follows by Proposition 2.21.

(iv) $\Rightarrow$ (i). Follows by Theorem 2.12.

(i) $\Rightarrow$ (ii). Follows from Lemma 2.64.

(ii) $\Rightarrow$ (iii). Follows from Lemma 2.65. □

We continue with technical lemmas, needed to establish Theorem 2.63.

Lemma 2.64 *If (2.1) is ISS, then it is WRS.*

Proof Let (2.1) be ISS. To prove WRS of (2.1) we are going to use Theorem B.22.

Since (2.1) is ISS, there exist $\beta \in \mathcal{KL}$ and $\gamma \in \mathcal{K}_\infty$ so that (2.4) holds for any t, x, u. Define $\alpha(r) := \beta(r, 0)$, for $r \in \mathbb{R}_+$. Substituting $u \equiv 0$ and $t = 0$ into (2.4) we see that $\alpha(r) \geq r$ for all $r \in \mathbb{R}_+$.

Pick any $\sigma \in \mathcal{K}_\infty$ so that $\sigma(r) \leq \gamma^{-1}\big(\frac{1}{4}\alpha^{-1}(\frac{2}{3}r)\big)$ for all $r \geq 0$. We may choose Lipschitz continuous maps $\varphi : \mathbb{R}^n \to \mathbb{R}_+$ and $\psi \in \mathcal{K}_\infty$ such that $\psi(|x|) \leq \varphi(x) \leq \sigma(|x|)$ (just pick a Lipschitz continuous on bounded balls $\psi \in \mathcal{K}_\infty$ and set $\varphi(x) := \psi(|x|)$ for all $x \in \mathbb{R}^n$, which guarantees that φ is also Lipschitz continuous on bounded balls).

We are going to show that for all $x \in \mathbb{R}^n$, all $d \in \mathcal{D}$, and almost all $t \geq 0$ it holds that:

$$\gamma\Big(\big|d(t)\varphi(\phi_\varphi(t, x, d))\big|\Big) \leq \frac{|x|}{2}. \tag{2.96}$$

First we show that (2.96) holds for almost all small enough times t. Since $\alpha^{-1}(r) \leq r$ for all $r > 0$, the following holds for a.e. $t \geq 0$

$$\begin{aligned}\gamma\Big(\big|d(t)\varphi(\phi_\varphi(t, x, d))\big|\Big) &\leq \gamma\big(\sigma(|\phi_\varphi(t, x, d)|)\big)\\ &\leq \frac{1}{4}\alpha^{-1}\Big(\frac{2}{3}|\phi_\varphi(t, x, d)|\Big)\\ &\leq \frac{1}{6}|\phi_\varphi(t, x, d)|.\end{aligned}$$

For any $d \in \mathcal{D}$ and any $x \in \mathbb{R}^n$ the last expression can be made smaller than $\frac{1}{2}|x|$ by choosing t small enough, since ϕ_φ is continuous in t. Now pick any $d \in \mathcal{D}$, $x \in \mathbb{R}^n$ and define $t^* = t^*(x, d)$ by

$$t^* := \operatorname{ess\,inf}\Big\{t \geq 0 : \gamma\Big(|d(t)|\varphi(\phi_\varphi(t, x, d))\Big) > \frac{|x|}{2}\Big\}.$$

Assume that $t^* < \infty$ (otherwise our claim is true). Then (2.96) holds for a.e. $t \in [0, t^*)$. Thus, for a.e. $t \in [0, t^*)$ it holds that

$$\begin{aligned}|\phi_\varphi(t,x,d)| &\le \beta(|x|,t) + \frac{|x|}{2}\\ &\le \beta(|x|,0) + \frac{1}{2}\alpha(|x|)\\ &= \frac{3}{2}\alpha(|x|).\end{aligned}$$

By continuity, this holds also for $t := t^*$. Using this estimate we find out that

$$\gamma\Big(|d(t^*)|\varphi(\phi_\varphi(t^*,x,d))\Big) \le \frac{1}{4}\alpha^{-1}\Big(\frac{2}{3}|\phi_\varphi(t^*,x,d)|\Big) \le \frac{1}{4}|x|.$$

But this contradicts to the definition of t^*. Thus, $t^* = +\infty$.

Now we see that for any $x \in \mathbb{R}^n$, any $d \in \mathcal{D}$, and all $t \ge 0$ we have that

$$|\phi_\varphi(t,x,d)| \le \beta(|x|,t) + \frac{|x|}{2}, \tag{2.97}$$

which shows global stability of (2.93). Since $\beta \in \mathcal{KL}$, there exists a $t_1 = t_1(|x|)$ so that $\beta(|x|,t_1) \le \frac{|x|}{4}$ and consequently

$$|\phi_\varphi(t,x,d)| \le \frac{3}{4}|x|,$$

for all $x \in \mathbb{R}^n$, any $d \in \mathcal{D}$, and all $t \ge \tau$.

By induction and using the cocycle property, we obtain that there exists a strictly increasing sequence of times $(t_k)_{k=1}^{\infty}$ that depends on the norm of $|x|$ but is independent of x and d so that

$$|\phi_\varphi(t,x,d)| \le \Big(\frac{3}{4}\Big)^k |x|,$$

for all $x \in \mathbb{R}^n$, any $d \in \mathcal{D}$ and all $t \ge t_k$.

But this means that for all $\varepsilon > 0$ and for all $\delta > 0$ there exists a time $\tau = \tau(\delta)$ so that for all $x \in \mathbb{R}^n$ with $|x| \le \delta$, for all $d \in \mathcal{D}$, and for all $t \ge \tau$ we have

$$|\phi_\varphi(t,x,d)| \le \varepsilon.$$

This shows uniform global attractivity of (2.93). Together with global stability of (2.93) this implies by Theorem B.22 UGAS of (2.93) and thus WRS of (2.1). □

Lemma 2.65 *If (2.1) is WRS, $f(0,0) = 0$, and Assumption 1.8 is satisfied then there exists an infinitely differentiable ISS Lyapunov function for (2.1).*

Proof Let (2.1) be WRS, which means that (2.93) is UGAS over $\mathcal{D}$ for suitable φ, ψ chosen in accordance with Definition 2.61.

Furthermore, g is continuous in both arguments and Lipschitz continuous in x uniformly w.r.t. d. Hence, by Theorem B.32, there is an infinitely differentiable strict Lyapunov function for (2.93), which we call $V : \mathbb{R}^n \to \mathbb{R}_+$, satisfying

$$\psi_1(|x|) \le V(x) \le \psi_2(|x|)$$

for certain $\psi_1, \psi_2 \in \mathcal{K}_\infty$ and whose Lie derivative along the solutions of (2.93) for all $x \in \mathbb{R}^n$ and for all $d \in \mathcal{D}$ satisfies the estimate

$$\dot{V}_d(x) \le -\alpha(V(x)), \tag{2.98}$$

for some $\alpha \in \mathcal{K}_\infty$. As V is infinitely differentiable, arguing as in Proposition 2.19, one can show that this estimate is equivalent to

$$\nabla V(x) \cdot f(x, \mathrm{d}\varphi(x)) \le -\alpha(V(x)), \quad x \in \mathbb{R}^n, \ |d| \le 1. \tag{2.99}$$

Hence, the following implication holds for all $u \in \mathbb{R}^m$, $x \in \mathbb{R}^n$:

$$|u| \le \varphi(x) \quad \Rightarrow \quad \nabla V(x) \cdot f(x, u) \le -\alpha(V(x)). \tag{2.100}$$

This automatically implies that

$$|u| \le \psi(|x|) \quad \Rightarrow \quad \nabla V(x) \cdot f(x, u) \le -\alpha(V(x)), \tag{2.101}$$

and by Proposition 2.19, V is an ISS Lyapunov function in implication form with a Lyapunov gain $\chi := \psi^{-1}$. □

2.7 ISS, Exponential ISS, and Nonlinear Changes of Coordinates

The same dynamical or control system can be represented in different coordinates. This section shows that ISS is equivalent to exponential ISS up to a nonlinear change of coordinates.

Definition 2.66 We say that $T : \mathbb{R}^n \to \mathbb{R}^n$ is a *change of coordinates*, if T is a homeomorphism satisfying $T(0) = 0$.

From the geometrical point of view, a natural stability notion for a nonlinear system should be invariant w.r.t. nonlinear changes of coordinates. Next, we show that this is the case for ISS:

Proposition 2.67 *Let (2.1) be ISS and let $T : \mathbb{R}^n \to \mathbb{R}^n$ be a change of coordinates. Then (2.1) is ISS in new coordinates $y = T(x)$, that is, there exist $\tilde{\beta} \in \mathcal{KL}$ and $\tilde{\gamma} \in \mathcal{K}$ so that for all $y \in \mathbb{R}^n$, $u \in \mathcal{U}$, and all $t \ge 0$ it holds that*

$$|\tilde{\phi}(t, y, u)| \leq \tilde{\beta}(|y|, t) + \tilde{\gamma}(\|u\|_\infty), \tag{2.102}$$

where $\tilde{\phi}(t, y, u)$ is a solution map of (2.1) in the new coordinates y.

Proof Let $T : \mathbb{R}^n \to \mathbb{R}^n$ be any homeomorphism with $T(0) = 0$. Then Lemma A.44 implies existence of $\psi_1, \psi_2 \in \mathcal{K}_\infty$ so that

$$\psi_1(|x|) \leq |T(x)| \leq \psi_2(|x|) \quad \forall x \in \mathbb{R}^n.$$

Let also (2.1) be ISS with corresponding $\beta \in \mathcal{KL}$ and $\gamma \in \mathcal{K}$.

Pick any $y \in \mathbb{R}^n$, any $u \in \mathcal{U}$ and let $x \in \mathbb{R}^n$ be so that $y = T(x)$. From (2.4), we have that

$$\begin{aligned}|\tilde{\phi}(t, y, u)| = \big|T(\phi(t, x, u))\big| &\leq \psi_2\big(|\phi(t, x, u)|\big)\\ &\leq \psi_2\big(\beta(|x|, t) + \gamma(\|u\|_\infty)\big)\\ &\leq \psi_2\big(2\beta(|x|, t)\big) + \psi_2\big(2\gamma(\|u\|_\infty)\big)\\ &\leq \psi_2\big(2\beta(\psi_1^{-1}(|y|), t)\big) + \psi_2\big(2\gamma(\|u\|_\infty)\big),\end{aligned}$$

which shows that (2.102) holds with $\tilde{\beta}(r, t) := \psi_2\big(2\beta(\psi_1^{-1}(r), t)\big)$ and $\tilde{\gamma}(r) := \psi_2\big(2\gamma(r)\big)$. □

Let us introduce the notion of exponential ISS:

Definition 2.68 System (2.1) is called *exponentially input-to-state stable (eISS)*, if (2.1) is forward complete and there exist $c, \lambda > 0$ and $\gamma \in \mathcal{K}$ such that for all $x \in \mathbb{R}^n$, $u \in \mathcal{U}$, and $t \geq 0$ the following holds

$$|\phi(t, x, u)| \leq ce^{-\lambda t}|x| + \gamma(\|u\|_\infty). \tag{2.103}$$

For nonlinear systems (in contrast to linear ones), exponential ISS is a much stronger notion than ISS, consider $\dot{x} = -x^3 + u$ as a simple example. Furthermore, although exponential ISS is invariant under linear changes of coordinates (see Exercise 2.31), it is not invariant under nonlinear changes of coordinates, in contrast to ISS. Despite this fact, we show in this section that for any sufficiently regular ISS system, there is a coordinate change that makes it eISS in new coordinates. Before proving this fact, we introduce the notion of an exponential ISS Lyapunov function:

Definition 2.69 $V \in C(\mathbb{R}^n, \mathbb{R}_+)$ is called an *exponential ISS Lyapunov function (in implication form)*, if there exist $\tilde{\psi}_1, \tilde{\psi}_2 \in \mathcal{K}_\infty$, $\mu > 0$, and $\chi \in \mathcal{K}$ such that

$$\tilde{\psi}_1(|x|) \leq V(x) \leq \tilde{\psi}_2(|x|) \quad \forall x \in \mathbb{R}^n, \tag{2.104}$$

and for any $x \in \mathbb{R}^n$, $u \in \mathcal{U}$, the following implication holds:

$$V(x) \geq \chi(\|u\|_\infty) \quad \Rightarrow \quad \dot{V}_u(x) \leq -\mu V(x). \tag{2.105}$$

The exponential decay of a Lyapunov function along the trajectory does not mean the exponential decay of a trajectory itself. For this implication, strong additional properties are needed, see Proposition 6.29. In fact, the existence of an ISS Lyapunov function is equivalent to the existence of an exponential ISS Lyapunov function:

Theorem 2.70 *Let V be a continuous ISS Lyapunov function for (2.1). Then for any $\mu > 0$ there exists a continuous exponential ISS Lyapunov function W for (2.1) with the decay coefficient μ in (2.105). Furthermore:*

(i) *If the decay rate α of V is a $\mathcal{K}$-function, then the Lyapunov gain of W can be chosen equal to the Lyapunov gain of V.*
(ii) *If V is infinitely differentiable on $\mathbb{R}^n\backslash\{0\}$, then W can be also chosen to be infinitely differentiable on $\mathbb{R}^n\backslash\{0\}$, and if V is Lipschitz continuous on bounded balls, then W can be chosen with the same property.*

Proof **Step 1.** Let V be an ISS Lyapunov function as in Definition 2.10 with corresponding functions $\psi_1, \psi_2, \chi, \alpha$. Given Proposition 2.17, we can assume without loss of generality that $\alpha \in \mathcal{K}_\infty$, by scaling of a Lyapunov function via suitable C^1-function that is infinitely differentiable on $(0, +\infty)$. In particular, if V is infinitely differentiable on $\mathbb{R}^n\backslash\{0\}$, then the above scaling procedure retains this property.

Pick any $\mu > 0$ and an arbitrary function $\omega \in \mathcal{K}$, infinitely differentiable on $(0, +\infty)$, of class $C^1(\mathbb{R}_+, \mathbb{R}_+)$, satisfying $\omega'(0) = 0$ and

$$\omega(r) \leq \min\left\{r, \tfrac{2}{\mu}\alpha(r)\right\}. \tag{2.106}$$

In particular, a possible choice is

$$\omega(r) = \frac{2}{\pi}\int_0^r \frac{\delta(s)}{1+s^2}ds,$$

where δ is a class $\mathcal{K}$ function, infinitely differentiable over $(0, +\infty)$ and satisfying $\delta(s) \leq \min\{s, \frac{2}{\mu}\alpha(s)\}$ for $s \geq 0$ (such δ always exists).

Step 2. Next define the scaling function $\rho : \mathbb{R}_+ \to \mathbb{R}_+$ by

$$\rho(r) := \begin{cases} e^{\int_1^r \frac{2}{\omega(s)}ds}, & \text{if } r > 0,\\ 0, & \text{if } r = 0. \end{cases} \tag{2.107}$$

By construction, ρ is infinitely differentiable over $(0, +\infty)$. Since $\omega(r) \leq r$ for $r > 0$, it holds for $r > 0$ that $-\frac{1}{\omega(r)} \leq -\frac{1}{r}$ and for $r \in (0, 1)$ we have

$$\rho(r) = e^{\int_1^r \frac{2}{\omega(s)}ds} = e^{-\int_r^1 \frac{2}{\omega(s)}ds} \leq e^{-\int_r^1 \frac{2}{s}ds} = e^{2\ln r} = r^2, \tag{2.108}$$

which implies that $\lim_{r\to+0}\rho(r) = 0$ and therefore ρ is continuous over $\mathbb{R}_+$.

Analogously,

$$\lim_{r\to+\infty}\rho(r)=e^{\int_1^\infty \frac{2}{\omega(s)}\mathrm{d}s}=\infty.$$

Clearly, ρ is increasing. Hence $\rho\in\mathcal{K}_\infty$.

Let us show that $\rho\in C^1(\mathbb{R}_+,\mathbb{R}_+)$. In view of (2.108), for $r>0$ it holds that $0\le\frac{1}{r}\big(\rho(r)-\rho(0)\big)\le r$ and hence

$$\rho'(0):=\lim_{r\to+0}\frac{1}{r}\big(\rho(r)-\rho(0)\big)=0.$$

At the same time, for $r>0$, we compute

$$\rho'(r)=\frac{2\rho(r)}{\omega(r)},\qquad \rho''(r)=\frac{2\rho(r)}{\omega^2(r)}\big(2-\omega'(r)\big).$$

Since $\omega\in C^1(\mathbb{R}_+,\mathbb{R}_+)$ with

$$\omega'(0):=\lim_{r\to+0}\frac{1}{r}\big(\omega(r)-\omega(0)\big)=0,$$

there exists $\varepsilon>0$ so that $0<\omega'(r)<1$ for $r\in(0,\varepsilon)$. Thus, for $r\in(0,\varepsilon)$ we obtain

$$\rho''(r)>\frac{2\rho(r)}{\omega^2(r)}=\frac{\rho'(r)}{\omega(r)}>0. \tag{2.109}$$

In particular, this means that ρ' is increasing and thus there exists $\lim_{r\to+0}\rho'(r)=c\ge 0$. Assume that $c>0$. Then for $r\in(0,\varepsilon]$ we have $\rho'(r)\ge c$ and by (2.109)

$$\rho''(r)>\frac{\rho'(r)}{\omega(r)}\ge\frac{\rho'(r)}{r}\ge\frac{c}{r}.$$

This implies for $r\in(0,\varepsilon]$ (remember that ρ is infinitely differentiable outside of the origin):

$$\rho'(r)=\rho'(\varepsilon)-\int_r^\varepsilon\rho''(s)\mathrm{d}s\le\rho'(\varepsilon)-\int_r^\varepsilon\frac{c}{s}\mathrm{d}s.$$

and hence $\rho'(r)\to-\infty$ as $r\to+0$, a contradiction. Consequently, $c=0$ and $\rho\in C^1(\mathbb{R}_+,\mathbb{R}_+)$.

Step 3. Now consider

$$W:=\rho\circ V. \tag{2.110}$$

Clearly, W is continuous and satisfies (2.104) with $\tilde{\psi}_1 := \rho \circ \psi_1$ and $\tilde{\psi}_2 := \rho \circ \psi_2$. Using Lemma A.30, we can compute the Lie derivative of W at $x \in \mathbb{R}^n$. For any $x \in \mathbb{R}^n$ and $u \in \mathbb{R}^m$ so that $|x| \geq \chi(|u|)$ we have

$$\begin{aligned}
\dot{W}(x) &= \frac{d\rho(r)}{dr}\Big|_{r=V(x)} \dot{V}(x) \\
&= \frac{2}{\omega(V(x))}\rho(V(x))\dot{V}(x) \\
&\leq -\frac{2}{\omega(V(x))} W(x)\alpha(V(x)) \\
&\leq -\frac{\mu}{\alpha(V(x))} W(x)\alpha(V(x)) \\
&= -\mu W(x).
\end{aligned}$$

This shows that W is an exponential ISS Lyapunov function for (2.1).

Since ρ is infinitely differentiable on $(0, +\infty)$, and in $C^1(\mathbb{R}_+, \mathbb{R}_+)$, (ii) immediately follows. □

Exponential ISS Lyapunov functions can be utilized to show that ISS is equivalent to exponential ISS up to a nonlinear change of coordinates. For this, we need additional machinery.

Definition 2.71 A change of coordinates $T : \mathbb{R}^n \to \mathbb{R}^n$ is called *smooth* if $T \in C^1(\mathbb{R}^n)$ and the restrictions of T and T^{-1} to $\mathbb{R}^n \backslash \{0\}$ are smooth (infinitely differentiable).

Let T be a change of coordinates. Consider a transformation $y(t) := T(x(t))$ of a system (2.1). We obtain

$$\frac{d}{dt}y(t) = \frac{d}{dt}T\big(x(t)\big) = T'(x(t))\dot{x}(t) = T'\big(T^{-1}(y(t))\big)f\big(T^{-1}(y(t)), u(t)\big),$$

where we use notation

$$T'(x(t)) := \begin{pmatrix} \big(\nabla T_1(x(t))\big)^T \\ \ldots \\ \big(\nabla T_n(x(t))\big)^T \end{pmatrix},$$

and $T'(x(t))\dot{x}(t)$ is a matrix product of $T'(x(t))$ and $\dot{x}(t)$.

Thus, after a change of coordinates T, the system (2.1) takes form

$$\dot{y}(t) = \tilde{f}(y(t), u(t)), \tag{2.111}$$

with $\tilde{f}(y, u) = T'\big(T^{-1}(y)\big)f\big(T^{-1}(y), u\big)$.

We are going to find changes of coordinates of (2.1) in which asymptotic dynamics of a system (2.1) have a particularly simple form. The following proposition ([13, Proposition 1]) provides us a key to solve this problem

Proposition 2.72 *Let $n \neq 4, 5$ and let $V \in C^1(\mathbb{R}^n, \mathbb{R}_+)$ satisfy (2.10) for certain $\psi_1, \psi_2 \in \mathcal{K}_\infty$. Let also V be infinitely differentiable on $\mathbb{R}^n\backslash\{0\}$ with nonvanishing gradient. Then there exist*

(i) $\zeta \in \mathcal{K}_\infty$, which is C^∞ outside of the origin and satisfies

$$\frac{\zeta(s)}{\zeta'(s)} \geq s \quad \forall s > 0, \tag{2.112}$$

(ii) and a corresponding smooth change of coordinates T with $T'(0) = 0$ so that for all y

$$\tilde{V}(y) := V(T^{-1}(y)) = \zeta(|y|). \tag{2.113}$$

Proof For the proof, please consult the original paper [13, Proposition 1] (where a somewhat stronger result is presented). □

Now we apply Proposition 2.72, Theorems 2.63 and 2.70 to show that an ISS system becomes exponentially ISS after a suitable smooth change of coordinates.

Theorem 2.73 *Let $n \neq 4, 5$, f be Lipschitz continuous on bounded balls of $\mathbb{R}^n \times \mathbb{R}^m$ and let (2.1) be ISS. Then there exists a smooth change of coordinates, which transforms (2.1) into an exponentially ISS system (2.111) with $c = \lambda = 1$.*

Proof Let (2.1) be ISS. According to Theorem 2.63, there exists an ISS Lyapunov function $V \in C^\infty(\mathbb{R}^n, \mathbb{R}_+)$ for (2.1), and due to Theorem 2.70, we can find a scaling ρ as in (2.107) so that $W = \rho \circ V$ is an infinitely differentiable on $\mathbb{R}^n\backslash\{0\}$ exponential ISS Lyapunov function with a rate coefficient $\mu = 1$. As W is an exponential ISS Lyapunov function, $\nabla W(x) \cdot f(x, 0) < 0$ for all $x \neq 0$, hence $\nabla W(x) \neq 0$ for $x \neq 0$. Let χ be a Lyapunov gain of W.

According to Proposition 2.72, we can find a nonlinear change of coordinates T with $T'(0) = 0$ and $\zeta \in \mathcal{K}_\infty$, which is C^∞ outside of the origin, satisfying (2.112) so that

$$\tilde{W}(y) := W(T^{-1}(y)) = \zeta(|y|). \tag{2.114}$$

Since $\tilde{W}$ is infinitely differentiable on $\mathbb{R}^n\backslash\{0\}$, the Lie derivative of $\tilde{W}$ w.r.t. a modified system (2.111) for $y \neq 0$ equals

$$\frac{d}{dt}\tilde{W}(y) = \big(\nabla\zeta(|y|)\big)^T \tilde{f}(y, u). \tag{2.115}$$

Let us compute $\nabla\zeta(|y|)$ for $y \neq 0$. Its ith component can be computed as follows:

$$\big(\nabla\zeta(|y|)\big)_i = \zeta'(|y|)\frac{\partial}{\partial y_i}\sqrt{y_1^2+\dots+y_n^2} = \frac{\zeta'(|y|)}{|y|}y_i.$$

Hence

$$\nabla\zeta(|y|) = \frac{\zeta'(|y|)}{|y|}y.$$

On the other hand, since W is an ISS Lyapunov function, there exists $\psi_2 \in \mathcal{K}_\infty$ so that $W(T^{-1}(y)) \leq \psi_2(|T^{-1}(y)|)$, for any $y \in \mathbb{R}^n$. Hence

$$|T^{-1}(y)| \geq \psi_2^{-1} \circ \zeta(|y|) \quad \forall y \in \mathbb{R}^n.$$

Define a Lyapunov gain for $\tilde{W}$ as $\tilde{\chi} := \zeta^{-1} \circ \psi_2 \circ \chi$. Then for $y \in \mathbb{R}^n$, $u \in \mathbb{R}^m$

$$|y| \geq \tilde{\chi}(|u|) \quad \Rightarrow \quad |x| = |T^{-1}(y)| \geq \psi_2^{-1} \circ \zeta(|y|) \geq \psi_2^{-1} \circ \zeta \circ \tilde{\chi}(|u|) = \chi(|u|),$$

which implies that

$$\begin{aligned}\frac{d}{dt}\tilde{W}(y) &= \frac{d}{dt}W(T^{-1}(y)) = \frac{d}{dt}W(x) \leq -W(x)\\ &= -W(T^{-1}(y)) = -\tilde{W}(y) = -\zeta(|y|).\end{aligned}$$

Substituting our findings into (2.115), we obtain for all $y \in \mathbb{R}^n$, $u \in \mathbb{R}^m$: $y \neq 0$ and $|y| \geq \tilde{\chi}(|u|)$

$$\frac{\zeta'(|y|)}{|y|}y^T\tilde{f}(y,u) \leq -\zeta(|y|),$$

which implies for the same y, u in view of (2.112) that

$$y^T\tilde{f}(y,u) \leq -\frac{\zeta(|y|)}{\zeta'(|y|)}|y| \leq -|y|^2.$$

Clearly, the last inequality holds also for $y = 0$.

Finally we compute the Lie derivative of $t \mapsto |y(t)|^2$ w.r.t. transformed system (2.111)

$$\frac{d}{dt}|y(t)|^2 = 2(y(t))^T\tilde{f}(y(t),u(t)) \leq -2|y(t)|^2,$$

and hence

$$|y(t)|^2 \leq e^{-2t}|y(0)|^2.$$

This implies that for all $y \in \mathbb{R}^n$, $u \in \mathbb{R}^m$

$$|y| \geq \tilde{\chi}(|u|) \quad \Rightarrow \quad |y(t)| \leq e^{-t}|y(0)|,$$

and (2.111) is exponentially ISS. □

Example 2.74 Consider a system

$$\dot{x} = -x^3. \tag{2.116}$$

This system is globally asymptotically stable, but it is not globally exponentially stable.

Let us show that there is no change of coordinates $y = T(x)$ in the class that we consider such that the transformed system looks $\dot{y} = -y$. Indeed, any such transformation $\tilde{T}$ must satisfy

$$\dot{y} = \tilde{T}'(x)\dot{x} = -\tilde{T}'(x)x^3 = -\tilde{T}(x) = -y,$$

and thus $\tilde{T}$ must satisfy

$$\tilde{T}'(x)x^3 = \tilde{T}(x).$$

For positive initial conditions, the solutions of this equation have the form

$$\tilde{T}_c(x) := ce^{\frac{1}{-2x^2}}, \quad x > 0,$$

for any positive $c > 0$. However, for any $c > 0$, $\tilde{T}(\mathbb{R}_+)$ is a bounded set, and thus $\tilde{T}$ cannot be a change of coordinates according to our definition.

However, it is possible to construct a change of coordinates, making the system exponentially stable. Take any $\alpha : \mathbb{R} \to \mathbb{R}$, such that $\alpha(0) = 0$, $\alpha'(x) > 0$ for all $x > 0$, $\alpha(x) = -\alpha(-x)$ for all $x \in \mathbb{R}$, and α is infinitely differentiable.

Now define

$$T(x) := \alpha(x)\tilde{T}_1(x), \quad x \in \mathbb{R}. \tag{2.117}$$

By definition, $T(0) = 0$, and $T \in C^\infty(\mathbb{R}\backslash\{0\})$. Furthermore, T is continuous at 0.

As $e^{\frac{1}{-2x^2}} \to 1$ as $|x| \to \infty$, any unbounded α as above makes T a monotone bijection. Hence T^{-1} is well-defined as well and is a continuous bijection. Hence, T is a homeomorphism.

Furthermore, for any $h \neq 0$, we have

$$\frac{1}{h}(T(h) - T(0)) = \frac{1}{h}T(h) = \frac{1}{|h|}\alpha(|h|)e^{\frac{1}{-2h^2}} \to 0, \quad h \to 0,$$

and thus $T'(0) = 0$, and it is easy to see that $T \in C^1(\mathbb{R})$. One can show that $T^{-1} \in C^\infty(\mathbb{R}\backslash\{0\})$, and thus T is a change of coordinates.

It remains to show that the system (2.116) is exponentially ISS in new coordinates. For any $y \neq 0$ we have

$$
\begin{aligned}
\dot{y} = \dot{T}(x) &= \alpha'(x)\tilde{T}_1(x)\dot{x} + \alpha(x)\dot{\tilde{T}}_1(x) \\
&= -\alpha'(x)x^3\tilde{T}_1(x) + \alpha(x)\tilde{T}_1(x)\Big(-\frac{1}{2}(-2)x^{-3}\dot{x}\Big) \\
&= -\alpha'(x)x^3\tilde{T}_1(x) - \alpha(x)\tilde{T}_1(x) \\
&= -\Big(1 + \frac{\alpha'(x)x^3}{\alpha(x)}\Big)T(x) \\
&\leq -y,
\end{aligned}
$$

and thus the system is exponentially ISS in new coordinates.

2.8 Integral Characterization of ISS

In this section, we characterize ISS in terms of notions involving integrals of the trajectories.

Definition 2.75 A system (2.1) is called

(i) *integral-to-integral ISS* if (2.1) is forward complete and there are $\alpha \in \mathcal{K}_\infty$ and $\psi \in \mathcal{K}_\infty$, $\sigma \in \mathcal{K}_\infty$ so that for all $x \in \mathbb{R}^n$, $u \in \mathcal{U}$ and $t \geq 0$ it holds that

$$\int_0^t \alpha(|\phi(s, x, u)|)ds \leq \psi(|x|) + \int_0^t \sigma(|u(s)|)ds, \tag{2.118}$$

(ii) *norm-to-integral ISS* if (2.1) is forward complete and there are $\alpha \in \mathcal{K}_\infty$ and $\psi \in \mathcal{K}_\infty$, $\sigma \in \mathcal{K}_\infty$ so that for all $x \in \mathbb{R}^n$, $u \in \mathcal{U}$ and $t \geq 0$ it holds that

$$\int_0^t \alpha(|\phi(s, x, u)|)ds \leq \psi(|x|) + t\sigma(\|u\|_\infty). \tag{2.119}$$

We relate ISS and integral ISS-like stability properties as follows:

Theorem 2.76 *Assume that f is Lipschitz continuous on bounded balls of $\mathbb{R}^n \times \mathbb{R}^m$. The following statements are equivalent:*

(i) (2.1) is ISS.
(ii) (2.1) is integral-to-integral ISS.
(iii) (2.1) is norm-to-integral ISS.

Proof (i) $\Rightarrow$ (ii). As (2.1) is ISS, and f is Lipschitz continuous on bounded balls of $\mathbb{R}^n \times \mathbb{R}^m$, Theorem 2.63 ensures existence of an infinitely differentiable ISS Lyapunov function for (2.1), which can be restated in a dissipative form as in Proposition 2.21. Thus, there is a function $V \in C^1(\mathbb{R}^n, \mathbb{R}_+)$ so that (2.10) holds for some $\psi_1, \psi_2 \in \mathcal{K}_\infty$ and for certain $\alpha, \gamma \in \mathcal{K}_\infty$ and all x and u we have that

$$\frac{d}{dt}V(\phi(t,x,u)) \leq -\alpha\big(V(\phi(t,x,u))\big) + \gamma(|u(t)|), \quad \text{for a.e. } t \geq 0.$$

As a composition of a C^1 and an absolutely continuous function, the function $t \mapsto V(\phi(t,x,u))$ is absolutely continuous (see Proposition 1.49), and thus its derivative is Lebesgue integrable, and by the fundamental theorem of calculus, we have that

$$V(\phi(t,x,u)) - V(\phi(0,x,u)) \leq -\int_0^t \alpha(V(\phi(s,x,u)))ds + \int_0^t \gamma(|u(s)|)ds.$$

As $V(\phi(t,x,u)) \geq 0$ and $\phi(0,x,u) = x$, this implies that

$$\int_0^t \alpha(V(\phi(s,x,u)))ds \leq V(x) + \int_0^t \gamma(|u(s)|)ds,$$

and by using (2.10), we obtain the integral-to-integral ISS estimate for all t, x, u:

$$\int_0^t \alpha \circ \psi_1(|\phi(s,x,u)|)ds \leq \psi_2(|x|) + \int_0^t \gamma(|u(s)|)ds.$$

(ii) $\Rightarrow$ (iii). Follows from $\int_0^t \sigma(|u(s)|)ds \leq \int_0^t \sigma(\|u\|_\infty)ds = t\sigma(\|u\|_\infty)$, for all $t \geq 0$ and $u \in \mathcal{U}$.

(iii) $\Rightarrow$ (i). Let (2.1) be norm-to-integral ISS with corresponding $\alpha, \psi, \sigma \in \mathcal{K}_\infty$. By ISS superposition Theorem 2.52, to prove ISS of (2.1), it is sufficient to show that (2.1) is LIM $\wedge$ 0-ULS.

Assume that (2.1) does not have the LIM property with $\gamma(r) := \alpha^{-1}(2\sigma(r))$, $r \geq 0$ in (2.82). Then there are $\varepsilon > 0$, $x \in \mathbb{R}^n$ and $u \in \mathcal{U}$ so that

$$|\phi(t,x,u)| \geq \varepsilon + \alpha^{-1}(2\sigma(\|u\|_\infty)), \quad t \geq 0.$$

From the norm-to-integral ISS, we obtain that

$$t\alpha\Big(\varepsilon + \alpha^{-1}\big(2\sigma(\|u\|_\infty)\big)\Big) \leq \int_0^t \alpha(|\phi(s,x,u)|)ds \leq \psi(|x|) + t\sigma(\|u\|_\infty). \tag{2.120}$$

On the other hand, Exercise A.4 implies that

$$t\alpha\big(\varepsilon + \alpha^{-1}(2\sigma(\|u\|_\infty))\big) \geq \frac{t}{2}\alpha(\varepsilon) + t\sigma(\|u\|_\infty). \tag{2.121}$$

Thus, $\frac{t}{2}\alpha(\varepsilon) \leq \psi(|x|)$ for all $t \geq 0$, a contradiction. Hence, (2.1) has the LIM property.

Let us show the 0-ULS property. Pick any $\varepsilon > 0$ and consider the set

$$K := \{x \in \mathbb{R}^n : |x| \in [\frac{\varepsilon}{2}, \varepsilon]\}.$$

Define also

$$c := \sup_{x \in K} |f(x, 0)|, \qquad \delta := \min\{\frac{\varepsilon}{2}, \psi^{-1}\big(\frac{\varepsilon}{2c}\alpha(\frac{\varepsilon}{2})\big)\}.$$

Assume that for some $x \in \mathbb{R}^n$ with $|x| < \delta$ there is a time t so that $|\phi(t, x, 0)| \geq \varepsilon$. As $t \mapsto \phi(t, x, 0)$ is a continuous map, and $|x| < \frac{\varepsilon}{2}$, there is an interval $[t_1, t_2] \in \mathbb{R}_+$ such that $|\phi(t_1, x, 0)| = \frac{\varepsilon}{2}$, $|\phi(t_2, x, 0)| = \varepsilon$, and $\phi(t, x, 0) \in K$ for all $t \in [t_1, t_2]$. We have:

$$\frac{\varepsilon}{2} = |\phi(t_2, x, 0)| - |\phi(t_1, x, 0)| \leq |\phi(t_2, x, 0) - \phi(t_1, x, 0)|. \tag{2.122}$$

As $t \mapsto \phi(t, x, 0)$ is an absolutely continuous function, we can apply the fundamental theorem of calculus to obtain

$$\begin{aligned}\frac{\varepsilon}{2} \leq \Big|\int_{t_1}^{t_2} \frac{d}{ds}\phi(s, x, 0)ds\Big| &= \Big|\int_{t_1}^{t_2} f(\phi(s, x, 0), 0)ds\Big| \\ &\leq \int_{t_1}^{t_2} |f(\phi(s, x, 0), 0)|ds \leq (t_2 - t_1)c. \end{aligned}\tag{2.123}$$

The norm-to-integral ISS property for $u := 0$ together with above estimate shows that

$$\psi(|x|) \geq \int_0^{+\infty} \alpha(|\phi(s, x, 0)|)ds \geq \int_{t_1}^{t_2} \alpha(|\phi(s, x, 0)|)ds \geq (t_2 - t_1)\alpha\big(\frac{\varepsilon}{2}\big) \geq \frac{\varepsilon}{2c}\alpha\big(\frac{\varepsilon}{2}\big).$$

Finally, by the definition of δ

$$\delta > |x| \geq \psi^{-1}\Big(\frac{\varepsilon}{2c}\alpha\big(\frac{\varepsilon}{2}\big)\Big) \geq \delta,$$

a contradiction. Hence $|\phi(t, x, 0)| \leq \varepsilon$ for all $t \geq 0$ and the system (2.1) is 0-ULS. Theorem 2.52 shows ISS of (2.1). □

The implication (i) $\Rightarrow$ (ii) was based on the smooth converse ISS Lyapunov theorem. This is why we require a quite strong assumption of Lipschitz continuity of f on bounded balls of $\mathbb{R}^n \times \mathbb{R}^m$. However, the implications ISS $\Rightarrow$ norm-to-integral ISS $\Rightarrow$ ULIM can be shown by assuming merely forward completeness of solutions, see Exercise 2.34. Furthermore, the following result holds:

Theorem 2.77 *Let (2.1) be a forward complete control system. Then (2.1) is ISS if and only if (2.1) is norm-to-integral ISS, has BRS, and has CEP property.*

Proof "$\Rightarrow$" ISS clearly implies BRS and CEP property. Norm-to-integral ISS follows by Exercise 2.34.

"$\Leftarrow$" By Exercise 2.34, norm-to-integral ISS implies ULIM property. Norm-to-integral ISS with CEP property implies ULS by [19, Proposition 3.4]. Now ISS superposition Theorem 2.51 implies ISS. □

2.9 Semiglobal Input-to-State Stability

Here we characterize ISS in terms of the following property, which ISS immediately implies.

Definition 2.78 A forward complete system (2.1) is called *semiglobally input-to-state stable (semiglobally ISS) w.r.t. inputs*, if for each $M > 0$ there exist $\beta_M \in \mathcal{KL}$ and $\gamma_M \in \mathcal{K}$ such that for all $x \in \mathbb{R}^n$, $u \in \mathcal{U}$: $\|u\|_\infty \le M$ and all $t \ge 0$ it holds that

$$|\phi(t, x, u)| \le \beta_M(|x|, t) + \gamma_M(\|u\|_\infty). \tag{2.124}$$

Without loss of generality we assume in this definition that $\beta_M(r, t) \ge \beta_N(r, t)$ for all $r, t \in \mathbb{R}_+$ and all M, N: $M \ge N$.

Proposition 2.79 *Let Σ be forward complete. Then (2.1) is ISS if and only if (2.1) is semiglobally ISS w.r.t. the inputs.*

Proof Clearly, any ISS system is semiglobally ISS w.r.t. inputs.

Conversely, let (2.1) be ISS w.r.t. the inputs with the corresponding sequence $(\gamma_M)_{M\in\mathbb{N}}$ of gain functions of class $\mathcal{K}_\infty$. Then (2.1) is ULS and has bounded reachability sets. Let us show that (2.1) has also bULIM property.

Take σ as in Lemma A.21 and define $\gamma(s) := \sigma(s + 1)\sigma(s)$, $s \in \mathbb{R}_+$.

For each $r \ge 0$ denote by $\bar{r} := \lceil r \rceil$ the ceiling function. Further, for each $r > 0$ and each $\varepsilon > 0$ define $\tau(\varepsilon, r)$ as the time τ for which $\beta_{\bar{r}}(r, \tau) = \varepsilon$, if such τ exists, and $\tau(\varepsilon, r) := 0$ otherwise.

Pick any $\varepsilon > 0$, any $r > 0$ and any $x \in \mathbb{R}^n$, $u \in \mathcal{U}$ with $|x| \le r$, $\|u\|_\infty \le r$. Denote also $M := \lceil \|u\|_\infty \rceil \le \bar{r}$. With (2.124) and (A.22) it holds that

$$\begin{aligned}|\phi(t,x,u)| &\le \beta_M(|x|,t) + \gamma_M(\|u\|_\infty)\\ &\le \beta_{\bar r}(r,t) + \sigma(M)\sigma(\|u\|_\infty)\\ &\le \beta_{\bar r}(r,t) + \sigma(\|u\|_\infty + 1)\sigma(\|u\|_\infty)\\ &= \beta_{\bar r}(r,t) + \gamma(\|u\|_\infty),\end{aligned}$$

and for all $t \ge \tau(\varepsilon, r)$ it holds that $|\phi(t,x,u)| \le \varepsilon + \gamma(\|u\|_\infty)$, which shows bULIM property for (2.1).

ISS superposition Theorem 2.51 finishes the proof. □

2.10 Input-to-State Stable Monotone Control Systems

Monotone control systems are an important subclass of control systems, which frequently arise in natural sciences, including chemistry and biology. In this section, we show that ISS for constant inputs implies ISS for arbitrary inputs for systems (2.1) that are monotone w.r.t. inputs.

For $x = (x_1, \dots, x_n), y = (y_1, \dots, y_n) \in \mathbb{R}^n$ we will say that $x \le y$ if $x_i \le y_i$ for all $i \in \{1, 2, \dots, n\}$. For $u, v \in \mathcal{U}$ we say that $u \le v$ if $u(t) \le v(t)$ for a.e. t.

Definition 2.80 We call a forward complete system (2.1) *monotone w.r.t. inputs*, if for all $u_1, u_2 \in \mathcal{U} : u_1 \le u_2$ it holds that $\phi(t,x,u_1) \le \phi(t,x,u_2)$ for all $t \ge 0$ and all $x \in \mathbb{R}^n$.

Definition 2.81 System (2.1) is called *input-to-state stable (ISS) w.r.t. inputs* $\tilde{\mathcal{U}} \subset \mathcal{U}$ if (2.1) is forward complete and there exist $\beta \in \mathcal{KL}$ and $\gamma \in \mathcal{K}$ such that for all $x \in \mathbb{R}^n$, all $u \in \tilde{\mathcal{U}}$, and all $t \ge 0$ the following holds

$$|\phi(t,x,u)| \le \beta(|x|,t) + \gamma(\|u\|_\infty). \tag{2.125}$$

We need a simple estimate:

Lemma 2.82 *If $a \le x \le b$ for some $a, b, x \in \mathbb{R}^n$, then $|x| \le |a| + |b|$.*

Proof $|x| = \sqrt{\sum_{i=1}^n x_i^2} \le \sqrt{\sum_{i=1}^n (a_i^2 + b_i^2)} \le \sqrt{\sum_{i=1}^n a_i^2} + \sqrt{\sum_{i=1}^n b_i^2} = |a| + |b|$.

□

Now we can show the following criterion for ISS of monotone control systems:

Proposition 2.83 *Let (2.1) be monotone with respect to inputs. Then (2.1) is ISS if and only if (2.1) is ISS w.r.t. constant inputs.*

Proof "⇒". Evident.

"⇐". Let (2.1) be a monotone control system. Pick any $u = (u_1, \dots, u_m) \in \mathcal{U}$, where $u_i \in L^\infty(\mathbb{R}_+, \mathbb{R})$. Define the vectors $v^-, v^+ \in \mathbb{R}^m$ by

$$v_i^- := \text{ess inf}_{s\geq 0} u_i(s), \qquad v_i^+ := \text{ess sup}_{s\geq 0} u_i(s),$$

for $i \in 1, \ldots, m$. Since u is globally essentially bounded, v^- and v^+ are well-defined. It holds that $v^- \leq u(s) \leq v^+$ for almost all $s \in \mathbb{R}_+$. Define the constant inputs

$$u^-(s) := (\inf_{i\in\{1,\ldots,m\}} v_i^-) \cdot (1, \ldots, 1)^T, \quad u^+(s) := (\sup_{i\in\{1,\ldots,m\}} v_i^+) \cdot (1, \ldots, 1)^T, \quad s \in \mathbb{R}^+.$$

Clearly, $u_- \leq u \leq u_+$. Moreover, it holds that:

$$\begin{aligned} \|u_-\|_\infty = \sqrt{m}|\inf_{i\in\{1,\ldots,m\}} v_i^-| &\leq \sqrt{m} \max_{i=1,\ldots,m} |v_i^-| \\ &\leq \sqrt{m} \max_{i=1,\ldots,m} \text{ess sup}_{s\geq 0}|u_i(s)| \\ &\leq \sqrt{m}\text{ess sup}_{s\geq 0}|u(s)| \\ &= \sqrt{m}\|u\|_\infty. \end{aligned} \tag{2.126}$$

Analogously,

$$\|u_+\|_\infty \leq \sqrt{m}\|u\|_\infty. \tag{2.127}$$

Due to monotonicity of (2.1) with respect to inputs, it holds that

$$\phi(t, x, u_-) \leq \phi(t, x, u) \leq \phi(t, x, u_+), \quad t \geq 0, \ x \in \mathbb{R}^n.$$

In view of Lemma 2.82, it holds that

$$|\phi(t, x, u)| \leq |\phi(t, x, u_-)| + |\phi(t, x, u_+)|, \quad t \geq 0, \ x \in \mathbb{R}^n,$$

which implies due to ISS of (2.1) w.r.t. constant inputs that there exist $\beta \in \mathcal{KL}$ and $\gamma \in \mathcal{K}_\infty$ (independent on x, u, t) so that

$$\begin{aligned} |\phi(t, x, u)| &\leq \beta\big(|x|, t\big) + \gamma\big(\|u_-\|_\infty\big) + \beta\big(|x|, t\big) + \gamma\big(\|u_+\|_\infty\big) \\ &\leq 2\beta\big(|x|, t\big) + 2\gamma\big(\sqrt{m}\|u\|_\infty\big), \end{aligned}$$

which shows ISS of (2.1). □

Having established Proposition 2.83, it is of interest to analyze whether ISS systems with constant inputs possess some special properties. In the exercises in Sect. 2.14, a reader may follow this path.

2.11 Input-to-State Stability, Dissipativity, and Passivity

Finally, we relate input-to-state stability to other important stability concepts for systems with inputs—dissipativity and passivity.

Consider a time-invariant nonlinear *system with inputs and outputs*

$$\dot{x} = f(x, u), \tag{2.128a}$$
$$y = h(x, u). \tag{2.128b}$$

The function $h : \mathbb{R}^n \times \mathbb{R}^m \to \mathbb{R}^p$ that we assume to be continuously differentiable is called the *output function* of the system (2.128).

A major concept, which brings Lyapunov methods to open systems, possessing inputs and outputs, is the notion of dissipativity.

Definition 2.84 The system (2.128) is called *dissipative* with a *supply rate* $s \in C(\mathbb{R}^m \times \mathbb{R}^p, \mathbb{R})$ if there exists $V : \mathbb{R}^n \to \mathbb{R}$ such that $V(0) = 0$, $V(x) \geq 0$ for all $x \neq 0$ and

$$V(\phi(t, x, u)) - V(x) \leq \int_0^t s(u(r), y(r))dr \tag{2.129}$$

holds for all $x \in \mathbb{R}^n$, $u \in \mathcal{U}$, and all $t \in [0, t_m(x, u))$.

The function V is called a *storage function.*

Dissipativity is an important structural property of control systems, providing a variety of additional features, which can be exploited in stability analysis and control.

Converse ISS Lyapunov theorems help to establish the link between ISS and dissipativity

Proposition 2.85 *Let f be Lipschitz continuous on bounded balls of $\mathbb{R}^n \times \mathbb{R}^m$ and assume that the output of the system is its state: $y(\cdot) = x(\cdot)$. If (2.128a) is ISS, then (2.128) is dissipative with a properly chosen storage function.*

Proof Since (2.128a) is ISS and Lipschitz continuous on bounded balls of $\mathbb{R}^n \times \mathbb{R}^m$, due to Theorem 2.63, there is an ISS Lyapunov function in a dissipative form $V \in C^1(\mathbb{R}^n, \mathbb{R}_+)$. Thus, V satisfies the inequality (2.10) for all $x \in \mathbb{R}^n$ and for all $x \in \mathbb{R}^n$ and all $u \in \mathbb{R}^m$ the following inequality holds:

$$\frac{d}{dt}V(\phi(t, x, u)) \leq -\alpha(|\phi(t, x, u)|) + \sigma(|u(t)|), \quad \text{for almost all } t \geq 0.$$

As $V \in C^1(\mathbb{R}^n, \mathbb{R}_+)$, V is Lipschitz continuous on bounded sets by Proposition 1.46. Since the solution map ϕ is absolutely continuous in time, the map $t \mapsto V(\phi(t, x, u))$ is absolutely continuous by Proposition 1.49.

Thus $\frac{d}{dt}V(\phi(t,x,u))$ is integrable and we can apply the fundamental theorem of calculus (Theorem 1.2) to yield

$$\begin{aligned} V(\phi(t,x,u)) - V(x) &\le \int_0^t -\alpha(|\phi(r,x,u)|) + \sigma(|u(r)|)dr \\ &= \int_0^t -\alpha(|y(r)|) + \sigma(|u(r)|)dr. \end{aligned}$$

Hence, (2.128a) is dissipative with a supply rate $s(u,y) = -\alpha(|y|) + \sigma(|u|)$. □

Another important subclass of dissipative systems is passive systems:

Definition 2.86 The system (2.128) is called *passive*, if it is dissipative w.r.t. the supply rate $s(u,y) := u^T y$.

In particular, the dissipativity inequality for the passive systems takes the form:

$$V(\phi(t,x,u)) - V(x) \le \int_0^t u^T(r)y(r)dr. \tag{2.130}$$

Passive systems are an essential tool for the stability analysis of interconnected systems. They are, in some sense, complementary to the ISS paradigm, see [4] (in particular, an interconnection of passive systems is again passive). The next examples show that already for linear systems the ISS-like notions and passivity are different concepts, even for time-invariant linear systems

$$\dot{x} = Ax + Bu, \tag{2.131}$$
$$y = Cx, \tag{2.132}$$

where $x(t) \in \mathbb{R}^n$, $u \in \mathbb{R}^m$, $y \in \mathbb{R}^p$, and A, B, C are real matrices of the appropriate dimension.

Example 2.87 Consider the following 1-dimensional system, which is called *"the integrator"*:

$$\dot{x} = u, \quad y(t) = x(t). \tag{2.133}$$

Clearly, this system is not ISS, and arbitrary small inputs may produce unbounded trajectories. However, it is passive, as we show next. Pick $V(x) = \frac{1}{2}x^2$. Computing the Lie derivative of V we obtain:

$$\dot{V}(x) = x(t)u(t),$$

which is equivalent to (2.130) since V is C^1.

We define the following extension of the input-to-state stability property to the systems with outputs (2.128).

Definition 2.88 System (2.128) is called *input-to-output stable (IOS)*, if (2.128) is forward complete and there exist $\beta \in \mathcal{KL}$ and $\gamma \in \mathcal{K}$ such that $\forall x \in \mathbb{R}^n$, $\forall u \in \mathcal{U}$, and $\forall t \geq 0$ the following holds

$$|y(\phi(t, x, u))| \leq \beta(|x|, t) + \gamma(\|u\|_\infty). \tag{2.134}$$

The next example shows that there are IOS systems, which are not passive.

Example 2.89 Consider the 2-dimensional system

$$\dot{x} = \begin{pmatrix} 0 & 1 \\ -a & -2a \end{pmatrix} x + \begin{pmatrix} 0 \\ 1 \end{pmatrix} u, \qquad y = (\mu \ \ 1)x \tag{2.135}$$

with $a, \mu \in \mathbb{R}$.

This system is IOS for any $a > 0$. However, if $\mu > 2a$, then this system is not passive, see [4, Section 1.4].

2.12 ISS and Regularity of the Right-Hand Side

In this chapter, we have developed an ISS theory for nonlinear systems (2.1) with a set of powerful tools, such as ISS superposition theorems, direct and converse ISS Lyapunov theorems, relations between ISS and integral-type stability properties. However, these results are shown under distinct requirements on the regularity of the right-hand side f. Here we would like to give an overview of the ISS toolbox, which is available for control theorists depending on the regularity of the system.

We can divide the regularity assumptions into four levels:

(L1) *Local existence and uniqueness at any point and BIC property.*
Direct Lyapunov theorems are our primary tool at this level. They ensure that the ISS estimate is valid at the interval of the existence of solutions, and the BIC property ensures that these solutions exist globally.

(L2) *Global existence and uniqueness (forward completeness).*
On this level, we have in addition the ISS superposition theorem (Theorem 2.51) and the characterization of ISS in terms of norm-to-integral ISS (Theorem 2.77). These results provide a firm basis for developing the small-gain theory for networks of ISS systems. Furthermore, these results can be extended to much more general system classes, as we discuss in Sect. 6.4. This means that the ISS characterizations for nonlinear forward complete ODE systems with such

mild regularity are almost as complex as the ISS criteria for general infinite-dimensional control systems, and specific traits of ODE systems cannot be exploited in most cases.

(L3) *Forward completeness and validity of Assumption* 1.8.
Assumption 1.8 (continuity of f on $\mathbb{R}^n \times \mathbb{R}^m$ together with Lipschitz continuity of f in x on bounded subsets) gives much more than just local existence and uniqueness. In combination with the specific structure of nonlinear ODE systems, this enables to show such strong properties as an equivalence between forward completeness and boundedness of reachability sets, and Lipschitz continuity of the flow map with respect to the initial state, as discussed in Sect. 1.3, 1.4, 1.5. This, in turn, allows us to obtain stronger versions of the ISS superposition theorem (Theorem 2.52), the equivalence between 0-UAS and LISS (Theorem 2.38), as well as the equivalence between LIM and ULIM properties (Proposition 2.49).

(L4) *Lipschitz continuity of f in both state and input variables.*
At this level, we can show the converse ISS Lyapunov theorems. This, in turn, allows us to "twist our eyes" at the geometric properties of ISS systems under the nonlinear changes of coordinates (Sect. 2.7) and show the equivalence between ISS and integral-to-integral ISS in Sect. 2.8.

2.13 Concluding Remarks

Input-to-state stability. The notions of input-to-state stability and of an ISS Lyapunov function have been introduced in the seminal paper [44]. In the same paper, it has been proved that the existence of an ISS Lyapunov function implies ISS (Proposition 2.12). For various extensions of CIUCS property (Proposition 2.7), a reader may consult [47].

ISS Lyapunov functions. Properties of the scalings of ISS and iISS Lyapunov functions have been investigated in [25, 43] in continuous-time setting and in [1, 11, 38] in discrete-time case. The scaling results were used for the characterization of the supply pairs for the ISS and ISS-like systems [1, 38, 43]. Lemma 2.14 is well-known, and much stronger results can be found in [17, Sect. 1.1]. Proposition 2.15 was shown in [43, Section II], and it was extended to more general settings in [11] and in [1]. Proposition 2.16 is a folklore and was formally stated, e.g., in [25, Proposition 5]. Exercises 2.9, 4.7 are taken from [25].

Proposition 2.17 was shown in [29, Remark 4.1], via a different argument. A discrete-time version of this result was shown in [21, Lemma 2.8]. A special case of Proposition 2.17, when the decay rate is a $\mathcal{K}$-function, was proved in [39, Lemma 11].

Proposition 2.21 is [48, Remark 2.4]. Section 2.2.4 is based on [9].

The model from Sect. 2.2.6 is adopted from [41]. Our analysis is, however, quite different from that in [41].

In [12], a deep neural network architecture and a training algorithm for computing approximate Lyapunov functions for nonlinear ODE systems have been proposed. Additionally, it is shown that nonlinear systems satisfying a small-gain condition admit compositional Lyapunov functions. This, in turn, ensures that the number of neurons needed for an approximation of a Lyapunov function with fixed accuracy grows only polynomially in the state dimension, which overcomes the curse of dimensionality.

Local ISS. Proposition 2.37 has been stated and proved, e.g., in [30, Theorem 1.4]. However, it was probably known before that and is reminiscent of the equivalence between asymptotic stability and so-called total stability (see [14, Theorems 56.3, 56.4]). [49, Lemma I.1] claiming that 0-GAS implies LISS is a precursor for Proposition 2.37. In turn, Proposition 2.37 is a special case of [31, Theorem 4], where this result has been generalized for infinite-dimensional systems using a proof technique previously exploited in [15, Corollary 4.2.3]. In our proof of Proposition 2.37, we follow the lines of [31, Theorem 4].

Although Theorem 2.38 tells us that a system that is asymptotically stable in the absence of disturbances is locally ISS in the presence of the disturbance inputs, it does not give precise estimates for the region in which is the system locally ISS. Estimating an attraction region of LISS systems is a challenging problem that is largely open. It has been approached by means of a numerical construction of a continuous piecewise affine LISS Lyapunov functions using the linear programming method [28].

The origins of the linearization method (see Corollary 2.39) go back to the father of stability theory—A. M. Lyapunov. The statement of a linearization method for nonlinear systems without controls can be found in any textbook on dynamical systems, see, e.g., [45, Theorem 19], or [57, p. 100].

Limit property and weak attractivity. The limit property was introduced in [49], and strong and uniform limit properties are due to [35]. Equivalence between LIM and ULIM properties for Lipschitz continuous systems (Proposition 2.49) was shown in [35] as well.

Uniform limit property with zero gain (aka uniform weak attractivity) has been introduced in [37] and studied in [32, 36]. This concept was also implicitly used in [52, Proposition 2.2] and [50, Proposition 3] to characterize UGAS of finite-dimensional hybrid and stochastic systems. Uniform weak attractivity is closely related to the notion of uniformly recurrent sets (see, e.g., [26, Lemma 6.5], [52, Section 2.4]) and is widely used in the theory of hybrid and especially stochastic systems [26, 50, 52].

The uniform weak attractivity notion is motivated by the weak attractivity, which was introduced (for systems without disturbances) in [7] and which is one of the pillars for the treatment of dynamical systems theory over locally compact metric spaces in [8]. In the language of [49], weak attractivity is precisely the limit property with zero gain. The concept of weak attractivity is intimately related to the classical notion of recurrent sets, see, e.g., [8, Definition 1.1 in Chap. 3], [26, 50, 52]. A reader may consult [32, Sect. 3.3, p. 97] for more on weak attractivity and recurrence. Equivalence of weak and uniform weak attractivity for ODEs with uniformly bounded disturbances is a consequence of [49, Corollary III.3].

ISS superposition theorem. The notions of an asymptotic gain and uniform asymptotic gain are due to [49]. The concept of strong asymptotic gain has been introduced in [35] and was used to obtain superposition theorems for the so-called strong ISS property. The uniform asymptotic gain on bounded sets (bUAG) property was introduced and studied in the context of ISS superposition theorems in [33]. However, the idea of using this kind of attractivity was also used previously, e.g., in [51]. In Example 2.46, we have seen that for nonlinear finite-dimensional ISS systems, the minimal asymptotic gain is not necessarily equal to the minimal ISS gain (which does not have to exist). In contrast to that, the minimal asymptotic gain equals the minimal ISS gain for a broad class of linear infinite-dimensional systems, see [22].

An ISS superposition theorem for systems that are Lipschitz continuous on bounded balls of $\mathbb{R}^n$ (Theorem 2.52) has been proved in [49]. ISS superposition theorem with mild regularity assumptions (Theorem 2.51) was shown in [33, 35] for a broad class of infinite-dimensional systems. Our proof technique for Theorems 2.51 and 2.52 is based on [33, 35].

Converse ISS Lyapunov theorems. In Sect. 2.6, we follow closely the method developed in [48]. In [48] it was assumed for simplicity that $f : \mathbb{R}^n \times \mathbb{R}^m \to \mathbb{R}^n$ is continuously differentiable and satisfies $f(0, 0) = 0$. For the construction of a Lipschitz continuous Lyapunov function in implication form it is enough, however, to require Lipschitz continuity of f on bounded balls of $\mathbb{R}^n \times \mathbb{R}^m$. If we aim for a merely continuous ISS Lyapunov function, then the conditions on f in (2.1) and on g in (2.93) can be possibly relaxed using the converse Lyapunov theorem [23, Theorem 3.4 at p. 135] (see [23, condition (REG1) at p. 130]). The relaxation will be, however, nontrivial, as relaxing assumptions on g we may lose the uniqueness of solutions for the system (2.93). The proof of Lemma 2.64 follows the lines of [48, Lemma 2.12].

Integral characterizations. Equivalence between integral-to-integral ISS and ISS (Theorem 2.76) is due to [45, Theorem 1]. Equivalence between norm-to-integral ISS and ISS for a broad class of infinite-dimensional systems has been shown in [19], where also the theory of noncoercive ISS Lyapunov functions has been developed. Further relationships among ISS, integral ISS, and integral-to-integral stability properties (as linear and nonlinear L^2-gain properties) have been established in [24].

Nonlinear coordinate changes. Main results in Sect. 2.7 have been proved in [13]. Definitions 2.66 and 2.71 are adopted from [13, p. 129]. The statement and proof of Theorem 2.70 is due to [39, Lemma 11]. This generalizes an earlier result [27, Theorem 3.6.10] for classical Lyapunov functions.

Monotone control systems have been firstly investigated in [3], inspired by the theory of monotone dynamical systems pioneered by M. Hirsch in the 1980s in a series of papers, beginning with [16]. For the introduction to this theory, a reader may consult [42]. The results of Sect. 2.10 have been shown in a more general context in [34] and applied for the ISS analysis of nonlinear parabolic systems.

Monotonicity is not the only **structural property** of nonlinear systems that significantly helps in the ISS analysis of control systems. For example, if a nonlinear ODE system is *homogeneous*, then 0-GAS of this system implies its ISS with respect to homogeneously involved perturbations [2, 6, 40]. ISS of *Lurie systems* was analyzed in [5].

Dissipativity. The notion of dissipativity was introduced to mathematical control theory by Willems in the seminal papers [55, 56]. It has transferred the Lyapunov methods, predominant in the study of systems without inputs ("closed systems") to the systems with inputs ("open systems"). Passive systems are a subclass of dissipative systems, which has attracted a lot of attention [54]. The methods of dissipative theory have become widely used in the analysis of interconnected systems, in part due to the property that interconnections of dissipative/passive systems are again dissipative/passive [4]. For example, port-Hamiltonian systems, widely used in modeling and analysis of finite and infinite-dimensional control systems [10, 20], are inherently passive.

2.14 Exercises

Basic definitions and results

Exercise 2.1 Let $\mathcal{KKL}$ be a class of functions $\omega \in C(\mathbb{R}_+ \times \mathbb{R}_+ \times \mathbb{R}_+, \mathbb{R}_+)$ that satisfy $\omega(\cdot, r, t) \in \mathcal{K}$ for any $r > 0$, $t \in \mathbb{R}_+$, $\omega(r, \cdot, t) \in \mathcal{K}$ for any $r > 0$, $t \in \mathbb{R}_+$, and $\omega(r_1, r_2, \cdot) \in \mathcal{L}$ for any $r_1 \neq 0$, $r_2 \neq 0$.

Let Assumption 2.1 hold. Prove that if there exists $\omega \in \mathcal{KKL}$ such that

$$x \in \mathbb{R}^n, \quad u \in \mathcal{U}, \quad t < t_m(x, u) \quad \Rightarrow \quad |\phi(t, x, u)| \leq \omega(|x|, \|u\|_\infty, t),$$

then (2.1) is ISS.

Exercise 2.2 We have defined ISS using the Euclidean norm in the state space $\mathbb{R}^n$. Show that the ISS notion is invariant w.r.t. the choice of the norm in $\mathbb{R}^n$. More precisely, if a forward complete system (2.1) is ISS according to Definition 2.2, then for any norm $\|\cdot\|$ in $\mathbb{R}^n$ there are $\tilde{\beta} \in \mathcal{KL}$, $\tilde{\gamma} \in \mathcal{K}_\infty$ that depend on the choice of the norm, so that for all $x \in \mathbb{R}^n$, all $u \in \mathcal{U}$, and all $t \geq 0$ the following holds:

$$\|\phi(t, x, u)\| \leq \tilde{\beta}(\|x\|, t) + \tilde{\gamma}(\|u\|_\infty). \tag{2.136}$$

Exercise 2.3 System (2.1) is called *input-to-state semiglobally exponentially stable*, if (2.1) is forward complete and there is $\gamma \in \mathcal{K}_\infty$ so that for each $r > 0$ there exist constants $\mu = \mu(r) > 0$ and $K = K(r) > 0$ s.t. $\forall x \in \overline{B_r}$, $\forall u \in \mathcal{U}$, and $\forall t \geq 0$ the following holds

$$|\phi(t, x, u)| \leq Ke^{-\mu t}|x| + \gamma(\|u\|_\infty). \tag{2.137}$$

Are input-to-state semiglobally exponentially stable systems necessarily ISS? Conversely, are ISS systems of the form (2.1) always input-to-state semiglobally exponentially stable?

Exercise 2.4 (Characterizations of ISS in terms of the norms of the trajectories) Show that (2.1) is ISS if and only if (2.1) is forward complete and there are $\beta \in \mathcal{KL}$ and $\gamma \in \mathcal{K}$ such that $\forall x \in \mathbb{R}^n$, $\forall u \in \mathcal{U}$, and $\forall t \geq 0$ the following holds

$$\|\phi(\cdot, x, u)\|_{C([t,+\infty),\mathbb{R}^n)} \leq \beta(|x|, t) + \gamma(\|u\|_\infty), \tag{2.138}$$

where $\|g\|_{C([t,+\infty),\mathbb{R}^n)} := \sup_{s\geq t} |g(s)|$ for any $g \in C([t, +\infty), \mathbb{R}^n)$.

ISS Lyapunov functions

Exercise 2.5 It is possible to define the concept of an ISS Lyapunov function without using the differentiation along the trajectory.

Let (2.1) satisfies the BIC property. Consider a (possibly discontinuous) function $V : \mathbb{R}^n \to \mathbb{R}_+$ such that there are $\psi_1, \psi_2 \in \mathcal{K}_\infty$:

$$\psi_1(|x|) \leq V(x) \leq \psi_2(|x|) \quad \forall x \in \mathbb{R}^n,$$

and there is $\beta \in \mathcal{KL}$ and $\gamma \in \mathcal{K}_\infty$, such that for all $x \in \mathbb{R}^n$, all $u \in \mathcal{U}$, and all $t \in [0, t_m(x, u))$ it holds that

$$V(\phi(t, x, u)) \leq \beta(V(x), t) + \gamma(\|u\|_\infty). \tag{2.139}$$

Show that existence of a discontinuous ISS Lyapunov function for (2.1) implies ISS.

Exercise 2.6 Show that replacing the inequality (2.28) in Definition 2.10 with the one below

$$|x| \geq \chi(|u|) \quad \Rightarrow \quad \nabla V(x) \cdot f(x, u) \leq -\alpha(|x|), \tag{2.140}$$

we obtain an equivalent definition of an ISS Lyapunov function.

Exercise 2.7 Let us replace the inequality (2.28) in Definition 2.10 with the following one

$$V(x) \geq \chi(|u|) \quad \Rightarrow \quad \nabla V(x) \cdot f(x, u) < 0. \tag{2.141}$$

Does the existence of such a function V imply ISS?

Exercise 2.8 Assume that there is $V \in C^1(\mathbb{R}^n, \mathbb{R}_+)$ and $\psi_1, \psi_2 \in \mathcal{K}_\infty$, $\alpha, \gamma \in \mathcal{K}$ such that

$$\psi_1(|x|) \leq V(x) \leq \psi_2(|x|) \quad \forall x \in \mathbb{R}^n \tag{2.142}$$

holds and for any $x \in \mathbb{R}^n$, $u \in \mathcal{U}$ the dissipation inequality holds:

$$\dot{V}_u(x) \leq -\alpha(V(x)) + \gamma(\|u\|_\infty). \tag{2.143}$$

Show that if $\sup_{s\geq 0} \alpha(s) > \sup_{s\geq 0} \gamma(s)$, then (2.1) is ISS.

Exercise 2.9 (Scalings of ISS Lyapunov functions)

(i) Show Proposition 2.16: Let V be an ISS Lyapunov function *in implication form* for (2.1). Show that $\mu \circ V$ is an ISS Lyapunov function in implication form for any nonlinear scaling μ.

(ii) Show that in Proposition 2.15 the requirement of convexity cannot be omitted. As an example you may consider the system

$$\dot{x} = -x + u, \tag{2.144}$$

and a corresponding ISS Lyapunov function $V(x) = \frac{1}{2}x^2$, which is an ISS Lyapunov function for (2.144) in both implication and dissipative form. Define a scaling $\mu(s) := \ln(1 + s)$ for $s \geq 0$.
Show that $W := \mu \circ V$ is an ISS Lyapunov function in implication form, but W is not an ISS Lyapunov function in a dissipative form.

Exercise 2.10 Show that if there is an ISS Lyapunov function for a system (2.1), then there is also a symmetric one W, satisfying $W(-x) = W(x)$ for all $x \in \mathbb{R}^n$.

Exercise 2.11 Let $x_i(t) \in \mathbb{R}$, $i = 1, 2$ and $u(t) \in \mathbb{R}$. Prove that the following system is ISS.

$$\dot{x}_1 = -x_1 + x_1^3 x_2, \tag{2.145a}$$

$$\dot{x}_2 = -x_2 - x_1^6 x_2 + u. \tag{2.145b}$$

Exercise 2.12 Let Q, P be as in item (iv) of Theorem 2.25 (and hence (2.2) is ISS). Show that $W(x) := (Ax)^T PAx$ is an ISS Lyapunov function for (2.2).

Exercise 2.13 (Krasovskii's Method) The method for construction of Lyapunov functions developed in Exercise 2.12 can be generalized to some classes of nonlinear systems. Consider an undisturbed nonlinear system

$$\dot{x} = f(x), \tag{2.146}$$

with locally Lipschitz $f : \mathbb{R}^n \to \mathbb{R}^n$. Denote $J(x) := \frac{\partial f}{\partial x}$ (the Jacobian of f).

(i) Assume that there are $Q, P > 0$ so that for any $x \in \mathbb{R}^n$

$$J^T(x)P + PJ(x) + Q \leq 0. \tag{2.147}$$

Define $W(x) := f^T(x)Pf(x)$, $x \in \mathbb{R}^n$. Show that if $W(x) \to \infty$ as $|x| \to \infty$, then W is a strict Lyapunov function for (2.146).

(ii) Apply Krasovskii's method to the system (2.146) with $x = (x_1, x_2) \in \mathbb{R}^2$ and

$$f(x) = \begin{pmatrix} -7x_1 + 4x_2 \\ x_1 - x_2 - x_2^5 \end{pmatrix}.$$

(iii) Think about ISS generalization of the Krasovskii's method.

Exercise 2.14 Consider the SIR epidemic model (see, e.g., [18])

$$\dot{S} = B - \mu S - \beta IS, \tag{2.148a}$$

$$\dot{I} = \beta IS - \gamma I - \mu I, \tag{2.148b}$$

$$\dot{R} = \gamma I - \mu R. \tag{2.148c}$$

Here S, I, R denote the number of susceptible, infected, and recovered individuals, respectively, and $B \in L_\infty(\mathbb{R}_+, \mathbb{R}_+)$ denotes the birth and/or immigration rate. The constant coefficients $\beta, \gamma, \mu > 0$ denote the contact, recovery, and mortality rates.

Show that this system is ISS with respect to the input B.

Linear systems and local input-to-state stability

Exercise 2.15 Prove Theorem 2.32.

Exercise 2.16 Prove that for linear systems

$$\dot{x} = Ax + Bu,$$

$A \in \mathbb{R}^{n\times n}, B \in \mathbb{R}^{n\times m}$, the notions of ISS and LISS are equivalent.

Exercise 2.17 Prove Proposition 2.24.

Exercise 2.18 Let $A \in \mathbb{R}^{n\times n}$. Show that

$$\dot{x} = Ax \tag{2.149}$$

is Lyapunov stable if and only if there are matrices $P, Q \in \mathbb{R}^{n\times n}$ so that $P > 0$ and $Q \geq 0$ satisfying the Lyapunov equation

$$A^T P + PA = -Q. \tag{2.150}$$

Exercise 2.19 Let $A = A^T \in \mathbb{R}^n$ and $B \in \mathbb{R}^{n\times m}$. Consider a linear equation

$$\dot{x} = Ax + Bu. \tag{2.151}$$

Assume that the system is ISS.

(i) Show that A is negative definite.

(ii) Using Lyapunov equation, find an ISS Lyapunov function for above system.

Asymptotic properties for control systems

Exercise 2.20 Finish the proof of Lemma 2.42.

Exercise 2.21 Formulate (and justify) a concept of a "UGS Lyapunov function", i.e., a Lyapunov-like function, whose existence implies UGS of a system (2.1). Can you propose both implicative and dissipative variants of the definition?

Exercise 2.22 Show that the solutions of (2.76) for inputs from $\mathcal{U} := L^\infty_{loc}(\mathbb{R}_+, \mathbb{R})$ (defined as in (2.8)) do not necessarily converge to 0 as $t \to \infty$.

Exercise 2.23 Prove that (2.76) is ISS by constructing an ISS Lyapunov function.

Exercise 2.24 Conduct an analysis as in Example 2.46 for the equation

$$\dot{x} = -\zeta(|u(t)|)x$$

for any $\zeta \in \mathcal{L}$.

Exercise 2.25 Consider the following properties:

$$u \in \mathcal{U},\ r > 0 \ \Rightarrow\ \overline{\lim_{t\to\infty}} \sup_{|x|\le r} |\phi(t, x, u)| \le \gamma(\|u\|_\infty)$$

and

$$r, s > 0 \ \Rightarrow\ \overline{\lim_{t\to\infty}} \sup_{|x|\le r} \sup_{\|u\|_\infty \le s} |\phi(t, x, u)| \le \gamma(s).$$

How are these properties related to sAG, UAG?

Exercise 2.26 (AG Characterization) Show the following statements:

(i) Let $f \in L^\infty(\mathbb{R}_+, \mathbb{R})$. Show that for every $\varepsilon > 0$ there exists $t_\varepsilon > 0$ so that

$$\text{ess sup}_{t>t_\varepsilon} f(t) \le \overline{\lim_{t\to+\infty}} f(t) + \varepsilon.$$

In this exercise, the superior limits should be understood in the "essential" sense (holding up to the measure zero).

(ii) Let $u \in L^\infty(\mathbb{R}_+, \mathbb{R}^m)$. Using item (i) show that

$$\overline{\lim_{t\to+\infty}} \gamma(\text{ess sup}_{s\ge t}|u(s)|) = \overline{\lim_{t\to+\infty}} \gamma(|u(t)|). \tag{2.152}$$

(iii) Prove that (2.1) is AG iff there exists $\gamma \in \mathcal{K}_\infty$ so that

$$x \in \mathbb{R}^n,\ u \in \mathcal{U} \ \Rightarrow\ \overline{\lim_{t\to+\infty}} |\phi(t, x, u)| \le \overline{\lim_{t\to+\infty}} \gamma(|u(t)|). \tag{2.153}$$

Exercise 2.27 (Open problem) Does AG imply sAG for systems (2.1)? Under which conditions?

Exercise 2.28 (Open problem) Does LIM imply AG for systems B.1? Under which conditions?

ISS superposition theorems

Exercise 2.29 Consider the following properties of a forward complete system (2.1):

(i) "Bounded deviation (BD)": $\exists \gamma \in \mathcal{K}_\infty$: $\forall t \geq 0$, $\forall x \in \mathbb{R}^n$, $\forall u \in \mathcal{U}$

$$|\phi(t, x, u) - \phi(t, x, 0)| \leq \gamma(\|u\|_\infty).$$

(ii) "Weak bounded deviation (WBD)": $\exists \gamma \in \mathcal{K}_\infty$: $\forall t \geq 0$, $\forall x \in \mathbb{R}^n$, $\forall u \in \mathcal{U}$

$$|\phi(t, x, u)| \leq |\phi(t, x, 0)| + \gamma(\|u\|_\infty).$$

Prove that 0-UGAS $\wedge$ WBD $\Rightarrow$ 0-UGAS $\wedge$ BD $\Rightarrow$ ISS. Do the converse implications hold?

Converse ISS Lyapunov theorem

Definition 2.90 A function $k : \mathbb{R}_+ \times \mathbb{R}^n \to \mathbb{R}^m$ is called a *feedback law* for (2.1), if k is measurable in the first argument, continuous in the second argument, and the substitution $u(t) = k(t, x(t))$ into (2.1) that has the form

$$\dot{x} = f(x, k(t, x)) \tag{2.154}$$

is well-posed.

System (2.1) is called *robustly stable*, if there exist $\rho \in \mathcal{K}_\infty$ and $\beta \in \mathcal{KL}$, such that for all feedback laws $k : \mathbb{R}_+ \times \mathbb{R}^n \to \mathbb{R}^m$ satisfying

$$|k(t, x)| \leq \rho(|x|), \quad \text{for all } x \in \mathbb{R}^n \quad \text{and almost all } t \geq 0, \tag{2.155}$$

every solution of the corresponding system (2.154) satisfies

$$|x(t)| \leq \beta(|x(0)|, t) \quad \forall t \geq 0.$$

Exercise 2.30 (ISS and robust stability) In this exercise, we establish equivalence between ISS and robust stability, which expands the ISS criteria shown in Theorem 2.63.

(i) Show that if there exists an ISS Lyapunov function for (2.1), then (2.1) is robustly stable.
(ii) Show that robust stability implies weak robust stability of (2.1).
(iii) Show that ISS is equivalent to robust stability provided that f is Lipschitz continuous on bounded balls of $\mathbb{R}^n \times \mathbb{R}^m$, and $f(0, 0) = 0$.

ISS, exponential ISS, and nonlinear changes of coordinates

Exercise 2.31 Show the following:

(i) Let $T : \mathbb{R}^n \to \mathbb{R}^n$ be a linear operator. Then T is a homeomorphism iff T is invertible.
(ii) Let (2.1) be eISS and let $T : \mathbb{R}^n \to \mathbb{R}^n$ be a linear homeomorphism. Then (2.1) is eISS in new coordinates $y = Tx$.

Integral characterization of ISS

Exercise 2.32 Given any $\alpha \in \mathcal{K}_\infty$ and $p \in \mathbb{N}$, there are changes of coordinates $T_l, T_u : \mathbb{R}^p \to \mathbb{R}^p$ so that

$$|T_l(y)| \le \alpha(|y|) \le |T_u(y)| \quad \forall y \in \mathbb{R}^p. \tag{2.156}$$

Exercise 2.33 (Relation to linear L^2-gains) Assume that f is Lipschitz continuous on bounded balls of $\mathbb{R}^n \times \mathbb{R}^m$. A forward complete system (2.1) has *linear L^2-gain with transient bound $\xi \in \mathcal{K}_\infty$ and gain $\gamma \ge 0$ (linear L^2-gain property)* if for all $x \in \mathbb{R}^n$, $u \in \mathcal{U}$, $t \ge 0$ the following holds:

$$\int_0^t |\phi(s, x, u)|^2 ds \le \max\Big\{\xi(|x|), \gamma^2 \int_0^t |u(s)|^2 ds\Big\}. \tag{2.157}$$

The inequality (2.157) is sometimes called *quadratic H_∞-inequality*. In this perspective, the inequality (2.118) can be seen as a *nonlinearization* of the quadratic H_∞ inequality.

Show that:

(i) The integral-to-integral stability property is invariant with respect to nonlinear changes of coordinates as introduced in Definition 2.66.
(ii) Every forward complete system (2.1) possessing linear L^2-gain property is ISS.
(iii) For every ISS system (2.1) there is a nonlinear change of coordinates for both the inputs and the state so that in new coordinates the system (2.1) has a linear L^2-gain property.
Hint: for (iii) use Exercise 2.32.

Exercise 2.34 Theorem 2.76 has been shown under assumption of Lipschitz continuity of f on bounded balls of $\mathbb{R}^n \times \mathbb{R}^m$.

Assume that (2.1) is forward complete (no further assumptions on f). Show that:

(i) ISS implies norm-to-integral ISS.
Hint: use Sontag's $\mathcal{KL}$-lemma (Proposition A.7).
(ii) Norm-to-integral ISS implies ULIM property.

Exercise 2.35 A continuous function $V : \mathbb{R}^n \to \mathbb{R}_+$ is called a *noncoercive ISS Lyapunov function* for (2.1), if there exist $\psi_2 \in \mathcal{K}_\infty$, $\alpha \in \mathcal{K}_\infty$, and $\xi \in \mathcal{K}$ such that

$$0 < V(x) \le \psi_2(|x|) \quad \forall x \in \mathbb{R}^n \backslash \{0\}, \tag{2.158}$$

and for any $x \in \mathbb{R}^n$, $u \in \mathcal{U}$ the following implication holds:

$$\dot{V}_u(x) \le -\alpha(V(x)) + \xi(\|u\|_\infty). \tag{2.159}$$

Show that for a forward complete system (2.1) existence of a noncoercive ISS Lyapunov function implies norm-to-integral ISS.

Input-to-state stable monotone control systems

Exercise 2.36 Assume that (2.1) is ISS. Show (under certain requirements on f) that for each constant input u and for each $T > 0$ there is some $x \in \mathbb{R}^n$ such that $\phi(T, x, u) = x$. You may use the following strategy:

(i) Fix any constant input u.
(ii) Show that for each ISS Lyapunov function V with a gain χ and for any input $u \in \mathcal{U}$ the set $K := \{x \in \mathbb{R}^n : V(x) \le \chi(\|u\|_\infty)\}$ is forward invariant.
(iii) Then for each $T \ge 0$ the map $M_T : x \mapsto \phi(T, x, u)$ is continuous (why?)
(iv) Pick any ball B, so that $K \subset \operatorname{int}(B)$. Consider a function $Q : B \to K$, which maps any $x \in B$ to $\phi(t_x, x, 0)$, where t_x is the first time t, so that $\phi(t, x, 0) \in K$. Show that Q is a continuous mapping (as $Q : B \to K$ is identity on $K \subset B$, this shows that Q is a *topological retraction*).
(v) Show that each continuous map $M : K \to K$ has a fixed point (to this end, apply Brouwer's Theorem A.39 to the map $M \circ Q : B \to B$). In particular, M_T has a fixed point.

Exercise 2.37 Assume that (2.1) is ISS.
Use Exercise 2.36 to show that:

(i) For each periodic input $u \in \mathcal{U}$ (i.e., $u(t + T) = u(t)$, for any $t \ge 0$) there is a $x \in \mathbb{R}^n$: $t \mapsto \phi(t, x, u)$ has the same period as u.
(ii) For each constant input u there is $x \in \mathbb{R}^n$ that is an equilibrium of (2.1), i.e., $f(x, u) = 0$.

Exercise 2.38 Consider the following (well-posed) interconnected system:

$$\begin{aligned} \dot{x} &= f(x, z, u), \\ \dot{z} &= g(z, u). \end{aligned} \tag{2.160}$$

Assume that both x-subsystem and z-subsystem are monotone with respect to the inputs. What can you say about the monotonicity of the whole system (2.160)?

References

1. Allan DA, Rawlings J, Teel AR (2021) Nonlinear detectability and incremental input/output-to-state stability. SIAM J Control Optim 59(4):3017–3039
2. Andrieu V, Praly L, Astolfi A (2008) Homogeneous approximation, recursive observer design, and output feedback. SIAM J Control Optim 47(4):1814–1850
3. Angeli D, Sontag ED (2003) Monotone control systems. IEEE Trans Autom Control 48(10):1684–1698
4. Arcak M, Meissen C, Packard A (2016) Networks of dissipative systems: compositional certification of stability, performance, and safety. Springer, New York
5. Arcak M, Teel A (2002) Input-to-state stability for a class of Lurie systems. Automatica 38(11):1945–1949
6. Bernuau E, Efimov D, Perruquetti W, Polyakov A (2014) On homogeneity and its application in sliding mode control. J Franklin Inst 351(4):1866–1901
7. Bhatia NP (1966) Weak attractors in dynamical systems. Boletin Sociedad Matematica Mexicana 11:56–64
8. Bhatia NP, Szegö GP (2002) Stability theory of dynamical systems. Springer, New York
9. Dashkovskiy S (2019) Practical examples of ISS systems. IFAC-PapersOnLine 52(16):1–6; 11th IFAC Symposium on Nonlinear Control Systems NOLCOS 2019
10. Duindam V, Macchelli A, Stramigioli S, Bruyninckx H (eds) (2009) Modeling and control of complex physical systems. The port-Hamiltonian approach, Springer, Berlin
11. Grimm G, Messina MJ, Tuna SE, Teel AR (2005) Model predictive control: for want of a local control Lyapunov function, all is not lost. IEEE Trans Autom Control 50(5):546–558
12. Grüne L (2021) Computing Lyapunov functions using deep neural networks. J Comput Dyn 8(2):131–152
13. Grüne L, Sontag E, Wirth F (1999) Asymptotic stability equals exponential stability, and ISS equals finite energy gain–if you twist your eyes. Syst Control Lett 38(2):127–134
14. Hahn W (1967) Stability of motion. Springer, New York
15. Henry D (1981) Geometric theory of semilinear parabolic equations. Springer, Berlin
16. Hirsch MW (1982) Systems of differential equations which are competitive or cooperative. I. Limit sets. SIAM J Math Anal 13(2):167–179
17. Hörmander L (2007) Notions of convexity. Springer, New York
18. Ito H (2020) Interpreting models of infectious diseases in terms of integral input-to-state stability. Math Control Signals Syst 32(4):611–631
19. Jacob B, Mironchenko A, Partington JR, Wirth F (2020) Noncoercive Lyapunov functions for input-to-state stability of infinite-dimensional systems. SIAM J Control Optim 58(5):2952–2978
20. Jacob B, Zwart HJ (2012) Linear port-Hamiltonian systems on infinite-dimensional spaces. Springer, Basel
21. Jiang Z-P, Wang Y (2002) A converse Lyapunov theorem for discrete-time systems with disturbances. Syst Control Lett 45(1):49–58
22. Karafyllis I (2021) On the relation of IOS-gains and asymptotic gains for linear systems. Syst Control Lett 152:104934
23. Karafyllis I, Jiang Z-P (2011) Stability and stabilization of nonlinear systems. Springer, London
24. Kellett CM, Dower PM (2016) Input-to-state stability, integral input-to-state stability, and L_2-gain properties: qualitative equivalences and interconnected systems. IEEE Trans Autom Control 61(1):3–17
25. Kellett CM, Wirth FR (2016) Nonlinear scaling of (i)ISS-Lyapunov functions. IEEE Trans Autom Control 61(4):1087–1092
26. Khasminskii R (2011) Stochastic stability of differential equations. Springer, New York
27. Lakshmikantham V, Leela S (1969) Differential and integral inequalities: Ordinary differential equations. Academic Press
28. Li H, Baier R, Grüne L, Hafstein S, Wirth F (2015) Computation of local ISS Lyapunov functions with low gains via linear programming. Available at https://hal.inria.fr/hal-01101284

29. Lin Y, Sontag ED, Wang Y (1996) A smooth converse Lyapunov theorem for robust stability. SIAM J Control Optim 34(1):124–160
30. Liu T, Jiang Z-P, Hill DJ (2014) Nonlinear control of dynamic networks. CRC Press
31. Mironchenko A (2016) Local input-to-state stability: characterizations and counterexamples. Syst Control Lett 87:23–28
32. Mironchenko A (2017) Uniform weak attractivity and criteria for practical global asymptotic stability. Syst Control Lett 105:92–99
33. Mironchenko A (2021) Small gain theorems for general networks of heterogeneous infinite-dimensional systems. SIAM J Control Optim 59(2):1393–1419
34. Mironchenko A, Karafyllis I, Krstic M (2019) Monotonicity methods for input-to-state stability of nonlinear parabolic PDEs with boundary disturbances. SIAM J Control Optim 57(1):510–532
35. Mironchenko A, Wirth F (2018) Characterizations of input-to-state stability for infinite-dimensional systems. IEEE Trans Autom Control 63(6):1602–1617
36. Mironchenko A, Wirth F (2019) Existence of non-coercive Lyapunov functions is equivalent to integral uniform global asymptotic stability. Math Control Signals Syst 31(4):1–26
37. Mironchenko A, Wirth F (2019) Non-coercive Lyapunov functions for infinite-dimensional systems. J Differ Equ 105:7038–7072
38. Nesic D, Teel AR (2001) Changing supply functions in input to state stable systems: the discrete-time case. IEEE Trans Autom Control 46(6):960–962
39. Praly L, Wang Y (1996) Stabilization in spite of matched unmodeled dynamics and an equivalent definition of input-to-state stability. Math Control Signals Syst 9(1):1–33
40. Ryan E (1995) Universal stabilization of a class of nonlinear systems with homogeneous vector fields. Syst Control Lett 26(3):177–184
41. Sanchez EN, Perez JP (1999) Input-to-state stability (ISS) analysis for dynamic neural networks. IEEE Trans Circuits Syst I Fundam Theory Appl 46(11):1395–1398
42. Smith HL (1995) Monotone dynamical systems. Am Math Soc
43. Sontag E, Teel A (1995) Changing supply functions in input/state stable systems. IEEE Trans Autom Control 40(8):1476–1478
44. Sontag ED (1989) Smooth stabilization implies coprime factorization. IEEE Trans Autom Control 34(4):435–443
45. Sontag ED (1998) Comments on integral variants of ISS. Syst Control Lett 34(1–2):93–100
46. Sontag ED (1998) Mathematical control theory: Deterministic finite-dimensional systems, 2nd edn. Springer, New York
47. Sontag ED (2003) A remark on the converging-input converging-state property. IEEE Trans Autom Control 48(2):313–314
48. Sontag ED, Wang Y (1995) On characterizations of the input-to-state stability property. Syst Control Lett 24(5):351–359
49. Sontag ED, Wang Y (1996) New characterizations of input-to-state stability. IEEE Trans Autom Control 41(9):1283–1294
50. Subbaraman A, Teel AR (2016) On the equivalence between global recurrence and the existence of a smooth Lyapunov function for hybrid systems. Syst Control Lett 88:54–61
51. Teel AR (1998) Connections between Razumikhin-type theorems and the ISS nonlinear small-gain theorem. IEEE Trans Autom Control 43(7):960–964
52. Teel AR (2013) Lyapunov conditions certifying stability and recurrence for a class of stochastic hybrid systems. Ann Rev Control 37(1):1–24
53. Teschl G (2012) Ordinary differential equations and dynamical systems. Am Math Soc
54. van der Schaft AJ (2000) L_2-gain and passivity techniques in nonlinear control. Springer
55. Willems JC (1972) Dissipative dynamical systems part I: General theory. Arch Ration Mech Anal 45(5):321–351
56. Willems JC (1972) Dissipative dynamical systems part II: Linear systems with quadratic supply rates. Arch Ration Mech Anal 45(5):352–393
57. Zabczyk J (2008) Mathematical control theory. Birkhäuser, Boston

Chapter 3
Networks of Input-to-State Stable Systems

Stability analysis and control of nonlinear systems are a rather complex task, which becomes even more challenging for large-scale systems. As there are no methods for constructing ISS Lyapunov functions for nonlinear systems, direct stability analysis (e.g., by constructing an ISS Lyapunov function) for large-scale systems is seldom possible. In this chapter, we prove powerful nonlinear small-gain theorems, which ensure input-to-state stability of large-scale networks, consisting of input-to-state stable nonlinear components, provided the interconnection structure of the network satisfies the small-gain condition. Broad applications of the small-gain approach include the design of robust controllers and observers for nonlinear systems, the design of decentralized observers for large-scale networks, and synchronization of multiagent systems.

3.1 Interconnections and Gain Operators

Consider a general feedback interconnection of n systems of the form

$$\dot{x}_i = f_i(x_1, \ldots, x_n, u), \quad i = 1, \ldots, n. \tag{3.1}$$

Here $u \in \mathcal{U} := L^\infty(\mathbb{R}_+, \mathbb{R}^m)$ is an *external input* that we without loss of generality assume to be the same for all subsystems. Otherwise, we can collect the inputs for individual subsystems u_k into a vector $(u_1, \ldots, u_n)$ that will be a common input for all subsystems. The state of x_i-subsystem belongs to the space $X_i := \mathbb{R}^{N_i}$, for some $N_i \in \mathbb{N}$. The dimension of the state space of the whole system we denote as $N := N_1 + \cdots + N_n$.

A feedback interconnection of 2 systems is depicted in Fig. 3.1.

A. Mironchenko, *Input-to-State Stability*, Communications and Control Engineering,
https://doi.org/10.1007/978-3-031-14674-9_3

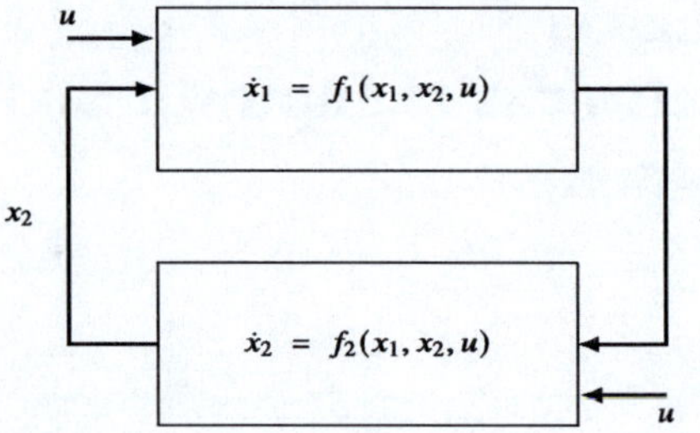

Fig. 3.1 Feedback interconnection

The state x_j, $j \neq i$ is *the internal input from j-th subsystem to the i-th subsystem*. The space $X_{\neq i}$ of internal input values of the i-th subsystem of (3.1) and the state space X of the coupling are given by

$$X_{\neq i} := \mathbb{R}^{N-N_i}, \quad X := \mathbb{R}^N.$$

The system (3.1) can be viewed as a system

$$\dot{x} = f(x, u), \tag{3.2}$$

with the state $x := (x_1, \dots, x_n)^T \in X := \mathbb{R}^N$, input $u \in L^\infty(\mathbb{R}_+, \mathbb{R}^m)$ and with the nonlinear dynamics

$$f(x, u) := (f_1^T(x, u), \dots, f_n^T(x, u))^T.$$

For any $i = 1, \dots, n$ consider also the i-th subsystem of the system (3.1), given by

$$\Sigma_i : \quad \dot{x}_i = \bar{f}_i(x_i, w_{\neq i}, u), \tag{3.3}$$

where

$$\bar{f}_i(x_i, w_{\neq i}, u) = f_i(w_1, \dots, w_{i-1}, x_i, w_{i+1}, \dots, w_n, u).$$

Here we denote by $w_{\neq i}$ the *total internal input to the i-th subsystem* that is defined by

$$w_{\neq i} := (w_1, \dots, w_{i-1}, w_{i+1}, \dots, w_n),$$

and belongs to the space of locally absolutely continuous functions $AC_{\text{loc}}(\mathbb{R}_+, X_{\neq i})$. We denote the internal inputs by w_i to underline that w_i are arbitrary inputs and not just those that are generated by the states of other subsystems.

Hence, the total input to the i-th subsystem belongs to $AC_{\text{loc}}(\mathbb{R}_+, X_{\neq i}) \times L^\infty(\mathbb{R}_+, \mathbb{R}^m)$. To make the notation easier, we consider as the *total input space to the i-th subsystem* a larger space $L^\infty_{\text{loc}}(\mathbb{R}_+, \mathbb{R}^{N+m-N_i})$.

By $\bar{\phi}_i\big(t, x_i, (w_{\neq i}, u)\big)$ we denote the state of the system (3.3) that we denote also by Σ_i, at time $t \geq 0$ corresponding to the initial condition $x_i \in \mathbb{R}^{N_i}$ and total input $(w_{\neq i}, u)$.

The central assumption for this chapter is:

Assumption 3.1 Suppose that the following holds:

(i) f is continuous on $\mathbb{R}^N \times \mathbb{R}^m$.
(ii) For each $i = 1, \dots, n$, f_i is Lipschitz continuous w.r.t. x_i on bounded subsets.
(iii) Each subsystem is forward complete.
(iv) (Well-posedness) For any initial condition $x \in \mathbb{R}^n$ and any input $u \in \mathcal{U}$, there is a unique maximal solution of (3.2).
(v) (3.2) satisfies the BIC property.

According to Definition 2.2, the i-th subsystem of the system (3.1) is ISS if there exist $\beta_i \in \mathcal{KL}$ and $\bar{\gamma}_i \in \mathcal{K}$ such that $w_{\neq i} \in L^\infty(\mathbb{R}_+, X_{\neq i})$, all external inputs $u \in \mathcal{U}$ and all $t \in \mathbb{R}_+$ the following holds

$$|\bar{\phi}_i\big(t, x_i, (w_{\neq i}, u)\big)| \leq \beta_i\left(|x_i|, t\right) + \bar{\gamma}_i\left(\left\|(w_{\neq i}, u)\right\|_\infty\right). \tag{3.4}$$

The estimate (3.4) is defined in terms of the norm of the whole input, and this is not suitable for stability analysis of coupled systems, as we are interested not only in the collective influence of all inputs over a subsystem but in the influence of particular subsystems over a given subsystem. Therefore, we reformulate the ISS property for a subsystem in the following form:

Lemma 3.2 (Summation formulation of ISS) *A forward complete system Σ_i is ISS if and only if there exist $\gamma_{ij}, \gamma_i \in \mathcal{K} \cup \{0\}$, $j = 1, \dots, n$ and $\beta_i \in \mathcal{KL}$, such that for all initial values $x_i \in X_i$, all internal inputs $w_{\neq i} := (w_1, \dots, w_{i-1}, w_{i+1}, \dots, w_n) \in L^\infty(\mathbb{R}_+, X_{\neq i})$, all external inputs $u \in \mathcal{U}$ and all $t \in \mathbb{R}_+$ the following estimate holds:*

$$|\bar{\phi}_i\big(t, x_i, (w_{\neq i}, u)\big)| \leq \beta_i\left(|x_i|, t\right) + \sum_{j \neq i} \gamma_{ij}\left(\left\|w_j\right\|_{[0,t]}\right) + \gamma_i\left(\|u\|_\infty\right). \tag{3.5}$$

Proof As Σ_i is ISS, there is a $\gamma \in \mathcal{K}_\infty$ so that for all $t, x_i, w_{\neq i}, u$ it holds that

$$\begin{aligned}
|\bar{\phi}_i\big(t, x_i, (w_{\neq i}, u)\big)| &\leq \beta_i\left(|x_i|, t\right) + \gamma(\|(w_{\neq i}, u)\|_{L^\infty(\mathbb{R}_+, X_{\neq i}) \times \mathcal{U}}) \\
&= \beta_i\big(|x_i|, t\big) + \gamma\Big(\Big(\sum_{j \neq i} \|w_j\|^2_\infty + \|u\|^2_\infty\Big)^{1/2}\Big) \\
&\leq \beta_i\big(|x_i|, t\big) + \gamma\Big(\sum_{j \neq i} \|w_j\|_\infty + \|u\|_\infty\Big) \\
&\leq \beta_i\big(|x_i|, t\big) + \sum_{j \neq i} \gamma(n\|w_j\|_\infty) + \gamma(n\|u\|_\infty),
\end{aligned}$$

where in the last estimate we have exploited the estimate $\gamma(s_1 + \dots + s_n) \le \gamma(ns_1) + \ldots + \gamma(ns_n)$, which holds for all $\gamma \in \mathcal{K}$ and all $s_1, \ldots, s_n \ge 0$.

Defining $\gamma_{ij}(r) := \gamma(nr)$ and $\gamma_i(r) := \gamma(nr)$, we obtain due to causality of Σ_i that

$$|\bar{\phi}_i\big(t, x_i, (w_{\neq i}, u)\big)| \le \beta_i\big(|x_i|, t\big) + \sum_{j\neq i} \gamma_{ij}\big(\|w_j\|_{[0,t]}\big) + \gamma_i\big(\|u\|_\infty\big).$$

Conversely, let the property in the statement of the lemma hold. Define $\gamma(r) := \gamma_i(r) + \sum_{j\neq i} \gamma_{ij}(r)$, $r \in \mathbb{R}_+$. We have for all $t, x_i, w_{\neq i}, u$ that

$$\begin{aligned}|\bar{\phi}_i\big(t, x_i, (w_{\neq i}, u)\big)| &\le \beta_i\left(|x_i|, t\right) + \sum_{j\neq i} \gamma_{ij}\left(\left\|w_j\right\|_{[0,t]}\right) + \gamma_i\left(\|u\|_\infty\right)\\ &\le \beta_i\left(|x_i|, t\right) + \sum_{j\neq i} \gamma_{ij}\left(\|(w_{\neq i}, u)\|_{C(\mathbb{R}_+, X_{\neq i})\times\mathcal{U}}\right)\\ &\quad + \gamma_i\left(\|(w_{\neq i}, u)\|_{C(\mathbb{R}_+, X_{\neq i})\times\mathcal{U}}\right)\\ &\le \beta_i\left(|x_i|, t\right) + \gamma\left(\|(w_{\neq i}, u)\|_{C(\mathbb{R}_+, X_{\neq i})\times\mathcal{U}}\right).\end{aligned}$$

This shows the claim. □

The functions γ_{ij} and γ_i in the statement of Lemma 3.2 are called *(asymptotic) gains*. For notational simplicity we allow the case $\gamma_{ij} \equiv 0$ and require $\gamma_{ii} \equiv 0$ for all i.

Analogously, one can restate the definitions of UGS and AG properties, which we leave without the proof:

Lemma 3.3 (Summation formulation of UGS) *Σ_i is UGS if and only if there exist $\gamma_{ij}, \gamma_i \in \mathcal{K} \cup \{0\}$, and $\sigma_i \in \mathcal{K}$, such that for all initial values $x_i \in X_i$, all internal inputs $w_{\neq i} := (w_1, \ldots, w_{i-1}, w_{i+1}, \ldots, w_n) \in L^\infty(\mathbb{R}_+, X_{\neq i})$, all external inputs $u \in \mathcal{U}$, and all $t \in \mathbb{R}_+$ the following inequality holds*

$$|\bar{\phi}_i\big(t, x_i, (w_{\neq i}, u)\big)| \le \sigma_i\left(|x_i|\right) + \sum_{j\neq i} \gamma_{ij}\big(\left\|w_j\right\|_{[0,t]}\big) + \gamma_i\left(\|u\|_\infty\right). \tag{3.6}$$

Lemma 3.4 (Summation formulation of AG property) *Σ_i is AG if and only if Σ_i is forward complete and there exist $\gamma_{ij}, \gamma_i \in \mathcal{K} \cup \{0\}$, such that for all $x_i \in X_i$, $u \in \mathcal{U}$, $w_{\neq i} := (w_1, \ldots, w_{i-1}, w_{i+1}, \ldots, w_n) \in L^\infty(\mathbb{R}_+, X_{\neq i})$, $\varepsilon > 0$ there is a time $\tau_i := \tau_i(x_i, u, w_{\neq i}, \varepsilon) < \infty$ so that*

$$t \ge \tau_i \quad\Rightarrow\quad |\bar{\phi}_i\big(t, x_i, (w_{\neq i}, u)\big)| \le \varepsilon + \sum_{j\neq i} \gamma_{ij}\left(\left\|w_j\right\|_\infty\right) + \gamma_i(\|u\|_\infty). \tag{3.7}$$

In the above definitions, w_j, $j = 1, \dots, n$ are arbitrary inputs that are not necessarily related to the states of other subsystems (i.e., we have considered all Σ_i as disconnected systems).

Pick arbitrary $x \in X$ and $u \in \mathcal{U}$ and define for $i = 1, \dots, n$ the modes of the interconnected system:

$$\phi_i := \phi_i(\cdot, x, u), \quad \phi_{\neq i} = (\phi_1, \dots, \phi_{i-1}, \phi_{i+1}, \dots, \phi_n). \tag{3.8}$$

According to the definition of the interconnection, for all $i = 1, \dots, n$ it holds for all s in the maximal interval of existence of $\phi_i(\cdot, x, u)$ that

$$\phi_i(s, x, u) = \bar{\phi}_i(s, x_i, (\phi_{\neq i}, u)). \tag{3.9}$$

We rewrite the definitions of ISS and UGS, specialized for the inputs $w_j := \phi_j$, in a shorter vectorized form, using the shorthand notation from [7], introduced next.

For vector functions $w = (w_1, \dots, w_n)^T : \mathbb{R}_+ \to X$ such that $w_i \in L^\infty(\mathbb{R}_+, X_i)$, $i = 1, \dots, n$ and times $0 \le t_1 \le t_2$ we define

$$\big| w_{[t_1,t_2]} \big| := \left(\|w_{i,[t_1,t_2]}\|_\infty\right)_{i=1}^n,$$

where $w_{i,[t_1,t_2]}$ is a restriction of w_1 to the interval $[t_1, t_2]$. Furthermore, we introduce for all t, u, and all $x = (x_1, \dots, x_n) \in X$ the following notation:

$$\big\| \bar{\phi}(t, x, u) \big\| := \left(\left|\bar{\phi}_i\left(t, x_i, (\phi_{\neq i}, u)\right)\right|\right)_{i=1}^n, \quad \vec{\gamma}(\|u\|_\infty) := \left(\gamma_i(\|u\|_\infty)\right)_{i=1}^n, \tag{3.10}$$

$$\vec{\sigma}(s) := \left(\sigma_i(s_i)\right)_{i=1}^n, \qquad \vec{\beta}(s, t) := \left(\beta_i(s_i, t)\right)_{i=1}^n, \quad \text{for all } s \in \mathbb{R}^n_+,\ t \ge 0. \tag{3.11}$$

We collect the internal gains in the matrix $\Gamma := (\gamma_{ij})_{i,j=1,\dots,n}$, called the (functional) *gain matrix*. If the gains are taken from the ISS restatement (3.5), we call the corresponding gain matrix Γ^{ISS}. Analogously, the gain matrices Γ^{UGS}, Γ^{AG} are defined.

For a given gain matrix Γ define the operator $\Gamma_\boxplus : \mathbb{R}^n_+ \to \mathbb{R}^n_+$ by

$$\Gamma_\boxplus(s) := \Big(\sum_{j=1}^n \gamma_{1j}(s_j), \dots, \sum_{j=1}^n \gamma_{nj}(s_j)\Big)^T, \quad s = (s_1, \dots, s_n)^T \in \mathbb{R}^n_+. \tag{3.12}$$

Again, to emphasize that the gains are from the ISS restatement (3.5), the corresponding gain operator will be denoted by $\Gamma^{ISS}_\boxplus$, with a similar convention for UGS and AG properties.

For two vectors $x, y \in \mathbb{R}^n$ we say that $x \geq y$ if $x_i \geq y_i$ for all $i = 1, \dots, n$. More notation concerning the partial order on $\mathbb{R}^n$ can be found in Sect. C.1.

Note that by the properties of γ_{ij} for $s_1, s_2 \in \mathbb{R}^n_+$ we have the implication

$$s_1 \geq s_2 \quad \Rightarrow \quad \Gamma_\boxplus(s_1) \geq \Gamma_\boxplus(s_2), \tag{3.13}$$

and hence $\Gamma_\boxplus$ defines a monotone (w.r.t. the partial order $\geq$ in $\mathbb{R}^n$) map.

With this notation, the ISS conditions (3.5) imply that for $t \geq 0$ it holds that

$$\left\| \vec{\phi}(t, x, u) \right\| \leq \vec{\beta}\left(\left\| \vec{\phi}(0, x, u) \right\|, t\right) + \Gamma^{ISS}_\boxplus\left(\left| \phi_{[0,t]} \right|\right) + \vec{\gamma}(\|u\|_\infty). \tag{3.14}$$

Analogously, the UGS conditions (3.6) imply that for $t \geq 0$ it holds that

$$\left\| \vec{\phi}(t, x, u) \right\| \leq \vec{\sigma}\left(\left\| \vec{\phi}(0, x, u) \right\|\right) + \Gamma^{UGS}_\boxplus\left(\left| \phi_{[0,t]} \right|\right) + \vec{\gamma}(\|u\|_\infty). \tag{3.15}$$

Finally, the AG conditions (3.7) imply that

$$\overline{\lim_{t\to\infty}} \left\| \vec{\phi}(t, x, u) \right\| \leq \Gamma^{AG}_\boxplus\left(\left| \phi_{[0,+\infty)} \right|\right) + \vec{\gamma}(\|u\|_\infty). \tag{3.16}$$

Clearly, ISS of individual subsystems does not imply any kind of stability of the network, even in the linear case, as shown by the following example:

Example 3.5 The subsystems of the system

$$\dot{x}_1 = -x_1 + 2x_2, \tag{3.17a}$$
$$\dot{x}_2 = 2x_1 - x_2. \tag{3.17b}$$

are exponentially ISS, but the whole system is not asymptotically stable, as $\sigma(A) = \{1, -3\}$, for $A = \begin{pmatrix} -1 & 2 \\ 2 & -1 \end{pmatrix}$.

To guarantee the stability of the interconnection Σ, the properties of the operators $\Gamma^{UGS}_\boxplus$ and $\Gamma^{ISS}_\boxplus$ will be crucial.

3.2 Small-Gain Theorem for Input-to-State Stability of Networks

We proceed directly to the main result of this section.

Theorem 3.6 (Trajectory-based ISS small-gain theorem) *Let Assumption 3.1 hold. Assume that all subsystems of* (3.1) *satisfy the ISS estimates as in Lemma 3.2.*

If $\Gamma^{ISS}_\boxplus$ *satisfies the* strong small-gain condition

$$(\mathrm{id} + \rho) \circ A(x) \not\geq x \quad \forall\, x \in \mathbb{R}^n_+ \setminus \{0\}, \tag{3.18}$$

then (3.1) *is ISS.*

Proof The proof is based on the UGS and AG small-gain theorems, shown in the following subsections.

Suppose additionally that f satisfies Assumption 1.8. As all subsystems of (3.1) satisfy the ISS estimates as in Lemma 3.2, they satisfy also the UGS and AG estimates as in Lemmas 3.3, 3.4, with the same gain operators, i.e., we can choose $\Gamma_{\boxplus}^{UGS} = \Gamma_{\boxplus}^{AG} := \Gamma_{\boxplus}^{ISS}$.

By Theorem 3.7, the network (3.1) is UGS and, in particular, forward complete. By Theorem 3.9, the interconnection (3.1) has AG property. The ISS superposition theorem (Theorem 2.52) implies (here we use that f satisfies Assumption 1.8) that (3.1) is ISS.

The proof of a general result, if f does not satisfy Assumption 1.8 but merely satisfies Assumption 2.1, is more involved, as in this case, the ISS superposition theorem for systems satisfying Assumption 1.8 (Theorem 2.52) cannot be used. In this case, one can establish the bUAG property instead of the AG property for the network and then use the ISS superposition Theorem 2.51 to infer the ISS of the system. We refer to [23] for the proof of this result (and to [24] for its extension to infinite networks). □

3.2.1 *Small-Gain Theorem for Uniform Global Stability of Networks*

We start with a small-gain theorem, which guarantees that an interconnection of UGS systems is a UGS system provided that the strong small-gain condition (3.18) holds.

Theorem 3.7 (Trajectory-based UGS small-gain theorem) *Let Assumption 3.1 hold. Assume that all subsystems of* (3.1) *are forward complete and satisfy the UGS estimates as in Lemma 3.3. If* $\Gamma_{\boxplus}^{UGS}$ *satisfies the strong small-gain condition* (3.18), *then* (3.1) *is UGS and thus is forward complete.*

Proof Pick any $u \in \mathcal{U}$ and any initial condition $x \in X$. As we assume that the interconnection (3.1) is a well-posed system, the solution of (3.1) exists on a certain interval $[0, t]$, where $t \in (0, +\infty]$. Define ϕ_i and $\phi_{\neq i}$ as in (3.8).

According to the definition of the interconnection, for all $i = 1, \ldots, n$ it holds that $\phi_i(s, x, u) = \bar{\phi}_i(s, x_i, (\phi_{\neq i}, u))$, $s \in [0, t]$ and hence

$$\begin{aligned}\Big(\sup_{s\in[0,t]} \big|\bar{\phi}_i\big(s, x_i, (\phi_{\neq i}, u)\big)\big|\Big)_{i=1}^{n} &= \Big(\sup_{s\in[0,t]} \big|\phi_i(s, x, u)\big|\Big)_{i=1}^{n} \\ &= \big(\|\phi_{i,[0,t]}\|_\infty\big)_{i=1}^{n} =: \big|\phi_{[0,t]}\big|\,.\end{aligned} \tag{3.19}$$

By assumptions, on $[0, t]$ the estimate (3.15) is valid. Taking in this estimate the supremum over $[0, t]$, and making use of (3.19), we see that

$$\left|\,\phi_{[0,t]}\,\right| \le \vec{\sigma}\big(\,\|\bar{\phi}(0,x,u)\|\,\big) + \Gamma^{UGS}_{\boxplus}\big(\,\left|\,\phi_{[0,t]}\,\right|\,\big) + \vec{\gamma}(\|u\|_\infty), \tag{3.20}$$

and thus

$$(I - \Gamma^{UGS}_{\boxplus})\big(\,\left|\,\phi_{[0,t]}\,\right|\,\big) \le \vec{\sigma}\big(\,\|\bar{\phi}(0,x,u)\|\,\big) + \vec{\gamma}(\|u\|_\infty). \tag{3.21}$$

As $\Gamma^{UGS}_{\boxplus}$ is a monotone gain operator satisfying the strong small-gain condition, by Theorem C.41, id $-\Gamma$ satisfies the MBI property, and there is $\xi \in \mathcal{K}_\infty$ so that

$$\begin{aligned}
\big|\left|\,\phi_{[0,t]}\,\right|\big| &\le \xi\big(\big|\vec{\sigma}\big(\,\|\bar{\phi}(0,x,u)\|\,\big) + \vec{\gamma}(\|u\|_\infty)\big|\big)\\
&\le \xi\big(\big|\vec{\sigma}\big(\,\|\bar{\phi}(0,x,u)\|\,\big)\big| + \big|\vec{\gamma}(\|u\|_\infty)\big|\big)\\
&\le \xi\big(2\big|\vec{\sigma}\big(\,\|\bar{\phi}(0,x,u)\|\,\big)\big|\big) + \xi\big(2\big|\vec{\gamma}(\|u\|_\infty)\big|\big).
\end{aligned} \tag{3.22}$$

As $\|\bar{\phi}(0,x,u)\| = (|x_1|,\ldots,|x_n|)$, the following holds:

$$\begin{aligned}
\big|\vec{\sigma}\big(\,\|\bar{\phi}(0,x,u)\|\,\big)\big| &= \sqrt{\sigma_1^2(|x_1|) + \cdots + \sigma_n^2(|x_n|)}\\
&\le \sqrt{\sigma_1^2(|x|) + \cdots + \sigma_n^2(|x|)} =: \tilde{\sigma}(|x|),
\end{aligned} \tag{3.23}$$

where $\tilde{\sigma} \in \mathcal{K}_\infty$ as all $\sigma_i \in \mathcal{K}_\infty$.

Furthermore,

$$|\phi(t,x,u)| = \Big(\sum_{i=1}^n |\phi_i(t,x,u)|^2\Big)^{1/2} \le \Big(\sum_{i=1}^n \|\phi_{i,[0,t]}\|^2_\infty\Big)^{1/2} = \big|\left|\,\phi_{[0,t]}\,\right|\big|. \tag{3.24}$$

Plugging (3.23) and (3.24) into (3.22), we obtain the UGS estimate

$$|\phi(t,x,u)| \le \big|\left|\,\phi_{[0,t]}\,\right|\big| \le \xi\big(2\tilde{\sigma}(|x|)\big) + \xi\big(2\big|\vec{\gamma}(\|u\|_\infty)\big|\big) \tag{3.25}$$

that is valid on the certain maximal interval of existence $[0,t^*)$ of $\phi(\cdot,x,u)$. As (3.1) possesses the BIC property by Assumption 3.1, (3.1) is forward complete and (3.25) is valid for all $t \in \mathbb{R}_+$. □

3.2.2 *Small-Gain Theorem for Asymptotic Gain Property*

For bounded functions $s : \mathbb{R}_+ \to \mathbb{R}^n_+$ one can define the upper limit as follows:

$$\overline{\lim_{t\to\infty}}\, s(t) := \inf_{t\ge 0} \sup_{\tau \ge t} s(\tau),$$

where the supremum and infimum are understood with respect to the order $\geq$, as in Definition C.3.

We need a technical lemma.

Lemma 3.8 *Let $s : \mathbb{R}_+ \to \mathbb{R}^n_+$ be continuous and globally bounded. Then*

$$\overline{\lim_{t\to\infty}}\, s(t) = \overline{\lim_{t\to\infty}} \left\| s_{[\frac{t}{2},\infty)} \right\|.$$

Proof As s is globally bounded, there exist finite

$$\overline{\lim_{t\to\infty}}\, s(t) =: a \in \mathbb{R}^n_+, \qquad \overline{\lim_{t\to\infty}} \left\| s_{[\frac{t}{2},\infty)} \right\| =: b \in \mathbb{R}^n_+.$$

For every $\varepsilon \in \mathbb{R}^n_+$, with $\varepsilon_i > 0$, $i = 1, \dots, n$ there are $t_a, t_b \geq 0$ such that

$$\forall t \geq t_a : \sup_{t\geq t_a} s(t) \leq a + \varepsilon \quad \text{and} \quad \forall t \geq t_b : \sup_{t\geq t_b} \left\| s_{[\frac{t}{2},\infty)} \right\| \leq b + \varepsilon. \tag{3.26}$$

Clearly we have

$$s(t) \leq \left\| s_{[\frac{t}{2},\infty)} \right\|$$

for all $t \geq 0$, i.e., $a \leq b$. On the other hand, the estimate $s(\tau) \leq a + \varepsilon$ for $\tau \geq t$ implies $\left\| s_{[\frac{\tau}{2},\infty)} \right\| \leq a + \varepsilon$ for $\tau \geq 2t$, i.e., $b \leq a$. This immediately gives $a = b$, and the claim is proved. □

Our next result is a small-gain theorem for the asymptotic gain property.

Theorem 3.9 (Trajectory-based AG small-gain theorem) *Let Assumption 3.1 hold. Assume that all subsystems of (3.1) are forward complete and satisfy the AG estimates as in Lemma 3.4. Assume that the whole interconnection (3.1) is forward complete. If $\Gamma^{AG}_{\boxplus}$ satisfies the strong small-gain condition (3.18), then (3.1) is AG.*

Proof Pick any $x \in X$ and $u \in \mathcal{U}$. In view of (3.9), for any $t \geq 0$ it holds that

$$\left\| \bar{\phi}(t, x, u) \right\| = \left(|\bar{\phi}_i(t, x_i, (\phi_{\neq i}, u))|\right)_{i=1}^n = (|\phi_i(t, x, u)|)_{i=1}^n = \left\| \phi(t, x, u) \right\| \in \mathbb{R}^n_+.$$

In view of (3.16) it holds for all $\tau > 0$ that

$$\overline{\lim_{t\to\infty}} \left\| \phi(t, x, u) \right\| \leq \Gamma^{AG}_{\boxplus}\left(\left\| \phi_{[\tau,+\infty)} \right\| \right) + \vec{\gamma}(\|u\|_\infty). \tag{3.27}$$

According to Lemma 3.8, we have that

$$\overline{\lim_{t\to\infty}} \left\| \phi(t, x, u) \right\| = \overline{\lim_{\tau\to\infty}} \left\| \phi_{[\tau,+\infty)} \right\| =: L, \tag{3.28}$$

and thus by computing the limit as $\tau \to \infty$ in (3.27) and noting that $\Gamma^{AG}_{\boxplus}$ is continuous, we obtain that

$$L \leq \Gamma_{\boxplus}^{AG}(L) + \vec{\gamma}(\|u\|_\infty), \tag{3.29}$$

and thus

$$(\mathrm{id} - \Gamma_{\boxplus}^{AG})L \leq \vec{\gamma}(\|u\|_\infty).$$

By Theorem C.41, the operator $\mathrm{id} - \Gamma_{\boxplus}^{AG}$ satisfies the MBI property, that is, there is $\xi \in \mathcal{K}_\infty$ so that

$$|L| \leq \xi(|\vec{\gamma}(\|u\|_\infty)|),$$

which is precisely the asymptotic gain property for the interconnection. □

Remark 3.10 A notable difference in the assumptions of the AG small-gain theorem in comparison with UGS and ISS small-gain theorems is that the forward completeness of the interconnection is required. The assumption of forward completeness cannot be dropped, as demonstrated in [7, Sect. 4.2].

3.2.3 Semimaximum Formulation of the ISS Small-Gain Theorem

In Lemma 3.2, we have reformulated the ISS property in a way that the total influence of subsystems is the sum of the internal gains. In some cases, this formulation is the most convenient, but other restatements can be more useful in other cases.

Another important restatement, which mixes summation and maximization, is given in the next lemma:

Lemma 3.11 *A forward complete system* Σ_i is ISS (in semimaximum formulation) *if there exist* $\gamma_{ij}, \gamma_i \in \mathcal{K} \cup \{0\}$, $j = 1, \ldots, n$ *and* $\beta_i \in \mathcal{KL}$, *such that for all initial values* $x_i \in X_i$, *all internal inputs* $w_{\neq i} := (w_1, \ldots, w_{i-1}, w_{i+1}, \ldots, w_n) \in L^\infty(\mathbb{R}_+, X_{\neq i})$, *all external inputs* $u \in \mathcal{U}$, *and all* $t \in \mathbb{R}_+$ *the following holds:*

$$\|\bar{\phi}_i(t, x_i, (w_{\neq i}, u))\|_{X_i} \leq \beta_i\left(\|x_i\|_{X_i}, t\right) + \max_{j \neq i}\left\{\gamma_{ij}\left(\|w_j\|_{[0,t]}\right)\right\} + \gamma_i\left(\|u\|_\infty\right). \tag{3.30}$$

Proof The proof is analogous to the proof of Lemma 3.2 and is omitted. □

We can collect all the internal gains γ_{ij} from the semimaximum formulation (3.30) of ISS again into the matrix Γ and introduce the operator $\Gamma_\otimes : \mathbb{R}^n_+ \to \mathbb{R}^n_+$ acting for $s = (s_1, \ldots, s_n)^T \in \mathbb{R}^n_+$ as

$$\Gamma_\otimes(s) := \left(\max_{j=1}^{n} \gamma_{1j}(s_j), \ldots, \max_{j=1}^{n} \gamma_{nj}(s_j)\right)^T. \tag{3.31}$$

Similarly to $\Gamma_{\boxplus}$, the operator $\Gamma_{\otimes}$ is a monotone w.r.t. the partial order $\geq$ in $\mathbb{R}^n$ map, i.e.,

$$s_1 \geq s_2 \quad \Rightarrow \quad \Gamma_{\otimes}(s_1) \geq \Gamma_{\otimes}(s_2). \tag{3.32}$$

A counterpart of the ISS small-gain Theorem 3.6 for the semimaximum formulation of the ISS property is given by the next result:

Theorem 3.12 (ISS small-gain theorem in semimaximum formulation) *Let Assumption 3.1 hold. Assume that all subsystems of* (3.1) *satisfy the ISS estimates as in Lemma 3.11. If $\Gamma_{\otimes}$ satisfies the strong small-gain condition* (3.18)*, then* (3.1) *is ISS.*

Proof Pick any $u \in \mathcal{U}$ and any $x \in X$. As the interconnection Σ is well posed, there is a certain $t_1 > 0$ so that $\phi(\cdot, x, u)$ is well defined on $[0, t_1]$. Using the notation and the arguments introduced in the UGS small-gain theorem in the summation form (Theorem 3.7), we obtain the following estimate for all $t \in [0, t_1]$:

$$\left|\, \phi_{[0,t]} \,\right| \leq \sigma(\, \| \bar{\phi}(0, x, u) \| \,) + \Gamma_{\otimes}^{ISS}(\, |\, \phi_{[0,t]} \,| \,) + \gamma(\|u\|_{\infty}),$$

and as $\Gamma_{\otimes}^{ISS}$ is a monotone operator satisfying the strong small-gain condition (3.18), we can again apply Proposition C.19 to show UGS property of the interconnection as it was done in Theorem 3.7.

The rest of the proof goes along the lines of the proof of Theorem 3.6 and is omitted. □

3.2.4 Small-Gain Theorem in the Maximum Formulation

Using the characterization of the small-gain condition for max-preserving operators, given in Theorem C.48, we can develop the small-gain theorems in a so-called maximum formulation. In this case, the small-gain condition can be slightly relaxed.

We start by reformulating the UGS, AG, and ISS properties in a maximum form.

Lemma 3.13 (Maximum formulation of UGS) *Σ_i is UGS if and only if there exist γ_{ij}, $\gamma_i \in \mathcal{K} \cup \{0\}$ and $\sigma_i \in \mathcal{K}$, such that for all initial values $x_i \in X_i$, all internal inputs $w_{\neq i} := (w_1, \ldots, w_{i-1}, w_{i+1}, \ldots, w_n) \in L^{\infty}(\mathbb{R}_+, X_{\neq i})$, all external inputs $u \in \mathcal{U}$, and all $t \in \mathbb{R}_+$ the following inequality holds*

$$|\bar{\phi}_i(t, x_i, (w_{\neq i}, u))| \leq \max_{j \neq i} \left\{ \sigma_i\left(\|x_i\|_{X_i}\right), \gamma_{ij}\left(\left\| w_j \right\|_{[0,t]}\right), \gamma_i\left(\|u\|_{\infty}\right) \right\}. \tag{3.33}$$

Proof The proof is analogous to the proof of Lemma 3.3 and is omitted. □

Lemma 3.14 (Maximum formulation of AG property) *Σ_i is AG if and only if Σ_i is forward complete and there exist γ_{ij}, $\gamma_i \in \mathcal{K} \cup \{0\}$, such that for all $x_i \in X_i$,*

$u \in \mathcal{U}$, $w_{\neq i} := (w_1, \ldots, w_{i-1}, w_{i+1}, \ldots, w_n) \in L^\infty(\mathbb{R}_+, X_{\neq i})$, $\varepsilon > 0$ *there is a time* $\tau_i := \tau_i(x_i, u, w_{\neq i}, \varepsilon) < \infty$ *so that it holds that*

$$t \geq \tau_i \quad \Rightarrow \quad |\bar{\phi}_i(t, x_i, (w_{\neq i}, u))| \leq \max_{j \neq i}\{\varepsilon, \gamma_{ij}\left(\|w_j\|_\infty\right), \gamma_i(\|u\|_\infty)\}. \tag{3.34}$$

Lemma 3.15 (Maximum formulation of ISS) *A forward complete system* Σ_i *is ISS if and only if there exist* $\gamma_{ij}, \gamma_i \in \mathcal{K} \cup \{0\}$, $j = 1, \ldots, n$ *and* $\beta_i \in \mathcal{KL}$, *such that for all initial values* $x_i \in X_i$, *all internal inputs* $w_{\neq i} := (w_1, \ldots, w_{i-1}, w_{i+1}, \ldots, w_n) \in L^\infty(\mathbb{R}_+, X_{\neq i})$, *all external inputs* $u \in \mathcal{U}$, *and all* $t \in \mathbb{R}_+$ *the following estimate holds:*

$$|\bar{\phi}_i(t, x_i, (w_{\neq i}, u))| \leq \max_{j \neq i}\left\{\beta_i(|x_i|, t), \gamma_{ij}\left(\|w_j\|_{[0,t]}\right), \gamma_i\left(\|u\|_\infty\right)\right\}. \tag{3.35}$$

Thanks to Theorem C.48, in the case of maximum formulation, one can relax the assumptions in UGS, AG, and ISS small-gain theorems and require only validity of the *small-gain condition*

$$A(x) \ngeq x \quad \forall\, x \in \mathbb{R}^n_+ \setminus \{0\}, \tag{3.36}$$

in contrast to the strong small-gain condition (3.18) in the case of summation and semimaximum formulations.

To stress that the gains are from the ISS restatement (3.35), the corresponding gain operator will be denoted by $\Gamma^{ISS}_\otimes$, with a similar convention for UGS and AG properties.

Theorem 3.16 (UGS small-gain theorem: maximum formulation) *Let Assumption 3.1 hold. Assume that all subsystems of* (3.1) *satisfy the UGS estimates as in Lemma 3.13. If* $\Gamma^{UGS}_\otimes$ *satisfies the "plain" small-gain condition* (3.36), *then* (3.1) *is UGS.*

Proof The proof is analogous to the proof of Theorem 3.7, thus we state only the differences. Pick any $u \in \mathcal{U}$ and any initial condition $x \in X$. Using the notation of Theorem 3.16, we obtain for $t \in [0, t_m(x, u))$ the estimate

$$\left|\phi_{[0,t]}\right| \leq \Gamma^{UGS}_\otimes(\left|\phi_{[0,t]}\right|) \vee \vec{\sigma}(\left\|\bar{\phi}(0, x, u)\right\|) \vee \vec{\gamma}(\|u\|_\infty), \tag{3.37}$$

where $a \vee b$ denotes the componentwise maximum of the vectors a, b.

Using Theorem C.48, there is $\xi \in \mathcal{K}_\infty$ such that the following holds:

$$\begin{aligned}
|\left|\phi_{[0,\tau]}\right|| &\leq \xi\left(|\vec{\sigma}(\left\|\bar{\phi}(0, x, u)\right\|) \vee \vec{\gamma}(\|u\|_\infty)|\right) \\
&\leq \xi\left(|\vec{\sigma}(\left\|\bar{\phi}(0, x, u)\right\|)| + |\vec{\gamma}(\|u\|_\infty)|\right) \\
&\leq \xi\left(2|\vec{\sigma}(\left\|\bar{\phi}(0, x, u)\right\|)|\right) + \xi\left(2|\vec{\gamma}(\|u\|_\infty)|\right).
\end{aligned} \tag{3.38}$$

Arguing as in Theorem 3.16, the claim follows. □

The AG small-gain theorem in a maximum formulation we state is as follows:

Theorem 3.17 (AG small-gain theorem: maximum formulation) *Let Assumption 3.1 hold. Assume that all subsystems of* (3.1) *satisfy the AG estimates as in Lemma 3.14. Assume that the whole interconnection* (3.1) *is forward complete. If* $\Gamma_{\otimes}^{AG}$ *satisfies the small gain condition* (3.36), *then* (3.1) *is AG.*

Proof Pick any $x \in X$ and $u \in \mathcal{U}$. Arguing as in Theorem 3.9 and recalling the definition of L in (3.28), we obtain the following counterpart of (3.29):

$$L \leq \Gamma_{\otimes}^{AG}(L) \vee \vec{\gamma}(\|u\|_{\infty}). \tag{3.39}$$

Using Theorem C.48, there is $\xi \in \mathcal{K}_{\infty}$ such that:

$$L = |L| \leq \xi(|\vec{\gamma}(\|u\|_{\infty})|),$$

which is precisely the AG property for the interconnection. □

Finally, the ISS small-gain theorem in the maximum formulation takes the form

Theorem 3.18 (ISS small-gain theorem: maximum formulation) *Let Assumption 3.1 hold. Assume that all subsystems of* (3.1) *satisfy the ISS estimates as in Lemma 3.15. If* $\Gamma_{\otimes}^{ISS}$ *satisfies the small-gain condition* (3.36), *then* (3.1) *is ISS.*

Proof (3.1) is UGS by Theorem 3.17, AG by Theorem 3.16 and ISS is implied by the ISS superposition theorem (Theorem 2.52). □

Remark 3.19 Note that in case if $\Gamma_{\otimes}^{ISS}$ satisfies the strong small-gain condition (3.18), the ISS small-gain theorem in the maximum formulation easily follows from the ISS small-gain theorem in the semimaximum formulation (Theorem 3.12). However, the tight form of Theorem 3.18 requires the use of the properties of max-preserving operators and cannot be derived from Theorem 3.12, which is not valid in general if $\Gamma_{\otimes}^{ISS}$ satisfies the condition (3.36). Theorem 3.18 was stated in [7, Sect. 4.3] (without proof), and in [17, Corollary 3.3] it was shown in a somewhat more general setting.

Remark 3.20 There are many other ways for characterizing the total influence of inputs over a given subsystem that can be formalized using so-called monotone aggregation functions; see Sect. C.7. Our approach, based on Proposition C.19, is not explicitly designed for a particular formulation of an ISS property and can be used for various formulations of ISS.

3.2.5 Interconnections of Two Systems

For interconnections of two systems the strong small-gain condition (3.18) takes a particularly simple cyclic form.

Proposition 3.21 *Let $n = 2$. Then the operators $\Gamma_\otimes$ and $\Gamma_\boxplus$ satisfy the strong small-gain condition* (3.18) *if and only if*

$$(\mathrm{id}+\rho)\circ\gamma_{12}\circ(\mathrm{id}+\rho)\circ\gamma_{21}(r) < r \quad \forall r > 0. \tag{3.40}$$

Proof For nonlinear gains and $n = 2$ the strong small-gain condition (3.18) for the operators $\Gamma_\otimes$ and $\Gamma_\boxplus$ takes the form: There exists $\rho \in \mathcal{K}_\infty$ such that

$$\begin{pmatrix}(\mathrm{id}+\rho)\circ\gamma_{12}(s_2)\\(\mathrm{id}+\rho)\circ\gamma_{21}(s_1)\end{pmatrix} \not\geq \begin{pmatrix}s_1\\s_2\end{pmatrix}, \quad (s_1, s_2)^T \in \mathbb{R}^2_+. \tag{3.41}$$

For the choice $(s_1, s_2) := (r, (\mathrm{id}+\rho)\circ\gamma_{21}(r)), r > 0$, the condition (3.41) implies (3.40).

Conversely, assume that (3.40) holds and (3.41) doesn't hold, that is, for some nonzero vector (s_1, s_2) it holds that

$$\begin{pmatrix}(\mathrm{id}+\rho)\circ\gamma_{12}(s_2)\\(\mathrm{id}+\rho)\circ\gamma_{21}(s_1)\end{pmatrix} \geq \begin{pmatrix}s_1\\s_2\end{pmatrix}.$$

This contradicts to (3.40) for the choice $r := s_1$. □

3.3 Cascade Interconnections

Cascade interconnections, i.e., connections without feedback loops, are a special case of feedback interconnections. See Fig. 3.2 for an illustration of a cascade interconnection of 2 systems, where x_2-subsystem is independent on x_1 but x_1 depends on x_2.

The next example shows that even the simplest cascade interconnections of two 0-GAS systems are not necessarily 0-GAS. Consider a planar system

$$\begin{aligned}\dot{x}_1 &= -x_1 + x_1^2 x_2,\\ \dot{x}_2 &= -x_2.\end{aligned} \tag{3.42}$$

Both subsystems of this system are 0-GAS, and it is easy to check that for $x_1(0) \neq 0$ and $x_2(0)$ large enough the whole system (3.42) exhibits a finite escape time. This indicates that the influence of inputs on subsystems should be taken into account

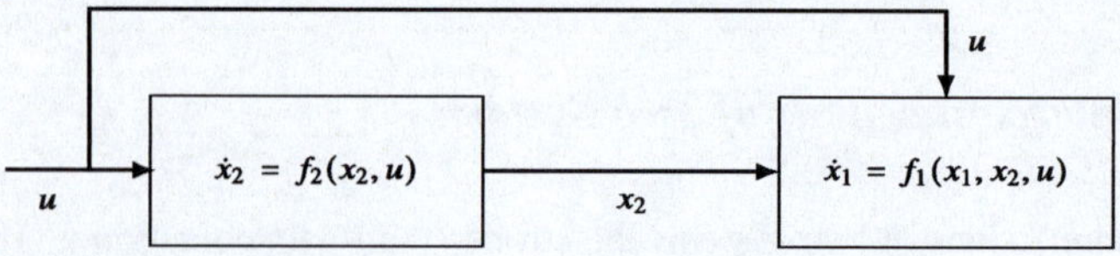

Fig. 3.2 Cascade interconnection

when studying coupled systems, which was one of the primary motivations for the introduction of the input-to-state stability property in [29].

The next result shows that cascade interconnections of ISS systems are always ISS.

Theorem 3.22 *Consider an interconnection of $n \in \mathbb{N}$ systems*

$$\dot{x}_i = f_i(x_i, x_{i+1}, \dots, x_n, u), \tag{3.43a}$$

$$i = 1, \dots, n. \tag{3.43b}$$

We assume that Assumption 3.1 holds and that all subsystems of (3.43) are ISS. Then the interconnection (3.43) is ISS.

Proof As all subsystems are ISS, there are $(\gamma_{ij})_{i,j=1}^n \subset \mathcal{K}_\infty$, such that all subsystems of (3.43) are ISS in maximum formulation as in (3.15). As f_i does not depend on x_j, $j < i$, this means that $\gamma_{ij} = 0$ for all $j \leq i$. Clearly, the cyclic small-gain condition in Theorem C.48(iii) holds, as all the cycles are identically zero. Thus, Theorem C.48 ensures that the small-gain condition holds, and the small-gain theorem in maximum formulation (Theorem 3.18) ensures that the cascade interconnection (3.43) is ISS. □

3.4 Example: Global Stabilization of a Rigid Body

The rotational dynamics of a rigid body in $\mathbb{R}^3$ is given by *Euler's rotation equations*, whose general form is:

$$J\dot{\omega} + \omega \times (J\omega) = M, \tag{3.44}$$

where M is a *vector of applied moments of force (torques)*, J is the (symmetric) *inertia matrix*, $\omega = (\omega_1, \omega_2, \omega_3)$ is the *vector of angular velocities about the principal axes*, and $\times$ is a vector product. Without the loss of generality (by *Sylvester's law of inertia*) we can assume that J is a diagonal matrix: $J = \text{diag}(J_1, J_2, J_3)$, where J_j are *principal moments of inertia. We assume that* $J_2 \neq J_3$.

We consider a model of a rigid body in $\mathbb{R}^3$, controlled by two torques acting along principal axes:

$$M(t) := (0, \tilde{u}_1(t), \tilde{u}_2(t))^T.$$

Defining $\tilde{u}(t) := (\tilde{u}_1(t), \tilde{u}_2(t))^T$, the model takes the form

$$J\dot{\omega} = \begin{pmatrix} 0 & \omega_3 & -\omega_2 \\ -\omega_3 & 0 & \omega_1 \\ \omega_2 & -\omega_1 & 0 \end{pmatrix} J\omega + \begin{pmatrix} 0 & 0 \\ 1 & 0 \\ 0 & 1 \end{pmatrix} \tilde{u}. \tag{3.45}$$

For example, this can be interpreted as a simple model of a satellite controlled by an opposing jet pair. In the coordinate form, the system (3.45) takes the form

$$J_1\dot{\omega}_1 = (J_2 - J_3)\omega_2\omega_3, \tag{3.46a}$$
$$J_2\dot{\omega}_2 = (J_3 - J_1)\omega_1\omega_3 + \tilde{u}_1, \tag{3.46b}$$
$$J_3\dot{\omega}_3 = (J_1 - J_2)\omega_1\omega_2 + \tilde{u}_2. \tag{3.46c}$$

We are going to globally stabilize (3.46) to the equilibrium position $\omega \equiv 0$. First, we simplify this system by means of a coordinate transformation (here we use that $J_2 \neq J_3$)

$$x_1 = \frac{J_1}{J_2 - J_3}\omega_1, \quad x_2 = \omega_2, \quad x_3 = \omega_3, \tag{3.47}$$

and by specializing feedback controls

$$\tilde{u}_1 := -(J_3 - J_1)\omega_1\omega_3 + J_2u_1, \tag{3.48a}$$
$$\tilde{u}_2 := -(J_1 - J_2)\omega_1\omega_2 + J_3u_2, \tag{3.48b}$$

where u_1 and u_2 are new controls. We obtain a new system:

$$\dot{x}_1 = x_2x_3, \tag{3.49a}$$
$$\dot{x}_2 = u_1, \tag{3.49b}$$
$$\dot{x}_3 = u_2. \tag{3.49c}$$

Next we show that the following feedback law globally stabilizes the system (3.49):

$$u_1 := -x_1 - x_2 - x_2x_3, \tag{3.50a}$$
$$u_2 := -x_3 + x_1^2 + 2x_1x_2x_3. \tag{3.50b}$$

To show this, we apply a one more transformation (as done in [2] for the corresponding local problem):

$$z_1 := x_1, \tag{3.51a}$$
$$z_2 := x_1 + x_2, \tag{3.51b}$$
$$z_3 := x_3 - x_1^2. \tag{3.51c}$$

The z_1-subsystem is transformed into:

$$\dot{z}_1 = x_2x_3 = (z_2 - z_1)(z_3 + z_1^2) = -z_1^3 + z_2z_3 - z_1z_3 + z_1^2z_2. \tag{3.52}$$

Some further computations lead us to:

$$\dot{z}_1 = -z_1^3 + \alpha(z_1, z_2, z_3), \tag{3.53a}$$

$$\dot{z}_2 = -z_2, \tag{3.53b}$$

$$\dot{z}_3 = -z_3, \tag{3.53c}$$

where

$$\alpha(z_1, z_2, z_3) = z_2 z_3 - z_1 z_3 + z_1^2 z_2.$$

As α is a polynomial of a degree 2 in z_1, it is easy to show that the z_1-subsystem is ISS w.r.t. z_2, z_3 as inputs. Clearly, z_2- and z_3-subsystems are UGAS. Since (3.53) is a cascade interconnection of ISS systems, (3.53) is UGAS. Analyzing the transformations (3.49) and (3.51), we see that the original system (3.46) is UGAS w.r.t. ω_1, ω_2, ω_3.

3.5 Lyapunov-Based Small-Gain Theorems

In most cases, the ISS of nonlinear systems is verified by constructing an appropriate ISS Lyapunov function. It sounds natural to use the information about the ISS Lyapunov functions for subsystems in the formulation of the small-gain theorems instead of using the ISS estimates for subsystems. In this section, we show that this is possible. Moreover, this results in a method for constructing ISS Lyapunov functions for the overall system, provided the ISS Lyapunov functions for subsystems are known.

As we have argued in a passage following the estimate (3.4), for the analysis of networks, we have to know the response of subsystems (or of corresponding Lyapunov functions) on the inputs from all other subsystems. To aggregate the individual inputs into the total input, we use the following formalism:

Definition 3.23 *(Monotone aggregation functions)* A function $\mu : \mathbb{R}^n_+ \to \mathbb{R}_+$ is called a monotone aggregation function (MAF) if μ is continuous and satisfies the following properties:

(i) *Positivity:* $\mu(v) \geq 0$, for all $v \in \mathbb{R}^n_+$.
(ii) *Strict increase:* if $v_i < w_i$ for all $i = 1, \dots, n$, then $\mu(v) < \mu(w)$.
(iii) *Unboundedness:* $\mu(v) \to \infty$ as $|v| \to \infty$.
(iv) *Subadditivity:* $\mu(v + w) \leq \mu(v) + \mu(w)$, for all $v, w \in \mathbb{R}^n_+$.

Two important examples of monotone aggregation functions are

$$\mu_{\boxplus} : s = (s_i)_{i=1}^n \mapsto s_1 + \cdots + s_n, \tag{3.54}$$

and

$$\mu_{\otimes} : s = (s_i)_{i=1}^n \mapsto \max\{s_1, \ldots, s_n\}. \tag{3.55}$$

Now, in addition to Assumption 3.1, let the following hold:

Assumption 3.24 For each $i = 1, \ldots, n$, there exists a continuous function $V_i : \mathbb{R}^{N_i} \to \mathbb{R}_+$ that is C^1 outside of $x_i = 0$, and satisfies the following properties:

(i) There exist $\psi_{i1}, \psi_{i2} \in \mathcal{K}_\infty$ such that

$$\psi_{i1}(|x_i|) \le V_i(x_i) \le \psi_{i2}(|x_i|) \quad \forall x_i \in \mathbb{R}^{N_i}. \tag{3.56}$$

(ii) There exists a MAF μ_i, together with $\chi_{ij} \in \mathcal{K}_\infty \cup \{0\}$, $j = 1, \ldots, n$, where $\chi_{ii} = 0$, and $\chi_{iu} \in \mathcal{K}$, $\alpha_i \in \mathcal{P}$ such that for all $x_j \in \mathbb{R}^{N_j}$, $j = 1, \ldots, n$, and all $u \in \mathbb{R}^m$ the following implication holds:

$$\begin{aligned} V_i(x_i) > \max\Big\{\mu_i\Big[\big(\chi_{ij}(V_j(x_j))\big)_{j=1}^n\Big], \chi_{iu}(|u|)\Big\} \\ \Rightarrow \nabla V_i(x_i) f_i(x_i, \bar{x}_i, u_i) \le -\alpha_i(V_i(x_i)). \end{aligned} \tag{3.57}$$

The functions χ_{ij}, $i, j = 1, \ldots, n$ are called *internal Lyapunov gains*, while the functions χ_{iu} are called external Lyapunov gains.

The function μ_i combines ("aggregates") the inputs from all x_j-subsystems into a total internal input to the x_i-subsystem from other elements of the network. If $\mu_i = \mu_{\otimes}$, the total input to the i-th subsystem is a maximum of the inputs from individual subsystems. If $\mu_i = \mu_{\boxplus}$, see (3.54), then the total input to the i-th subsystem is a sum of the inputs from individual subsystems. Allowing for such flexibility, we can obtain sharper conditions for the stability of the network in a wide variety of applications.

We collect the Lyapunov gains $\chi_{ij} \in \mathcal{K}_\infty \cup \{0\}$ for $i, j \in \{1, 2, \ldots, n\}$ into the matrix $\Gamma := (\chi_{ij})_{i,j=1}^n$. All MAFs μ_i we combine into the vector of monotone aggregation functions $\mu := (\mu_i)_{i=1}^n$.

Then a pair (Γ, μ) gives rise to the following monotone and continuous operator

$$\Gamma_\mu(s) := \begin{pmatrix} \mu_1\big(\chi_{11}(s_1), \ldots, \chi_{1n}(s_n)\big) \\ \vdots \\ \mu_n\big(\chi_{n1}(s_1), \ldots, \chi_{nn}(s_n)\big) \end{pmatrix}, \quad s = (s_i)_{i=1}^n \in \mathbb{R}^n_+. \tag{3.58}$$

If $\mu_i = \mu_{\boxplus}$ for all i, then Γ_μ corresponds to the operator $\Gamma_{\boxplus}$, defined in (3.12). If $\mu_i = \mu_{\otimes}$ for all i, then Γ_μ is precisely $\Gamma_{\otimes}$, introduced in (3.31).

We will need the following technical lemma:

Lemma 3.25 *Let $(x_k^1)_{k=1}^\infty$, ..., $(x_k^m)_{k=1}^\infty$ be sequences of real numbers. Let the limit $\lim_{k\to\infty} \max_{1\le i\le m} x_k^i$ exist. Then*

$$\lim_{k\to\infty}\max_{1\le i\le m} x_k^i = \max_{1\le i\le m}\overline{\lim_{k\to\infty}} x_k^i. \tag{3.59}$$

Proof For all $k \in \mathbb{N}$ define $i(k) := \arg\max_{1\le i\le m} x_k^i$ - the index of a maximal element of $\{x_k^i : i = 1, \dots, m\}$ (if there is more than one maximal element, then take for definiteness the index of the first maximizing element). Then $\max_{1\le i\le m} x_k^i = x_k^{i(k)}$ for all $k \in \mathbb{N}$. Extract from the sequence $(x_k^{i(k)})$ the maximal subsequences of the form $(x^j_{n_k^j})$, $j = 1, \dots, m$, where (n_k^j) is the monotone increasing sequence of indexes. At least some of $(x^j_{n_k^j})$, $j = 1, \dots, m$ are infinite. Without loss of generality let it be $(x^1_{n_k^1})$.

By assumption, the sequence $(x_k^{i(k)})$ is convergent. Hence all its (infinite) subsequences are convergent and have the same limit value. Thus we obtain

$$\lim_{k\to\infty}\max_{1\le i\le m} x_k^i = \lim_{k\to\infty} x_k^{i(k)} = \lim_{k\to\infty} x^1_{n_k^1} \le \overline{\lim_{k\to\infty}} x_k^1 \le \max_{1\le i\le m}\overline{\lim_{k\to\infty}} x_k^i. \tag{3.60}$$

To obtain the reverse inequality, denote $a_i := \overline{\lim}_{k\to\infty} x_k^i$. Then for a certain index j, we have

$$\max_{1\le i\le m}\overline{\lim_{k\to\infty}} x_k^i = \max_{1\le i\le m} a_i = a_j = \overline{\lim_{k\to\infty}} x_k^j.$$

Hence, there is a sequence $(x^j_{n_k})_{k\in\mathbb{N}}$, such that

$$\lim_{k\to\infty} x^j_{n_k} = \max_{1\le i\le m}\overline{\lim_{k\to\infty}} x_k^i.$$

We have that $\max_{1\le i\le m} x^i_{n_k} \ge x^j_{n_k}$. Furthermore, as the limit $\lim_{k\to\infty}\max_{1\le i\le m} x_k^i$ exists, then also the limit $\lim_{k\to\infty}\max_{1\le i\le m} x^i_{n_k}$ exists, and

$$\lim_{k\to\infty}\max_{1\le i\le m} x_k^i = \lim_{k\to\infty}\max_{1\le i\le m} x^i_{n_k} \ge \lim_{k\to\infty} x^j_{n_k} = \max_{1\le i\le m}\overline{\lim_{k\to\infty}} x_k^i. \tag{3.61}$$

From (3.60) and (3.61) we obtain (3.59). □

Corollary 3.26 *Let $f_i : \mathbb{R} \to \mathbb{R}$ be defined and bounded in some neighborhood D of $t = 0$. The following holds:*

$$\overline{\lim_{t\to 0}}\max_{1\le i\le m} f_i(t) = \max_{1\le i\le m}\overline{\lim_{t\to 0}} f_i(t). \tag{3.62}$$

Proof Under made assumptions, the upper limits in both parts of the equation (3.62) exist and are both finite. As for all $j = 1, \dots, m$ and all $t \in D$ it holds that $\max_{1\le i\le m} f_i(t) \ge f_j(t)$, we obtain for any j that

$$\overline{\lim_{t\to 0}}\max_{1\le i\le m} f_i(t) \ge \overline{\lim_{t\to 0}} f_j(t),$$

and hence, taking the maximum over j and calling the index i, we have

$$\overline{\lim_{t\to 0}} \max_{1\le i\le m} f_i(t) \ge \max_{1\le i\le m} \overline{\lim_{t\to 0}} f_i(t).$$

To prove the converse inequality, we use Lemma 3.25.

$$\begin{aligned}\overline{\lim_{t\to 0}} \max_{1\le i\le m} f_i(t) &= \sup_{(t_{n_k})_k\to 0} \lim_{k\to\infty} \max_{1\le i\le m} f_i(t_{n_k}) \\ &= \sup_{(t_{n_k})_k\to 0} \max_{1\le i\le m} \overline{\lim_{k\to\infty}} f_i(t_{n_k}) \le \max_{1\le i\le m} \overline{\lim_{t\to 0}} f_i(t),\end{aligned}$$

where the supremum is taken over all convergent to 0 sequences t_{n_k}. □

In Appendix C we have studied paths of strict decay (Definition C.38) with respect to monotone operators on $\mathbb{R}^n_+$. Such paths possessing a certain kind of regularity will be ultimately helpful for construction of ISS Lyapunov functions for interconnected systems with ISS components.

Definition 3.27 Let $\Gamma : \mathbb{R}^n_+ \to \mathbb{R}^n_+$ be a nonlinear operator. A function $\sigma = (\sigma_1, \dots, \sigma_n)^T : \mathbb{R}^n_+ \to \mathbb{R}^n_+$, where $\sigma_i \in \mathcal{K}_\infty$, $i = 1, \dots, n$ is called a *regular path of nonstrict decay* with respect to the operator Γ if it possesses the following properties:

(i) For every compact interval $K \subset (0, \infty)$, there exist $0 < c \le C < \infty$ such that for all $r_1, r_2 \in K$ and $i \in \mathbb{N}$

$$c|r_1 - r_2| \le |\sigma_i^{-1}(r_1) - \sigma_i^{-1}(r_2)| \le C|r_1 - r_2|. \tag{3.63}$$

(ii) Γ is nonincreasing on $\sigma(\mathbb{R}_+)$:

$$\Gamma(\sigma(r)) \le \sigma(r) \quad \forall r \ge 0. \tag{3.64}$$

If the map σ satisfies also $\Gamma(\sigma(r)) \ll \sigma(r)$, $\forall r > 0$, then σ is called a *regular path of strict decay*.

The next theorem provides a construction of an ISS Lyapunov function for an interconnection of ISS subsystems provided a path of nonstrict decay for the gain operator is available.

Theorem 3.28 (Lyapunov-based ISS small-gain theorem) *Let Assumptions 3.1 and 3.24 hold. If there exists a regular path of nonstrict decay $\sigma = (\sigma_1, \dots, \sigma_n)^T$ corresponding to the operator Γ_μ defined by (3.58), then an ISS Lyapunov function for (3.1) can be constructed as*

$$V(x) := \max_{i=1}^{n} \sigma_i^{-1}(V_i(x_i)), \tag{3.65}$$

and the Lyapunov gain of the whole system is given by

$$\chi(r) := \max_{i=1}^{n} \sigma_i^{-1}(\chi_i(r)). \quad (3.66)$$

Proof We divide the proof into four parts.

Step 1. Partition of X. Pick any $u \in \mathcal{U}$. To prove that V is an ISS Lyapunov function, it is useful to divide its domain of definition into subsets, on which V takes a simpler form. Thus, for all $i \in \{1, \dots, n\}$ define a set

$$M_i = \left\{ x \in \mathbb{R}^N : \sigma_i^{-1}(V_i(x_i)) > \sigma_j^{-1}(V_j(x_j))\ \forall j = 1, \dots, n,\ j \neq i \right\}.$$

From the continuity of V_i and σ_i^{-1}, $i = 1, \dots, n$ it follows that all M_i are open. Also note that $\mathbb{R}^N = \bigcup_{i=1}^n \overline{M_i}$ and for all i, j: $i \neq j$ holds $M_i \bigcap M_j = \emptyset$.

Step 2. Estimates for $\dot{V}_u(x)$ on $\bigcup_{i=1}^n M_i$. Take some $i \in \{1, \dots, n\}$ and pick any $x \in M_i$. Assume that $V(x) \geq \chi(\|u\|_\infty)$ holds, where χ is given by (3.66). We obtain

$$\sigma_i^{-1}(V_i(x_i)) = V(x) \geq \chi(\|u\|_\infty) = \max_{j=1}^{n} \sigma_j^{-1} \circ \chi_j(\|u\|_\infty) \geq \sigma_i^{-1}(\chi_i(\|u\|_\infty)).$$

Since $\sigma_i^{-1} \in \mathcal{K}_\infty$, it holds that

$$V_i(x_i) \geq \chi_i(\|u\|_\infty). \quad (3.67)$$

On the other hand, from the condition of nonstrict decay (3.64), and using the monotonicity of the MAF μ_i, we obtain that

$$\begin{aligned} V_i(x_i) = \sigma_i(V(x)) &\geq \Big[\Gamma_\mu\Big(\sigma\big(V(x)\big)\Big)\Big]_i \\ &= \mu_i\Big[\Big(\chi_{ij}(\sigma_j(V(x)))\Big)_{j=1}^n\Big] = \mu_i\Big[\Big(\chi_{ij}\big(\sigma_j\big(\sigma_i^{-1}(V_i(x_i))\big)\big)\Big)_{j=1}^n\Big] \\ &\geq \mu_i\Big[\Big(\chi_{ij}\big(\sigma_j\big(\sigma_j^{-1}(V_j(x_j))\big)\big)\Big)_{j=1}^n\Big] = \mu_i\Big[\Big(\chi_{ij}(V_j(x_j))\Big)_{j=1}^n\Big]. \end{aligned}$$

Combining with (3.67), we obtain

$$V_i(x_i) \geq \max\Big\{\mu_i\Big[\Big(\chi_{ij}(V_j(x_j))\Big)_{j=1}^n\Big], \chi_i(\|u\|_\infty)\Big\}. \quad (3.68)$$

By (3.57), we have that

$$\dot{V}_i(x_i) = \nabla V_i(x_i) f_i(x_i, \bar{x}_i, u_i) \leq -\alpha_i(V_i(x_i)). \quad (3.69)$$

As $\sigma_i^{-1}(V_i(x_i)) > \sigma_j^{-1}(V_j(x_j))$ for all $j \neq i$, and since the trajectories of all subsystems are continuous, there is a time $t > 0$ such that for all $s \in [0, t]$ it holds that $\sigma_i^{-1}\big(V_i(\phi_i(s, x, u))\big) > \sigma_j^{-1}\big(V_j(\phi_j(s, x, u))\big)$ for all $j \neq i$. Thus,

$$V(\phi(s, x, u)) = \sigma_i^{-1}\big(V_i(\phi_i(s, x, u))\big), \quad s \in [0, t].$$

Note also that $V_i(\phi_i(s, x, u)) \leq V_i(x_i)$ in view of (3.69). Now we have

$$\begin{aligned}\frac{1}{s}\Big(V(\phi(s, x, u)) - V(x)\Big) &= \frac{1}{s}\Big(\sigma_i^{-1}\big(V_i(\phi_i(s, x, u))\big) - \sigma_i^{-1}\big(V_i(x_i)\big)\Big)\\ &= -\frac{1}{s}\Big|\sigma_i^{-1}\big(V_i(\phi_i(s, x, u))\big) - \sigma_i^{-1}\big(V_i(x_i)\big)\Big|. \quad (3.70)\end{aligned}$$

Define (pointwise) $\sigma_{\min} := \min_{i=1}^n \sigma_i$ and $\sigma_{\max} := \max_{i=1}^n \sigma_i$.

Thanks to continuity of $s \mapsto \sigma_i^{-1}\big(V_i(\phi_i(s, x, u))\big)$, we choose t small enough, such that for all $s \in [0, t]$ it holds that

$$\frac{1}{2}V_i(x_i) \leq V_i(\phi_i(s, x, u)) \leq V_i(x_i) = \sigma_i(V(x)) \leq \sigma_{\max}(V(x)).$$

Similarly, we estimate

$$\frac{1}{2}V_i(x_i) = \frac{1}{2}\sigma_i(V(x)) \geq \frac{1}{2}\sigma_{\min}(V(x)).$$

Define for any $r \geq 0$

$$K(r) := \Big[\frac{1}{2}\sigma_{\min}(r), \sigma_{\max}(r)\Big],$$

and let $c = c(K(r)) > 0$ be the maximal constant such that

$$|\sigma_i^{-1}(r_1) - \sigma_i^{-1}(r_2)| \geq c|r_1 - r_2|, \quad \forall r_1, r_2 \in K(r).$$

For all $s \in (0, t)$, we obtain from (3.70) that

$$\begin{aligned}\frac{1}{s}\Big(V(\phi(s, x, u)) - V(x)\Big) &\leq -c(K(V(x)))\frac{1}{s}\Big|V_i(\phi_i(s, x, u)) - V_i(x_i)\Big|\\ &= c(K(V(x)))\frac{1}{s}\big(V_i(\phi_i(s, x, u)) - V_i(x_i)\big).\end{aligned}$$

Taking the limit superior, we obtain

$$\begin{aligned}\dot{V}_u(x) &= \overline{\lim_{s \to +0}}\, \frac{1}{s}\Big(V(\phi(s, x, u)) - V(x)\Big)\\ &\leq c(K(V(x)))\dot{V}_i(x_i)\\ &\leq -c(K(V(x)))\alpha_i(V_i(x_i))\\ &= -c(K(V(x)))\alpha_i(\sigma_i(V(x))).\end{aligned}$$

Define now

$$\hat{\alpha}(r) := c(K(r)) \min_{i=1}^{n} \alpha_i(\sigma_i(r)), \quad r \geq 0.$$

Step 3. Lower bound for $\hat{\alpha}$. Next we lower bound $\hat{\alpha}$ by a positive definite function. For each $r > 0$, define $K_2(r) := \bigcup_{q \in K(r)} \mathrm{str}(1, q)$, where $\mathrm{str}(1, q)$ equals $[1, q]$ for $q \geq 1$ and $[q, 1]$ for $q < 1$. Clearly, $K_2(r)$ is a compact subset of $(0, \infty)$. Further, we introduce

$$\hat{\alpha}_2(r) := c(K_2(r)) \min_{i=1}^{n} \alpha_i(\sigma_i(r)), \quad \forall r > 0.$$

As $K(r) \subset K_2(r)$ for any $r > 0$, $\hat{\alpha}(r) \geq \hat{\alpha}_2(r)$ for all $r > 0$. Furthermore, there is $r_{\min}$ such that $K_2(r_1) \supset K_2(r_2)$ for all $r_1, r_2 \in (0, r_{\min})$ with $r_1 < r_2$. This implies that $\hat{\alpha}_2$ is a nondecreasing positive function on $(0, r_{\min})$, and $\lim_{r \downarrow 0} \hat{\alpha}_2(r) = 0$. Moreover, for all $r \in (0, r_{\min})$, we have

$$\hat{\alpha}_2(r) = \frac{2}{r} \int_{r/2}^{r} \hat{\alpha}_2(r)\, ds \geq \frac{2}{r} \int_{r/2}^{r} \hat{\alpha}_2(s)\, ds,$$

where $\hat{\alpha}_2$ is integrable on $(0, r_{\min})$ as it is monotone on this interval. Hence, $\hat{\alpha}_2$ and thus $\hat{\alpha}$ can be lower bounded by a continuous function on $[0, r_{\min}]$. Similarly, there is $r_{\max}$ such that $K_2(r_1) \supset K_2(r_2)$ for all $r_1, r_2 \in (r_{\max}, \infty)$ with $r_1 > r_2$. This implies that $\hat{\alpha}_2$ is a nonincreasing positive function on $(r_{\max}, \infty)$. Consequently, for $r \in (r_{\max}, \infty)$, we have

$$\hat{\alpha}_2(r) = \frac{1}{r} \int_{r}^{2r} \hat{\alpha}_2(r)\, ds \geq \frac{1}{r} \int_{r}^{2r} \hat{\alpha}_2(s)\, ds.$$

Hence, $\hat{\alpha}_2$, and thus $\hat{\alpha}$, can be lower bounded by a continuous function on $[r_{\max}, \infty)$. As $\hat{\alpha}$ assumes positive values and is bounded away from zero on every compact interval in $(0, \infty)$, $\hat{\alpha}$ can be lower bounded by a positive definite function, which we denote by α.

Overall, for all $x \in \bigcup_{i=1}^{n} M_i$ it holds that

$$\dot{V}_u(x) \leq -\alpha(V(x)).$$

Step 4. Estimates for $\dot{V}_u(x)$ on X. Now let $x \notin \bigcup_{i=1}^{n} M_i$. From $\mathbb{R}^N = \bigcup_{i=1}^{n} \overline{M_i}$ it follows that $x \in \bigcap_{i \in I(x)} \partial M_i$ for some index set $I(x) \subset \{1, \ldots, n\}$, $|I(x)| \geq 2$.

$$\bigcap_{i\in I(x)} \partial M_i = \big\{x \in \mathbb{R}^N : \forall i \in I(x),\ \forall j \notin I(x)\ \sigma_i^{-1}(V_i(x_i)) > \sigma_j^{-1}(V_j(x_j))$$
$$\forall i, j \in I(x)\ \sigma_i^{-1}(V_i(x_i)) = \sigma_j^{-1}(V_j(x_j))\big\}.$$

Due to the continuity of ϕ, there exists $t > 0$, such that for all $s \in [0, t)$

$$\phi(s, x, u) \in \Big(\bigcap_{i\in I(x)} \partial M_i\Big) \cup \Big(\bigcup_{i\in I(x)} M_i\Big).$$

Then, by definition of the Lie derivative we obtain

$$\begin{aligned}\dot V_u(x) &= \overline{\lim_{t\to+0}} \frac{1}{t}\Big(V(\phi(t,x,u)) - V(x)\Big)\\ &= \overline{\lim_{t\to+0}} \frac{1}{t}\Big(\max_{i\in I(x)}\{\sigma_i^{-1}(V_i(\phi_i(t,x,u)))\} - \max_{i\in I(x)}\{\sigma_i^{-1}(V_i(x_i))\}\Big). \quad (3.71)\end{aligned}$$

From the definition of $I(x)$ it follows that

$$\sigma_i^{-1}(V_i(x_i)) = \sigma_j^{-1}(V_j(x_j)) \quad \forall i, j \in I(x),$$

and therefore the index i, on which the maximum of $\max_{i\in I(x)}\{\sigma_i^{-1}(V_i(x_i))\}$ is reached, may be always set equal to the index, on which the maximum $\max_{i\in I(x)}\{\sigma_i^{-1}(V_i(\phi_i(t,x,u)))\}$ is reached.
We continue the estimates (3.71)

$$\dot V_u(x) = \overline{\lim_{t\to+0}} \max_{i\in I(x)} \Big\{\frac{1}{t}\left(\sigma_i^{-1}(V_i(\phi_i(t,x,u))) - \sigma_i^{-1}(V_i(x_i))\right)\Big\}.$$

Using Lemma 3.25, we obtain

$$\begin{aligned}\dot V_u(x) &= \max_{i\in I(x)} \Big\{\overline{\lim_{t\to+0}} \frac{1}{t}\left(\sigma_i^{-1}(V_i(\phi_i(t,x,u))) - \sigma_i^{-1}(V_i(x_i))\right)\Big\}\\ &\le -\alpha(V(x)).\end{aligned}$$

Overall, we have that for all $x \in \mathbb{R}^N$ and $u \in \mathcal{U}$ it holds that

$$V(x) \ge \chi(\|u\|_\infty) \quad \Rightarrow \quad \dot V_u(x) \le -\alpha(V(x)),$$

and the ISS Lyapunov function in implication form for the whole interconnection is constructed. ISS of the whole system follows by Theorem 2.12. □

Remark 3.29 We have shown that the network of ISS systems is ISS, provided that there is a path of nonstrict decay for the operator Γ_μ. The existence of a path of strict (and thus also of a nonstrict) decay (at least up to its regularity) is guaranteed provided that Γ_μ satisfies a so-called strong small-gain condition; see Theorem C.41.

3.5.1 Small-Gain Theorem for Homogeneous and Subadditive Gain Operators

Definition 3.30 We call a function $\mu : \mathbb{R}^n_+ \to \mathbb{R}_+$ *homogeneous (of degree one)*, if for all $s \in \mathbb{R}^n_+$ and all $c \geq 0$ it holds that $\mu(cs) = c\mu(s)$.

We immediately have the following:

Lemma 3.31 *Let all MAFs μ_i in Assumption 3.24 be homogeneous, and let all internal gains χ_{ij} be linear functions.*

Then the operator Γ_μ defined by (3.58) is homogeneous and subadditive (as defined in Definition C.56).

Now we can state a constructive small-gain theorem for the case of homogeneous gain operators.

Theorem 3.32 (Lyapunov-based ISS small-gain theorem: homogeneous case) *Let Assumptions 3.1 and 3.24 hold. Assume further that all MAFs μ_i in Assumption 3.24 are homogeneous, and that all internal gains χ_{ij} are linear functions.*

If $r(\Gamma_\mu) < 1$, then there are $\lambda \in (0, 1)$ and $z \in \operatorname{int}(\mathbb{R}^n_+)$ such that

$$\Gamma_\mu(z) \leq \lambda z. \tag{3.72}$$

Furthermore, an ISS Lyapunov function for (3.1) can be constructed as

$$V(x) := \max_{i=1}^{n} \frac{1}{z_i} V_i(x_i). \tag{3.73}$$

Proof As $r(\Gamma_\mu) < 1$, Theorem C.62 ensures that there are $\lambda \in (0, 1)$ and $z \in \operatorname{int}(\mathbb{R}^n_+)$ such that (3.72) holds. Hence, a linear path of (strict and thus also nonstrict) decay for Γ_μ can be constructed as $\sigma(r) := rx^0$, $r \geq 0$. Applying Theorem 3.28, we obtain the claim. □

3.5.2 Examples on the Max-Formulation of the Small-Gain Theorem

In Corollary C.54, it is shown that if $\Gamma_\otimes$ satisfies the small-gain condition

$$\Gamma_\otimes(x) \not\geq x, \quad x \neq 0, \tag{3.74}$$

then one can explicitly construct a path of nonstrict decay with respect to $\Gamma_{\otimes}$. However, this path of nonstrict decay may fail to satisfy the regularity requirements of Definition 3.27.

Nevertheless, one can show the following existence result due to [5, Theorem 5.2(iii)].

Theorem 3.33 *If $\Gamma_{\otimes}$ satisfies the small-gain condition (3.74), then there is a regular path of strict decay with respect to $\Gamma_{\otimes}$.*

We continue with two examples.

Example 3.34 Consider the system

$$\begin{aligned}\dot{x}_1 &= -x_1 + x_2^2,\\ \dot{x}_2 &= -x_2 + a\sqrt{|x_1|}.\end{aligned} \tag{3.75}$$

We analyze UGAS of this system by means of small-gain theorems. Pick

$$V_1(x_1) := x_1^2, \quad x_1 \in \mathbb{R}, \tag{3.76}$$

as an ISS Lyapunov function candidate for the first subsystem. We obtain:

$$\dot{V}_1(x_1) = 2x_1\dot{x}_1 = 2x_1(-x_1 + x_2^2) \le -2x_1^2 + 2|x_1|x_2^2. \tag{3.77}$$

The Lyapunov gain χ_{12} we choose as $\chi_{12}(r) := \frac{1}{1-\varepsilon}r^2$ for some $\varepsilon \in (0,1)$ and all $r \ge 0$. Then $|x_1| \ge \chi_{12}(|x_2|)$ implies $x_2^2 \le (1-\varepsilon)|x_1|$, which immediately leads to

$$\dot{V}_1(x_1) \le -2\varepsilon x_2^2 = -2\varepsilon V_1(x_1). \tag{3.78}$$

This shows that the first subsystem is ISS with V_1 as an ISS Lyapunov function.

Now take the following ISS Lyapunov function candidate for the second subsystem:

$$V_2(x_2) := x_2^2, \quad x_2 \in \mathbb{R}. \tag{3.79}$$

We have

$$\dot{V}_2(x_2) = 2x_2\dot{x}_2 = 2x_2(-x_2 + a\sqrt{|x_1|}) \le -2x_2^2 + 2|a||x_2|\sqrt{|x_1|}. \tag{3.80}$$

Pick $\chi_{21}(r) := \frac{|a|}{1-\varepsilon_2}\sqrt{r}$, for some $\varepsilon_2 \in (0,1)$ and all $r \ge 0$. Then $|x_2| \ge \chi_{21}(|x_1|)$ implies that $|a|\sqrt{|x_1|} \le (1-\varepsilon_2)|x_2|$, and thus

$$\dot{V}_2(x_2) \le -2\varepsilon_2 x_2^2 = -2\varepsilon_2 V_2(x_2), \tag{3.81}$$

and the second subsystem is also ISS.

Hence, $\mu_1, \mu_2 = \mu_\otimes$, and the gain operator (3.58) induced by the Lyapunov gains χ_{12} and χ_{21} takes form

$$\Gamma_\otimes(x) := \begin{pmatrix} \chi_{12}(x_2) \\ \chi_{21}(x_1) \end{pmatrix} = \begin{pmatrix} \frac{1}{1-\varepsilon} x_2^2 \\ \frac{|a|}{1-\varepsilon_2} \sqrt{x_1} \end{pmatrix}, \quad x = (x_1, x_2) \in \mathbb{R}^2_+. \tag{3.82}$$

In view of Theorem 3.33, a regular path of strict decay for $\Gamma_\otimes$ exists provided that the cyclic small-gain condition

$$\chi_{12} \circ \chi_{21}(r) < r, \quad r > 0$$

holds.

In our case, this reduces to

$$\chi_{12} \circ \chi_{21}(r) = \frac{1}{1-\varepsilon} \frac{a^2}{(1-\varepsilon_2)^2} r < r,$$

for some $\varepsilon, \varepsilon_2 > 0$ and all $r > 0$. Clearly, this is equivalent to $|a| < 1$.

For our gain operator, the path provided by Corollary C.54 will be Lipschitz continuous. However, it is not hard to explicitly construct also a continuously differentiable path of strict decay. Take any $b > 0$ and define

$$\sigma(r) := \begin{pmatrix} (1+b)r^2 \\ r \end{pmatrix}, \quad r \geq 0.$$

We have

$$\Gamma(\sigma(r)) = \begin{pmatrix} \frac{1}{1-\varepsilon} r^2 \\ \frac{|a|}{1-\varepsilon_2} \sqrt{(1+b)r} \end{pmatrix}.$$

Choosing ε_2 and b small enough such that

$$\frac{|a|}{1-\varepsilon_2} \sqrt{(1+b)} < 1,$$

and choosing $\varepsilon > 0$ small enough, such that $\frac{1}{1-\varepsilon} < 1 + b$, we see that σ is a path of strict decay for $\Gamma_\otimes$.

Applying the small-gain theorem, we obtain that (3.75) is ISS if $|a| < 1$, and using σ we can construct a Lyapunov function for (3.75).

If $a \in \{1, -1\}$, then the points satisfying algebraic equations

$$x_1 = x_2^2,$$
$$x_2 = a\sqrt{|x_1|},$$

are the stationary points of (3.75), which shows that for $a \in \{1, -1\}$ the system (3.75) is not ISS. Using monotonicity arguments, one can show that for $a : |a| > 1$ the system (3.75) is again not ISS.

To conclude our investigations, (3.75) is ISS iff $|a| < 1$. □

Example 3.35 Consider the following control system that arises, e.g., in modeling of chemical reactions or production networks [3].

$$\dot{x}_i(t) = \sum_{j=1, j\neq i}^{n} c_{ij}(x(t))\tilde{f}_j(x_j(t)) - \tilde{c}_{ii}(x(t))\tilde{f}_i(x_i(t)) + u_i(t), \quad i = 1, \dots, n. \tag{3.83}$$

As the physical sense of x_i in this model is usually the mass of a chemical substance or some other nonnegative quantity, we will assume from the outset that $x_j(0) \geq 0$ for all j, $f_i \in \mathcal{K}_\infty$, and $u_i(t) \geq 0$ for all i and a.e. $t \geq 0$.

Furthermore, we assume that for all $x \in \mathbb{R}^n_+ \backslash \{0\}$ it holds that $\tilde{c}_{ii}(x) > 0$ and $c_{ij}(x) \geq 0,\ i \neq j$.

Denoting $c_{ii} := -\tilde{c}_{ii}$ we can rewrite above equations in a vector form

$$\dot{x}(t) = C(x(t))\tilde{f}(x(t)) + u(t), \tag{3.84}$$

where $\tilde{f}(x(t)) = (\tilde{f}_1(x_1(t)), \dots, \tilde{f}_n(x_n(t)))^T$, $u(t) = (u_1(t), \dots, u_n(t))^T$ and $C(x(t)) = (c_{ij}(x(t)))_{ij}$.

Since $\tilde{f}_i \in \mathcal{K}_\infty$ and for all $x > 0$ $c_{ii}(x) < 0$ and $c_{ij}(x) \geq 0,\ i \neq j$, and as $x(0) \geq 0$ and $u(t) \geq 0$, for all $t > 0$, we have that $x(t) \geq 0$ for all $t > 0$. Thus, the solution of (3.83) remains nonnegative on the whole domain of existence.

To analyze the stability of the system (3.84), we exploit Theorem 3.28. We construct ISS Lyapunov functions V_i and corresponding gains χ_{ij} for each subsystem (which ensures that the subsystems are ISS) and seek for conditions, guaranteeing that the small-gain condition (3.36) holds.

Step 1. ISS Lyapunov functions and corresponding gains for subsystems of (3.83). Pick

$$V_i(x_i) = |x_i|$$

as an ISS Lyapunov function for the i-th subsystem of (3.83). Evidently, $V_i(x_i)$ satisfies the condition (3.56). Take the Lyapunov gains χ_{ij}, χ_i as

$$\chi_{ij}(s) := \tilde{f}_i^{-1}\left(\tfrac{a_i}{a_j}\tfrac{1}{1+\delta_j}\tilde{f}_j(s)\right), \ \chi_i(s) := \tilde{f}_i^{-1}\left(\tfrac{1}{r_i}s\right), \tag{3.85}$$

where δ_j, a_j, $j = 1, \dots, n$ and r_i are positive reals. It follows from (3.85) that

$$x_i \geq \chi_{ij}\left(x_j\right) \Rightarrow \tilde{f}_j(x_j) \leq \tfrac{a_j}{a_i}(1+\delta_j)\tilde{f}_i(x_i),$$

$$x_i \geq \chi_i\left(|u_i|\right) \Rightarrow |u_i| \leq r_i \tilde{f}_i(x_i).$$

Using the inequalities from the right-hand side of the implications above and assuming that the following condition holds

$$\sum_{j=1,\ j\neq i}^{n} c_{ij}(x)\tfrac{a_j}{a_i}(1+\delta_j) + c_{ii}(x) + r_i \leq -h_i \quad \forall x \in \mathbb{R}^n_+, \text{ for some } h_i > 0, \tag{3.86}$$

we obtain for all $x_i \in \mathbb{R}_+$ satisfying

$$V_i(x_i) \geq \max\left\{\max_{j\neq i} \chi_{ij}\left(V_j(x_j)\right), \chi_i\left(|u_i|\right)\right\}$$

that

$$\begin{aligned}\frac{dV_i(x_i(t))}{dt} &= \sum_{j=1}^{n} c_{ij}(x(t))\tilde{f}_j(x_j(t)) + u_i(t)\\ &\leq \left(\sum_{j=1,j\neq i}^{n} c_{ij}(x(t))\tfrac{a_j}{a_i}(1+\delta_j) + c_{ii}(x(t)) + r_i\right)\tilde{f}_i\left(x_i(t)\right) \leq -\alpha_i(V_i(x_i(t))),\end{aligned}$$

where $\alpha_i(r) := h_i\tilde{f}_i(r)$. Thus, under the condition (3.86), $V_i(x_i) = |x_i|$ is an ISS Lyapunov function for the i-th subsystem with the gains, given by (3.85).

Step 2. Small-gain analysis of the whole network. For all $i = 1, \ldots, n$ the MAFs have a form $\mu_i = \mu_\otimes$, and the gain operator for the network is given by (3.31). Let us check the cyclic small-gain condition as in Theorem C.48.

Consider a composition $\chi_{k_1k_2} \circ \chi_{k_2k_3}$. It holds that

$$\begin{aligned}\chi_{k_1k_2} \circ \chi_{k_2k_3} &= \tilde{f}_{k_1}^{-1}\left(\tfrac{a_{k_1}}{a_{k_2}}\tfrac{1}{1+\delta_{k_2}}\tilde{f}_{k_2}\left(\tilde{f}_{k_2}^{-1}\left(\tfrac{a_{k_2}}{a_{k_3}}\tfrac{1}{1+\delta_{k_3}}\tilde{f}_{k_3}(s)\right)\right)\right)\\ &= \tilde{f}_{k_1}^{-1}\left(\tfrac{a_{k_1}}{a_{k_3}}\tfrac{1}{(1+\delta_{k_3})(1+\delta_{k_2})}\tilde{f}_{k_3}(s)\right).\end{aligned}$$

In the same way, we obtain the expression for the cycle condition in (C.26) (here we use that $k_1 = k_p$):

$$\chi_{k_1k_2} \circ \chi_{k_2k_3} \circ \ldots \circ \chi_{k_{p-1}k_p}(s) = \tilde{f}_{k_1}^{-1}\left(\tfrac{1}{\prod_{i=2}^{p}(1+\delta_{k_i})}\tilde{f}_{k_1}(s)\right) < s.$$

Thus, the small-gain condition (C.26) holds for all $\delta_i > 0$. Theorem C.48 ensures that the condition (3.74) holds, and in view of Theorem 3.33, there is a path of strict decay for the system (3.84). By Theorem 3.28, the network (3.84) is ISS.

If c_{ij} are globally bounded, i.e., there exists $M > 0$ such that for all $x \in \mathbb{R}^n_+$: $c_{ij}(x) \le M$ for all $i, j = 1, \ldots, n,\ i \neq j$, then the inequality (3.86) can be simplified. To this end, note that

$$\forall w_i > 0\, \exists \delta_j > 0,\, j = 1, \ldots, n: \sum_{j=1, j\neq i}^{n} c_{ij}(x) \tfrac{a_j}{a_i} \delta_j \le M\Bigl(\sum_{j=1, j\neq i}^{n} \tfrac{a_j}{a_i} \delta_j \Bigr) < w_i.$$

Using these estimates, we can instead of (3.86) require the following assumption:

$$\sum_{j=1,\, j\neq i}^{n} c_{ij}(x) a_j \le -c_{ii}(x) a_i - \epsilon_i, \quad x \in \mathbb{R}^n_+,$$

where $\epsilon_i = a_i(r_i + h_i + w_i)$. In matrix notation, with $a = (a_1, \ldots, a_n)^T$, $\epsilon = (\epsilon_1, \ldots, \epsilon_n)^T$, it takes the form

$$C(x)a < -\epsilon. \tag{3.87}$$

We summarize our investigations in the following proposition.

Proposition 3.36 *Consider a network as in* (3.83) *and assume that the* c_{ij} *are globally bounded for all* $i, j = 1, \ldots, n,\ i \neq j$. *If there exist* $a \in \mathbb{R}^n$, $\epsilon \in \mathbb{R}^n$, $a_i > 0$, $\epsilon_i > 0$, $i = 1, \ldots, n$ *such that the condition* $C(x)a < -\epsilon$ *holds for all* $x \in \mathbb{R}^n_+$, *then the whole network* (3.84) *is ISS.*

Remark 3.37 If C is a constant matrix, then the condition $Ca < -\epsilon$ is equivalent to $Ca < 0$ (with a, ϵ as in the proposition above).

Remark 3.38 Assume that C is a constant matrix with negative elements on the main diagonal and all other elements being nonnegative. Then C is *diagonal dominant* (see, e.g., [1]) if it holds $c_{ii} + \sum_{j\neq i} c_{ij} < 0$ for all $i = 1, \ldots, n$. In this case, one can easily prove with help of Gershgorin circle theorem (see [1, Fact 4.10.17]) that C is Hurwitz. Similarly, the previous condition can be replaced with another one: There are n numbers $a_i > 0$ such that $c_{ii}a_i + \sum_{j\neq i} c_{ij}a_j < 0$ for all $i = 1, \ldots, n$ (which is equivalent to the existence of a positive vector a such that $Ca < 0$). In this case the matrix is also Hurwitz (see, e.g., [20]).

3.5.3 *Interconnections of Linear Systems*

Consider the following interconnected system

$$\dot{x}_i = A_i x_i(t) + \sum_{j=1}^{n} B_{ij} x_j(t) + C_i u(t),\ i = 1, \ldots, n, \tag{3.88}$$

where $x_i(t) \in \mathbb{R}^{N_i}$, $A_i \in \mathbb{R}^{N_i \times N_i}$, $i = 1, \dots, n$, $B_{ij} \in \mathbb{R}^{N_i \times N_j}$, $i, j \in \{1, \dots, n\}$ and $u \in \mathcal{U}$. We assume that $B_{ii} = 0$, $i = 1, \dots, n$. Otherwise we can always substitute $\tilde{A}_i = A_i + B_{ii}$.

We define $N := N_1 + \dots + N_n$ and introduce the matrices $A := \operatorname{diag}(A_1, \dots, A_n) \in \mathbb{R}^{N \times N}$, $B := (B_{ij})_{i,j=1,\dots,n} \in \mathbb{R}^{N \times N}$ and $C := (C_1, \dots, C_n)^T$, a column matrix, consisting of C_i. Then (3.88) can be rewritten in the form

$$\dot{x}(t) = (A + B)x(t) + Cu(t). \tag{3.89}$$

We are going to apply Lyapunov technique developed in this section to the system (3.88). Assume that x_i-subsystem of (3.88) is ISS for any $i = 1, \dots, n$. This is the same as to say that there exists $P_i > 0$ so that

$$A_i^T P_i + P_i A_i = -I \tag{3.90}$$

holds. Here I denotes the identity matrix. As we know by Theorem 2.25,

$$V_i(x_i) = x_i^T P_i x_i \tag{3.91}$$

is an ISS Lyapunov function for the i-th subsystem of (3.88). Since P_i is a positive matrix, the following holds for certain $a_i > 0$:

$$a_i^2 |x_i|^2 \le V_i(x_i) \le \|P_i\| |x_i|^2, \quad x_i \in \mathbb{R}^{N_i}. \tag{3.92}$$

Differentiating V_i w.r.t. the i-th subsystem of (3.88), we obtain for all $x_i \in \mathbb{R}^{N_i}$

$$\begin{aligned}\dot{V}_i(x_i) &= \dot{x}_i^T P_i x_i + x_i^T P_i \dot{x}_i \\ &\le (A_i x_i)^T P_i x_i + x_i^T P_i A_i x_i + 2|x_i| \|P_i\| \Big(\sum_{i \ne j} \|B_{ij}\| |x_j| + \|C_i\| |u| \Big).\end{aligned}$$

Using (3.90), we obtain

$$\dot{V}_i(x_i) \le -|x_i|^2 + 2|x_i| \|P_i\| \Big(\sum_{i \ne j} \|B_{ij}\| |x_j| + \|C_i\| |u| \Big).$$

Now take $\varepsilon \in (0, 1)$ and let

$$|x_i| \ge \frac{2\|P_i\|}{1 - \varepsilon} \Big(\sum_{i \ne j} \|B_{ij}\| |x_j| + \|C_i\| |u| \Big). \tag{3.93}$$

Then we obtain for all $x_i \in \mathbb{R}^{N_i}$

$$\dot{V}_i(x_i) \le -\varepsilon |x_i|^2. \tag{3.94}$$

It is easy to see that the following condition together with (3.92) implies (3.93).

$$V_i(x_i) \geq \|P_i\|^3\Big(\frac{2}{1-\varepsilon}\Big)^2\Big(\sum_{i\neq j}\frac{\|B_{ij}\|}{a_j}\sqrt{V_j(x_j)} + \|C_i\||u|\Big)^2. \tag{3.95}$$

In turn, this condition is implied by

$$V_i(x_i) \geq \max\Big\{\Big(\sum_{i\neq j}\|P_i\|^{3/2}\frac{4}{1-\varepsilon}\frac{\|B_{ij}\|}{a_j}\sqrt{V_j(x_j)}\Big)^2, \|P_i\|^3\Big(\frac{4}{1-\varepsilon}\Big)^2\|C_i\|^2|u|^2\Big\}. \tag{3.96}$$

Define the gains by:

$$\chi_{ij}(s) = \Big(\frac{2\|P_i\|^{3/2}}{1-\varepsilon}\frac{\|B_{ij}\|}{a_j}\Big)\sqrt{s}, \tag{3.97}$$

for all $i \neq j, i = 1, \dots, n$.

We also choose MAFs

$$\mu_i(s) := (s_1 + s_2 + \cdots + s_n)^2.$$

The gains χ_{ij} together with MAFs (μ_i) generate the following gain operator

$$\Gamma_\mu(s) := \Big(\mu_i\big((\chi_{ij}(s_j))_{j=1}^n\big)\Big)_{i=1}^n.$$

If there is a regular path of nonstrict decay for this operator, we can conclude the ISS of the system (3.89).

3.6 Tightness of Small-Gain Conditions

The trajectory-based ISS small-gain theorem (Theorem 3.6) states that for a network of ISS systems, the strong small-gain condition (3.18) implies ISS of the interconnection. The following example due to [7, Example 18] shows that the ISS small-gain theorem (Theorem 3.6) doesn't hold in general if we assume that Γ satisfies merely small-gain condition (3.36) instead of the strong small-gain condition.

Example 3.39 Consider a coupled system

$$\dot{x}_1 = -x_1 + x_2(1 - e^{-x_2}) + u(t), \tag{3.98a}$$

$$\dot{x}_2 = -x_2 + x_1(1 - e^{-x_1}) + u(t), \tag{3.98b}$$

where $x_i(t) \in \mathbb{R}$ and $u(t) \in \mathbb{R}$.

The ISS estimate for the first subsystem (treating x_1 as the state and $x_2(\cdot)$ as an input) can be obtained as

$$\begin{aligned}\phi_1(t, x_1, (x_2, u)) &= e^{-t}x_1(0) + \int_0^t e^{-(t-s)}\big(x_2(s)(1 - e^{-x_2(s)}) + u(s)\big)\,ds \\ &\leq e^{-t}x_1(0) + \|x_2\|_\infty(1 - e^{-\|x_2\|_\infty}) + \|u\|_\infty.\end{aligned}$$

As the second subsystem has the same structure as the first one, we see that both subsystems are ISS with internal gains $\gamma_{12}(s) = \gamma_{21}(s) = s(1 - e^{-s}) < s$. Both gains are less than identity, and in particular, the small-gain condition (3.36) holds.

However, there is a solution $x_1 = x_2 = const$, given by

$$\dot{x}_1 = -x_2 e^{-x_2} + u, \quad \text{with } u = x_2 e^{-x_2},$$

and $x_1 = x_2$ can be chosen arbitrarily large with $u \to 0$ as $x_1 \to \infty$. In particular, (3.98) is not ISS.

The corresponding gain operator is analyzed in Example C.49, where it is shown that it does not satisfy the MBI property and the strong small-gain condition.

Although the strong small-gain condition is necessary for ISS of a network, it may still be that the network is 0-GAS. Whether this holds or not is an open question (as far as the author is concerned). However, the following result shows that the failure of the small-gain condition (3.36) may restrain the system from being 0-GAS.

Theorem 3.40 *Let a gain matrix* $\Gamma := (\gamma_{ij})$, $\gamma_{ij} \in \mathcal{K}_\infty$ *for all* $i, j = 1, \ldots, n$, $\gamma_{ii} = 0$ *be given. If the condition* (3.36) *is not satisfied, then there exists a continuous function* $f : \mathbb{R}^n \times \mathbb{R}^m \to \mathbb{R}^n$ *so that for all* $i = 1, \ldots, n$ *the estimates* (3.5) *hold for all* $t \geq 0$*, but the whole system* (3.1) *is not 0-GAS.*

Proof For arbitrary gain matrix Γ satisfying the assumptions of the theorem, we are going to construct a corresponding system satisfying (3.5), which is not 0-GAS.

Assume that Γ does not satisfy (3.36). According to Theorem C.48, there exists some cycle such that the condition (C.26) is violated. Let $s > 0$ be such that

$$\gamma_{12} \circ \gamma_{23} \circ \ldots \circ \gamma_{r-1r} \circ \gamma_{r1}(s) \geq s, \tag{3.99}$$

where $2 \leq r \leq n$ (violation of the small-gain condition on other cycles can be treated in the same way).

Due to continuity of γ_{ij}, there exist constants $\varepsilon_j \in [0, 1)$, $j = 2, \ldots, r$, such that for functions $\chi_{ij} := (1 - \varepsilon_j)\gamma_{ij}$ and the same s it holds that

$$\chi_{12} \circ \chi_{23} \circ \ldots \circ \chi_{r-1r} \circ \chi_{r1}(s) = s. \tag{3.100}$$

Let us enlarge the domain of definition of functions χ_{ij} to $\mathbb{R}$, defining $\chi_{ij}(-p) = -\chi_{ij}(p)$ for all $p > 0$, and all $i, j = 1, \dots n$, $i \neq j$.

Consider the following system:

$$\begin{cases} \dot{x}_1(t) = -x_1(t) + \chi_{12}\left(x_2(t)\right), \\ \dot{x}_2(t) = -x_2(t) + \chi_{23}\left(x_3(t)\right), \\ \dots \\ \dot{x}_r(t) = -x_r(t) + \chi_{r1}\left(x_1(t)\right), \\ \dot{x}_{r+1}(t) = -x_{r+1}(t), \\ \dots \\ \dot{x}_n(t) = -x_n(t). \end{cases} \tag{3.101}$$

For the first equation, using variation of constants formula, we obtain the following estimates:

$$\begin{aligned} |x_1(t)| &\leq |x_1(0)|\, e^{-t} + \left| \int_0^t e^{\tau - t} \chi_{12}(x_2(\tau)) d\tau \right| \\ &\leq |x_1(0)|\, e^{-t} + e^{-t} \int_0^t e^{\tau}\, |\chi_{12}(x_2(\tau))|\, d\tau \\ &= |x_1(0)|\, e^{-t} + e^{-t} \int_0^t e^{\tau} \chi_{12}(|x_2(\tau)|) d\tau \\ &\leq |x_1(0)|\, e^{-t} + e^{-t} \int_0^t e^{\tau} d\tau\, \chi_{12}\left(\|x_2\|_\infty\right) \\ &\leq |x_1(0)|\, e^{-t} + \chi_{12}\left(\|x_2\|_\infty\right). \end{aligned}$$

Similar estimates can be made for all equations. Thus, inequalities (3.5) are satisfied. Now we are going to prove that the system (3.101) is not 0-GAS.

Fixed points of the system (3.101) are the solutions $(x_1, \dots, x_n)$ of the following system:

$$\begin{cases} x_1 = \chi_{12}\left(x_2\right), \\ x_2 = \chi_{23}\left(x_3\right), \\ \dots \\ x_{r-1} = \chi_{r-1r}\left(x_r\right), \\ x_r = \chi_{r1}\left(x_1\right), \\ x_i = 0, \ i = r+1, \dots, n. \end{cases} \tag{3.102}$$

Substituting the i-th equation of (3.102) into the $(i-1)$-th, $i = r, \dots, 2$, we obtain the equivalent system:

$$\begin{cases} x_1 = \chi_{12} \circ \chi_{23} \circ \dots \circ \chi_{r1}(x_1), \\ x_2 = \chi_{23} \circ \chi_{34} \circ \dots \circ \chi_{r1}(x_1), \\ \dots \\ x_{r-1} = \chi_{r-1r} \circ \chi_{r1}(x_1), \\ x_r = \chi_{r1}(x_1), \\ x_i = 0, \; i = r+1, \dots, n. \end{cases} \tag{3.103}$$

For all solutions $s > 0$ of the equation (3.100), the first equation of the system (3.103) is satisfied with $x_1 = s$, and a point

$$(x_1, \dots, x_{r-1}, x_r, \dots, x_n)$$
$$= (s, \chi_{23} \circ \chi_{34} \circ \dots \circ \chi_{r-1r} \circ \chi_{r1}(s), \dots, \chi_{r-1r} \circ \chi_{r1}(s), \chi_{r1}(s), 0, \dots, 0)$$

is a fixed point for (3.101). Hence the system (3.101) has a nonzero fixed point, and therefore it is not 0-GAS. □

The counterpart of this result can also be proved for the Lyapunov-type small-gain theorem.

Theorem 3.41 *Let a matrix of Lyapunov gains* $\Gamma := (\gamma_{ij})$, $i, j = 1, \dots, n$, $\gamma_{ii} = 0$ *be given. Let there exist* $s > 0$, *such that for some cycle in* Γ *it holds that*

$$\gamma_{12} \circ \gamma_{23} \circ \dots \circ \gamma_{r-1r} \circ \gamma_{r1}(s) > s, \tag{3.104}$$

where $2 \le r \le n$ *(we can always renumber the nodes to obtain the cycle of the needed form). Then there exist a continuous function* $f : \mathbb{R}^n \times \mathbb{R}^m \to \mathbb{R}^n$ *and Lyapunov functions* V_i *for subsystems (in maximum formulation), so that for all* $i = 1, \dots, n$ *it holds (3.57) with* $\mu = \mu_\otimes$, *but the whole system (3.1) is not 0-GAS.*

Proof Take constants $\varepsilon_i \in (0, 1)$, $i = 2, \dots, r$, such that for the functions $\chi_{ij} := (1-\varepsilon_i)\gamma_{ij}$ and some $s > 0$ it holds

$$\chi_{12} \circ \chi_{23} \circ \dots \circ \chi_{r-1r} \circ \chi_{r1}(s) = s.$$

Consider the system (3.101). Take

$$V_i(x_i) = |x_i|, \quad x_i \in \mathbb{R}$$

as Lyapunov functions for i-th subsystem. For $i = 1, \dots, r-1$ if

$$V_i(x_i) \ge \gamma_{ii+1}(x_{i+1}) = \frac{1}{1-\varepsilon_i}\chi_{ii+1}(x_{i+1})$$

holds, then

$$\dot{V}_i(x_i) \leq -V_i(x_i) + (1 - \varepsilon_i)V_i(x_i) = -\varepsilon_i V_i(x_i).$$

Thus, for $i = 1, \ldots, r-1$ V_i is an ISS Lyapunov function for i-th subsystem. In fact, it holds also for all $i = 1, \ldots, n$. Moreover, Γ is a matrix of Lyapunov gains for the system (3.101). According to the proof of Theorem 3.40, (3.101) is not a 0-GAS system. □

We discuss the obtained results at the end of the next section.

3.7 Concluding Remarks

The study of interconnections plays a significant role in the mathematical systems theory, as it allows one to establish stability for a complex system based on the properties of its less complex components. In this context, small-gain theorems prove to be useful and general in analyzing feedback interconnections, which are ubiquitous in the control literature. An overview of classical small-gain theorems involving input–output gains of linear systems can be found in [10], [19, Chapter 5]. In [13, 22], the small-gain technique was extended to nonlinear feedback systems within the input–output context.

Nonlinear small-gain theorems for general feedback interconnections of two ISS systems were introduced in [15, 16]. Their generalization to networks composed of any finite number of ISS systems was reported in [5–7], with several variations summarized in [8]. Applications of small-gain theorems have been studied in [21].

Small-gain theorems have also been extended to interconnections of systems possessing the ISS-like properties. In [17, 26] small-gain theorems for networks of input-to-output stable (IOS) systems have been developed. Small-gain theorems for networks using vector Lyapunov functions have been proposed in [18]. In Sect. 4.12, we briefly analyze small-gain theorems for couplings of strongly integral ISS systems. References to the corresponding literature can be found in Sect. 4.13.

This chapter is mostly based on the works [5–7, 23, 24].

Small-gain theorems. Lemma 3.8 is precisely [7, Lemma 7]. UGS and AG small-gain Theorems 3.7, 3.9 and ISS small-gain Theorem 3.6 for Lipschitz continuous on bounded balls of $\mathbb{R}^n \times \mathbb{R}^m$ right-hand sides have been shown in [7, Theorems 8, 9, 10]. Extension of Theorem 3.6 to ODEs without bi-Lipschitz property of f is a special case of ISS small-gain theorems shown in [23] for networks of infinite-dimensional systems.

Monotone aggregation functions have been considered in [5, 27]. Lyapunov-based small-gain Theorem 3.28 was originally proved (in a somewhat different formulation) in [5, 6]. In the small-gain theorems shown in these papers, the existence of paths of *strict* decay was assumed, but in fact, the proofs in [5, 6] remain the same if nonstrict decay is assumed. To treat nondifferentiability of the composite Lyapunov function for the networks, in the proofs of [5, 6] the subgradient formalism is used. In contrast to that, we use the Dini derivatives, which makes it possible to extend the proof to the case of finite networks of infinite-dimensional systems.

While in most works on small-gain theory, couplings of globally ISS systems are considered, in [9] small-gain theorems for couplings of locally ISS systems have been proposed, which give LISS criteria for the composite system as well as estimates of the stability region of the network.

The Lyapunov function that we have constructed for the interconnection is obtained as a maximization of the scaled ISS Lyapunov function for subsystems. There are, however, also small-gain theorems where an ISS Lyapunov function for interconnection is obtained as a sum of scaled ISS Lyapunov functions for subsystems; see [4, 14, 21]. In this form, small-gain theorems have been applied to distributed control design [21], compositional construction of (in)finite-state abstractions [25, 28], cyber-security of networked systems [11], and networked control systems with asynchronous communication [12]. In Sect. 6.6, we will present a sum-form small-gain theorem for infinite networks with linear gains; see Theorem 6.42.

Cascades. Theorem 3.22 has been proved in [29] using a direct argument, without the use of a more general small-gain technique. The example presented in Sect. 3.4 is taken from [30, Sect. 4.1]. Later, in Sect. 4.9, we perform a much more detailed analysis of cascade interconnections.

3.8 Exercises

Exercise 3.1 Let $\gamma_1, \gamma_2 \in \mathcal{K}$. Show that

$$\gamma_1 \circ \gamma_2 < \mathrm{id} \quad \Leftrightarrow \quad \gamma_2 \circ \gamma_1 < \mathrm{id}\,.$$

What equivalent representations do exist for $\gamma_1 \circ \gamma_2 \circ \ldots \circ \gamma_n < \mathrm{id}$, where $\gamma_i \in \mathcal{K}_\infty$ for all $i = 1, \ldots, n$?

Exercise 3.2 Let $\Gamma = \begin{pmatrix} 0 & \gamma_{12} \\ \gamma_{21} & 0 \end{pmatrix} \in \mathbb{R}^{2\times 2}$ be a linear gain matrix. Prove directly that it holds that $\gamma_{12} \cdot \gamma_{21} < 1 \Leftrightarrow r(\Gamma) < 1$.

Exercise 3.3 Investigate under which conditions a cascade interconnection of two (nonlinear) exponentially ISS systems is exponentially ISS?

Exercise 3.4 Investigate robustness (in the ISS sense) of the controller (3.50) with respect to additive disturbances.

Exercise 3.5 Show that the system (3.98) is 0-GAS (although the strong small-gain condition doesn't hold).

References

1. Bernstein D (2009) Matrix mathematics: theory, facts, and formulas. Princeton University Press, Princeton
2. Brockett R (1983) Asymptotic stability and feedback stabilization. Diff Geom Control Theory 27(1):181–191
3. Dashkovskiy S, Görges M, Kosmykov M, Mironchenko A, Naujok L (2011) Modeling and stability analysis of autonomously controlled production networks. Logistics Res 3(2):145–157
4. Dashkovskiy S, Ito H, Wirth F (2011) On a small gain theorem for ISS networks in dissipative Lyapunov form. Euro J Control 17(4):357–365
5. Dashkovskiy S, Rüffer B, Wirth F (2010) Small gain theorems for large scale systems and construction of ISS Lyapunov functions. SIAM J Control Optim 48(6):4089–4118
6. Dashkovskiy S, Rüffer BS, Wirth FR (2006) On the construction of ISS Lyapunov functions for networks of ISS systems. In: Proceedings of 17th International Symposium on Mathematical Theory of Networks and Systems, pp 77–82
7. Dashkovskiy S, Rüffer BS, Wirth FR (2007) An ISS small gain theorem for general networks. Math Control Sign Syst 19(2):93–122
8. Dashkovskiy SN, Efimov DV, Sontag ED (2011) Input to state stability and allied system properties. Autom Remote Control 72(8):1579–1614
9. Dashkovskiy SN, Rüffer BS (2010) Local ISS of large-scale interconnections and estimates for stability regions. Syst Control Lett 59(241–247)
10. Desoer CA, Vidyasagar M (2009) Feedback systems: input-output properties. SIAM, Philadelphia, PA
11. Feng S, Tesi P, De Persis C (2017) Towards stabilization of distributed systems under denial-of-service. In: Proceedings of 56th IEEE Conference on Decision and Control, pp 5360–5365, Melbourne
12. Heemels WPMH, Borgers DP, van de Wouw N, Nešić D, Teel AR (2013) Stability analysis of nonlinear networked control systems with asynchronous communication: a small-gain approach. In: Proceedings of 52nd IEEE Conference on Decision and Control, pp 4631–4637, Florence
13. Hill DJ (1991) A generalization of the small-gain theorem for nonlinear feedback systems. Automatica 27(1043–1045)
14. Ito H, Jiang Z-P, Dashkovskiy S, Rüffer B (2013) Robust stability of networks of iISS systems: Construction of sum-type Lyapunov functions. IEEE Trans Autom Control 58(5):1192–1207
15. Jiang Z-P, Mareels IMY, Wang Y (1996) A Lyapunov formulation of the nonlinear small-gain theorem for interconnected ISS systems. Automatica 32(8):1211–1215
16. Jiang Z-P, Teel AR, Praly L (1994) Small-gain theorem for ISS systems and applications. Math Control Sign Syst 7(2):95–120
17. Jiang Z-P, Wang Y (2008) A generalization of the nonlinear small-gain theorem for large-scale complex systems. In: Proceedings of 7th World Congress on Intelligent Control and Automation, pp 1188–1193
18. Karafyllis I, Jiang Z-P (2011) A vector small-gain theorem for general non-linear control systems. IMA J Math Control Inform 28:309–344
19. Khalil HK (2002) Nonlinear systems. Prentice Hall, New York
20. Leenheer P, Angeli D, Sontag E (2007) Monotone chemical reaction networks. J Math Chem 41(3):295–314
21. Liu T, Jiang Z-P, Hill DJ (2014) Nonlinear control of dynamic networks. CRC Press
22. Mareels IMY, Hill DJ (1992) Monotone stability of nonlinear feedback systems. J Math Syst Estimation Control 2:275–291
23. Mironchenko A (2021) Small gain theorems for general networks of heterogeneous infinite-dimensional systems. SIAM J Control Optim 59(2):1393–1419
24. Mironchenko A, Kawan C, Glück J (2021) Nonlinear small-gain theorems for input-to-state stability of infinite interconnections. Math Control Sign Syst 33:573–615

25. Pola G, Pepe P, Di Benedetto MD (2016) Symbolic models for networks of control systems. IEEE Trans Autom Control 61(11):3663–3668
26. Rüffer B (2007) Monotone dynamical systems, graphs, and stability of large-scale interconnected systems. PhD thesis, Fachbereich 3 (Mathematik & Informatik) der Universität Bremen
27. Rüffer BS (2010) Monotone inequalities, dynamical systems, and paths in the positive orthant of Euclidean n-space. Positivity 14(2):257–283
28. Rungger M, Zamani M (2016) Compositional construction of approximate abstractions of interconnected control systems. IEEE Trans Control Network Syst 5(1):116–127
29. Sontag ED (1989) Smooth stabilization implies coprime factorization. IEEE Trans Autom Control 34(4):435–443
30. Sontag ED (2008) Input to state stability: basic concepts and results. In: Nonlinear and optimal control theory, Chap. 3. Springer, Heidelberg, pp 163–220

Chapter 4
Integral Input-to-State Stability

Despite all advantages of the ISS framework, for some practical systems, input-to-state stability is too restrictive. ISS excludes systems whose state stays bounded as long as the magnitude of applied inputs and initial states remains below a specific threshold but becomes unbounded when either the magnitude of an input or the magnitude of an initial state exceeds the threshold. Due to saturation and limitations in actuators and processing rates, such behavior is frequent in biochemical processes, population dynamics, and traffic flows. The idea of integral input-to-state stability (iISS) is to capture such nonlinearities.

We develop the integral ISS theory along the path that we have used for the ISS theory. In Sect. 4.2, we introduce the concept of an iISS Lyapunov function and show that its existence implies the iISS of a system. In Sect. 4.3, we derive the characterizations of the 0-GAS property for systems with inputs that are used in Sect. 4.4 to present a converse iISS Lyapunov theorem and characterize the iISS in terms of several dissipativity notions. In Sect. 4.6, we present the superposition theorems for the iISS property, as well as integral-to-integral characterizations of iISS, which are the counterparts of the corresponding results developed in Sect. 2.5. In Sect. 4.7, we discuss several important distinctions between ISS and iISS theories. In particular, cascades of iISS systems are not ISS, which makes "pure" iISS less useful for the study of interconnections. To remove this drawback, in Sect. 4.8, we introduce the strong iISS property. While it is still considerably less restrictive than ISS, strong iISS has appealing Lyapunov sufficient conditions, and at the same time, cascade interconnections of strongly iISS systems are strongly iISS. We close the chapter by presenting Lyapunov-based small-gain theorems for strongly iISS systems.

A. Mironchenko, *Input-to-State Stability*, Communications and Control Engineering,
https://doi.org/10.1007/978-3-031-14674-9_4

4.1 Basic Properties of Integrally ISS Systems

In this chapter, we continue to study systems with inputs of the form

$$\dot{x} = f(x, u), \tag{4.1}$$

where $x(t) \in \mathbb{R}^n$, an input u belongs to the set $\mathcal{U} := L^\infty(\mathbb{R}_+, \mathbb{R}^m)$, and $f : \mathbb{R}^n \times \mathbb{R}^m \to \mathbb{R}^n$.

In the whole chapter, we assume that Assumption 1.8 *holds.* This ensures that (4.1) is well posed (Theorem 1.16) and has the BIC property (Proposition 1.20).

Definition 4.1 System (4.1) is called *integral input-to-state stable (iISS)* if there exist $\theta \in \mathcal{K}_\infty$, $\mu \in \mathcal{K}$, and $\beta \in \mathcal{KL}$ such that for all $x \in \mathbb{R}^n$, $u \in \mathcal{U}$, and all $t \in [0, t_m(x, u))$ it holds that

$$|\phi(t, x, u)| \leq \beta(|x|, t) + \theta\left(\int_0^t \mu(|u(s)|)ds\right). \tag{4.2}$$

The ISS estimate provides an upper bound of the system's response to the maximal magnitude of the applied input. In contrast to that, *integral ISS gives an upper bound of the response of the system with respect to a kind of the energy fed into the system, described by the integral in the right-hand side of* (4.2).

The following concepts formalize the robustness with respect to disturbances of finite energy.

Definition 4.2 We say that a forward complete system (4.1) has

(i) *(uniform) Bounded energy-convergent state (BECS) property if there is* $\xi \in \mathcal{K}$ such that

$$\int_0^{+\infty} \xi(|u(s)|)ds < +\infty, \quad r \geq 0 \quad \Rightarrow \quad \lim_{t\to\infty} \sup_{|x|\leq r} |\phi(t, x, u)| = 0. \tag{4.3}$$

(ii) *Bounded energy-bounded state (BEBS) property if there is* $\xi \in \mathcal{K}$ *such that*

$$\int_0^{+\infty} \xi(|u(s)|)ds < +\infty, \quad r \geq 0 \quad \Rightarrow \quad \sup_{|x|\leq r,\ t\geq 0} |\phi(t, x, u)| < +\infty. \tag{4.4}$$

We start with a collection of basic properties of iISS systems.

Proposition 4.3 *Let Assumption 1.8 hold. Let (4.1) be an iISS control system. Then (4.1) is forward complete, has BRS, is 0-UGAS, LISS, and satisfies BECS and BEBS properties with* $\xi := \mu$.

Proof **FC.** Let $\theta \in \mathcal{K}_\infty$, $\mu \in \mathcal{K}$ and $\beta \in \mathcal{KL}$ be as in the definition of iISS. Pick any $x \in \mathbb{R}^n$, $u \in \mathcal{U}$. Assume that the maximal existence time $t_m(x, u)$ of a solution $\phi(\cdot, x, u)$ is finite. Integral ISS implies that for all $t \in [0, t_m(x, u))$ it holds that

$$|\phi(t, x, u)| \leq \beta(|x|, 0) + \theta\big(t_m(x, u)\mu(\|u\|_\infty)\big), \tag{4.5}$$

and thus the solution is bounded as $t \to t_m(x, u) - 0$, and in view of the BIC property of (4.1) (here we use Assumption 1.8 to invoke Proposition 1.20), the solution $\phi(\cdot, x, u)$ exists and is unique on a larger time interval than $[0, t_m(x, u))$, which contradicts to the definition of $t_m(x, u)$. Hence $t_m(x, u) = +\infty$.

BRS. The estimate (4.5) shows BRS.

0-UGAS. Clear.

LISS. As (4.1) is iISS, we have that (4.1) is 0-UAS and $f(0, 0) = 0$. Hence (4.1) is LISS thanks to LISS criterion (Theorem 2.38).

BECS. Let (4.1) be an iISS system with corresponding functions $\theta \in \mathcal{K}_\infty$ and $\mu \in \mathcal{K}$. It is forward complete as shown above. Pick any $u \in \mathcal{U}$ so that $\int_0^\infty \mu(|u(s)|)ds < \infty$.

To show the claim of the proposition, we need to show that for all $\varepsilon > 0$ and $r > 0$ there is a time $t^* = t^*(\varepsilon, r) > 0$ so that

$$|x| \leq r, \quad t \geq t^* \;\Rightarrow\; |\phi(t, x, u)| \leq \varepsilon. \tag{4.6}$$

Pick any $r, \varepsilon > 0$ and choose $t_1 > 0$ so that $\int_{t_1}^\infty \mu(|u(s)|)ds \leq \theta^{-1}(\frac{\varepsilon}{2})$.

As $u(\cdot + t_1) \in \mathcal{U}$, we use the cocycle property of (4.1) to obtain for all $t > 0$ and all $x \in B_r$ that

$$\begin{aligned}
|\phi(t + t_1, x, u)| &= \big|\phi\big(t, \phi(t_1, x, u), u(\cdot + t_1)\big)\big| \\
&\leq \beta\big(|\phi(t_1, x, u)|, t\big) + \theta\left(\int_0^t \mu(|u(s + t_1)|)ds\right) \\
&\leq \beta\left(\beta(|x|, t_1) + \theta\left(\int_0^{t_1} \mu(|u(s)|)ds\right), t\right) \\
&\quad + \theta\left(\int_{t_1}^{t+t_1} \mu(|u(s)|)ds\right) \\
&\leq \beta\left(\beta(r, 0) + \theta\left(\int_0^{+\infty} \mu(|u(s)|)ds\right), t\right) + \frac{\varepsilon}{2}.
\end{aligned}$$

Pick any t_2 in a way that

$$\beta\left(\beta(r,0)+\theta\left(\int_0^{+\infty}\mu(|u(s)|)ds\right),t_2\right)\leq\frac{\varepsilon}{2}.$$

This ensures that for all $t\geq 0$ and all $x\in\overline{B_r}$

$$\begin{aligned}|\phi(t+t_2+t_1,x,u)| &\leq \beta\left(\beta(r,0)+\theta\left(\int_0^{+\infty}\mu(|u(s)|)ds\right),t+t_2\right)+\frac{\varepsilon}{2}\\ &\leq \beta\left(\beta(r,0)+\theta\left(\int_0^{+\infty}\mu(|u(s)|)ds\right),t_2\right)+\frac{\varepsilon}{2}\\ &\leq \varepsilon.\end{aligned}$$

Since $\varepsilon>0$ and $r>0$ are arbitrary, the claim of the proposition follows.

BEBS. By BECS property, for any $u\in\mathcal{U}$ so that $\int_0^{\infty}\mu(|u(s)|)ds<\infty$ and for any $r>0$, there is time t^* such that (4.6) holds with $\varepsilon:=1$. By BRS, $\sup_{|x|\leq r,\ t\in[0,t^*]}|\phi(t,x,u)|<+\infty$. Combining these two estimates, we obtain BEBS property. □

Remark 4.4 For nonlinear systems, iISS is stronger than forward completeness together with 0-UGAS, as shown by an example in [5, Sect. V].

As we have argued in Sect. 2.7, for nonlinear systems, it is natural to require invariance of stability concepts under (nonlinear) changes of coordinates. In Proposition 2.67, we have shown that ISS remains valid under nonlinear changes of coordinates. A similar result holds for iISS.

Proposition 4.5 *Let (4.1) be iISS and let $T:\mathbb{R}^n\to\mathbb{R}^n$ be a homeomorphism with $T(0)=0$. Then (4.1) is iISS in new coordinates $y=T(x)$, that is, there exist $\tilde\beta\in\mathcal{KL}$ and $\tilde\theta,\tilde\mu\in\mathcal{K}$ so that for all $y\in\mathbb{R}^n$, $u\in\mathcal{U}$ and all $t\geq 0$ it holds that*

$$|\tilde\phi(t,y,u)|\leq\tilde\beta(|y|,t)+\tilde\theta\left(\int_0^t\tilde\mu(|u(s)|)ds\right),\tag{4.7}$$

where $\tilde\phi(t,y,u)$ is a solution map of (4.1) in the new coordinates y.

Proof Let $T:\mathbb{R}^n\to\mathbb{R}^n$ be any homeomorphism with $T(0)=0$. According to Lemma A.44, there are $\psi_1,\psi_2\in\mathcal{K}_\infty$ such that

$$\psi_1(|x|)\leq|T(x)|\leq\psi_2(|x|)\quad\forall x\in\mathbb{R}^n.$$

Let also (4.1) be iISS with corresponding $\beta\in\mathcal{KL}$ and $\theta,\mu\in\mathcal{K}$.

Pick any $y \in \mathbb{R}^n$, $u \in \mathcal{U}$ and let $x \in \mathbb{R}^n$ be so that $y = T(x)$. From (4.2), we have that

$$\begin{aligned}|\tilde{\phi}(t, y, u)| = \big|T(\phi(t, x, u))\big| &\le \psi_2\big(|\phi(t, x, u)|\big)\\ &\le \psi_2\Big(\beta(|x|, t) + \theta\Big(\int_0^t \mu(|u(s)|)ds\Big)\Big)\\ &\le \psi_2\big(2\beta(|x|, t)\big) + \psi_2\Big(2\theta\Big(\int_0^t \mu(|u(s)|)ds\Big)\Big)\\ &\le \psi_2\big(2\beta(\psi_1^{-1}(|y|), t)\big) + \psi_2\Big(2\theta\Big(\int_0^t \mu(|u(s)|)ds\Big)\Big),\end{aligned}$$

which shows that (4.7) holds with $\tilde{\beta}(r, t) := \psi_2\big(2\beta(\psi_1^{-1}(r), t)\big)$ and $\tilde{\theta}(r) := \psi_2\big(2\theta(r)\big)$, $r, t \ge 0$. □

Let us introduce an important subclass of integral ISS systems.

Definition 4.6 Let $q \ge p \ge 1$ be given and let (4.1) be well posed with inputs from $\mathcal{U} := L^p_{\text{loc}}(\mathbb{R}_+, \mathbb{R}^m)$ (i.e., the solutions of (4.1) exist locally and are unique for any initial state from $\mathbb{R}^n$ and for any input u from $\mathcal{U}$). The system (4.1) is called *L^q-input-to-state stable* (L^q-ISS), if it is forward complete and there exist $\beta \in \mathcal{KL}$ and $\gamma \in \mathcal{K}$ such that for all $x \in \mathbb{R}^n$, $u \in L^q(\mathbb{R}_+, \mathbb{R}^m)$ and $t \ge 0$ it holds that

$$|\phi(t, x, u)| \le \beta(|x|, t) + \gamma(\|u\|_{L^q(\mathbb{R}_+,\mathbb{R}^m)}). \tag{4.8}$$

If additionally there are $M, \lambda, G > 0$ so that (4.8) holds with $\beta(r, t) = Me^{-\lambda t}r$, and $\gamma(r) = Gr$, for all $t, r \ge 0$, then we call (4.1) *exponentially L^q-input-to-state stable.*

We close the section with a brief analysis of linear systems

$$\dot{x} = Ax + Bu \tag{4.9}$$

for which the solutions are well defined for all $u \in L^1_{\text{loc}}(\mathbb{R}_+, \mathbb{R}^m)$.

Proposition 4.7 *The following statements are equivalent:*

(i) (4.9) *is ISS.*
(ii) (4.9) *is 0-GAS.*
(iii) (4.9) *is L^p-ISS for all $p \in [1, +\infty]$.*
(iv) (4.9) *is L^p-ISS for some $p \in [1, +\infty)$.*
(v) (4.9) *is iISS.*

Proof (i) ⇔ (ii). Shown in Theorem 2.25.
(iii) ⇒ (iv) ⇒ (v) ⇒ (ii). Clear.

(ii) ⇒ (iii). As (4.9) is 0-GAS, A is Hurwitz, that is, there are $M, \lambda > 0$ such that $\|e^{At}\| \le Me^{-\lambda t}$ for all $t \ge 0$. For any t, x, u we have

$$\begin{aligned}|\phi(t,x,u)| &= \Big|e^{At}x + \int_0^t e^{A(t-r)}Bu(r)dr\Big|\\ &\le \|e^{At}\||x| + \int_0^t \|e^{A(t-r)}\|\|B\||u(r)|dr,\\ &\le Me^{-\lambda t}|x| + M\|B\|\int_0^t e^{-\lambda(t-r)}|u(r)|dr.\end{aligned}$$

Estimating $e^{-\lambda(t-r)} \le 1$, $r \le t$ we obtain that (4.9) is ISS w.r.t. $L^1(\mathbb{R}_+, \mathbb{R}^m)$.

To prove the claim for $p > 1$, pick any $q \ge 1$ so that $\frac{1}{p} + \frac{1}{q} = 1$. We continue the above estimates, using the Hölder's inequality

$$\begin{aligned}|\phi(t,x,u)| &\le Me^{-\lambda t}|x| + M\|B\|\int_0^t e^{-\frac{\lambda}{2}(t-r)}e^{-\frac{\lambda}{2}(t-r)}|u(r)|dr\\ &\le Me^{-\lambda t}|x| + M\|B\|\Big(\int_0^t e^{-\frac{q\lambda}{2}(t-r)}dr\Big)^{\frac{1}{q}}\cdot\Big(\int_0^t e^{-\frac{p\lambda}{2}(t-r)}|u(r)|^p dr\Big)^{\frac{1}{p}}\\ &\le Me^{-\lambda t}|x| + M\|B\|\Big(\frac{2}{q\lambda}\Big)^{\frac{1}{q}}\Big(\int_0^t |u(r)|^p dr\Big)^{\frac{1}{p}}\\ &= Me^{-\lambda t}|x| + w\|u\|_{L^p(\mathbb{R}_+,\mathbb{R}^m)},\end{aligned}$$

where $w := M\|B\|\big(\frac{2}{q\lambda}\big)^{\frac{1}{q}}$. This means that (4.9) is ISS w.r.t. $L^p(\mathbb{R}_+, \mathbb{R}^m)$ also for $p > 1$. As L^∞-ISS is the same as ISS, which is implied by 0-GAS, the claim follows. □

4.2 iISS Lyapunov Functions

As in ISS theory, Lyapunov functions play an essential role for the analysis of integral ISS of nonlinear systems. We restrict ourselves in this section to the analysis of continuously differentiable iISS Lyapunov functions.

Definition 4.8 A map $V \in C(\mathbb{R}^n, \mathbb{R}_+) \cap C^1(\mathbb{R}^n\backslash\{0\}, \mathbb{R}_+)$ is called an *iISS Lyapunov function* if there exist $\psi_1, \psi_2 \in \mathcal{K}_\infty$, $\sigma \in \mathcal{K}$, and $\alpha \in \mathcal{P}$ satisfying

$$\psi_1(|x|) \le V(x) \le \psi_2(|x|), \quad x \in \mathbb{R}^n, \tag{4.10}$$

and

$$\nabla V(x) \cdot f(x,u) \le -\alpha(|x|) + \sigma(|u|), \quad x \in \mathbb{R}^n \backslash \{0\},\ u \in \mathbb{R}^m. \tag{4.11}$$

Please note that the decay rate α of iISS Lyapunov functions is merely a positive definite function and not a $\mathcal{K}_\infty$-function as this is the case for ISS Lyapunov functions. This means that an iISS Lyapunov function may increase along the whole trajectory provided that the input is sufficiently large.

Proposition 4.9 *Let Assumption 1.8 hold. If there exists an iISS Lyapunov function for (4.1), then (4.1) is iISS and LISS.*

Proof **LISS** follows by Theorem 2.32 (note that for LISS it is enough for decay rate α to be a positive definite function, see Remark 2.31).

Integral ISS. As $\alpha \in \mathcal{P}$, by Proposition A.13, there are $\omega \in \mathcal{K}_\infty$ and $\sigma \in \mathcal{L}$ such that $\alpha(r) \ge \omega(r)\sigma(r)$ for all $r \ge 0$. Pick any $\tilde{\alpha} \in \mathcal{P}$ satisfying

$$\tilde{\alpha}(r) \ge \omega(\psi_2^{-1}(r)) \cdot \sigma(\psi_1^{-1}(r)), \quad r \in \mathbb{R}_+. \tag{4.12}$$

By (4.11) we have for any $x \in \mathbb{R}^n$, $w \in \mathbb{R}^m$ that

$$\begin{aligned} \nabla V(x) \cdot f(x,w) &\le -\omega(|x|)\sigma(|x|) + \sigma(|w|) \\ &\le -\omega \circ \psi_2^{-1}(V(x)) \cdot \sigma \circ \psi_1^{-1}(V(x)) + \sigma(|w|) \\ &\le -\tilde{\alpha}(V(x)) + \sigma(|w|). \end{aligned} \tag{4.13}$$

Now take any input $u \in \mathcal{U}$, any initial state $x \in \mathbb{R}^n$ and consider the corresponding solution $\phi(\cdot, x, u)$ that is defined on a maximal interval $[0, t_m(x,u))$.

As $V \in C^1(\mathbb{R}^n, \mathbb{R}_+)$ and ϕ is absolutely continuous w.r.t. t as a solution of an ODE, the map $y : s \mapsto V(\phi(s,x,u))$ is absolutely continuous and differentiable on $[0, t_m(x,u))$ almost everywhere (possibly outside of the points where $y(s) = 0$), and at all such points it holds by (4.13) that

$$\begin{aligned} \frac{d}{ds} V(\phi(s,x,u)) &= \nabla V(\phi(s,x,u)) \cdot f(\phi(s,x,u), u(s)) \\ &\le -\tilde{\alpha}\big(V(\phi(s,x,u))\big) + \sigma(|u(s)|). \end{aligned}$$

By Proposition A.37, there is $\beta \in \mathcal{KL}$ such that for all $x \in \mathbb{R}^n$ and $u \in \mathcal{U}$

$$V(\phi(t,x,u)) \le \beta(V(x), t) + \int_0^t 2\sigma(|u(s)|)ds, \quad t \in [0, t_m(x,u)),$$

and thus by (4.10)

$$\psi_1(|\phi(t,x,u)|) \le \beta(\psi_2(|x|),t) + \int_0^t 2\sigma(|u(s)|)ds, \quad t \in [0, t_m(x,u)).$$

Using that $\psi_1^{-1}(a+b) \le \psi_1^{-1}(2a) + \psi_1^{-1}(2b)$ for all $a, b \ge 0$, we finally obtain for $t \in [0, t_m(x,u))$ that

$$\begin{aligned} |\phi(t,x,u)| &\le \psi_1^{-1}\Big(\beta(\psi_2(|x|),t) + \int_0^t 2\sigma(|u(s)|)ds\Big) \\ &\le \psi_1^{-1}\Big(2\beta\big(\psi_2(|x|),t\big)\Big) + \psi_1^{-1}\Big(4\int_0^t \sigma(|u(s)|)ds\Big). \end{aligned} \tag{4.14}$$

This shows iISS. □

Remark 4.10 Usually, in the context of the iISS property, the iISS Lyapunov functions belonging to the class $C^1(\mathbb{R}^n, \mathbb{R}_+)$ are considered. We allow iISS Lyapunov functions to be only continuous at 0 as many standard Lyapunov functions like $x \mapsto |x|$ are not differentiable at 0.

Example 4.11 Consider a single integrator

$$\dot{x} = u. \tag{4.15}$$

Clearly, this system is globally stable if $u \equiv 0$, and a controller $u := -x + d$ makes (4.15) ISS with respect to the actuator disturbance d. However, in applications of control technology, physical inputs (like force, torque, thrust, stroke, etc.) are often limited in size. Therefore the (undisturbed) feedback $u(t) := -x(t)$ is not physical for large x. Instead, a realistic controller could be

$$u(t) := -\xi(x(t)) + d(t),$$

where $\xi : \mathbb{R} \to \mathbb{R}$ is a Lipschitz continuous strictly increasing globally bounded function with $\xi(0) = 0$. The resulting closed-loop system takes form

$$\dot{x} = -\xi(x) + d. \tag{4.16}$$

We show that such a controller makes the closed-loop system iISS by verification that $V : \mathbb{R} \to \mathbb{R}_+$, defined by

$$V(x) := \int_0^x \xi(s)ds, \quad x \in \mathbb{R},$$

is an iISS Lyapunov function for (4.16).

Clearly, $V(0) = 0$, $V \in C^1(\mathbb{R}, \mathbb{R}_+)$, and V is positive definite and radially unbounded function. By Proposition A.14 and Corollary A.11, V satisfies the sandwich estimates (4.10).

As ξ is continuous, $V \in C^1(\mathbb{R}, \mathbb{R}_+)$. Furthermore, for any $\varepsilon > 0$ it holds that

$$\begin{aligned}\dot{V}(x) &= \xi(x)\dot{x} = \xi(x)(-\xi(x) + d)\\ &= -(\xi(x))^2 + \xi(x)d\\ &\le -(\xi(x))^2 + \varepsilon(\xi(x))^2 + \frac{1}{\varepsilon}d^2,\end{aligned}$$

and choosing $\varepsilon \in (0, 1)$, we see that V is an iISS Lyapunov function with the decay rate $(1 - \varepsilon)(\xi(x))^2$.

As we have shown, existence of an iISS Lyapunov function with a decay rate $\alpha \in \mathcal{P}$ ensures integral ISS of (4.1). At the same time, if $\alpha \notin \mathcal{K}_\infty$, we cannot expect in general that (4.1) is ISS, unless the function σ in (4.11) satisfies certain additional properties.

Proposition 4.12 *If there exists an iISS Lyapunov function for* (4.1) *and*

$$\lim_{\tau\to\infty} \alpha(\tau) = \infty \ \text{or} \ \underline{\lim}_{\tau\to\infty} \alpha(\tau) \ge \lim_{\tau\to\infty} \sigma(\tau) \tag{4.17}$$

holds, then (4.1) *is ISS.*

Proof Suppose that V is an iISS Lyapunov function in a dissipative form, i.e., (4.11) holds with some $\alpha \in \mathcal{P}$ and $\sigma \in \mathcal{K}$. As $\sigma \in \mathcal{K}$, (4.17) implies that $\underline{\lim}_{\tau\to\infty} \alpha(\tau) > 0$. Lemma A.28 ensures that there is $\hat{\alpha} \in \mathcal{K}$ such that $\hat{\alpha}(s) \le \alpha(s)$ for all $s \in \mathbb{R}_+$ and $\lim_{\tau\to\infty} \hat{\alpha}(\tau) = \underline{\lim}_{\tau\to\infty} \alpha(\tau)$. Hence, from (4.11) we derive

$$\nabla V(x) \cdot f(x, u) \le -\hat{\alpha}(|x|) + \sigma(|u|), \quad x \in \mathbb{R}^n,\ u \in \mathbb{R}^m.$$

If $\underline{\lim}_{\tau\to\infty} \alpha(\tau) = \infty$, then $\hat{\alpha} \in \mathcal{K}_\infty$, and V is an ISS Lyapunov function in a dissipative form, and hence the system (4.1) is ISS.

Otherwise, let $\lim_{\tau\to\infty} \hat{\alpha}(\tau) = \underline{\lim}_{\tau\to\infty} \alpha(\tau) \ge \lim_{\tau\to\infty} \sigma(\tau)$. Take $\rho \in \mathcal{K}_\infty$ such that $\rho \circ \hat{\alpha}(s) > \hat{\alpha}(s)$ for all $s \in \mathbb{R}_+$ and $\lim_{s\to\infty} \rho \circ \hat{\alpha}(s) - \hat{\alpha}(s) = 0$.

Define $\gamma := \hat{\alpha}^{-1} \circ \rho \circ \sigma$. For each $s \in \mathbb{R}_+$ it holds that

$$\rho \circ \sigma(s) < \lim_{s\to\infty} \rho \circ \sigma(s) \le \lim_{s\to\infty} \rho \circ \hat{\alpha}(s) = \lim_{s\to\infty} \hat{\alpha}(s).$$

Thus, $\rho \circ \sigma(\mathbb{R}_+) \subset \hat{\alpha}(\mathbb{R}_+)$, and hence γ is well defined on $\mathbb{R}_+$. Overall, $\gamma \in \mathcal{K}$.

We have for all $x \in \mathbb{R}^n$ and all $u \in \mathbb{R}^m$ that

$$|x| \geq \gamma(|u|) \quad \Rightarrow \quad \nabla V(x) \cdot f(x,u) \leq -\hat{\alpha}(|x|) + \rho^{-1} \circ \hat{\alpha}(|x|) \\ = -(\mathrm{id} - \rho^{-1}) \circ \hat{\alpha}(|x|).$$

By construction of ρ it holds that $-\hat{\alpha}(|x|) + \rho^{-1} \circ \hat{\alpha}(|x|) < 0$ for all $x \in \mathbb{R}^n$: $x \neq 0$, which means that $x \mapsto -(\mathrm{id} - \rho^{-1}) \circ \hat{\alpha}(|x|)$ is a positive definite function, and V is an ISS Lyapunov function in implication form. □

4.3 Characterization of 0-GAS Property

Here we derive Lyapunov-based criteria for the 0-GAS property for systems with inputs whose right-hand side depends continuously on the input. These criteria are interesting in their own right but also they will be helpful in deriving Lyapunov-based criteria for integral ISS in Sect. 4.4.

Proposition 4.13 *Consider the system (4.1) satisfying Assumption 1.8. Let also $f(0,0) = 0$. The following statements are equivalent:*

(i) (4.1) is 0-GAS.

(ii) There is an infinitely differentiable strict Lyapunov function (as introduced in Definition B.24).

(iii) There are $V \in C^1(\mathbb{R}^n, \mathbb{R}_+)$, $\psi_1, \psi_2 \in \mathcal{K}_\infty$, $\alpha \in \mathcal{K}_\infty$, and $\delta, \lambda \in \mathcal{K}$ such that

$$\psi_1(|x|) \leq V(x) \leq \psi_2(|x|), \quad x \in \mathbb{R}^n, \tag{4.18}$$

and

$$\nabla V(x) \cdot f(x,u) \leq -\alpha(|x|) + \lambda(|x|)\delta(|u|), \quad x \in \mathbb{R}^n,\ u \in \mathbb{R}^m. \tag{4.19}$$

(iv) There are $V \in C^1(\mathbb{R}^n, \mathbb{R}_+)$, $\psi_1, \psi_2 \in \mathcal{K}_\infty$, $\rho \in \mathcal{P}$, and $\pi, \sigma \in \mathcal{K}$, such that (4.18) holds and

$$\nabla(\pi \circ V)(x) \cdot f(x,u) \leq -\rho(|x|) + \sigma(|u|), \quad x \in \mathbb{R}^n,\ u \in \mathbb{R}^m. \tag{4.20}$$

Proof (i) ⇔ (ii). Follows from Theorem B.32.

(ii) ⇒ (iii). In view of (ii), take $V \in C^1(\mathbb{R}^n, \mathbb{R}_+)$, $\psi_1, \psi_2 \in \mathcal{K}_\infty$, and $\alpha \in \mathcal{K}_\infty$, such that (4.18) holds and

$$\nabla V(x) \cdot f(x,0) \leq -\alpha(|x|), \quad x \in \mathbb{R}^n. \tag{4.21}$$

Define

$$\tilde{\zeta}(r,s) := \max_{|x|\le r,\ |u|\le s} |f(x,u) - f(x,0) - f(0,u)|. \tag{4.22}$$

As f is continuous in both arguments, and $f(0,0) = 0$, it holds that $\tilde{\zeta}$ is continuous, nondecreasing in each argument, and $\tilde{\zeta}(r,0) = \tilde{\zeta}(0,r) = 0$ for all $r \ge 0$. Hence, $\zeta : (r,s) \mapsto rs + \tilde{\zeta}(r,s)$ upperbounds $\tilde{\zeta}$ on $\mathbb{R}_+ \times \mathbb{R}_+$, and $\zeta(r,\cdot) \in \mathcal{K}$ for each $r > 0$ and $\zeta(\cdot,s) \in \mathcal{K}$ for each $s > 0$. By Lemma A.20, there is $\sigma \in \mathcal{K}_\infty$, such that

$$\zeta(r,s) \le \sigma(r)\sigma(s), \quad r,s \ge 0.$$

Now we have for all $x \in \mathbb{R}^n,\ u \in \mathbb{R}^m$ that

$$\begin{aligned}\nabla V(x) \cdot f(x,u) &= \nabla V(x) \cdot \big(f(x,u) - f(0,u) - f(x,0)\big)\\ &\quad + \nabla V(x) \cdot f(0,u) + \nabla V(x) \cdot f(x,0)\\ &\le |\nabla V(x)|\zeta(|x|,|u|) + |\nabla V(x)||f(0,u)| - \alpha(|x|). \end{aligned}\tag{4.23}$$

As $x = 0$ is a global minimum for V, and $V \in C^1(\mathbb{R}^n, \mathbb{R}_+)$, then $\nabla V(0) = 0$, and thus the function

$$\kappa(r) := r + \max_{|x|\le r} |\nabla V(x)|, \quad r \ge 0,$$

is a $\mathcal{K}_\infty$-function. As f is continuous, and $f(0,0) = 0$, then $|f(0,u)| \le \chi(|u|)$ for some $\chi \in \mathcal{K}$ and all $u \in \mathbb{R}^m$.

Continuing the estimates (4.23), we obtain

$$\begin{aligned}\nabla V(x) \cdot f(x,u) &\le \kappa(|x|)\sigma(|x|)\sigma(|u|) + \kappa(|x|)\chi(|u|) - \alpha(|x|)\\ &\le -\alpha(|x|) + \lambda(|x|)\delta(|u|),\end{aligned}$$

with $\lambda(r) := \kappa(r)\sigma(r) + \kappa(r)$ and $\delta(r) := \sigma(r) + \chi(r)$, $r \ge 0$.

(iii) $\Rightarrow$ (iv). Take V, λ, δ as in (iii), and consider

$$\pi(r) := \int_0^r \frac{1}{1+\chi(s)} ds, \quad r \ge 0, \tag{4.24}$$

where $\chi := \lambda \circ \psi_1^{-1}$. We have for all $x \in \mathbb{R}^n,\ u \in \mathbb{R}^m$ that

$$\nabla(\pi \circ V)(x) \cdot f(x,u) = \frac{\nabla V(x) \cdot f(x,u)}{1+\chi(V(x))}$$
$$\leq \frac{-\alpha(|x|)}{1+\chi(V(x))} + \frac{\lambda(|x|)\delta(|u|)}{1+\chi(V(x))}$$
$$\leq \frac{-\alpha(|x|)}{1+\chi(V(x))} + \delta(|u|),$$

where we have used that $\chi(V(x)) \geq \lambda(|x|)$.

(iv) $\Rightarrow$ (i). The estimate (4.20) implies by setting $u = 0$ that

$$\nabla V(x) \cdot f(x,0) < 0, \quad x \in \mathbb{R}^n,$$

which shows that V is a strict Lyapunov function for the undisturbed system (4.1), and hence (4.1) is 0-GAS. □

4.4 Lyapunov-Based Characterizations of iISS Property

Motivated by the dissipativity framework, briefly considered in Sect. 2.11, we characterize the integral ISS property in terms of dissipativity notions, introduced next.

Definition 4.14 Consider a (well-posed) system (4.1) and let $p \in \mathbb{N}$. A continuous map $h : \mathbb{R}^n \to \mathbb{R}^p$, with $h(0) = 0$ we will call an *output* of (4.1).

The map $y : (t, x, u) \mapsto h(\phi(t, x, u))$, defined for each initial state $x \in \mathbb{R}^n$, each input $u \in \mathcal{U}$, and each $t \in [0, t_m(x, u))$ is called the *output map* of (4.1).

Definition 4.15 The system (4.1) with output h is called:

(i) *Weakly zero-detectable* if, for each $x \in \mathbb{R}^n$, such that $t_m(x, 0) = +\infty$ and $y(\cdot, x, 0) \equiv 0$ it holds that $\phi(t, x, 0) \to 0$ as $t \to +\infty$.
(ii) *Dissipative* if there exist $V \in C^1(\mathbb{R}^n, \mathbb{R}_+)$, $\psi_1, \psi_2 \in \mathcal{K}_\infty$, $\alpha \in \mathcal{P}$, and $\sigma \in \mathcal{K}$, satisfying
$$\psi_1(|x|) \leq V(x) \leq \psi_2(|x|), \quad x \in \mathbb{R}^n, \tag{4.25}$$
and
$$\nabla V(x) \cdot f(x,u) \leq -\alpha(|h(x)|) + \sigma(|u|), \quad x \in \mathbb{R}^n, \ u \in \mathbb{R}^m. \tag{4.26}$$
(iii) *Smoothly dissipative*, if it is dissipative with an infinitely differentiable storage function V.

Definition 4.16 The system (4.1) is called *zero-output dissipative* if there exist $V \in C^1(\mathbb{R}^n, \mathbb{R}_+)$, $\psi_1, \psi_2 \in \mathcal{K}_\infty$, and $\sigma \in \mathcal{K}$, satisfying (4.25) and

$$\nabla V(x) \cdot f(x,u) \leq \sigma(|u|), \quad x \in \mathbb{R}^n, \ u \in \mathbb{R}^m. \tag{4.27}$$

If V above can be chosen to be infinitely differentiable (aka smooth), then (4.1) is called *zero-output smoothly dissipative*.

We characterize the iISS as follows:

Theorem 4.17 *Consider a system* (4.1) *and let* f *be Lipschitz continuous on bounded balls of* $\mathbb{R}^n \times \mathbb{R}^m$. *The following assertions are equivalent:*

(i) (4.1) *is iISS.*
(ii) (4.1) *admits an infinitely differentiable iISS Lyapunov function.*
(iii) *There is an output that makes* (4.1) *smoothly dissipative and weakly zero-detectable.*
(iv) (4.1) *is 0-GAS and zero-output smoothly dissipative.*

Proof (i) $\Rightarrow$ (ii). We omit the proof of this implication and refer to the paper [5].

(ii) $\Rightarrow$ (i). Follows by Proposition 4.9.

(ii) $\Rightarrow$ (iii). Let $V \in C^\infty(\mathbb{R}^n, \mathbb{R}_+)$ be an iISS Lyapunov function for (4.1), such that for certain $\sigma \in \mathcal{K}$ and $\alpha \in \mathcal{P}$ the estimate (4.11) holds. Choose the output $h(x) := \alpha(|x|)$, $x \in \mathbb{R}^n$. With this output the system (4.11) is smoothly dissipative, and the estimate (4.26) clearly holds with $\alpha := \mathrm{id}$.

Furthermore, if $y(t, x, 0) = 0$ for all $t \geq 0$, then $\phi(t, x, 0) = 0$ for all $t \geq 0$, since α is a positive definite function. Hence (4.1) is weakly zero-detectable.

(iii) $\Rightarrow$ (iv). As (4.1) is smoothly dissipative, it is also zero-output smoothly dissipative. Furthermore, we can take V from the definition of dissipativity as a Lyapunov function for the undisturbed system. Together with weak zero-detectability, we can use La Salle principle to show 0-GAS of (4.1).

(iv) $\Rightarrow$ (ii). As (4.1) is 0-GAS, by Proposition 4.13(iv), there are $V \in C^1(\mathbb{R}^n, \mathbb{R}_+)$, $\psi_1, \psi_2 \in \mathcal{K}_\infty$, $\rho \in \mathcal{P}$, and $\pi, \sigma \in \mathcal{K}$, such that (4.18) holds and

$$\nabla(\pi \circ V)(x) \cdot f(x, u) \leq -\rho(|x|) + \sigma(|u|), \quad x \in \mathbb{R}^n,\ u \in \mathbb{R}^m.$$

Furthermore, as (4.1) is zero-output smoothly dissipative, there are $V_1 \in C^1(\mathbb{R}^n, \mathbb{R}_+)$, $\tilde{\psi}_1, \tilde{\psi}_2 \in \mathcal{K}_\infty$, $\tilde{\sigma} \in \mathcal{K}$, such that

$$\tilde{\psi}_1(|x|) \leq V_1(x) \leq \tilde{\psi}_2(|x|), \quad x \in \mathbb{R}^n,$$

and

$$\nabla V_1(x) \cdot f(x, u) \leq \tilde{\sigma}(|u|), \quad x \in \mathbb{R}^n,\ u \in \mathbb{R}^m.$$

Now $W := V_1 + \pi \circ V$ satisfies

$$\tilde{\psi}_1(|x|) \leq W(x) \leq \tilde{\psi}_2(|x|) + \pi \circ \psi_2(|x|), \quad x \in \mathbb{R}^n,$$

and

$$\nabla W(x) \cdot f(x, u) \leq -\rho(|x|) + \sigma(|u|) + \tilde{\sigma}(|u|), \quad x \in \mathbb{R}^n,\ u \in \mathbb{R}^m,$$

which shows that W is an iISS Lyapunov function for (4.1).

As a corollary of Theorem 4.17, one can show also the following characterizations of the iISS property. □

Proposition 4.18 *Consider a system* (4.1) *with f being Lipschitz continuous on bounded balls of $\mathbb{R}^n \times \mathbb{R}^m$. The following statements are equivalent:*

(i) The system (4.1) *is iISS.*

(ii) There exist a positive definite proper C^∞ function $W : \mathbb{R}^n \to \mathbb{R}$, ρ, $\gamma \in \mathcal{K}_\infty$, and $b \in \mathcal{P}$, satisfying $\int_0^{+\infty} \frac{1}{1+b(r)} dr = +\infty$ such that for all $x \in \mathbb{R}^n$, $u \in \mathbb{R}^m$ we have

$$|x| \geq \rho(|d|) \quad \Rightarrow \quad \nabla W(x) f(x, u) \leq \gamma(|d|) b(W(x)). \tag{4.28}$$

(iii) There exist a positive definite proper C^∞ function $W : \mathbb{R}^n \to \mathbb{R}$, ρ, $\gamma \in \mathcal{K}_\infty$, and α, $b \in \mathcal{P}$, satisfying $\int_0^{+\infty} \frac{1}{1+b(r)} dr = +\infty$ such that for all $x \in \mathbb{R}^n$, $u \in \mathbb{R}^m$ we have

$$|x| \geq \rho(|d|) \quad \Rightarrow \quad \nabla W(x) f(x, u) \leq -\alpha(|x|) + \gamma(|d|) b(W(x)). \tag{4.29}$$

Proof The proof is left as Exercise 4.9. □

4.5 Example: A Robotic Manipulator

Consider a manipulator with one rotational and one linear actuator; see Fig. 4.1. We model the arm as a segment with mass M and length L and the hand as a point particle with mass m. Denote the *linear extension* of the manipulator by r and its *angular position* by θ. The second-order equations of motion have the form

$$\Big(mr^2 + \frac{ML^2}{3}\Big)\ddot{\theta} + 2mr\dot{r}\dot{\theta} = \tau, \tag{4.30a}$$

$$m\ddot{r} - mr\dot{\theta}^2 = F. \tag{4.30b}$$

Here we consider the *torque* τ and the *force* F as control inputs. We denote $q := (\theta, r)$, and note that the equations (4.30) can be written as a 4-dimensional system of the form (4.1) with the state variables $(q, \dot{q})$. We are going to use the controls τ and F to track the desired position $q_d = (\theta_d, r_d)$.

Assuming that we can measure the whole state $(q, \dot{q}) = (\theta, r, \dot{\theta}, \dot{r})$, we use the following controller design

$$\tau := -k_{11}(\dot{\theta} + d_{11}) - k_{12}(\theta + d_{12} - \theta_d) = -k_{11}\dot{\theta} - k_{12}\theta + u_1, \tag{4.31a}$$

$$F := -k_{21}(\dot{r} + d_{21}) - k_{22}(r + d_{22} - r_d) = -k_{21}\dot{r} - k_{22}r + u_2, \tag{4.31b}$$

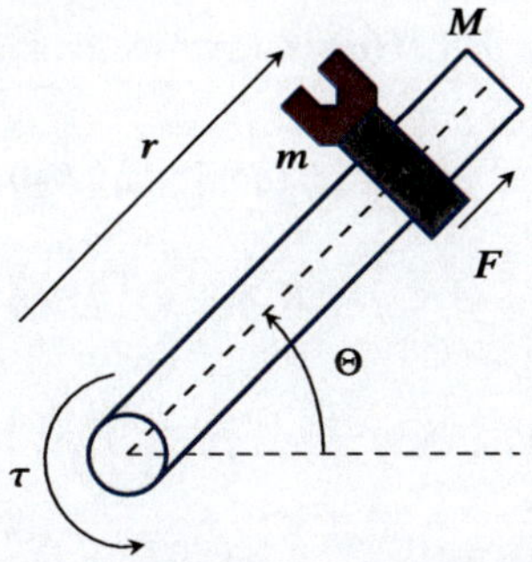

Fig. 4.1 Manipulator, as modeled by (4.30). Here r denotes the linear extension, θ is the angular position, τ is a torque, and F is a force. The figure is reprinted/adapted by permission from Springer Nature Customer Service Centre GmbH: Springer, Nonlinear and Optimal Control Theory, Chapter title: 'Input to State Stability: Basic Concepts and Results' by E. Sontag [30, Fig. 15] ©2008

where

$$u_1 = -k_{11}d_{11} - k_{12}(d_{12} - \theta_d), \tag{4.32a}$$

$$u_2 = -k_{21}d_{21} - k_{22}(d_{22} - r_d) \tag{4.32b}$$

are the "total" external inputs acting on the system.

With controllers (4.31) the closed-loop system takes the form

$$(mr^2 + M\frac{L^2}{3})\ddot{\theta} + 2mr\dot{r}\dot{\theta} = -k_{11}\dot{\theta} - k_{12}\theta + u_1, \tag{4.33a}$$

$$m\ddot{r} - mr\dot{\theta}^2 = -k_{21}\dot{r} - k_{22}r + u_2. \tag{4.33b}$$

Let us show integral ISS of this system.

Define for all $(q, \dot{q})$ the *inertia matrix* $H(q)$ and the *Coriolis torques* $C(q, \dot{q})$ as

$$H(q) := \begin{pmatrix} mr^2 + M\frac{L^2}{3} & 0 \\ 0 & m \end{pmatrix}, \quad C(q, \dot{q}) := mr \begin{pmatrix} \dot{r} & \dot{\theta} \\ -\dot{\theta} & 0 \end{pmatrix}. \tag{4.34}$$

With this notation, (4.33) can be represented in the matrix form as

$$H(q)\ddot{q} + C(q, \dot{q})\dot{q} = -K_1\dot{q} - K_2q + u, \tag{4.35}$$

with $u := (u_1, u_2)^T$, $K_1 = \text{diag}(k_{11}, k_{21})$, and $K_2 = \text{diag}(k_{12}, k_{22})$.

To analyze the system (4.35), consider the full mechanical energy of the system

$$V(q, \dot{q}) := \frac{1}{2}\dot{q}^T H(q)\dot{q} + \frac{1}{2}q^T K_2 q. \tag{4.36}$$

Differentiating V, and using that $H(q)$ is a symmetric matrix, we obtain

$$\begin{aligned}
\frac{d}{dt}V(q,\dot q) &= \ddot q^T H(q)\dot q + \frac{1}{2}\dot q^T \frac{d}{dt}(H(q))\dot q + q^T K_2 \dot q \\
&= (-C(q,\dot q)\dot q - K_1\dot q - K_2 q + u)^T \dot q + \dot q^T \begin{pmatrix} mr\dot r & 0 \\ 0 & 0 \end{pmatrix} \dot q + q^T K_2 \dot q \\
&= (-C(q,\dot q)\dot q - K_1\dot q + u)^T \dot q + \begin{pmatrix} mr\dot r\dot\theta \\ 0 \end{pmatrix}^T \dot q \\
&= (-K_1\dot q + u)^T \dot q + \left(-mr\begin{pmatrix} \dot r & \dot\theta \\ -\dot\theta & 0 \end{pmatrix}\dot q\right)^T \dot q + \begin{pmatrix} mr\dot r\dot\theta \\ 0 \end{pmatrix}^T \dot q \\
&= (-K_1\dot q + u)^T \dot q + \left(-mr\begin{pmatrix} 2\dot r\dot\theta \\ -\dot\theta^2 \end{pmatrix} + mr\begin{pmatrix} \dot r\dot\theta \\ 0 \end{pmatrix}\right)^T \dot q \\
&= (-K_1\dot q + u)^T \dot q + \left(-mr\begin{pmatrix} \dot r\dot\theta \\ -\dot\theta^2 \end{pmatrix}\right)^T \dot q \\
&= (-K_1\dot q + u)^T \dot q \\
&\le -c_1|\dot q|^2 + c_2|u|^2,
\end{aligned} \tag{4.37}$$

for certain $c_1, c_2 > 0$. Here the last estimate is due to Young's inequality.

The estimate (4.37) shows that (4.35) is dissipative with the output $h(q,\dot q) = \dot q$. Assume that $u \equiv 0$ and $h(q(t),\dot q(t)) = \dot q(t) \equiv 0$ on the interval of existence of a solution. Looking onto equations (4.35), we immediately see that this implies that $q \equiv 0$, and thus the system (4.35) is weakly zero-detectable with the output h.

It is not hard to see that the system (4.35) can be transformed into the form (4.1) with f being Lipschitz continuous on bounded balls of $\mathbb{R}^n \times \mathbb{R}^m$. Hence Theorem 4.17 ensures iISS of (4.35).

4.6 Integral ISS Superposition Theorems

Having characterized integral ISS in terms of iISS Lyapunov functions, we proceed to the characterization of iISS in terms of weaker stability properties (superposition theorems).

In Sect. 4.1, we have introduced the bounded energy-convergent state and bounded energy-bounded state properties. Let us introduce somewhat weaker counterparts of these properties:

Definition 4.19 We say that a forward complete system (4.1) has:

(i) *(Uniform) bounded energy-weakly convergent state (BEWCS)* property if there is $\xi \in \mathcal{K}$ such that

$$\int_0^{+\infty} \xi(|u(s)|)ds < +\infty \quad \Rightarrow \quad \underline{\lim}_{t\to\infty} |\phi(t,x,u)| = 0. \tag{4.38}$$

This means that for each $x \in \mathbb{R}^n$ and each $u \in \mathcal{U}$ such that $\int_0^{+\infty} \xi(|u(s)|)ds < +\infty$ it holds that $t_m(x, u) = +\infty$ and $\underline{\lim}_{t\to\infty} |\phi(t, x, u)| = 0$.

(ii) *Bounded energy-frequently bounded state (BEFBS)* property if there is $\xi \in \mathcal{K}$ such that

$$\int_0^{+\infty} \xi(|u(s)|)ds < +\infty \quad \Rightarrow \quad \underline{\lim}_{t\to\infty} |\phi(t, x, u)| < +\infty. \tag{4.39}$$

More precisely, for each $x \in \mathbb{R}^n$ and each $u \in \mathcal{U}$ such that $\int_0^{+\infty} \xi(|u(s)|)ds < +\infty$ it holds that $t_m(x, u) = +\infty$ and $\underline{\lim}_{t\to\infty} |\phi(t, x, u)| < +\infty$.

(iii) *Uniformly bounded energy-bounded state (UBEBS)* property if there exist $\theta \in \mathcal{K}_\infty$, $\mu \in \mathcal{K}$, $\sigma \in \mathcal{K}_\infty$ and $c > 0$ such that for all $x \in \mathbb{R}^n$, $u \in \mathcal{U}$, and all $t \in [0, t_m(x, u))$ it holds that

$$|\phi(t, x, u)| \le \sigma(|x|) + \theta\left(\int_0^t \mu(|u(s)|)ds\right) + c. \tag{4.40}$$

Now we can state a superposition principle for the iISS property.

Theorem 4.20 *Consider the system* (4.1) *and let f be Lipschitz continuous on bounded balls of $\mathbb{R}^n \times \mathbb{R}^m$. The following assertions are equivalent:*

(i) (4.1) *is iISS.*
(ii) (4.1) *is BEWCS and 0-ULS.*
(iii) (4.1) *is BEFBS and 0-GAS.*
(iv) (4.1) *is UBEBS and 0-GAS.*
(v) There exist $\theta \in \mathcal{K}_\infty$, $\mu, \gamma \in \mathcal{K}$ and $\beta \in \mathcal{KL}$ such that for all $x \in \mathbb{R}^n$, $u \in \mathcal{U}$, and all $t \in [0, t_m(x, u))$ it holds that

$$|\phi(t, x, u)| \le \beta(|x|, t) + \theta\left(\int_0^t \mu(|u(s)|)ds\right) + \gamma\big(\sup_{0\le s\le t} |u(s)|\big). \tag{4.41}$$

Proof The equivalences between items (i)–(iii) are due to [3, Theorem 3]. Integral ISS clearly implies the properties in items (iv), (v).

(iv) $\Rightarrow$ (iii). Let (4.1) be UBEBS. Pick any $x \in \mathbb{R}^n$ and any $u \in \mathcal{U}$ such that $\int_0^{+\infty} \mu(|u(s)|)ds < \infty$. The BIC property that is ensured by Proposition 1.20 together with (4.40) shows that for such inputs $t_m(x, u) = +\infty$. BEFBS property now follows from (4.40).

(v) $\Rightarrow$ (iii). By (v), (4.1) is 0-GAS. Using BIC property, we see that for any $x \in \mathbb{R}^n$ and any $u \in \mathcal{U}$ such that $\int_0^{+\infty} \mu(|u(s)|)ds < \infty$, the corresponding solution exists on $\mathbb{R}_+$. As $u \in \mathcal{U}$, the last term in (4.41) is globally bounded, and thus (4.41) ensures BEFBS property. □

Assume that f is Lipschitz continuous on bounded balls of $\mathbb{R}^n \times \mathbb{R}^m$. In Theorem 2.76, we have shown that (4.1) is ISS if and only if (4.1) is forward complete, and there are $\alpha \in \mathcal{K}$ and $\psi \in \mathcal{K}_\infty$, $\sigma \in \mathcal{K}_\infty$ it holds that

$$\int_0^t \alpha(|\phi(s,x,u)|)ds \leq \psi(|x|) + \int_0^t \sigma(|u(s)|)ds, \quad (x,u,t) \in \mathbb{R}^n \times \mathcal{U} \times \mathbb{R}_+.$$

Next, we state the characterization of integral ISS of this type.

Theorem 4.21 *Consider the system (4.1) with f being Lipschitz continuous on bounded balls of $\mathbb{R}^n \times \mathbb{R}^m$. Then (4.1) is iISS if and only if (4.1) is forward complete and there exists $\alpha, \psi, \gamma, \sigma \in \mathcal{K}_\infty$ such that solutions satisfy*

$$\int_0^t \alpha(|\phi(s,x,u)|)ds \leq \psi(|x|) + \gamma\left(\int_0^t \sigma(|u(s)|)ds\right), \quad (x,u,t) \in \mathbb{R}^n \times \mathcal{U} \times \mathbb{R}_+. \tag{4.42}$$

Proof Please consult [4, Theorem 1] for the proof. □

4.7 Integral ISS Versus ISS

As we have seen, integral ISS is a property that is weaker than ISS (and hence shared by a larger class of systems) and allows for several useful characterizations. Next, we illustrate by means of examples that some essential properties of ISS systems do not hold for iISS systems in general.

Our first example shows that trajectories of iISS systems subject to the uniformly bounded inputs that are vanishing at infinity may be unbounded.

Example 4.22 Consider a scalar system

$$\dot{x} = -\frac{x}{1+x^2} + u. \tag{4.43}$$

To show that this system is iISS, consider a function $V : \mathbb{R} \to \mathbb{R}_+$, $V(x) = |x|$. We have that

$$\frac{d}{dt}V(x) \leq -\frac{|x|}{1+x^2} + |u|,$$

and thus V is an iISS Lyapunov function, and hence (4.43) is iISS.

At the same time, consider a feedback $u(t) := \frac{2x(t)}{1+x^2(t)}$. Substituting it into (4.43), we obtain a closed-loop system

$$\dot{x} = \frac{x}{1+x^2}. \tag{4.44}$$

For any initial condition the trajectory diverges to ∞, and at the same time $u(t) \to 0$ as $t \to \infty$. This violates both *convergent input-uniformly convergent state property (CIUCS)* (compare to Proposition 2.7) and *bounded input-bounded state (BIBS)* property (cf. Lemma 2.42).

Another downside of iISS is that even the simplest cascade interconnections of two iISS systems are not necessarily iISS, in contrast to cascade interconnections of ISS systems.

***Example 4.23** (Cascade of 2 iISS systems that is not iISS)* Consider a cascade interconnection

$$\dot{x} = -\mathrm{sat}(x) + xz, \tag{4.45a}$$
$$\dot{z} = -z^3. \tag{4.45b}$$

Here $\mathrm{sat}(x) := \mathrm{sgn}(x)\min\{1, |x|\}$, $x \in \mathbb{R}$.

Define $V(x) := \frac{1}{2}\ln(1+x^2)$, $x \in \mathbb{R}$. Then

$$\begin{aligned}\dot{V}(x) &= \frac{x}{1+x^2}\dot{x} = \frac{x}{1+x^2}\big(-\mathrm{sat}(x) + xz\big)\\ &= -\mathrm{sat}(x)\frac{x}{1+x^2} + \frac{x^2}{1+x^2}z\\ &\le -\mathrm{sat}(x)\frac{x}{1+x^2} + |z|.\end{aligned}$$

This shows that V is an iISS Lyapunov function for (4.45a), and hence (4.45a) is iISS. Clearly, z-subsystem is 0-UGAS, iISS, and ISS.

Let us show that the interconnection (4.45) is unstable. Let $z(0) = 1$. The corresponding solution of z-subsystem is $z(t) = \frac{1}{\sqrt{1+2t}}$. Assuming that $x(0) > 1$, we have that as long as the solution of the x-subsystem satisfies $x(t) > 1$, it holds that

$$\dot{x}(t) = -1 + \frac{1}{\sqrt{1+2t}}x(t),$$

and thus

$$x(t) = e^{\sqrt{1+2t}-1}\Big(x(0) - \int_0^t e^{1-\sqrt{1+2\tau}}d\tau\Big). \tag{4.46}$$

Using the change of variables $s = -1 + \sqrt{1+2\tau}$, we obtain

$$\int_0^t e^{1-\sqrt{1+2\tau}} d\tau \le \int_0^\infty e^{1-\sqrt{1+2\tau}} d\tau = \int_0^\infty e^{-s}(s+1)ds = 2,$$

and thus for $x(0) \ge 3$ from (4.46) we obtain that $x(t) \ge e^{\sqrt{1+2t}-1}$, which means that $x(t) \ge 1$ for all $t \ge 0$ and $x(t) \to \infty$ as $t \to \infty$.

4.8 Strong Integral Input-to-State Stability

This section introduces the strong iISS property, which offers a halfway between the strength of input-to-state stability and the generality of integral input-to-state stability. Based on this concept, we will develop powerful tools for the stability analysis of interconnections of such systems.

We start with

Definition 4.24 System (4.1) is called input-to-state stable with respect to small inputs, if there exist $\beta \in \mathcal{KL}$, $\gamma \in \mathcal{K}_\infty$, and $R > 0$ such that for all $x \in \mathbb{R}^n$, $u \in \mathcal{U}$: $\|u\|_\infty \le R$ and $t \in [0, t_m(x, u))$ it holds that

$$|\phi(t, x, u)| \le \beta(|x|, t) + \gamma(\|u\|_\infty). \tag{4.47}$$

In this case the BIC property implies that $t_m(x, u) = +\infty$ for all such solutions.

Note that "nice" behavior for small inputs does not guarantee that the system is well behaved for large u. In fact, it does not have even to be forward complete.

Example 4.25 Consider the scalar system

$$\dot{x} = -x + u + \xi(u)x^2,$$

where $\xi : \mathbb{R} \to \mathbb{R}$ denotes any locally Lipschitz function satisfying $\xi(u) = 0$ for all $u \in [-1, 1]$ and $|\xi(u)| \ge 1$ whenever $|u| \ge 2$. Clearly, this system is ISS w.r.t. inputs in $B_{1,\mathcal{U}}$. However, for any constant input satisfying $u \ge 2$ pointwise, the solution satisfies the estimate

$$\dot{x}(t) \ge -x(t) + x^2(t),$$

which has a finite escape time for any initial condition greater than 1.

Obstructions demonstrated in Examples 4.22, 4.23 motivate the following strengthening of the iISS property:

Definition 4.26 System (4.1) is called *strongly integral input-to-state stable (strongly iISS)*, if (4.1) is iISS and ISS with respect to small inputs.

The system (4.43) studied in Example 4.22 is not ISS with respect to small inputs, and hence it is not strongly iISS. Strong iISS can be verified using the following Lyapunov sufficient condition.

Proposition 4.27 *If there exists an iISS Lyapunov function for* (4.1) *with the decay rate* $\alpha \in \mathcal{P}$*, satisfying* $\underline{\lim}_{s\to\infty} \alpha(s) > 0$*, then* (4.1) *is strongly iISS.*

Proof By Proposition 4.9, the system (4.1) is iISS. As $\alpha \in \mathcal{P}$ satisfies $\underline{\lim}_{s\to\infty} \alpha(s) > 0$, then by Lemma A.28 there is $\rho \in \mathcal{K}$ such that $\rho(s) \le \alpha(s)$ for all $s \ge 0$. From now on, we assume that the dissipative inequality in the iISS definition is satisfied with ρ instead of α.

Let $\sigma(|u|) \le \frac{1}{2}\rho(|x|)$, i.e., $|x| \ge \rho^{-1}(2\sigma(|u|)) =: \chi(|u|)$, where we assume that $\|u\|_\infty \le R$, and R is chosen such that $2\sigma(R) < \rho(+\infty)$. For such x, u it holds that

$$\nabla V(x) \cdot f(x, u) \le -\frac{1}{2}\rho(|x|),$$

which shows that V is an ISS Lyapunov function for (4.1) with the input space $\mathcal{U}_R := \{u \in L^\infty(\mathbb{R}, \mathbb{R}^m) : \|u\|_\infty \le R)\}$, and thus (4.1) is ISS w.r.t. small inputs and hence strongly iISS. □

We close this section with a property that holds for strongly iISS systems but may fail for iISS systems.

Proposition 4.28 *If (2.1) is strongly iISS, then (2.1) has CIUCS property.*

Proof The proof is left as Exercise 4.11. □

4.9 Cascade Interconnections Revisited

Consider a cascade interconnection

$$\begin{aligned} \dot{x} &= f(x, z), \\ \dot{z} &= g(z, u), \end{aligned} \tag{4.48}$$

where $x(t) \in \mathbb{R}^{n_1}$, $z(t) \in \mathbb{R}^{n_2}$, $u \in \mathcal{U} := L^\infty(\mathbb{R}_+, \mathbb{R}^m)$, and f, g are such that both subsystems are forward complete.

We have already seen in Sect. 3.3 that if both subsystems are ISS, then the interconnection is ISS as well. On the other hand, Example 4.23 indicates that if the x-subsystem ("driven subsystem") is only iISS, then the coupled system is not integral ISS in general. In this section, based on the concept of strong iISS, we perform a more precise stability analysis of cascade interconnections.

We start with cascades of systems, which are ISS w.r.t. small inputs.

Proposition 4.29 *Assume that Assumption 1.8 holds and consider the cascade interconnection (4.48). If both x-subsystem and z-subsystem are ISS with respect to small inputs, then the cascade interconnection is again ISS with respect to small inputs.*

Proof is left as Exercise 4.12. □

Now we turn our attention to the case when the driven system is ISS.

Theorem 4.30 *Let Assumption 1.8 hold. Consider the cascade interconnection (4.48), where x-subsystem is ISS. If z-subsystem is ISS / strongly iISS / iISS / 0-GAS, then the cascade interconnection (4.48) has the same property.*

Proof **z-subsystem is iISS ⇒ cascade is iISS.** We denote the flow maps of the x-subsystem and z-subsystem by ϕ_1 and ϕ_2, respectively. By the cocycle property (Proposition 1.17) and due to input-to-state stability of x-subsystem we have the following estimate (here we denote by v the input to the x-subsystem):

$$\begin{aligned}|\phi_1(t,x,v)| &= \Big|\phi_1\Big(\tfrac{t}{2},\phi_1\big(\tfrac{t}{2},x,v\big),v\big(\cdot+\tfrac{t}{2}\big)\Big)\Big|\\ &\leq \beta_1\Big(\big|\phi_1\big(\tfrac{t}{2},x,v\big)\big|,\tfrac{t}{2}\Big)+\gamma\big(\sup_{\frac{t}{2}\leq s\leq t}|v(s)|\big)\\ &\leq \beta_1\Big(\beta_1\big(|x|,\tfrac{t}{2}\big)+\gamma\big(\sup_{0\leq s\leq\frac{t}{2}}|v(s)|\big),\tfrac{t}{2}\Big)+\gamma\big(\sup_{\frac{t}{2}\leq s\leq t}|v(s)|\big)\\ &\leq \beta_1\big(2\beta_1\big(|x|,\tfrac{t}{2}\big),\tfrac{t}{2}\big)+\beta_1\Big(2\gamma\big(\sup_{0\leq s\leq\frac{t}{2}}|v(s)|\big),\tfrac{t}{2}\Big)+\gamma\big(\sup_{\frac{t}{2}\leq s\leq t}|v(s)|\big).\end{aligned} \tag{4.49}$$

As z-subsystem is iISS, there exist $\beta_2\in\mathcal{KL}$ and $\theta,\mu\in\mathcal{K}_\infty$ so that for all $z\in\mathbb{R}^{n_2}$, all $u\in\mathcal{U}$ and for all $t\geq 0$ it holds that

$$|\phi_2(t,z,u)|\leq\beta_2(|z|,t)+\theta\Big(\int_0^t\mu(|u(s)|)ds\Big).$$

This implies that

$$\sup_{0\leq s\leq t/2}|\phi_2(s,z,u)|\leq\beta_2(|z|,0)+\theta\Big(\int_0^{t/2}\mu(|u(r)|)dr\Big). \tag{4.50}$$

Substituting this expression as the input v into (4.49), we obtain that

$$
\begin{aligned}
|\phi_1(t,x,z)| &\le \beta_1\big(2\beta_1(|x|,\tfrac{t}{2}),\tfrac{t}{2}\big)+\beta_1\Big(2\gamma\Big(\beta_2(|z|,0)+\theta\Big(\int_0^{t/2}\mu(|u(r)|)dr\Big)\Big),\tfrac{t}{2}\Big)\\
&\quad+\gamma\Big(\sup_{\frac{t}{2}\le s\le t}\Big(\beta_2(|z|,s)+\theta\Big(\int_0^{s}\mu(|u(r)|)dr\Big)\Big)\Big)\\
&\le \beta_1\big(2\beta_1(|x|,\tfrac{t}{2}),\tfrac{t}{2}\big)+\beta_1\Big(2\gamma\big(2\beta_2(|z|,0)\big)+2\gamma\Big(2\theta\Big(\int_0^{t/2}\mu(|u(r)|)dr\Big)\Big),\tfrac{t}{2}\Big)\\
&\quad+\gamma\Big(\beta_2(|z|,\tfrac{t}{2})+\theta\Big(\int_0^{t}\mu(|u(r)|)dr\Big)\Big)\\
&\le \beta_1\big(2\beta_1(|x|,\tfrac{t}{2}),\tfrac{t}{2}\big)+\beta_1\Big(4\gamma\big(2\beta_2(|z|,0)\big),\tfrac{t}{2}\Big)+\beta_1\Big(4\gamma\Big(2\theta\Big(\int_0^{t}\mu(|u(r)|)dr\Big)\Big),0\Big)\\
&\quad+\gamma\big(2\beta_2(|z|,\tfrac{t}{2})\big)+\gamma\Big(2\theta\Big(\int_0^{t}\mu(|u(r)|)dr\Big)\Big).
\end{aligned}
$$

Denote

$$
\bar{\beta}_1(r,t):=\beta_1\big(2\beta_1(r,\tfrac{t}{2}),\tfrac{t}{2}\big)+\beta_1\Big(4\gamma\big(2\beta_2(r,0)\big),\tfrac{t}{2}\Big)+\gamma\big(2\beta_2(r,\tfrac{t}{2})\big),
$$

and

$$
\bar{\gamma}(r):=\beta_1\big(4\gamma(2\theta(r)),0\big)+\gamma\big(2\theta(r)\big).
$$

Denote the state of (4.48) by $w:=(x,z)^T$ and the flow map by ϕ_w. Since $|x|\le|w|$ and $|z|\le|w|$, we have that

$$
|\phi_1(t,x,z)|\le\bar{\beta}_1(|w|,t)+\bar{\theta}\Big(\int_0^{t}\mu(|u(r)|)dr\Big)
$$

Overall,

$$|\phi_w(t, w, u)| \le \bar{\beta}_1(|w|, t) + \bar{\theta}\Big(\int_0^t \mu(|u(r)|)dr\Big)$$
$$+\beta_2(|w|, t) + \theta\Big(\int_0^t \mu(|u(s)|)ds\Big),$$

which shows iISS of the overall system (4.48).

z**-subsystem is 0-GAS** $\Rightarrow$ **cascade is 0-GAS,** follows from the iISS case.

z**-subsystem is strongly iISS** $\Rightarrow$ **cascade is strongly iISS.** By previous arguments, the cascade is iISS. By Proposition 4.29 the cascade is also ISS w.r.t. small inputs, and thus by definition, it is strongly iISS.

z**-subsystem is ISS** $\Rightarrow$ **cascade is ISS,** follows by Theorem 3.22. □

Finally, if f is Lipschitz continuous on bounded balls of $\mathbb{R}^n \times \mathbb{R}^m$, we have the following result for the stability of cascades with a strongly iISS driven system:

Theorem 4.31 *Assume that both f and g are Lipschitz continuous on bounded balls of $\mathbb{R}^n \times \mathbb{R}^m$. Consider the cascade interconnection* (4.48). *Let x-subsystem be strongly iISS. If z-subsystem is strongly iISS / iISS / 0-GAS, then the cascade interconnection* (4.48) *has the same property.*

Proof z**-subsystem is 0-GAS** $\Rightarrow$ **cascade is 0-GAS.** As z-system is 0-GAS, the system $\dot{z} = g(z, 0)$ is ISS (as it does not depend on inputs). Since x-subsystem is ISS w.r.t. small inputs, the coupling of x-subsystem with $\dot{z} = g(z, 0)$ is also ISS w.r.t. small inputs by Proposition 4.29, and thus the coupling (4.48) is 0-GAS.

z**-subsystem is iISS** $\Rightarrow$ **cascade is iISS.** As x-subsystem is strongly iISS, it is ISS over the set of inputs of magnitude at most $R > 0$.

As z-subsystem is iISS, there exist $\beta_2 \in \mathcal{KL}$ and $\theta, \mu \in \mathcal{K}_\infty$ so that for all $z \in \mathbb{R}^{n_2}$, all $u \in \mathcal{U}$ and for all $t \ge 0$ it holds that

$$|\phi_2(t, z, u)| \le \beta_2(|z|, t) + \theta\Big(\int_0^t \mu(|u(s)|)ds\Big).$$

Assume that $\int_0^{+\infty} \mu(|u(s)|)ds < +\infty$. By Proposition 4.3 it holds that $\phi_2(t, z, u) \to 0$, as $t \to +\infty$.

Take $\tau > 0$ such that $|\phi_2(t, z, u)| \le R$ for $t \ge \tau$.

By the cocycle property, and using that x-subsystem is ISS with respect to the inputs of the magnitude at most R, we have for certain $\beta_1 \in \mathcal{KL}$ and $\gamma \in \mathcal{K}_\infty$ the following estimate for the trajectory of the x-subsystem:

$$\begin{aligned}|\phi_1(t, x, z)| &\le \beta_1(|\phi_1(\tau, x, z)|, t - \tau) + \gamma(\operatorname{ess\,sup}_{s \ge \tau}|z(s)|)\\ &\le \beta_1(|\phi_1(\tau, x, z)|, 0) + \gamma(R).\end{aligned}$$

This shows that the cascade interconnection (4.48) has BEFBS property. As the cascade (4.48) is also 0-GAS (by the first part of the proof), Theorem 4.20 shows iISS of (4.48).

z-subsystem is strongly iISS $\Rightarrow$ cascade is strongly iISS. By previous arguments, the cascade is iISS. By Proposition 4.29 the cascade is also ISS w.r.t. small inputs, and thus by definition, it is strongly iISS. □

4.10 Relationships Between ISS-Like Notions

We summarize the relationships between different ISS-like stability notions in the following theorem.

Proposition 4.32 *Let f be Lipschitz continuous on bounded balls of $\mathbb{R}^n \times \mathbb{R}^m$. Then for the system (2.1) the implications depicted in Fig. 4.2 hold.*

Furthermore, Lipschitz continuity on bounded balls of $\mathbb{R}^n \times \mathbb{R}^m$ is needed only for the implication "ISS $\Rightarrow$ strong iISS". For other implications, it is enough to require that Assumption 1.8 holds.

Proof **ISS $\Rightarrow$ strong iISS.** As f is Lipschitz continuous on bounded balls of $\mathbb{R}^n \times \mathbb{R}^m$ and (2.1) is ISS, smooth converse ISS Lyapunov theorem (Theorem 2.63) implies existence of an ISS Lyapunov function in dissipative form $V \in C^1(\mathbb{R}^n, \mathbb{R}_+)$ for (2.1). By Proposition 2.20, V satisfies (2.31), with $\alpha \in \mathcal{K}_\infty$. Thus, V is an iISS Lyapunov function, and by Proposition 4.9, (2.1) is iISS.

As ISS implies ISS w.r.t. small inputs, strong iISS follows.

Strong iISS $\Rightarrow$ iISS $\Rightarrow$ 0-GAS $\wedge f(0,0) = 0$ $\Rightarrow$ **0-UAS** $\wedge f(0,0) = 0$. Clear.

Strong iISS $\Rightarrow$ ISS w.r.t. small inputs $\Rightarrow$ 0-GAS $\wedge f(0,0) = 0$. Clear.

0-UAS $\wedge f(0,0) = 0 \Leftrightarrow$ **LISS,** follows by Proposition 4.3. □

We also visualize the relationships between ISS-like notions via the Euler diagram in Fig. 4.3 (motivated by [8, Fig. 1]).

ISS $\Rightarrow$ strong iISS $\Rightarrow$ iISS $\Rightarrow$ 0-GAS $\wedge f(0,0) = 0$ $\Rightarrow$ 0-UAS $\wedge f(0,0) = 0 \Leftrightarrow$ LISS

$\searrow$ $\nearrow$

ISS w.r.t. small inputs

Fig. 4.2 Relationships between ISS-like notions for the case of f being Lipschitz continuous on bounded balls of $\mathbb{R}^n \times \mathbb{R}^m$

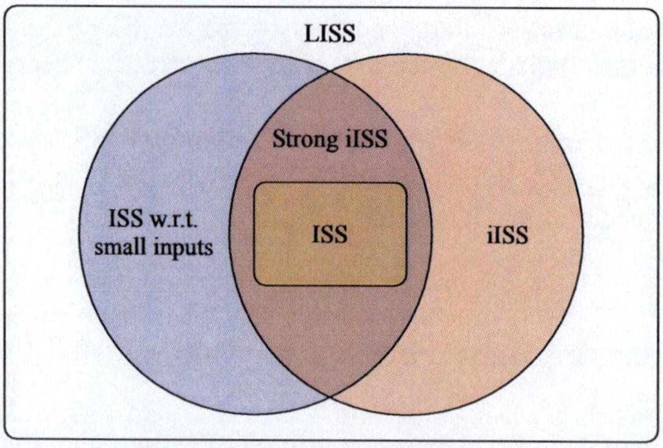

Fig. 4.3 The hierarchy of ISS-related concepts for the case of f being Lipschitz continuous on bounded balls of $\mathbb{R}^n \times \mathbb{R}^m$. © 2014 IEEE. This figure is a derivative of the original figure from "A. Chaillet, D. Angeli, and H. Ito. Combining iISS and ISS With Respect to Small Inputs: The Strong iISS Property. IEEE Transactions on Automatic Control, 59(9):2518–2524, 2014." This derivative figure is created by A. Mironchenko, and is reprinted with permission from IEEE.

4.11 Bilinear Systems

One of the simplest classes of nonlinear control systems is bilinear systems, which form a bridge between the linear and the nonlinear theories and are essential in numerous applications such as biochemical reactions, quantum-mechanical processes [7, 28], reaction-diffusion-convection processes controlled by means of catalysts [21], etc. Here we characterize the strong iISS of such systems.

Consider the system

$$\dot{x} = Ax + g(x, u), \tag{4.51}$$

where $A \in \mathbb{R}^{n\times n}$ and $g : \mathbb{R}^n \times \mathbb{R}^m \to \mathbb{R}^n$ is such that the solution of (4.51) exists and is unique for all initial conditions and for all inputs $u \in \mathcal{U} := L^\infty(\mathbb{R}_+, \mathbb{R}^m)$. Furthermore, assume that g satisfies for some $a, b \in \mathbb{R}_+$ the following estimate:

$$|g(x, v)| \le (a|x| + b)|v|, \qquad x \in \mathbb{R}^n, \quad v \in \mathbb{R}^m. \tag{4.52}$$

A special case of (4.51) is bilinear systems of the form

$$\dot{x} = Ax + \sum_{j=1}^{m} u_j A_j x + Bu\,, \tag{4.53}$$

where $x \in \mathbb{R}^n$, $u = (u_1, \dots, u_m) \in \mathbb{R}^m$, $A, A_1, \dots, A_m \in \mathbb{R}^{n\times n}$, $B \in \mathbb{R}^{n\times m}$.

Bilinear systems are seldom ISS, but they are strongly iISS provided A is a Hurwitz matrix.

Proposition 4.33 *Let* (4.51) *satisfy the assumption* (4.52). *The following statements are equivalent:*

(i) (4.51) *is strongly iISS.*
(ii) (4.51) *is iISS.*
(iii) (4.51) *is 0-UGAS.*
(iv) *A is Hurwitz.*
(v) *A function* $W : \mathbb{R}^n \to \mathbb{R}_+$*, defined by*

$$W(x) = \ln\Big(1 + x^T P x\Big), \quad x \in \mathbb{R}^n, \tag{4.54}$$

where $P \in \mathbb{R}^{n\times n}$ *is a positive definite matrix satisfying*

$$A^T P + PA = -Q$$

for a certain negative definite matrix Q*, is an iISS Lyapunov function for* (4.51).

Proof (i) $\Rightarrow$ (ii) $\Rightarrow$ (iii) $\Rightarrow$ (iv). Clear.

(iv) $\Rightarrow$ (v) $\wedge$ (i). As A is Hurwitz, there exist symmetric, positive definite matrices $P, Q \in \mathbb{R}^{n\times n}$ such that $A^\top P + PA = -Q$. For these P, Q consider the function W, defined by (4.54).

For all x, u it holds that

$$\begin{aligned}
\dot{W}_u(x) &= \frac{1}{1+x^T Px}\Big(2x^T P(Ax + g(x,u))\Big)\\
&\le \frac{1}{1+x^T Px}\Big(- x^T Qx + 2|x|\|P\||g(x,u)|\Big)\\
&\le \frac{1}{1+x^T Px}\Big(- \lambda_{\min}(Q)|x|^2 + 2|x|\|P\|(a|x|+b)|u|\Big),
\end{aligned} \tag{4.55}$$

where $\lambda_{\min}(Q)$ is the minimal eigenvalue of a positive symmetric matrix Q. Now integral ISS can be shown as follows:

$$\begin{aligned}
\dot{W}(x) &\le -\frac{\lambda_{\min}(Q)|x|^2}{1+x^T Px} + \frac{2|x|\|P\|(a|x|+b)}{1+x^T Px}|u|\\
&\le -\frac{\lambda_{\min}(Q)|x|^2}{1+x^T Px} + C|u|,
\end{aligned}$$

for $C := \max_{x\in\mathbb{R}^n} \frac{2|x|\|P\|(a|x|+b)}{1+x^T Px} < \infty$. This shows that W is an iISS Lyapunov function for (4.51), and thus (4.51) is iISS.

To show ISS w.r.t. small inputs, we infer from (4.55) that

$$\begin{aligned}
\dot{W}_u(x) &\le \frac{1}{1+x^T Px}\Big(- \lambda_{\min}(Q)|x|^2 + 2a\|P\||x|^2|u| + 2b|x||u|\Big)\\
&\le \frac{1}{1+x^T Px}\Big(- \lambda_{\min}(Q)|x|^2 + 2a\|P\||x|^2|u| + \varepsilon b^2|x|^2 + \frac{1}{\varepsilon}|u|^2\Big),
\end{aligned}$$

where we have applied Young's inequality for the last estimate. Assuming that $|u| \leq R$ for some fixed R that we specify later, we obtain that

$$\dot{W}_u(x) \leq \frac{1}{1+x^T Px}\Big(- \lambda_{\min}(Q)|x|^2 + 2a\|P\|R|x|^2 + \varepsilon b^2|x|^2\Big) + \frac{1}{1+x^T Px}\frac{1}{\varepsilon}|u|^2 \quad (4.56)$$

$$\leq \Big(2a\|P\|R + \varepsilon b^2 - \lambda_{\min}(Q)\Big)\frac{|x|^2}{1+x^T Px} + \frac{1}{\varepsilon}|u|^2. \quad (4.57)$$

Choosing ε, R small enough to ensure that $2a\|P\|R + \varepsilon b^2 - \lambda_{\min}(Q) < 0$, we see that the system (4.51) is ISS w.r.t. the inputs satisfying $\|u\|_\infty \leq R$.

This shows (v) and (i). As (v) implies (ii), the proposition is proved. □

4.12 Small-Gain Theorems for Couplings of Strongly iISS Systems

We finish this chapter with a small-gain theorem for a coupling of two strongly iISS systems of the form

$$\dot{x}_i(t) = f_i(x_1, x_2, u), \quad i = 1, 2, \quad (4.58)$$

where $x_i \in X_i := \mathbb{R}^{N_i}$, $i = 1, 2$, $u \in \mathcal{U} := L^\infty(\mathbb{R}_+, \mathbb{R}^m)$, and f_1, f_2 are such that both subsystems are well posed and forward complete.

Assume that both subsystems possess iISS Lyapunov functions $V_i \in C^1(\mathbb{R}^{N_i}, \mathbb{R}_+)$, with corresponding $\psi_{i1}, \psi_{i2} \in \mathcal{K}_\infty$, $\alpha_i \in \mathcal{K}$, $\sigma_i \in \mathcal{K}$, and $\kappa_i \in \mathcal{K} \cup \{0\}$ for $i = 1, 2$ such that

$$\psi_{i1}(|x_i|) \leq V_i(x_i) \leq \psi_{i2}(|x_i|), \quad x_i \in \mathbb{R}^{N_i}, \quad (4.59)$$

and the dissipation inequalities

$$\dot{V}_i(x_i) \leq -\alpha_i(|x_i|) + \sigma_i(|x_{3-i}|) + \kappa_i(|u|) \quad (4.60)$$

hold for all $x_i \in X_i$, $x_{3-i} \in X_{3-i}$, and $u \in \mathbb{R}^m$.

As $\alpha_i \in \mathcal{K}$, $i = 1, 2$, Proposition 4.27 shows that our assumptions imply strong iISS of both subsystems.

To present a small-gain criterion for the interconnected system (4.58) whose components are not necessarily ISS, we make use of a generalized expression of inverse mappings on the set of extended nonnegative numbers $\overline{\mathbb{R}_+} := [0, \infty]$. For $\omega \in \mathcal{K}$, define the function $\omega^\ominus \colon \overline{\mathbb{R}_+} \to \overline{\mathbb{R}_+}$ as

$$\omega^\ominus(s) = \sup\{v \in \mathbb{R}_+ : s \geq \omega(v)\} = \begin{cases} \omega^{-1}(s) & \text{, if } s \in \operatorname{Im}\omega, \\ +\infty & \text{, otherwise.} \end{cases}$$

A function $\omega \in \mathcal{K}$ is extended to ω: $\overline{\mathbb{R}_+} \to \overline{\mathbb{R}_+}$ as $\omega(s) := \sup_{v \in \{y \in \mathbb{R}_+ : y \leq s\}} \omega(v)$. These notations are useful for presenting the following result succinctly.

Theorem 4.34 *Suppose that*

$$\lim_{s\to\infty} \alpha_i(s) = \infty \text{ or } \lim_{s\to\infty} \sigma_{3-i}(s)\kappa_i(1) < \infty \tag{4.61}$$

is satisfied for $i = 1, 2$. *If there exists* $c > 1$ *such that*

$$\psi_{11}^{-1} \circ \psi_{12} \circ \alpha_1^{\ominus} \circ c\sigma_1 \circ \psi_{21}^{-1} \circ \psi_{22} \circ \alpha_2^{\ominus} \circ c\sigma_2(s) \leq s, \quad s \in \mathbb{R}_+, \tag{4.62}$$

then system (4.58) *is iISS. Moreover, if additionally* $\alpha_i \in \mathcal{K}_\infty$ *for* $i = 1, 2$, *then system* (4.58) *is ISS. Furthermore,*

$$V(x) = \int_0^{V_1(x_1)} \lambda_1(s)ds + \int_0^{V_2(x_2)} \lambda_2(s)ds \tag{4.63}$$

is an iISS (ISS) Lyapunov function for (4.58), *where* $\lambda_i \in \mathcal{K}$ *is given for* $i = 1, 2$ *by*

$$\lambda_i(s) = [\alpha_i(\psi_{i2}^{-1}(s))]^{\psi} [\sigma_{3-i}(\psi_{3-i\,1}^{-1}(s))]^{\psi+1} \quad \forall s \in \mathbb{R}_+, \tag{4.64}$$

with an arbitrary $q \geq 0$ *satisfying*

$$\begin{array}{ll} q = 0 & \text{, if } c > 2, \\ q^{-\frac{q}{q+1}} < \frac{c}{q+1} \leq 1 & \text{, otherwise.} \end{array} \tag{4.65}$$

Proof We refer to [25] for the proof of this result. □

It is straightforward to see that there always exists $q \geq 0$ satisfying (4.65). It is also worth mentioning that the Lyapunov function (4.63) is not in the maximization form employed in Chap. 3 for establishing ISS. The use of the summation form (4.63) for systems that are not necessarily ISS is motivated by the limitation of the maximization form as clarified in [14].

Remark 4.35 It can be verified as in [16, p. 1204] that condition (4.65) in Theorem 4.34 can be replaced by

$$\exists \tau \in (1, c] \text{ s.t. } \left(\frac{\tau}{c}\right)^{q+1} < \tau - 1.$$

In fact, this property is implied by condition (4.65). For each arbitrarily chosen $\tau \in (1, c]$, one obtains solutions to the above inequality explicitly whenever $c > 1$. Furthermore, it is worth noting that as in [11, Theorem 4], it can be verified that one can get rid of (4.65) if there exist $c_i > 0$, $i = 1, 2$, and $k > 0$ such that

$$c_1\sigma_2(\psi_{11}^{-1}(s))^k \le \alpha_1(\psi_{12}^{-1}(s)) \quad \forall s \in \mathbb{R}_+,$$
$$c_2\sigma_1(\psi_{21}^{-1}(s)) \le \alpha_2(\psi_{22}^{-1}(s))^k \quad \forall s \in \mathbb{R}_+$$

hold. Notice that $c_1c_2 > 1$ implies the existence of $c > 1$ satisfying (4.62) in this case.

Remark 4.36 Condition (4.62) can be called an iISS small-gain condition. On the first glance it may seem a bit technical, but it simplifies considerably if $V_i(x_i) = \psi_i(|x_i|)$, for some $\psi_1, \psi_2 \in \mathcal{K}_\infty$. In this case $\psi_{i1} = \psi_{i2} = \psi_i$, $i = 1, 2$ and (4.62) takes the form

$$\alpha_1^{\ominus} \circ c\sigma_1 \circ \alpha_2^{\ominus} \circ c\sigma_2(s) \le s, \quad s \in \mathbb{R}_+. \tag{4.66}$$

The term $\alpha_i^{\ominus} \circ \sigma_i$ can be interpreted as a Lyapunov gain of the i-th subsystem. This interpretation justifies the name "small-gain condition" for (4.62).

Example 4.37 Consider the following planar system:

$$\dot{x}_1 = -x_1 + x_1x_2^4, \tag{4.67a}$$

$$\dot{x}_2 = -ax_2 - bx_2^3 + \Big(\frac{x_1^2}{1+x_1^2}\Big)^{1/2}. \tag{4.67b}$$

Let us find for which values of $a, b > 0$ this system is GAS.

Motivated by the analysis of bilinear systems, we consider the following iISS Lyapunov function for the x_1-subsystem:

$$V_1(x_1) := \ln(1 + x_1^2), \quad x_1 \in \mathbb{R}. \tag{4.68}$$

The dissipation inequality for V_1 is easy to derive:

$$\dot{V}_1(x_1) = \frac{1}{1+x_1^2} 2x_1(-x_1 + x_1x_2^4) \le -\frac{2x_1^2}{1+x_1^2} + 2x_2^4. \tag{4.69}$$

The x_2-subsystem is ISS with the ISS Lyapunov function

$$V_2(x_2) := x_2^2, \quad x_2 \in \mathbb{R}. \tag{4.70}$$

The dissipation inequality for V_2 is

$$\begin{aligned} \dot{V}_2(x_2) &= -2ax_2^2 - 2bx_2^4 + 2x_2\Big(\frac{x_1^2}{1+x_1^2}\Big)^{1/2} \\ &\le -2\Big(a - \frac{\omega}{2}\Big)x_2^2 - 2bx_2^4 + \frac{1}{\omega}\frac{x_1^2}{1+x_1^2}, \end{aligned} \tag{4.71}$$

where in the last estimate we have used Young's inequality, which is valid for any $\omega > 0$.

Hence, the inequalities (4.59) are satisfied with the class $\mathcal{K}_\infty$ functions $\psi_{11} = \psi_{12} : s \mapsto \ln(1 + s^2)$ and $\psi_{21} = \psi_{22} : s \mapsto s^2$. Due to (4.69) and (4.71), we have (4.60) for

$$\alpha_1(s) = \frac{2s^2}{1+s^2}, \quad \sigma_1(s) = 2s^4, \quad \kappa_1(s) = 0, \tag{4.72}$$

$$\alpha_2(s) = 2\left(a - \frac{\omega}{2}\right)s^2 + 2bs^4, \;\; \sigma_2(s) = \frac{1}{\omega}\left(\frac{s^2}{1+s^2}\right), \;\; \kappa_2(s) = 0, \tag{4.73}$$

defined with $\omega \in (0, 2a]$.

First of all, in our case the small-gain condition (4.62) reduces to (4.66). Secondly, since $\alpha_2^{-1}(s) \le \left(\frac{s}{2b}\right)^{\frac{1}{4}}$ for all $s \ge 0$, we have that $\sigma_1 \circ \alpha_2^{-1}(s) \le \frac{s}{b}$ for all $s \ge 0$.

Thus, Theorem 4.34 ensures GAS of the interconnection provided that the following condition holds:

$$\alpha_1^{\ominus} \circ \frac{c^2}{b}\sigma_2(s) \le s, \quad s \ge 0.$$

This is the same as to require

$$\frac{c^2}{\omega b}\frac{s^2}{1+s^2} \le 2\frac{s^2}{1+s^2}, \quad s \ge 0.$$

There is $c > 1$ such that the above condition holds if and only if

$$\frac{1}{\omega b} < 2.$$

As $\omega \in (0, 2a]$, the optimal choice is $\omega := 2a$, and we arrive at sufficient condition for the stability of the network as

$$ab > \frac{1}{4}. \tag{4.74}$$

4.13 Concluding Remarks

The concepts of integral input-to-state stability (integral ISS) and iISS Lyapunov function have been introduced in [29]. Integral ISS is one of the most important ISS-like concepts with a powerful theory and numerous applications. A detailed treatment of the integral ISS theory deserves a separate book, and thus this chapter should be considered rather as an introduction to the iISS theory. In particular, we have omitted the proofs of several characterizations of the iISS property, and some of the important topics have been not discussed, in particular, qualitative equivalence (up to a nonlinear change of coordinates) between iISS and nonlinear L^2-gain property, shown in [20].

Basic properties. Proposition 4.3 is due to [29, Proposition 6]. Proposition 4.7 is a folklore. We follow the argument in [26, Proposition 4]; see also [27, Theorem 3.5]. The relationship between ISS and integral ISS for general infinite-dimensional linear systems is much more complex; see [18] for the detailed analysis and [27, Section 3.3] for a brief overview. Lyapunov concepts for the L^p-ISS property have been introduced in [24].

Criteria for iISS. Characterizations of iISS property have been developed in a series of papers [3–5, 29]. In particular, Proposition 4.13 is due to [5, Proposition II.5 and Lemma IV.10]. Theorem 4.17 is the main result in [5]. Proposition 4.18 is due to [22, Proposition 1]. The example of a manipulator from Sect. 4.5 is due to [5], and we followed [5] closely in our description. Another controller for the considered manipulator that achieves ISS has been derived in [1].

Theorem 4.20 was shown in [3], though some parts of it appeared already in [4]. Theorem 4.21 is due to [4, Theorem 1].

Strong iISS and cascades. The notion of strong integral ISS has been introduced in [8] and further studied in [9]. Example 4.23 is due to [6, Example 1]. Theorem 4.30 was shown in [23, Proposition 2]. Proposition 4.29 and Theorem 4.31 are due to [9, Theorem 1] (see also [12] for some precursors of the latter result).

In Proposition 4.32 the restrictions to the right-hand side f are imposed to invoke the smooth converse ISS Lyapunov theorem. Recently, an alternative argument has been proposed in [10], where the implication ISS $\Rightarrow$ iISS has been shown under less demanding assumptions on the nonlinearity (and for a larger class of control systems encompassing some classes of time-variant ODEs) and without the usage of Lyapunov functions.

Bilinear systems. Integral ISS of bilinear systems has been shown in [29, Theorem 5], strong iISS of such systems (Proposition 4.33) has been proved in [8, Corollary 2]. This result has been generalized to certain classes of infinite-dimensional systems in [27, Proposition 3.34].

Small-gain theorems. For an overview of the obstacles encountered in addressing interconnections of iISS systems, we refer to [13]. Further notable results in iISS small-gain theory include [2, 11, 15, 19]. In contrast to ISS subsystems, iISS subsystems that are not ISS usher the issue of incompatibility of spaces in the time domain for trajectory-based approaches, as well as insufficiency of max-type constructions that we have used in Chap. 3. This motivated researchers to look for the sum-type constructions of iISS Lyapunov functions for the interconnections. The small-gain Theorem 4.34 for couplings of strongly iISS systems is one of the results of this kind. It was also extended to the couplings of two infinite-dimensional strongly iISS control systems and used for the couplings of 2 nonlinear parabolic systems in [25]. Small-gain theorems for couplings of systems that are iISS but are not strongly iISS have been obtained in [17].

4.14 Exercises

Exercise 4.1 Show that a forward complete system (4.1) is iISS if and only if there are $\kappa \in \mathcal{K}_\infty$, $\mu \in \mathcal{K}$, and $\beta \in \mathcal{KL}$ such that for all $x \in \mathbb{R}^n$, $u \in \mathcal{U}$, and all $t \geq 0$ it holds that

$$\kappa(|\phi(t, x, u)|) \leq \beta(|x|, t) + \int_0^t \mu(|u(s)|)ds. \tag{4.75}$$

Exercise 4.2 Show that performing nonlinear changes of coordinates for x and u variables for L^p-ISS systems, we obtain that such a system will be iISS in new variables. In particular, L^p-ISS property (in contrast to iISS) is not preserved under nonlinear coordinate changes.

Exercise 4.3 For given $\xi \in \mathcal{K}_\infty$ and $r > 0$ denote

$$\mathcal{Q}_r(\xi) := \{u \in \mathcal{U} : \int_0^{+\infty} \xi(|u(s)|)ds \leq r\}.$$

We say that a forward complete system (4.1) has:

(i) *Uniform bounded energy-convergent state (UBECS) property:* There is $\xi \in \mathcal{K}$ such that

$$r > 0 \;\Rightarrow\; \lim_{t\to\infty} \sup_{|x|\leq r,\; u\in\mathcal{Q}_r(\xi)} |\phi(t, x, u)| = 0. \tag{4.76}$$

(ii) *Uniform bounded energy-bounded state (UBEBS) property:* There is $\xi \in \mathcal{K}$ such that

$$r > 0 \;\Rightarrow\; \sup_{|x|\leq r,\; u\in\mathcal{Q}_r(\xi),\; t\geq 0} |\phi(t, x, u)| < +\infty. \tag{4.77}$$

Does any iISS system (4.1) satisfy UBECS and UBEBS properties?

Exercise 4.4 Show that $V \in C^1(\mathbb{R}^n, \mathbb{R}_+)$ is an iISS Lyapunov function if and only if there exist $\psi_1, \psi_2 \in \mathcal{K}_\infty$, $\sigma \in \mathcal{K}$, and $\alpha \in \mathcal{P}$ satisfying

$$\psi_1(|x|) \leq V(x) \leq \psi_2(|x|), \quad x \in \mathbb{R}^n, \tag{4.78}$$

and

$$\nabla V(x) \cdot f(x, u) \leq -\alpha(V(x)) + \sigma(|u|), \quad x \in \mathbb{R}^n,\; u \in \mathbb{R}^m. \tag{4.79}$$

Exercise 4.5 Show that the following scalar system is iISS:

$$\dot{x} = -\arctan(x) + u.$$

Exercise 4.6 Consider the equations of motion of a nonlinear oscillator

$$\dot{x}_1 = x_2, \tag{4.80a}$$

$$\dot{x}_2 = -k(x_1) - \mu x_2 + d, \tag{4.80b}$$

where $(x_1, x_2) \in \mathbb{R}^2$, $\mu > 0$, $d \in L^\infty(\mathbb{R}_+, \mathbb{R})$, and $k : \mathbb{R} \to \mathbb{R}$ is a monotone function, satisfying $k(0) = 0$ and $rk(r) > 0$ for all $r \neq 0$.

Show that the system is iISS. Under which conditions does it become strongly ISS / ISS?

Exercise 4.7 In Proposition 2.16, we have shown that nonlinear scalings of ISS Lyapunov functions in implication form are again ISS Lyapunov functions in implication form.

Show by means of a counterexample that this does not hold for iISS Lyapunov functions. Consider a system

$$\dot{x} = -x + xu,$$

and a corresponding iISS Lyapunov function $V(x) = \ln(1 + x^2)$. Find a scaling $\xi \in \mathcal{K}_\infty \cap C^1(\mathbb{R}_+, \mathbb{R}_+)$, such that $\xi \circ V$ is not an iISS Lyapunov function.

Exercise 4.8 A positive definite map $W : \mathbb{R}^n \to \mathbb{R}_+$ is called *semiproper* if for each $r \in \mathrm{Im}(W)$ the set $\{x \in \mathbb{R}^n : W(x) \leq r\}$ is compact.

Show that a positive definite map $W \in C(\mathbb{R}^n, \mathbb{R}_+)$ is semiproper if and only if there are $\xi \in \mathcal{K}$ and a proper (see Definition A.9) function $W_1 \in C(\mathbb{R}^n, \mathbb{R}_+)$ such that $W = \xi \circ W_1$.

Exercise 4.9 Show Proposition 4.18.

To show the implication (ii) $\Rightarrow$ (i), one may consider the Lyapunov function

$$V(x) := \int_0^{W(x)} \frac{1}{1 + b(r)} dr, \quad x \in \mathbb{R}^n.$$

For (i) $\Rightarrow$ (iii) one may recall that iISS is equivalent to existence of an iISS Lyapunov function V and consider $W(x) := e^{V(x)} - 1$.

Exercise 4.10 Show that under certain conditions on f BEWCS and BEFBS properties imply forward completeness.

Exercise 4.11 Show Proposition 4.28.

Exercise 4.12 Show Proposition 4.29.

References

1. Angeli D (1999) Intrinsic robustness of global asymptotic stability. Syst Control Lett 38(4–5):297–307
2. Angeli D, Astolfi A (2007) A tight small gain theorem for not necessarily ISS systems. Syst Control Lett 56:87–91
3. Angeli D, Ingalls B, Sontag ED, Wang Y (2004) Separation principles for input-output and integral-input-to-state stability. SIAM J Control Optim 43(1):256–276
4. Angeli D, Sontag E, Wang Y (2000) Further equivalences and semiglobal versions of integral input to state stability. Dyn Control 10(2):127–149
5. Angeli D, Sontag ED, Wang Y (2000) A characterization of integral input-to-state stability. IEEE Trans Autom Control 45(6):1082–1097
6. Arcak M, Angeli D, Sontag E (2002) A unifying integral ISS framework for stability of nonlinear cascades. SIAM J Control Optim 40(6):1888–1904
7. Bruni C, Dipillo G, Koch G (1974) Bilinear systems: an appealing class of "nearly linear" systems in theory and applications. IEEE Trans Autom Control 19(4):334–348
8. Chaillet A, Angeli D, Ito H (2014) Combining iISS and ISS with respect to small inputs: the strong iISS property. IEEE Trans Autom Control 59(9):2518–2524
9. Chaillet A, Angeli D, Ito H (2014) Strong iISS is preserved under cascade interconnection. Automatica 50(9):2424–2427
10. Haimovich H, Mancilla-Aguilar JL (2019) ISS implies iISS even for switched and time-varying systems (if you are careful enough). Automatica 104:154–164
11. Ito H (2006) State-dependent scaling problems and stability of interconnected iISS and ISS systems. IEEE Trans Autom Control 51:1626–1643
12. Ito H (2010) A Lyapunov approach to cascade interconnection of integral input-to-state stable systems. IEEE Trans Autom Control 55(3):702–708
13. Ito H (2013) Utility of iISS in composing Lyapunov functions. In: Proceedings of 9th IFAC Symposium on Nonlinear Control Systems, pp 723–730
14. Ito H, Dashkovskiy S, Wirth F (2012) Capability and limitation of max- and sum-type construction of Lyapunov functions for networks of iISS systems. Automatica 48(6):1197–1204
15. Ito H, Jiang Z-P (2009) Necessary and sufficient small gain conditions for integral input-to-state stable systems: a Lyapunov perspective. IEEE Trans Autom Control 54(10):2389–2404
16. Ito H, Jiang Z-P, Dashkovskiy S, Rüffer B (2013) Robust stability of networks of iISS systems: construction of sum-type Lyapunov functions. IEEE Trans Autom Control 58(5):1192–1207
17. Ito H, Kellett CM (2018) A small-gain theorem in the absence of strong iISS. IEEE Trans Autom Control 64(9):3897–3904
18. Jacob B, Nabiullin R, Partington JR, Schwenninger FL (2018) Infinite-dimensional input-to-state stability and Orlicz spaces. SIAM J Control Optim 56(2):868–889
19. Karafyllis I, Jiang Z-P (2012) A new small-gain theorem with an application to the stabilization of the chemostat. Int J Robust Nonlinear Control 22(14):1602–1630
20. Kellett CM, Dower PM (2016) Input-to-state stability, integral input-to-state stability, and L_2-gain properties: qualitative equivalences and interconnected systems. IEEE Trans Autom Control 61(1):3–17
21. Khapalov AY (2003) Controllability of the semilinear parabolic equation governed by a multiplicative control in the reaction term: a qualitative approach. SIAM J Control Optim 41(6):1886–1900
22. Liberzon D, Sontag ED, Wang Y (1999) On integral-input-to-state stabilization. In: Proceedings of 1999 American Control Conference, pp 1598–1602
23. Liberzon D, Sontag ED, Wang Y (2002) Universal construction of feedback laws achieving ISS and integral-ISS disturbance attenuation. Syst Control Lett 46(2):111–127
24. Mironchenko A (2020) Lyapunov functions for input-to-state stability of infinite-dimensional systems with integrable inputs. IFAC-PapersOnLine 53:5336–5341
25. Mironchenko A, Ito H (2015) Construction of Lyapunov functions for interconnected parabolic systems: an iISS approach. SIAM J Control Optim 53(6):3364–3382

26. Mironchenko A, Ito H (2016) Characterizations of integral input-to-state stability for bilinear systems in infinite dimensions. Math Control Rel Fields 6(3):447–466
27. Mironchenko A, Prieur C (2020) Input-to-state stability of infinite-dimensional systems: recent results and open questions. SIAM Rev 62(3):529–614
28. Pardalos PM, Yatsenko V (2008) Optimization and control of bilinear systems. Springer, Boston, MA
29. Sontag ED (1998) Comments on integral variants of ISS. Syst Control Lett 34(1–2):93–100
30. Sontag ED (2008) Input to state stability: basic concepts and results. In: Nonlinear and optimal control theory, Chap. 3. Springer, Heidelberg, pp 163–220

Chapter 5
Robust Nonlinear Control and Observation

An objective of a control theory is to influence the dynamics of a system to guarantee its desired behavior. Challenges arising in this way are manifold and include *nonlinearity* of a system, a need to ensure *robustness* (or *reliability*) of designed controllers in spite of actuator and observation errors, hidden (unmodeled) dynamics of a system, and external disturbances acting on the system. Furthermore, networks to be controlled may be huge with a nontrivial topological structure. Input-to-state stability theory allows to tackle these challenges by delivering a systematic theory for the design of robust controllers and observers for nonlinear control systems, and networks of such systems. Lyapunov and small-gain methods play a central role in this development. The literature on ISS-based nonlinear control is vast, and in this chapter, we discuss only some of the most characteristic results in this field.

We start with the description of the problem of input-to-state stabilization by full state feedback. Then, in Sects. 5.2, 5.3, 5.4, we propose several methods for robust stabilization: ISS feedback redesign, universal formulas for stabilizing controllers and ISS backstepping. In Sect. 5.5, we use the ISS backstepping together with small-gain theorems (via gain assignment technique) to achieve the stabilization of an axial compressor model.

In many applications, it is not possible to update the controller at each time instant. Hence, in Sect. 5.6, we study the event-based control, which strives to solve the control problem by updating the controller at discrete time instants.

In most cases, it is impossible to measure the whole state, and only a certain function of the state, called the system's output, is available for control purposes. In the second part of this chapter, starting with Sect. 5.7, we will study how to recover the knowledge about the state of the system and to use it to achieve the robust stabilization.

At the end of the chapter, we give an overview of obstructions on the way to stabilization and briefly discuss the methods how to overcome them, including references to the corresponding literature.

A. Mironchenko, *Input-to-State Stability*, Communications and Control Engineering,
https://doi.org/10.1007/978-3-031-14674-9_5

5.1 Input-to-state Stabilization

Consider a system

$$\dot{x} = f(x, u, d), \tag{5.1}$$

where $x(t) \in \mathbb{R}^n$, $u \in \mathcal{U} := L^\infty(\mathbb{R}_+, \mathbb{R}^m)$, and $d \in \mathcal{D} := L^\infty(\mathbb{R}_+, \mathbb{R}^p)$ for some $n, m, p \in \mathbb{N}$.

This system possesses two inputs:

- a *control input* u that we can design in order to achieve certain objectives.
- a *disturbance input* d: an unknown signal that we cannot influence and which may induce a destabilizing effect.

We assume that f is continuous on $\mathbb{R}^n \times \mathbb{R}^m \times \mathbb{R}^p$, and Lipschitz continuous with respect to (x, u) on bounded subsets of $\mathbb{R}^n \times \mathbb{R}^m$. For a while, assume that we can measure the whole state of a system at each moment. Thus, the knowledge of $x(t)$ can be used to design u at each moment.

In this chapter, we use *"smooth"* synonymously with "infinitely differentiable".

Definition 5.1 The system (5.1) is called *input-to-state stabilizable* if there exists a Lipschitz continuous $k : \mathbb{R}^n \to \mathbb{R}^m$, such that

$$\dot{x}(t) = f\big(x(t), k(x(t)), d(t)\big), \tag{5.2}$$

is ISS w.r.t. the disturbance d. The function k is called an *input-to-state stabilizing feedback control.*

If k above can be chosen to be smooth, then (5.1) is called *smoothly input-to-state stabilizable.*

Definition 5.2 The system (5.1) is called *globally asymptotically stabilizable* if there exists a Lipschitz continuous $k : \mathbb{R}^n \to \mathbb{R}^m$, so that

$$\dot{x}(t) = f\big(x(t), k(x(t)), 0\big),$$

is GAS. The function k is called a *stabilizing feedback control.*

If k above can be chosen smooth, then (5.1) is called *smoothly globally asymptotically stabilizable.*

The system (5.2) is frequently called a *closed-loop system*, and the system (5.1) is called an *open-loop system.* As we have assumed that f is Lipschitz continuous in (x, u), after substitution of a Lipschitz continuous feedback $u = k(x)$, the closed-loop system (5.2 is well-posed), which can be shown similarly to (Lemma 2.60).

In what follows, we develop several basic methods of robust stabilization of control systems.

5.2 ISS Feedback Redesign

Assume that disturbances enter the system (5.1) in a special way, namely a disturbance d distorts the control signal sent to the system but does not change the structure of the system:

$$\dot{x}(t) = f\big(x(t), u(t) + d(t)\big). \tag{5.3}$$

Such disturbances are called additive *actuator disturbances*. The schematic description for such a system in the absence and in the presence of disturbances is depicted in Figs. 5.1, 5.2.

The next example shows that controllers that asymptotically stabilize the system in the absence of disturbances may have poor performance if the disturbances are present.

Example 5.3 **(Nonrobust stabilization)** Consider a stabilization problem for the following system:

$$\dot{x} = x + (x^2 + 1)(u + d). \tag{5.4}$$

First assume that $d \equiv 0$ and pick $k(x) := -\frac{2x}{1+x^2}$. The feedback control $u(t) := k(x(t))$ globally (asymptotically) stabilizes the system (5.4), and the closed-loop system takes form:

$$\dot{x} = -x. \tag{5.5}$$

The stabilization problem for the undisturbed system (5.4) is solved.

Now assume that actuator disturbances are present. Using the same stabilizing controller, we obtain after substitution of $u(t) := k(x(t))$ into (5.4) the following equation

$$\dot{x} = -x + (x^2 + 1)d. \tag{5.6}$$

This system is 0-GAS, but for the disturbance $d \equiv 1$, the state of (5.6) blows up in finite time for any initial condition.

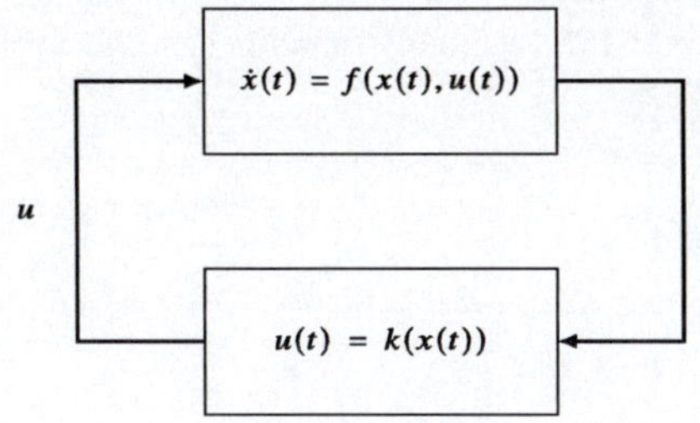

Fig. 5.1 Stabilization problem without actuator disturbances

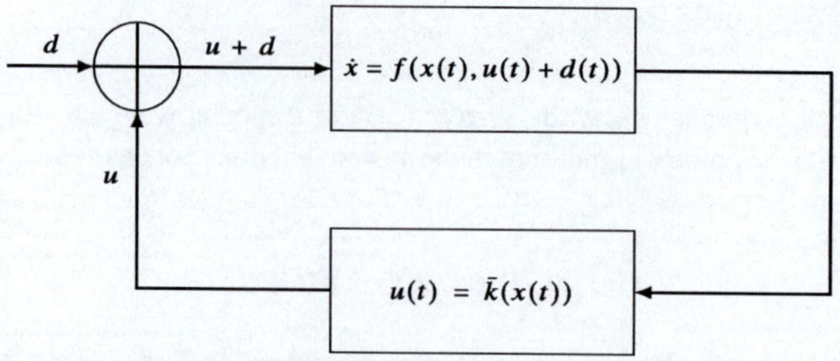

Fig. 5.2 Stabilization problem with actuator disturbances

We see that a feedback controller that stabilizes an undisturbed system may have a poor performance in presence of actuator disturbances. This motivates a natural question, whether it is possible to modify the stabilizing feedback in a way to make the controller input-to-state stabilizing in presence of disturbances. Next, we positively answer this question for the special class of *control-affine systems*

$$\dot{x}(t) = g_0(x) + g_1(x)(u + d), \tag{5.7}$$

where $x(t) \in \mathbb{R}^n$, $u(t) \in \mathbb{R}$, and $g_0, g_1 : \mathbb{R}^n \to \mathbb{R}^n$ are smooth functions.

We have the following result:

Theorem 5.4 (5.7) *is smoothly globally asymptotically stabilizable if and only if* (5.7) *is smoothly input-to-state stabilizable.*

Proof Let (5.7) be smoothly globally asymptotically stabilizable. Then there exists a smooth feedback $k = k(x(t))$ such that

$$\dot{x}(t) = g_0(x(t)) + g_1(x(t))k(x(t)) \tag{5.8}$$

is GAS. Due to the converse Lyapunov theorem (Theorem B.32), there exists a smooth Lyapunov function V for (5.8), satisfying for some $\alpha \in \mathcal{K}_\infty$ and all $x \neq 0$

$$\dot{V}_{d=0}(x(t)) = \nabla V(x(t)) \cdot \Big(g_0(x(t)) + g_1(x(t))k(x(t))\Big) \leq -\alpha(x(t)). \tag{5.9}$$

Let us design a controller $\bar{k} : \mathbb{R}^n \to \mathbb{R}$ that input-to-state stabilizes (5.7). We will use V as an ISS Lyapunov function candidate for (5.7). The Lie derivative of V w.r.t. the system

$$\dot{x}(t) = g_0(x(t)) + g_1(x(t))\big(\bar{k}(x(t)) + d\big)$$

equals

$$\dot{V}_d(x) = \nabla V(x) \cdot \Big(g_0(x) + g_1(x)\big(\bar{k}(x) + d\big)\Big)$$
$$= \nabla V(x) \cdot \Big(g_0(x) + g_1(x)k(x)\Big) + \nabla V(x) \cdot g_1(x)\Big(\bar{k}(x) - k(x) + d\Big).$$

Choose

$$\bar{k}(x) := k(x) - \nabla V(x) \cdot g_1(x).$$

We continue estimates using this $\bar{k}$ and inequality (5.9) to obtain

$$\dot{V}_d(x) \le -\alpha(x) - \Big(\nabla V(x) \cdot g_1(x)\Big)^2 + \nabla V(x) \cdot g_1(x)d$$
$$\le -\alpha(x) - \Big(\nabla V(x) \cdot g_1(x)\Big)^2 + \frac{1}{2}\Big(\nabla V(x) \cdot g_1(x)\Big)^2 + \frac{1}{2}|d|^2$$
$$= -\alpha(x) - \frac{1}{2}\Big(\nabla V(x) \cdot g_1(x)\Big)^2 + \frac{1}{2}|d|^2.$$

This shows that V is an ISS Lyapunov function in a dissipative form and $\bar{k}$ is an input-to-state stabilizing controller for (5.7). □

Example 5.5 Let us apply (Theorem 5.4) for the system (5.5). The GAS Lyapunov function for (5.5) can be chosen as $V(x) := \frac{1}{2}x^2$. According to (Theorem 5.4) the following controller

$$\bar{k}(x) = -\frac{2x}{1+x^2} - x(1+x^2) \tag{5.10}$$

input-to-state stabilizes (5.4).

5.3 ISS Control Lyapunov Functions

Until now, we have used Lyapunov functions to verify and characterize the stability properties of the systems. This section shows that it is possible to solve the input-to-state stabilization problem using so-called ISS control Lyapunov functions. Moreover, for the case of control-affine systems, this leads to explicit formulas for the input-to-state stabilizing controllers.

Consider input-affine systems

$$\dot{x} = f(x, d) + G(x)u, \tag{5.11}$$

where $f : \mathbb{R}^n \times \mathbb{R}^p \to \mathbb{R}^n$ and $G : \mathbb{R}^n \to \mathbb{R}^{n\times m}$ are smooth functions.

A great deal of activities for the input-to-state stabilization of nonlinear systems was centered around the following concept:

Definition 5.6 A smooth function $V : \mathbb{R}^n \to \mathbb{R}$ is called an *ISS control Lyapunov function (ISS-CLF)* for (5.1), if there exist $\psi_1, \psi_2 \in \mathcal{K}_\infty$ such that

$$\psi_1(|x|) \le V(x) \le \psi_2(|x|) \quad \forall x \in \mathbb{R}^n$$

holds and there exist $\alpha, \chi \in \mathcal{K}_\infty$ such that for any $x \in \mathbb{R}^n$ and any $d \in \mathbb{R}^p$ the following holds:

$$\begin{aligned} \inf_{u\in\mathbb{R}^m} \nabla V(x) \cdot \big(f(x,d) + G(x)u\big) &= \inf_{u\in\mathbb{R}^m} \big(a(x,d) + b(x)u\big) \\ &\le -\alpha(|x|) + \chi(|d|), \end{aligned} \tag{5.12}$$

where

$$a(x,d) := \nabla V(x) \cdot f(x,d), \qquad b(x) := \nabla V(x) \cdot G(x).$$

Note that a is a scalar function, and $b(x)$ is an m-dimensional *row* vector at any x.

Define for each $x \in \mathbb{R}^n$

$$\omega(x) := \max_{d\in\mathbb{R}^p} \big(a(x,d) - \chi(|d|)\big), \tag{5.13}$$

and note that we can always make $\omega(x) < 0$ for all nonzero $x \in \mathbb{R}^n$ by enlarging χ, e.g., by considering instead of χ the function $r \mapsto \chi(r) + \hat{\chi}(r)$, where

$$\hat{\chi}(r) := \max_{|x|\le r,\ |d|\le r} |a(x,d)|, \quad r \ge 0.$$

The function $\chi + \hat{\chi}$ belongs to $\mathcal{K}_\infty$ as a is a continuous function and $a(0,0) = 0$ (since $\nabla V(0) = 0$ as 0 is a strict minimum of a smooth function V).

Furthermore, for any $x \in \mathbb{R}^n$ it holds that

$$\begin{aligned} \omega(x) \ge a(x,0) - \chi(0) = \nabla V(x) \cdot f(x,0) &\ge -|\nabla V(x) \cdot f(x,0)| \\ &\ge -h(|x|), \end{aligned} \tag{5.14}$$

where $h(r) := \sup_{|x|\le r} |\nabla V(x) \cdot f(x,0)|$, $r \ge 0$.

Pick any $\rho \in \mathcal{K}_\infty$, such that $\rho > \mathrm{id}$. Using $\chi + \hat{\chi}$ instead of χ in (5.13), for any fixed $x \in \mathbb{R}^n$ we see that

$$\begin{aligned}\max_{d:|d|>\rho(|x|)} &\big(a(x,d) - \hat{\chi}(|d|) - \chi(|d|)\big)\\ &\le \max_{d:|d|>\rho(|x|)} \big(a(x,d) - \max_{|\tilde{x}|\le|d|,\ |\tilde{d}|\le|d|} |a(\tilde{x},\tilde{d})| - \chi(|d|)\big)\\ &\le \max_{d:|d|>\rho(|x|)} \big(-\chi(|d|)\big) \le -\chi\circ\rho(|x|).\end{aligned}$$

Here we have used that for $|d| > \rho(|x|)$ we have

$$\max_{|\tilde{x}|\le|d|,\ |\tilde{d}|\le|d|} |a(\tilde{x},\tilde{d})| \ge \max_{|\tilde{x}|\le|x|,\ |\tilde{d}|\le|d|} |a(\tilde{x},\tilde{d})| \ge |a(x,d)|.$$

Now taking ρ large enough, we see that the maximum in (5.13) cannot be obtained on the set $|d| > \rho(|x|)$ in view of (5.14).

Summarizing the above considerations, we see that for sufficiently large $\rho, \chi \in \mathcal{K}_\infty$ we have

$$\omega(x) = \max_{|d|\le\rho(|x|)} \big(a(x,d) - \chi(|d|)\big). \tag{5.15}$$

Assumption 5.7 We further assume that V satisfies the *small control property*: for each $\varepsilon > 0$ there is $\delta > 0$ such that whenever $0 < |x| < \delta$ there exists some $u \in \mathbb{R}^m$: $|u| \le \varepsilon$, for which

$$\omega(x) + b(x)u \le -\alpha(|x|),$$

where without loss of generality, we assume that α is the same as in (5.12).

Definition 5.8 We say that $k : \mathbb{R}^n \to \mathbb{R}^n$ is *almost smooth* if k is continuous on $\mathbb{R}^n$ and smooth in $\mathbb{R}^n\backslash\{0\}$.

The function ω in (5.13) is Lipschitz continuous, but it is not necessarily smooth. However, we can pick another function $\bar{\omega}$, which is almost smooth and satisfies

$$\omega(x) + \frac{1}{3}\alpha(|x|) \le \bar{\omega}(x) \le \omega(x) + \frac{2}{3}\alpha(|x|) \quad \forall x. \tag{5.16}$$

The following result gives an explicit "universal" construction of stabilizing feedback laws for a control-affine system possessing an ISS-CLF. This simplifies the stabilization problem considerably since the construction of an ISS-CLF is usually an easier problem than a direct construction of a stabilizing feedback. Indeed, in the latter case, one needs anyway to provide an ISS Lyapunov function for the closed-loop system to show that the control indeed input-to-state stabilizes the system.

Theorem 5.9 (Universal stabilizing laws for affine systems) *Consider the system* (5.11). *Let V be a smooth ISS control Lyapunov function for a control-affine system* (5.11) *satisfying the small control property. Then the feedback*

$$u(t) := k(x(t)) := K\big(\bar{\omega}(x(t)), b^T(x(t))\big), \tag{5.17}$$

where $\bar{\omega}$ is given as above (see 5.16), input-to-state stabilizes the system (5.11) and k is almost smooth, where $K : \mathbb{R} \times \mathbb{R}^m \to \mathbb{R}^m$ is given by

$$K(a, b) := \begin{cases} -\frac{a+\sqrt{a^2+|b|^4}}{|b|^2} b, & b \neq 0, \\ 0, & b = 0. \end{cases} \tag{5.18}$$

The expression (5.17) for the stabilizing feedback k is also called Sontag's formula.

Proof **k is smooth on $\mathbb{R}^n \backslash \{0\}$.** Define an open set

$$S := \{(a, b) \in \mathbb{R}^2 : a < 0 \text{ or } b > 0\} \tag{5.19}$$

and the function $\varphi : S \to \mathbb{R}$ defined by

$$\varphi(a, b) := \begin{cases} 0, & (a, 0) \in S, \\ \frac{a+\sqrt{a^2+b^2}}{b}, & \text{otherwise}. \end{cases} \tag{5.20}$$

Further define $F : S \times \mathbb{R} \to \mathbb{R}$ by

$$F(a, b, w) := bw^2 - 2aw - b. \tag{5.21}$$

For each $(a, b) \in S$ it holds that

$$F(a, b, \varphi(a, b)) = \frac{a^2 + 2a\sqrt{a^2+b^2} + a^2 + b^2}{b} - 2a\frac{a + \sqrt{a^2+b^2}}{b} - b = 0.$$

Furthermore, $\frac{\partial F}{\partial w} = 2bw - 2a$, and thus for all $(a, b) \in S$

$$\frac{\partial F}{\partial w}(a, b, \varphi(a, b)) = 2\sqrt{a^2+b^2} > 0.$$

By the analytic implicit function theorem, φ is a real-analytic function.

It is easy to see that k can be represented using function φ as

$$k(x) = \begin{cases} b(x)\varphi(\bar{\omega}(x), |b(x)|^2), & x \neq 0, \\ 0, & x = 0. \end{cases} \tag{5.22}$$

Here note that if $|b(x)| = 0$, then by (5.12) and (5.13) it holds that $\omega(x) \leq -\alpha(|x|)$, and thus $\bar{\omega}(x) \leq -\frac{1}{3}\alpha(|x|) < 0$, provided $x \neq 0$, and thus $(\bar{\omega}(x), |b(x)|^2) \in S$ for any $x \neq 0$, and in particular the representation (5.22) makes sense.

By construction, $\bar{\omega}$ is almost smooth, and b is smooth since G and V are. Thus, by real-analyticity of φ, we see from the representation (5.22) that k is smooth outside of the origin.

k is continuous on $\mathbb{R}^n$. It suffices to prove continuity of k at zero. As V has a minimum at 0 and V is smooth, $\nabla V(0) = 0$. Thus, also $b(0) = 0$. Since b is smooth and V satisfies the small control property, for any $\varepsilon > 0$ there is $\delta > 0$ such that for all $x \in \mathbb{R}^n$: $0 < |x| \leq \delta$, it holds that $|b(x)| \leq \frac{\varepsilon}{2}$ and there exists some $u \in \mathbb{R}^m$: $|u| \leq \frac{\varepsilon}{2}$, for which

$$\omega(x) + b(x)u \leq -\alpha(|x|).$$

Hence

$$\bar{\omega}(x) + b(x)u \leq \omega(x) + \frac{2}{3}\alpha(|x|) + b(x)u \leq -\frac{1}{3}\alpha(|x|)$$

and by Cauchy–Schwarz inequality, it holds that

$$\bar{\omega}(x) \leq |b(x)||u| \leq \frac{\varepsilon}{2}|b(x)|.$$

If $\bar{\omega}(x) > 0$, then also $|\bar{\omega}(x)| \leq \frac{\varepsilon}{2}|b(x)|$, and according to the formula (5.22) we have (recall that we assumed that $b(x) \neq 0$) that

$$\begin{aligned} |k(x)| &= |b(x)\varphi(\bar{\omega}(x), |b(x)|^2)| \leq |b(x)|\frac{|\bar{\omega}(x)| + \sqrt{\bar{\omega}(x)^2 + |b(x)|^4}}{|b(x)|^2} \\ &\leq \frac{|\bar{\omega}(x)| + |\bar{\omega}(x)| + |b(x)|^2}{|b(x)|} \leq \frac{|b(x)|\varepsilon + |b(x)|^2}{|b(x)|} \leq \frac{3}{2}\varepsilon. \end{aligned} \tag{5.23}$$

If $\bar{\omega}(x) \leq 0$, then

$$0 < \bar{\omega}(x) + \sqrt{\bar{\omega}(x)^2 + |b(x)|^4} \leq \bar{\omega}(x) + |\bar{\omega}(x)| + |b(x)|^2 = |b(x)|^2,$$

and thus

$$|k(x)| = |b(x)|\varphi(\bar{\omega}(x), |b(x)|^2) \leq |b(x)| \leq \frac{1}{2}\varepsilon. \tag{5.24}$$

Thus, k is continuous at 0 and as k is smooth on $\mathbb{R}^n \backslash \{0\}$, k is continuous on $\mathbb{R}^n$ and almost smooth.

k is a stabilizing feedback. Consider the system (5.11) with the controller (5.17):

$$\dot{x} = f(x, d) + G(x)k(x). \tag{5.25}$$

The Lie derivative of V equals

$$\dot{V}(x) = a(x, d) + b(x)k(x) \leq \omega(x) + b(x)k(x) + \chi(|d|).$$

By (5.16) we have

$$\begin{aligned}\dot{V}(x) &\le \bar{\omega}(x) - \frac{1}{3}\alpha(|x|) + b(x)k(x) + \chi(|d|)\\ &\le \bar{\omega}(x) - \frac{1}{3}\alpha(|x|) - \left(\bar{\omega}(x) + \sqrt{|\bar{\omega}(x)|^2 + |b(x)|^4}\right) + \chi(|d|)\\ &\le -\frac{1}{3}\alpha(|x|) + \chi(|d|).\end{aligned}$$

This shows that V is an ISS Lyapunov function for (5.25), and hence (5.25) is ISS. □

5.4 ISS Backstepping

We discuss the ISS backstepping method for recursive design of stabilizing controllers for control systems of the following form:

$$\dot{x} = f(x) + G_1(x)z + G_2(x)d, \tag{5.26a}$$

$$\dot{z} = u + F(x, z)d, \tag{5.26b}$$

where $x \in \mathbb{R}^n$ and $z \in \mathbb{R}^m$ denote the states of the subsystems (5.26a), (5.26b), $u \in \mathbb{R}^m$ is a control input, and $d \in \mathbb{R}^p$ is a disturbance acting on the system.

With the control u, we can directly control the z-subsystem, and it is easy to check that the feedback $u = -z - (1 + |F(x, z)|^2)$ stabilizes the z-subsystem in the ISS sense w.r.t. the disturbance d. But at the same time, we cannot directly influence the x-subsystem. Nevertheless, in the following result, we show that if x-subsystem can be input-to-state stabilized by using z as a control, then the whole system can be input-to-state stabilized w.r.t. d. Moreover, suppose the ISS Lyapunov function for the x-subsystem is known. Then it is possible to explicitly construct an input-to-state stabilizing controller for the whole system and construct an ISS Lyapunov function for the whole system.

Theorem 5.10 *Assume that all the nonlinearities involved in the system (5.26) are smooth and $f(0) = 0$. Further assume that (5.26a) is input-to-state stabilizable by a smooth control law $z = k(x)$ satisfying $k(0) = 0$. Then the overall system (5.26) is input-to-state stabilizable by a certain smooth control law $u = \bar{k}(x, z)$.*

Proof The closed-loop system (5.26a) with a smooth controller $z = k(x)$ takes form

$$\dot{x} = f(x) + G_1(x)k(x) + G_2(x)d. \tag{5.27}$$

By assumption, it is ISS. As all the nonlinearities in (5.26) are smooth, Theorem 2.63 ensures the existence of a smooth ISS Lyapunov function W for the system (5.27), satisfying for certain $\alpha, \sigma \in \mathcal{K}_\infty$ and all $x \in \mathbb{R}^n$, $d \in \mathbb{R}^p$

$$\nabla W(x) \cdot \big(f(x) + G_1(x)k(x) + G_2(x)d\big) \le -\alpha(|x|) + \sigma(|d|). \tag{5.28}$$

Consider the function $V : \mathbb{R}^n \times \mathbb{R}^m \to \mathbb{R}_+$, defined by

$$V(x, z) := W(x) + \frac{1}{2}|k(x) - z|^2, \quad x \in \mathbb{R}^n,\ z \in \mathbb{R}^m. \tag{5.29}$$

As $k(0) = 0$, we have that $V(0, 0) = 0$. Furthermore, V is a positive definite and radially unbounded function. By Proposition A.14 and Corollary A.11, V satisfies the sandwich estimates (4.10).

Let us obtain the dissipative estimate for V.

$$\begin{aligned}
\dot V_d(x, z) &= \nabla W(x) \cdot \big(f(x) + G_1(x)z + G_2(x)d\big) \\
&\qquad + (k(x) - z)^T \Big(\nabla k(x) \cdot \big(f(x) + G_1(x)z + G_2(x)d\big) - u - F(x, z)d\Big) \\
&= \nabla W(x) \cdot \big(f(x) + G_1(x)k(x) + G_2(x)d\big) + \nabla W(x) \cdot G_1(x)\big(z - k(x)\big) \\
&\qquad + (k(x) - z)^T \Big(\nabla k(x) \cdot \big(f(x) + G_1(x)z + G_2(x)d\big) - u - F(x, z)d\Big).
\end{aligned}$$

Using (5.28), we proceed to

$$\begin{aligned}
\dot V_d(x, z) \le& -\alpha(|x|) + \sigma(|d|) + (k(x) - z)^T \Big(- u - (\nabla W(x) \cdot G_1(x))^T \\
&+ \nabla k(x) \cdot \big(f(x) + G_1(x)z\big) + \nabla k(x) \cdot G_2(x)d - F(x, z)d\Big).
\end{aligned} \tag{5.30}$$

Now choose $u = \bar k(x, z)$, with

$$\begin{aligned}
\bar k(x, z) :=& - (\nabla W(x) \cdot G_1(x))^T + \nabla k(x) \cdot \big(f(x) + G_1(x)z\big) \\
&+ (k(x) - z)\Big(1 + |\nabla k(x) \cdot G_2(x)|^2 + |F(x, z)|^2\Big).
\end{aligned} \tag{5.31}$$

With this controller, we proceed from (5.30) to

$$\begin{aligned}
\dot V_d(x, z) \le& -\alpha(|x|) + \sigma(|d|) - |k(x) - z|^2\Big(1 + |\nabla k(x) \cdot G_2(x)|^2 + |F(x, z)|^2\Big) \\
&+ (k(x) - z)^T \nabla k(x) \cdot G_2(x)d - (k(x) - z)^T F(x, z)d.
\end{aligned} \tag{5.32}$$

By Cauchy–Schwarz inequality, we proceed to

$$\begin{aligned}
\dot V_d(x, z) &\le -\alpha(|x|) + \sigma(|d|) - |k(x) - z|^2\Big(1 + |\nabla k(x) \cdot G_2(x)|^2 + |F(x, z)|^2\Big) \\
&\qquad + \frac{1}{2}|k(x) - z|^2|\nabla k(x) \cdot G_2(x)|^2 + \frac{1}{2}|d|^2 + \frac{1}{2}|(k(x) - z)^T F(x, z)|^2 + \frac{1}{2}|d|^2 \\
&\le -\alpha(|x|) + \sigma(|d|) - |k(x) - z|^2 + |d|^2.
\end{aligned} \tag{5.33}$$

This shows that V is a smooth ISS Lyapunov function in a dissipative form for the system (5.26) with the feedback law (5.31). □

Example 5.11 We are going to construct a global input-to-state stabilizing feedback for a planar control system

$$\dot{x} = z^3 + x^2 d, \tag{5.34a}$$
$$\dot{z} = u + d. \tag{5.34b}$$

We follow the method developed in Theorem 5.10, with some simplifications due to the specific form of a system.

First of all, x-subsystem can be input-to-state stabilized by the feedback $z = k(x) := -x$. The corresponding closed-loop x-subsystem takes the form

$$\dot{x} = -x^3 + x^2 d,$$

and the ISS Lyapunov function for this system can be chosen as

$$W(x) := x^2.$$

Following the proof of Theorem 5.10, we consider the function V defined by (5.29), which takes the form

$$V(x,z) := x^2 + \frac{1}{2}(x+z)^2. \tag{5.35}$$

We obtain:

$$\begin{aligned}
&\nabla V(x,z)\cdot(z^3 + x^2 d, u + d)\\
&\quad = 2x(z^3 + x^2 d) + (x+z)(\dot{x} + \dot{z})\\
&\quad = 2x(-x^3 + x^2 d) + 2x(x^3 + z^3) + (x+z)(z^3 + x^2 d + u + d)\\
&\quad = 2x(-x^3 + x^2 d) + 2x(x+z)(x^2 - xz + z^2) + (x+z)(z^3 + x^2 d + u + d)\\
&\quad = 2x(-x^3 + x^2 d) + (x+z)(2x^3 - 2x^2 z + 2xz^2 + z^3 + x^2 d + u + d)\\
&\quad = 2x(-x^3 + x^2 d) + (x+z)(2x^3 - 2x^2 z + 2xz^2 + z^3 + u) + (x+z)(x^2+1)d.
\end{aligned}$$

With the feedback controller

$$u := \Big(-(2x^3 - 2x^2 z + 2xz^2 + z^3) - (x+z) \Big)(x^2+1)^2 \tag{5.36}$$

we proceed to

$$\begin{aligned}
&\nabla V(x,z)\cdot(z^3 + x^2 d, u + d)\\
&\quad = 2x(-x^3 + x^2 d) - (x+z)^2(x^2+1)^2 + (x+z)(x^2+1)d\\
&\quad \le -2x^4 + 2x^3 d - (x+z)^2(x^2+1)^2 + \frac{1}{2}(x+z)^2(x^2+1)^2 + \frac{1}{2}|d|^2.
\end{aligned}$$

Using Young's inequality $x^3 d \leq \frac{|d|^4}{4} + \frac{(|x|^3)^{4/3}}{4/3} = \frac{|d|^4}{4} + \frac{3x^4}{4}$, we obtain that

$$\begin{aligned}
&\nabla V(x,z)\cdot(z^3 + x^2 d, u + d) \\
&\quad \leq -2x^4 + \frac{|d|^4}{2} + \frac{3x^4}{2} - \frac{1}{2}(x+z)^2(x^2+1)^2 + \frac{1}{2}|d|^2 \\
&\quad \leq -\frac{1}{2}x^4 - \frac{1}{2}(x+z)^2(x^2+1)^2 + \frac{1}{2}|d|^2 + \frac{|d|^4}{2},
\end{aligned}$$

which shows that V is a smooth ISS Lyapunov function for the closed-loop system (5.34) with the smooth controller (5.36). In particular, this closed-loop system is ISS.

The recursive nature of the backstepping allows using the method also for couplings of an arbitrary finite number of systems if the interconnection has the lower triangular form.

5.5 Global Stabilization of Axial Compressor Model. Gain Assignment Technique

This section shows how ISS backstepping can be used together with small-gain theorems to globally asymptotically stabilize important practical nonlinear systems.

Jet engine compression systems have been studied actively during the last decades. One of the key objectives was to prevent two types of instability: *rotating stall* and *surge*. Consider an *axial compressor model*, which is the three-state Galerkin approximation of a nonlinear PDE model derived in [71].

$$\dot{\phi} = -\psi + \frac{3}{2}\phi + \frac{1}{2} - \frac{1}{2}(\phi+1)^3 - 3(\phi+1)R, \tag{5.37a}$$

$$\dot{\psi} = \frac{1}{\beta^2}(\phi + 1 - u), \tag{5.37b}$$

$$\dot{R} = \sigma R\big(-2\phi - \phi^2 - R\big). \tag{5.37c}$$

Here ϕ and ψ are the deviations of the mass flow and the pressure rise from their set points. R is the normalized stall cell squared amplitude, and σ, β are positive constants. The control input u is the flow through the throttle.

In view of the physical sense of R, we assume that $R(0) \geq 0$. This guarantees that $R(t) \geq 0$ for all $t \geq 0$.

Our aim is to construct a controller u that globally asymptotically stabilizes the system (5.37).

Assuming that we can measure the full state (ϕ, ψ, R), one can construct a controller $u = u(\phi, \psi, R)$ using 2 steps of backstepping technique that we have introduced in Sect. 5.4. Indeed, using ϕ as a control input, one can stabilize the R-system.

Using ψ as a control, one can stabilize ϕ-subsystem. Finally, the ISS backstepping allows stabilizing the whole system using u as an input.

This section shows that one can construct a globally asymptotically stabilizing controller u using measurements of ϕ, ψ only, using an alternative approach.

Proposition 5.12 *The controller* $u = u(\phi, \psi)$ *given by* (5.45) *globally asymptotically stabilizes the system* (5.37) *using the measurements of* ϕ, ψ.

Proof We view the system (5.37) as an interconnection of the (ϕ, ψ)-subsystem and the R-subsystem. We show that R-subsystem is ISS with respect to the input ϕ from the other subsystem. Furthermore, using ISS backstepping, we design u to input-to-state stabilize the (ϕ, ψ)-subsystem with R as in input. We choose the gain small enough to guarantee that the small-gain condition holds (this step is called *gain assignment*). Finally, we employ the small-gain theorem to show that the whole system (5.37) is ISS.

Step 1. ISS Lyapunov function for$R$-subsystem. Let us show ISS of R-subsystem (5.37c). Take an ISS Lyapunov function candidate

$$Z(R) := R^2, \quad R \geq 0.$$

Lie derivative of Z has the form:

$$\begin{aligned}\dot{Z}_\phi(R) &= 2\sigma R^2\big(-2\phi - \phi^2 - R\big)\\ &\leq -2\sigma R^3 + 4\sigma|\phi|R^2.\end{aligned}$$

Now pick any $\delta > 0$. The following implication holds

$$R \geq (2+\delta)|\phi| \quad \Rightarrow \quad \dot{Z}_\phi(R) \leq -2\sigma R^3 + \frac{4}{2+\delta}\sigma R^3 = -\frac{2\delta\sigma}{2+\delta}R^3. \tag{5.38}$$

This shows that Z is an ISS Lyapunov function for R-subsystem.

Step 2. Stabilization of ϕ-subsystem using ψ as a control input. Decomposing $(\phi + 1)^3$, we can equivalently represent the ϕ-subsystem as

$$\dot{\phi} = -\psi - \frac{3}{2}\phi^2 - \frac{1}{2}\phi^3 - 3(\phi+1)R.$$

The ϕ-subsystem can be input-to-state stabilized by treating ψ as an input, using the feedback

$$\psi := c_1\phi, \tag{5.39}$$

for a suitable $c_1 > 0$. The corresponding ISS Lyapunov function can be chosen as

$$W(\phi) := \frac{1}{2}\phi^2, \quad \phi \in \mathbb{R}.$$

We have:

$$\begin{aligned}\dot{W}_R(\phi) &= -\phi\Big(c_1\phi + \frac{3}{2}\phi^2 + \frac{1}{2}\phi^3 + 3(\phi+1)R\Big)\\ &= -c_1\phi^2 - \frac{3}{2}\phi^3 - \frac{1}{2}\phi^4 - 3\phi^2 R - 3\phi R.\end{aligned}$$

As $R \geq 0$, we have

$$\dot{W}_R(\phi) \leq -c_1\phi^2 + \frac{3}{2}|\phi|^3 - \frac{1}{2}\phi^4 + 3|\phi|R. \tag{5.40}$$

Recall Cauchy's inequality

$$ab \leq \frac{1}{4\varepsilon}a^2 + \varepsilon b^2, \quad a, b \geq 0, \quad \varepsilon > 0.$$

To get rid of the cubic term, we use this inequality with $\varepsilon = \frac{1}{2}$, to obtain

$$\frac{3}{2}|\phi|^3 = \frac{3}{2}|\phi||\phi|^2 \leq \frac{1}{2}\Big(\frac{3}{2}|\phi|\Big)^2 + \frac{1}{2}\phi^4 = \frac{9}{8}\phi^2 + \frac{1}{2}\phi^4. \tag{5.41}$$

Further, we take any $\varepsilon > 0$ and use the Cauchy's inequality for the mixed term $3|\phi|R$ in (5.40) as follows:

$$3|\phi|R \leq \frac{1}{4\varepsilon}9|\phi|^2 + \varepsilon R^2.$$

Substituting the above findings into (5.40), we obtain that

$$\dot{W}_R(\phi) \leq -\Big(c_1 - \frac{9}{8} - \frac{9}{4\varepsilon}\Big)|\phi|^2 + \varepsilon R^2. \tag{5.42}$$

Take $c_1, \varepsilon > 0$ such that

$$c_1 - \frac{9}{8} - \frac{9}{4\varepsilon} = \frac{c(\varepsilon)}{2},$$

for a certain $c(\varepsilon) > 0$. With this choice, we have

$$\dot{W}_R(\phi) \leq -c(\varepsilon)W(\phi) + \varepsilon R^2. \tag{5.43}$$

This shows that the feedback (5.39) for a proper choice of c_1, ε makes the ϕ-subsystem ISS, and W is a corresponding ISS Lyapunov function.

Step 3. ISS backstepping for (ϕ, ψ)-subsystem. Following the ISS backstepping methodology, we construct an ISS Lyapunov function for the (ϕ, ψ)-subsystem (for a suitable input) as

$$V(\phi,\psi) := W(\phi) + |c_1\phi - \psi|^2. \tag{5.44}$$

We have:

$$\begin{aligned}
\dot{V}_R(\phi,\psi) &= -\phi\Big(\psi + \frac{3}{2}\phi^2 + \frac{1}{2}\phi^3 + 3(\phi+1)R\Big) + (c_1\phi - \psi)(c_1\dot{\phi} - \dot{\psi}) \\
&= -\phi\Big(c_1\phi + \frac{3}{2}\phi^2 + \frac{1}{2}\phi^3 + 3(\phi+1)R\Big) + \phi(c_1\phi - \psi) \\
&\quad + (c_1\phi - \psi)\Big(-c_1\Big(\psi + \frac{3}{2}\phi^2 + \frac{1}{2}\phi^3 + 3(\phi+1)R\Big) - \frac{1}{\beta^2}(\phi + 1 - u)\Big) \\
&= \dot{W}_R(\phi) - 3c_1(c_1\phi - \psi)(\phi+1)R \\
&\quad + (c_1\phi - \psi)\Big(\phi - c_1\Big(\psi + \frac{3}{2}\phi^2 + \frac{1}{2}\phi^3\Big) - \frac{\phi+1}{\beta^2} + \frac{1}{\beta^2}u\Big).
\end{aligned}$$

Using Cauchy's inequality once again for the mixed term, we have for the above ε:

$$-3c_1(c_1\phi - \psi)(\phi+1)R \le \varepsilon R^2 + \frac{1}{4\varepsilon}9c_1^2(c_1\phi - \psi)^2(\phi+1)^2.$$

Now choose the controller u as:

$$\begin{aligned}
u(\phi,\psi) := &-\beta^2\Big(\phi - c_1\Big(\psi + \frac{3}{2}\phi^2 + \frac{1}{2}\phi^3\Big) - \frac{\phi+1}{\beta^2}\Big) \\
&- \beta^2\frac{1}{4\varepsilon}9c_1^2(c_1\phi - \psi)(\phi+1)^2 \\
&- c(\varepsilon)(c_1\phi - \psi).
\end{aligned} \tag{5.45}$$

With this controller, and recalling (5.43), the Lie derivative of V takes the form

$$\begin{aligned}
\dot{V}_R(\phi,\psi) &\le -c(\varepsilon)W(\phi) + \varepsilon R^2 - c(\varepsilon)|c_1\phi - \psi|^2 + \varepsilon R^2 \\
&= -c(\varepsilon)V(\phi,\psi) + 2\varepsilon R^2.
\end{aligned}$$

Now take any $a \in (0,1)$. Then

$$V(\phi,\psi) \ge \frac{2\varepsilon}{c(\varepsilon)(1-a)}Z(R) \quad \Rightarrow \quad \dot{V}_R(\phi,\psi) \le -ac(\varepsilon)V(\phi,\psi).$$

Hence, V is an ISS Lyapunov function for the (ϕ,ψ)-subsystem with the controller u defined by (5.45).

Step 4. Small-gain analysis. The condition $R \ge (2+\delta)|\phi|$ in (5.38) is equivalent to $Z(R) = R^2 \ge 2(2+\delta)^2\frac{1}{2}|\phi|^2$. Thus (5.38) ensures that

$$Z(R) \ge 2(2+\delta)^2\, V(\phi,\psi) \quad \Rightarrow \quad \dot{Z}_\phi(R) \le -\frac{2\delta\sigma}{2+\delta}R^3. \tag{5.46}$$

The coupled system is ISS provided the small-gain condition holds:

$$2(2+\delta)^2 \cdot \frac{2\varepsilon}{c(\varepsilon)(1-a)} < 1.$$

However, for any given $\varepsilon, \delta > 0$ and $a \in (0, 1)$ one can find $c_1 > 0$ large enough such that this condition holds.

Hence, (5.37) is globally asymptotically stabilizable by using the controller (5.45). □

5.6 Event-based Control

In previous sections, in our design of stabilizing laws, we have assumed that we can update a controller continuously. However, for digital controllers, this assumption is not valid, as the value of the controller can be updated only at some discrete moments. It is necessary to ensure in advance that by updating the control law at reasonably chosen triggering times, we can stabilize the system. As we will see, using the ISS methodology, we can ensure this and even more.

Consider a control system

$$\dot{x} = f(x, u), \tag{5.47}$$

with f that is locally Lipschitz continuous in both arguments. Assume that there is a locally Lipschitz continuous control law

$$u := k(x), \tag{5.48}$$

such that the closed-loop system

$$\dot{x} = f(x, k(x+e)), \tag{5.49}$$

is ISS with respect to the *measurement error* e.

As we assume that f is Lipschitz continuous on bounded balls of $\mathbb{R}^n \times \mathbb{R}^m$, and k is Lipschitz continuous, then $(x, e) \mapsto f(x, k(x+e))$ is locally Lipschitz continuous in both x and e. This implies the well-posedness of (5.49) and furthermore by Theorem 2.63 there is a smooth ISS Lyapunov function $V : \mathbb{R}^n \to \mathbb{R}_+$ in a dissipative form for (5.49). In view of Proposition 2.20, this means that there exist $\psi_1, \psi_2 \in \mathcal{K}_\infty$, $\alpha \in \mathcal{K}_\infty$, and $\xi \in \mathcal{K}$ such that the sandwich estimate holds

$$\psi_1(|x|) \le V(x) \le \psi_2(|x|) \quad \forall x \in \mathbb{R}^n, \tag{5.50}$$

and for all $x \in \mathbb{R}^n$, $e \in \mathbb{R}^n$ the following dissipation inequality is valid:

$$\nabla V(x) \cdot f(x, k(x+e)) \leq -\alpha(|x|) + \xi(|e|). \tag{5.51}$$

In the *digital implementation* of the control law (5.48) one usually follows the so-called *sample-and-hold technique*, according to which, one performs the measurements $(x(t_k))_{k \in \mathbb{N}}$ of the state at time instants $(t_k)_{k \in \mathbb{N}}$ and employs the corresponding "digital" control law

$$u_d(t) := u(t_k) = k(x(t_k)), \quad t \in [t_k, t_{k+1}). \tag{5.52}$$

In the *sample-data control* (see Sect. 5.11.2) the sequence (t_k) is chosen to be periodic, and such a control can then be called a *time-triggered control*. Instead, we would like to update the control value only if the current one "doesn't work properly". To formalize this idea, we introduce the *error function*

$$e(t) := x(t_k) - x(t), \quad t \in [t_k, t_{k+1}). \tag{5.53}$$

The system (5.47) endowed with the digital controller (5.52) takes the form

$$\dot{x}(t) = f\big(x(t), k\big(x(t_k)\big)\big), \quad t \in [t_k, t_{k+1}), \tag{5.54}$$

which is precisely (5.49) with the error e as in (5.53).

In view of (5.51), if for some $\sigma \in (0, 1)$

$$\xi(|e|) \leq \sigma\alpha(|x|), \tag{5.55}$$

then

$$\nabla V(x) \cdot f\big(x, k(x+e)\big) \leq -(1-\sigma)\alpha(|x|). \tag{5.56}$$

To ensure that (5.55) is always satisfied, it suffices to update the controller provided the following condition (or *event*) holds:

$$\xi(|e|) \geq \sigma\alpha(|x|). \tag{5.57}$$

With the event-triggering condition (5.57), the estimate (5.56) holds along all trajectories of (5.54), V is a strict Lyapunov function for (5.54) and thus (5.54) is UGAS.

However, the above analysis makes sense if the event-triggered control is *feasible*, i.e., the sequence of triggering times (t_k) does not have a *Zeno behavior* (i.e., provided that it is unbounded). The following theorem shows that this is the case if we assume that α^{-1} and ξ are locally Lipschitz on bounded sets.

Theorem 5.13 *Consider a control system (5.47) with f being Lipschitz continuous on bounded balls of $\mathbb{R}^n \times \mathbb{R}^m$, and with a locally Lipschitz controller given by (5.48) making the closed-loop system (5.49) ISS w.r.t. the measurement disturbance e, and* *let V be an ISS Lyapunov function for the closed-loop system as defined above with α^{-1} and ξ locally Lipschitz on bounded sets.*

Then for every compact set $S \subset \mathbb{R}^n$ with $0 \in S$ there is a time $\tau > 0$ such that for any initial condition in S the interexecution times $(t_{i+1} - t_i)_{i \in \mathbb{N}}$ implicitly defined by the execution rule (5.57) with $\sigma \in (0, 1)$ are lower bounded by τ, i.e., $t_{i+1} - t_i \geq \tau$ for any $i \in \mathbb{N}$.

Proof Take any compact set $S \subset \mathbb{R}^n$ with $0 \in S$, and consider a compact set

$$R := \{x \in \mathbb{R}^n : V(x) \leq \mu\}, \tag{5.58}$$

where μ is chosen large enough such that $S \subset R$.

Define also another compact set

$$E := \Big\{e \in \mathbb{R}^n : \alpha^{-1}\Big(\frac{1}{\sigma}\xi(|e|)\Big) \leq \sup_{x \in R} |x|\Big\}. \tag{5.59}$$

As α^{-1} and ξ are assumed to be locally Lipschitz continuous, then $r \mapsto \alpha^{-1}(\frac{1}{\sigma}\xi(r))$ is locally Lipschitz continuous as well. Let P be a Lipschitz constant for this map corresponding to the compact set E. Then it holds that

$$\Big|\alpha^{-1}\Big(\frac{1}{\sigma}\xi(r)\Big) - \alpha^{-1}\Big(\frac{1}{\sigma}\xi(s)\Big)\Big| \leq P|r - s|, \quad r, s \in E. \tag{5.60}$$

In particular, it holds that

$$\alpha^{-1}\Big(\frac{1}{\sigma}\xi(|e|)\Big) \leq P|e|, \quad e \in E.$$

Now instead of the execution rule (5.57) consider a more conservative rule $P|e| \geq |x|$. Updating a controller according to this rule, we enforce the validity of a condition $P|e| \leq |x|$ along all trajectories, which in view of (5.60) implies (5.55).

We want to show that the triggering times for the high-frequency execution rule $P|e| \geq |x|$ are uniformly bounded from below. This will imply the claim of the theorem.

First note that

$$\frac{\mathrm{d}}{\mathrm{d}t}|e| = \frac{\mathrm{d}}{dt}\sqrt{e^T e} = \frac{d}{dt}\sqrt{e^T e} = \frac{1}{2\sqrt{e^T e}}\frac{\mathrm{d}}{dt}(e^T e) = \frac{1}{2\sqrt{e^T e}} 2e^T \dot{e} = \frac{e^T \dot{e}}{|e|}. \tag{5.61}$$

Furthermore, as f is Lipschitz continuous on bounded balls of $\mathbb{R}^n \times \mathbb{R}^m$ and k is Lipschitz continuous, then $(x, e) \mapsto f(x, k(x + e))$ is Lipschitz continuous in both x and e. As (5.49) is ISS, $f(0, k(0)) = 0$, and thus there is $L > 0$ such that for any $(x, e) \in R \times E$ it holds that

$$|f(x, k(x + e)| \leq L|x| + L|e|. \tag{5.62}$$

Now we compute

$$\begin{aligned}\frac{\mathrm{d}}{\mathrm{d}t}\frac{|e|}{|x|} &= \frac{1}{|x|^2}\Big(\frac{e^T\dot{e}}{|e|}|x| - \frac{x^T\dot{x}}{|x|}|e|\Big) = \frac{e^T\dot{e}}{|e||x|} - \frac{|e|x^T\dot{x}}{|x|^3} = -\frac{e^T\dot{x}}{|e||x|} - \frac{|e|x^T\dot{x}}{|x|^3}\\ &\leq \frac{|\dot{x}|}{|x|} + \frac{|e||\dot{x}|}{|x|^2} = \frac{|\dot{x}|}{|x|}\Big(1+\frac{|e|}{|x|}\Big)\\ &\leq \frac{L|x|+L|e|}{|x|}\Big(1+\frac{|e|}{|x|}\Big) \leq L\Big(1+\frac{|e|}{|x|}\Big)^2. \end{aligned} \tag{5.63}$$

Defining $y := \frac{|e|}{|x|}$, we see that

$$\dot{y} \leq L(1+y)^2,$$

and for the initial condition $y(0) = 0$ (corresponding to the zero error after the update of the controller), we have:

$$y(t) \leq \phi(t),$$

where ϕ is the solution of the equation $\dot{\phi} = L(1+\phi)^2$, with the initial condition $\phi(0) = 0$.

Thus, the interexecution times are bounded by the time it takes for ϕ to evolve from 0 to $\frac{1}{P}$. □

5.7 Outputs and Output Feedback

In Sect. 5.2, we investigated a problem of an input-to-state stabilization of nonlinear systems (5.1) by means of a *static state feedback* $u(t) = k(x(t))$. However, in most cases, the measurements of the whole state of a system are not available due to the high costs of sensors or the technical impossibility of such measurements. Instead, we can measure only a certain function of the state $y = h(x)$ that is called the *output* of the system. The system (5.1) together with the output law forms a *control system with outputs*

$$\dot{x} = f(x,u,d), \tag{5.64a}$$

$$y = h(x), \tag{5.64b}$$

where $x(t) \in \mathbb{R}^n$, $u \in \mathcal{U} := L^\infty(\mathbb{R}_+, \mathbb{R}^m)$, $d \in \mathcal{D} := L^\infty(\mathbb{R}_+, \mathbb{R}^q)$, $y(t) \in \mathbb{R}^p$.

A fundamental problem of control theory is to stabilize the state of the system using solely the knowledge of its output. A natural approach to this problem would be to use the *static output feedback* $u(t) = k(y(t))$. However, this method is too restrictive even for linear systems that are controllable and observable, as the next example shows.

Example 5.14 Consider the *double integrator* system

$$\dot{x} = \begin{pmatrix} 0 & 1 \\ 0 & 0 \end{pmatrix} x + \begin{pmatrix} 0 \\ 1 \end{pmatrix} u, \qquad y = \begin{pmatrix} 1 & 0 \end{pmatrix} x. \tag{5.65}$$

By Kalman matrix criterion, it easily follows that this system is controllable and observable (the definitions can be found in any textbook on control theory, see, e.g., [95]). Nevertheless, (5.65) is not even locally stabilizable by means of a continuous static output feedback. Pick any continuous feedback $k : \mathbb{R} \to \mathbb{R}$, denote $x := (x_1, x_2)^T$, and consider the closed-loop system

$$\dot{x}_1 = x_2, \tag{5.66a}$$
$$\dot{x}_2 = k(x_1). \tag{5.66b}$$

Since k is continuous, the function

$$V : y = (y_1, y_2)^T \mapsto y_2^2 - 2\int_0^{y_1} k(s)\,ds, \quad y_1, y_2 \in \mathbb{R},$$

is continuously differentiable on $\mathbb{R}^2$.

Pick any $x_0 \in \mathbb{R}^2$ and let $x(\cdot)$ be a solution of (5.66) with an initial condition x_0, which exists since k is continuous. It holds that

$$\frac{d}{dt}V(x(t)) = \nabla V(x(t)) \cdot \dot{x}(t) = 2x_2(t)k(x_1(t)) - 2k(x_1(t))x_2(t) = 0,$$

and thus $t \mapsto V(x(t))$ is a constant function. By definition of V we have $V((0, 1)) = 1$, and thus $V(\phi(\cdot, (0, 1))) \equiv 1$. However, $V((0, 0)) = 0$, and thus $\phi(t, (0, 1))$ does not converge to $(0, 0)$ and hence (5.65) is not stabilizable by means of a continuous static output feedback.

5.8 Robust Nonlinear Observers

To overcome the limitations of continuous static state feedback controllers, one may try to restore the state of (5.64) using so-called observers. This additional information can then be used to design a "dynamic" output feedback.

Definition 5.15 (*Robust observer*) Consider a system of the form

$$\dot{\bar{x}} = g(\bar{x}, u + \tilde{u}, y + \tilde{y}), \quad \bar{x}(t) \in \mathbb{R}^n, \tag{5.67}$$

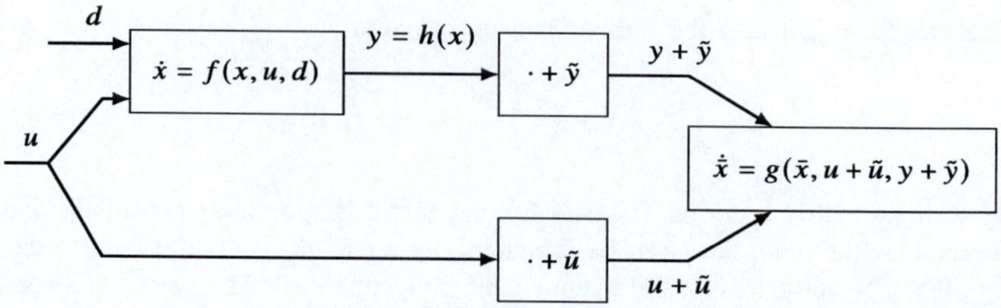

Fig. 5.3 Illustration for the definition of an observer

which receives the input signal u and the output signal y of the system (5.64), which may be distorted by unknown disturbances $\tilde{u}$ and $\tilde{y}$. Here $u, \tilde{u} \in \mathcal{U}$, $y, \tilde{y} \in \mathcal{Y} := L^\infty(\mathbb{R}_+, \mathbb{R}^p)$, and $g : \mathbb{R}^n \times \mathbb{R}^m \times \mathbb{R}^p \to \mathbb{R}^n$ is such that the solution of (5.67) exists and is unique for all initial conditions, inputs, and disturbances. This solution at time t subject to the initial state $\bar{x}$ and the inputs and disturbances $u, y, \tilde{u}, \tilde{y}$ we denote by $\bar{\phi}(t, \bar{x}, u, y, \tilde{u}, \tilde{y})$.

Define the *error* between the states of (5.64) and (5.67) for all $t \geq 0$, $x \in \mathbb{R}^n$, $\bar{x} \in \mathbb{R}^n$, any known input u and any unknown disturbances $\tilde{u} \in \mathcal{U}, d \in \mathcal{D}, \tilde{y} \in \mathcal{Y}$ by

$$e(t, x, \bar{x}, u, d, \tilde{u}, \tilde{y}) := \phi(t, x, u, d) - \bar{\phi}(t, \bar{x}, u, y, \tilde{u}, \tilde{y}),$$

where $y = h(\phi(t, x, u, d))$.

The system (5.67) is called a *(full-order) robust observer* of a nonlinear system (5.64) if the error dynamics is ISS with respect to disturbances acting on the system, in the sense that there are $\beta \in \mathcal{KL}$ and $\gamma_1, \gamma_2, \gamma_3 \in \mathcal{K}_\infty$ such that for all $t, x, \bar{x}, u, d, \tilde{u}, \tilde{y}$

$$|e(t, x, \bar{x}, u, d, \tilde{u}, \tilde{y})| \leq \beta(|x - \bar{x}|, t) + \gamma_1(\|d\|_\infty) + \gamma_2(\|\tilde{u}\|_\infty) + \gamma_3(\|\tilde{y}\|_\infty). \tag{5.68}$$

If in addition γ_1 can be chosen equal to 0, we call (5.67) a *(full-order) 0-gain robust observer* of a nonlinear system (5.64).

In particular, if $d \equiv 0$ (i.e., there are no disturbances acting on the system), and if $\tilde{u} \equiv 0$ and $\tilde{y} \equiv 0$ (there are no disturbances acting on the observer (5.67)), then $\bar{x}(t) - x(t) \to 0$ as $t \to \infty$, and thus the robust observer asymptotically converges to the unknown state of the system. Robust 0-gain observer does the same also for nonzero disturbances d. Graphically an observer is described in Fig. 5.3.

Proposition 5.16 *The equations governing the dynamics of any robust observer* (5.67) *for the system* (5.64) *must have the* "output injection" *form:*

$$\dot{\bar{x}} = f(\bar{x}, u + \tilde{u}, 0) + L\big(\bar{x}, u + \tilde{u}, y - h(\bar{x}) + \tilde{y}\big), \tag{5.69}$$

where $L(a, b, 0) = 0$ *for all* a, b.

Proof Let a robust observer (5.67) for the system (5.64) be given. Consider the special case when the disturbances are acting neither on (5.64), nor on the observer system, i.e., $d \equiv 0$, $\tilde{u} \equiv 0$ and $\tilde{y} \equiv 0$ and the initial states of the observer and of the system itself are identical. In view of the estimate (5.68), the trajectories of the system (5.64) and of the observer (5.67) coincide. Hence also the right-hand sides of (5.64) and (5.67) coincide as well. Consequently, for all x, u, it holds that

$$f(x, u, 0) = g(x, u, h(x)).$$

Let us rewrite g as

$$g(\bar{x}, u + \tilde{u}, y + \tilde{y}) = f(\bar{x}, u + \tilde{u}, 0) + g(\bar{x}, u + \tilde{u}, y + \tilde{y}) - f(\bar{x}, u + \tilde{u}, 0).$$

Define

$$R(a, b, c) := g(a, b, c) - f(a, b, 0)$$

and consider the invertible transformation

$$T : (a, b, c) \mapsto (a, b, c - h(a)),$$

with the inverse given by

$$T^{-1} : (a, b, c) \mapsto (a, b, c + h(a)).$$

Now define

$$L(a, b, c) := R \circ T^{-1}(a, b, c) = R(a, b, c + h(a)).$$

Then

$$R(a, b, c) = L \circ T(a, b, c) = L(a, b, c - h(a)).$$

This implies the formula (5.69).

Finally, $L(a, b, 0) = R(a, b, h(a)) = g(a, b, h(a)) - f(a, b, 0) = 0.$ □

The formula (5.69) can be interpreted in a way that the dynamics of the observer is a copy of the dynamics of the "observed" system plus the term, depending on the difference $y - h(\bar{x})$ between the output of the system (5.64) and the output of (5.64) that would be produced if $\bar{x}$ were the state of (5.64).

In the rest of this section, we assume that $\mathcal{D} = \{0\}$, that is, no external disturbances are acting on our system. The output of (5.64) with initial condition $x \in \mathbb{R}^n$, input $u \in \mathcal{U}$, and disturbance $d = 0$, at time $t \geq 0$ we denote by $y(t, x, u) := h(\phi(t, x, u, 0))$.

Definition 5.17 Let $\mathcal{D} = \{0\}$. The system (5.64) is called *incrementally input/output-to-state stable (i-IOSS)* if there are $\beta \in \mathcal{KL}$ and $\gamma_1, \gamma_2 \in \mathcal{K}_\infty$ such that, for every two initial states $x_1, x_2 \in \mathbb{R}^n$, and any controls $u_1, u_2 \in \mathcal{U}$, it holds that

$$|\phi(t, x_2, u_2, 0) - \phi(t, x_1, u_1, 0)| \leq \beta(|x_2 - x_1|, t) + \gamma_1(\|u_2 - u_1\|_{[0,t]}) \\ +\gamma_2(\|y(t, x_2, u_2) - y(t, x_1, u_1)\|_{[0,t]}),$$

for all t in the common domain of definition of $\phi(\cdot, x_1, u_1, 0)$ and $\phi(\cdot, x_2, u_2, 0)$.

The following proposition relates the question about the existence of observers to the ISS theory:

Proposition 5.18 *Let* $\mathcal{D} = \{0\}$. *If there is a robust (0-gain) observer for* (5.64), *then* (5.64) *is incrementally IOSS.*

Proof Pick any $x_1, x_2 \in \mathbb{R}^n$ and any $u_1, u_2 \in \mathcal{U}$. We consider x_1 as the initial state of (5.64), u_1 as the input entering the system (5.64), and define $y(t) := y(t, x_1, u_1)$.

By Proposition 5.16, any observer for (5.64) has the form (5.69). We consider the dynamics of the observer subject to the input disturbance $\tilde{u} := u_2 - u_1$ and the output disturbance $\tilde{y}(t) := y(t, x_2, u_2) - y(t, x_1, u_1)$. For $u := u_1$, and $y, \tilde{u}, \tilde{y}$, the observer takes form

$$\dot{\bar{x}} = f(\bar{x}, u_2, 0) + L\big(\bar{x}, u_2, y(\cdot, x_2, u_2) - h(\bar{x})\big). \tag{5.70}$$

Recall that $L\big(\bar{x}, u_2, 0\big) = 0$ for all $\bar{x}, u_2$. Thus, the map $t \mapsto \phi(t, x_2, u_2)$ that solves (5.64) with the initial state x_2, input u_2 and disturbance $d = 0$, solves also (5.70) with initial condition $\bar{x}(0) = x_2$.

Now the estimate (5.70) is precisely the estimate (5.68) for observer error dynamics. □

Let us relate the obtained results to the classical observability concepts.

Definition 5.19 Let $\mathcal{D} = \{0\}$. We say that the control $u \in \mathcal{U}$:

(i) *Distinguishes the states* $x_1, x_2 \in \mathbb{R}^n$ *of the system* (5.64) *on* $[0, T]$, if there is $t \in [0, T]$ so that

$$y(t, x_1, u) \neq y(t, x_2, u).$$

(ii) *Distinguishes the states* $x_1, x_2 \in \mathbb{R}^n$ of the system *(5.64)*, if there $T > 0$ such that u distinguishes the states $x_1, x_2 \in \mathbb{R}^n$ on $[0, T]$.

Definition 5.20 Let $\mathcal{D} = \{0\}$. The system (5.64) is called:

(i) *Observable (by any input)*, if all pairs (x_1, x_2) with $x_1 \neq x_2$ are distinguishable by any input $u \in \mathcal{U}$.

(ii) *Detectable* or *asymptotically observable* (by any input), if for any $u \in \mathcal{U}$ and any $x_1, x_2 \in \mathbb{R}^n$ it holds that

$$y(t, x_1, u) = y(t, x_2, u) \quad \forall t \in [0, \infty) \;\Rightarrow\; |\phi(t, x_1, u) - \phi(t, x_2, u)| \to 0 \text{ for } t \to \infty.$$

Directly from the definition of incremental IOSS it follows that

Proposition 5.21 *Let $\mathcal{D} = \{0\}$. Any incrementally IOSS system* (5.64) *is detectable by any input.*

5.9 Observers and Dynamic Feedback for Linear Systems

This section shows a classical result that a linear system with outputs can be stabilized using dynamic output feedback, provided that it is stabilizable (by full state feedback) and detectable.

We start with a well-known criterion of the asymptotic observability for linear systems

Proposition 5.22 *Consider the system*

$$\dot{x} = Ax + Bu, \qquad y = Cx, \tag{5.71}$$

where $A \in \mathbb{R}^{n\times n}$, $B \in \mathbb{R}^{n\times m}$, $C \in \mathbb{R}^{p\times n}$, $n, m, p \in \mathbb{N}$, and $u \in \mathcal{U}$.

The following holds:

(i) (5.71) is detectable by any input $\Leftrightarrow$ there is $L \in \mathbb{R}^{n\times p}$ so that $A + LC$ is Hurwitz.

(ii) (5.71) is observable by any input $\Leftrightarrow$ rank $O[A, C] = n$, where

$$O[A, C] = \begin{pmatrix} C \\ CA \\ \vdots \\ CA^{n-1} \end{pmatrix}. \tag{5.72}$$

Proof See, e.g., [110, Theorem 1.6]. □

In the next result, we present a well-known *Luenberger design of (robust) observers* for linear control systems

$$\dot{x} = Ax + Bu + Dd, \tag{5.73a}$$

$$y = Cx, \tag{5.73b}$$

where $A \in \mathbb{R}^{n\times n}$, $B \in \mathbb{R}^{n\times m}$, $D \in \mathbb{R}^{n\times q}$, $C \in \mathbb{R}^{p\times n}$, where $n, m, q, p \in \mathbb{N}$, $u \in \mathcal{U}$, and $d \in \mathcal{D}$.

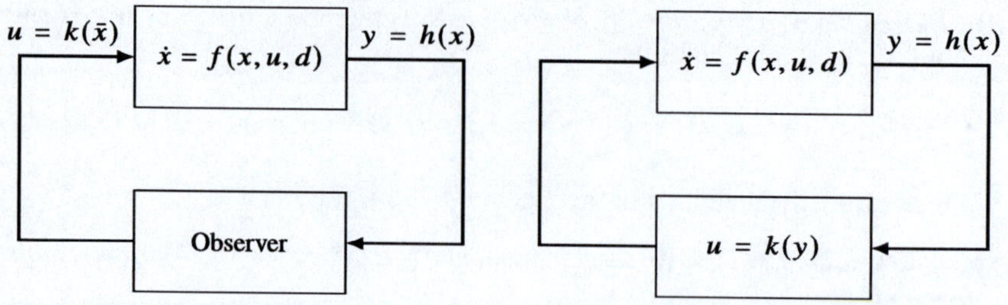

(a) Dynamic output feedback (b) Static output feedback

Fig. 5.4 Dynamic and static output feedback

Theorem 5.23 *The following statements are equivalent:*

(i) There is a robust observer for the system (5.73).
(ii) There is a linear robust observer for the system (5.73).
(iii) (Luenberger's observer) *The following system is a robust observer for* (5.73):

$$\dot{\bar{x}} = A\bar{x} + L(C\bar{x} - (y + \tilde{y})) + B(u + \tilde{u}). \tag{5.74}$$

(iv) The system (5.71) *is asymptotically observable (i.e., detectable) for any input.*

Proof Clearly, (iii) ⇒ (ii) ⇒ (i). Furthermore, (i) implies (iv) by Propositions 5.18, 5.21.

(iv) ⇒ (iii). As (5.71) is detectable, by Proposition 5.22 there exists L so that $A + LC$ is Hurwitz. Consider the system (5.74) and define the error between the states of the system x and the state of the observer $\bar{x}$ by $e := x - \bar{x}$. Subtracting the equation (5.74) from (5.73a) we get the following system for the error e:

$$\dot{e} = Ae - L(C\bar{x} - Cx) + Dd + L\tilde{y} - B\tilde{u},$$

which can be rewritten as

$$\dot{e} = (A + LC)e + Dd + L\tilde{y} - B\tilde{u}. \tag{5.75}$$

Since $A + LC$ is Hurwitz, (5.75) is 0-GAS and due to Theorem 2.25 it is ISS. This shows that (5.74) is a robust observer for (5.73). □

Next, we show that one can overcome obstructions encountered in Example 5.14 by employing the *dynamic feedback*, whose structure is depicted at Fig. 5.4a.

Theorem 5.24 *Let* (5.73) *be stabilizable and detectable. Then the coupled system* (5.73)-(5.74) *is input-to-state stabilizable (with respect to disturbances d, $\tilde{u}$, $\tilde{y}$) by means of a feedback $u = F\bar{x}$, for some $F \in \mathbb{R}^{m\times n}$ where $\bar{x}$ comes from* (5.74).

Proof As (5.73) is stabilizable and detectable, there exist F, L so that $A + BF$ and $A + LC$ are Hurwitz matrices. Consider the system (5.73) coupled with the observer (5.74) by means of the feedback law $u = F\bar{x}$. We obtain a coupled system

$$\begin{aligned}\dot{x} &= Ax + BF\bar{x} + Dd,\\ \dot{\bar{x}} &= A\bar{x} + L(C\bar{x} - Cx) + BF\bar{x} - L\tilde{y} + B\tilde{u}.\end{aligned} \tag{5.76}$$

This system can be rewritten as

$$\begin{aligned}\begin{pmatrix}\dot{x}\\ \dot{\bar{x}}\end{pmatrix} &= \begin{pmatrix}A & BF\\ -LC & A+LC+BF\end{pmatrix}\begin{pmatrix}x\\ \bar{x}\end{pmatrix} + \begin{pmatrix}D\\ 0\end{pmatrix}d + \begin{pmatrix}0\\ -L\end{pmatrix}\tilde{y} + \begin{pmatrix}0\\ B\end{pmatrix}\tilde{u}\\ &= T^{-1}\begin{pmatrix}A+BF & BF\\ 0 & A+LC\end{pmatrix}T\begin{pmatrix}x\\ \bar{x}\end{pmatrix} + \begin{pmatrix}D\\ 0\end{pmatrix}d + \begin{pmatrix}0\\ -L\end{pmatrix}\tilde{y} + \begin{pmatrix}0\\ B\end{pmatrix}\tilde{u}.\end{aligned}$$

Here $T = \begin{pmatrix}I & 0\\ -I & I\end{pmatrix}$ and $T^{-1} = \begin{pmatrix}I & 0\\ I & I\end{pmatrix}$.

Since $A + BF$ and $A + LC$ are Hurwitz matrices, $\begin{pmatrix}A+BF & BF\\ 0 & A+LC\end{pmatrix}$ as well its equivalent transformation $T^{-1}\begin{pmatrix}A+BF & BF\\ 0 & A+LC\end{pmatrix}T$ are also Hurwitz matrices. This shows that (5.76) is 0-GAS and due to Theorem 2.25, (5.76) is ISS. □

Example 5.25 As we know from Example 5.14, the following double integrator system is not stabilizable by continuous static output feedback:

$$\dot{x} = \begin{bmatrix}0 & 1\\ 0 & 0\end{bmatrix}x + \begin{bmatrix}0\\ 1\end{bmatrix}u, \quad y = \begin{bmatrix}1 & 0\end{bmatrix}x. \tag{5.77}$$

However, by Kalman matrix condition, this system is controllable and observable, and thus, by Theorem 5.24, it can be stabilized by a dynamic feedback, which we construct next.

Following the proof of Theorem 5.24, to solve the problem of dynamic stabilization of a double integrator (5.77), we have to find F and L so that $A + BF$ and $A + LC$ are Hurwitz matrices. A general form of the matrices $A + BF$ and $A + LC$ is given by

$$A + BF = \begin{pmatrix}0 & 1\\ f_1 & f_2\end{pmatrix}, \qquad A + LC = \begin{pmatrix}l_1 & 1\\ l_2 & 0\end{pmatrix}.$$

Choosing, e.g., $f_2 := -2$, $f_1 := -1$, $l_1 := -2$, $l_2 := -1$, we obtain that $\sigma(A + BF) = \{-1\}$ and $\sigma(A + LC) = \{-1\}$.

5.10 Observers for Nonlinear Systems

As we have seen, Luenberger's observer constructed in the previous section is automatically robust due to the linearity of the error dynamics. For nonlinear systems, the robustness of the observer comes to the forefront of research. Consider

$$\dot{x} = Ax + G\big(\gamma(Hx) + \Delta(Hx)d\big) + \rho(y, u), \tag{5.78a}$$

$$y = Cx, \tag{5.78b}$$

where $A \in \mathbb{R}^{n\times n}$, $C \in \mathbb{R}^{p\times n}$, $H \in \mathbb{R}^{r\times n}$, $G \in \mathbb{R}^{n\times r}$ for certain $n, p, r \in \mathbb{N}$ and the state-dependent nonlinearities $\gamma(Hx)$ and $\Delta(Hx)$ are r-dimensional vector functions of the form

$$\gamma(Hx) := \Big(\gamma_i\big((Hx)_i\big)\Big)_{i=1}^{r}, \quad \Delta(Hx) := \Big(\Delta_i\big((Hx)_i\big)\Big)_{i=1}^{r}.$$

We assume that γ, Δ, and ρ are Lipschitz continuous on bounded balls.

In (5.78) the scalar disturbance d does not distort the input signal (as actuator disturbances do) but brings the additional *unmodeled dynamics* $\Delta(Hx)$ into play. Next, we construct a robust observer for the system (5.78), using certain *linear matrix inequalities (LMIs)*.

Theorem 5.26 *Assume that for all* $i = 1, \ldots, r$ *there exist* $\sigma_i \in \mathcal{K}$ *such that for all* $a, b, d \in \mathbb{R}$ *it holds that*

$$(a - b)(\gamma_i(a) - \gamma_i(b) + \Delta_i(a)d) \geq -\sigma_i(|d|). \tag{5.79}$$

Let also there exist a matrix $P = P^T > 0$, *a diagonal matrix* $\Lambda > 0$ *and* $\nu > 0$ *so that an LMI*

$$\begin{pmatrix} (A + LC)^T P + P(A + LC) + \nu I & PG + (H + KC)^T \Lambda \\ G^T P + \Lambda(H + KC) & 0 \end{pmatrix} \leq 0 \tag{5.80}$$

holds for certain matrices $K \in \mathbb{R}^{r\times p}$, $L \in \mathbb{R}^{n\times p}$.

Then

$$\dot{\bar{x}} = A\bar{x} + L(C\bar{x} - y) + G\gamma\big(H\bar{x} + K(C\bar{x} - y)\big) + \rho(y, u) \tag{5.81}$$

is a robust observer for the system (5.78).

Proof Let the assumptions of the theorem hold. Denote the error between the states of the system and the observer by

$$e(t) := x(t) - \bar{x}(t),$$

and let

$$v := Hx, \qquad w := H\bar{x} + K(C\bar{x} - y).$$

Consider the system governing the error dynamics

$$\dot{e} = (A + LC)e + G\big(\gamma(v) - \gamma(w) + \Delta(v)d\big). \tag{5.82}$$

It is of virtue to introduce a new variable

$$z := v - w = (H + KC)e.$$

Also, we denote

$$\varphi(v, w, d) := \gamma(v) - \gamma(w) + \Delta(v)d.$$

Condition (5.79) for $a := v$ and $b := w$ leads for all $i = 1, \ldots, r$ to

$$z_i\varphi_i(v, w, d) \geq -\sigma_i(|d|), \quad z_i \in \mathbb{R},\ t \geq 0,\ d \in \mathbb{R}. \tag{5.83}$$

We can rewrite (5.82) as

$$\dot{e} = (A + LC)e + G\varphi(v, w, d). \tag{5.84}$$

Consider the following ISS Lyapunov function candidate:

$$V(e) = e^T Pe. \tag{5.85}$$

As $P > 0$, V satisfies the sandwich estimates

$$\lambda_{\min}(P)|e|^2 \leq V(e) \leq \lambda_{\max}(P)|e|^2, \quad e \in \mathbb{R}^n.$$

Let us compute Lie derivative of V:

$$\begin{aligned}
\dot{V}(e) &= \dot{e}^T Pe + e^T P\dot{e} \\
&= \big((A + LC)e + G\varphi(v, w, d)\big)^T Pe + e^T P\big((A + LC)e + G\varphi(v, w, d)\big) \\
&= e^T\big((A + LC)^T P + P(A + LC)\big)e + \varphi^T(v, w, d)G^T Pe \\
&\quad + e^T PG\varphi(v, w, d) = e^T\big((A + LC)^T P + P(A + LC) + \nu I\big)e - \nu e^T e \\
&\quad + \varphi^T(v, w, d)\big(G^T P + \Lambda(H + KC)\big)e - \varphi^T(v, w, d)\Lambda(H + KC)e \\
&\quad + e^T\big(PG + (H + KC)^T\Lambda\big)\varphi(v, w, d) - e^T(H + KC)^T\Lambda\varphi(v, w, d).
\end{aligned}$$

Applying the inequality (5.80) and recalling that $z = (H + KC)e$, we see that

$$\begin{aligned}\dot{V}(e) &\leq -\nu|e|^2 - \varphi^T(v, w, d)\Lambda z - z^T\Lambda\varphi(v, w, d)\\ &= -\nu|e|^2 - 2z^T\Lambda\varphi(v, w, d)\\ &= -\nu|e|^2 - 2\sum_{i=1}^{r} z_i\lambda_i\varphi_i(v, w, d).\end{aligned}$$

Since $\Lambda > 0$ and due to (5.83) we finally arrive at

$$\begin{aligned}\dot{V}(e) &\leq -\nu|e|^2 + 2\sum_{i=1}^{r}\lambda_i\sigma_i(|d|)\\ &\leq -\frac{\nu}{\lambda_{\max}(P)}V(e) + \sigma(|d|),\end{aligned} \tag{5.86}$$

with $\sigma(|d|) := 2\sum_{i=1}^{r}\lambda_i\sigma_i(|d|)$. This shows that V is an exponential ISS Lyapunov function for (5.82) and thus (5.81) is a robust observer for (5.78). □

Remark 5.27 There are many efficient tools for the solution of LMIs, including MATLAB routines. The feasibility (existence of solutions) of the LMI (5.80) is not guaranteed a priori and has to be determined numerically.

5.11 Concluding Remarks and Extensions

The nonlinear control design is a topic of many research monographs, including [16, 25, 49, 89], see also the survey [41].

5.11.1 Concluding Remarks

ISS feedback redesign. Recall that control-affine systems have the following form:

$$\dot{x} = f(x) + G(x)u, \tag{5.87}$$

where $f : \mathbb{R}^n \to \mathbb{R}^n$, $G : \mathbb{R}^n \to \mathbb{R}^{n\times m}$ are smooth functions.

Theorem 5.4 on ISS feedback redesign that achieves robustness of the closed-loop systems with respect to the additive actuator disturbances was proved by Sontag in the seminal paper [93], where he introduced the ISS concept.

Achieving the ISS with respect to measurement disturbances is a harder problem. In [23] an example of a planar control-affine system (5.87) has been constructed that can be GAS-stabilized by means of a continuous feedback controller $u = \tilde{k}(x)$ but for which no continuous controller $u = k(x + d)$ can prevent finite escape times from every initial condition in the presence of small sensor disturbances d. At the same time, in [22] it was shown that for one-dimensional GAS-stabilizable control-affine systems (5.87), it is possible to design a continuous periodic time-varying feedback making the closed-loop system (5.87) ISS with respect to measurement disturbances (however, the author argues that time variance is essential). For higher-dimensional control-affine systems, feedback controllers for GAS-stabilizable systems achieving ISS stabilization w.r.t. measurement disturbances have been reported in [25]. Still, many questions remain open in this direction.

In [80], ISS and iISS feedback redesign are analyzed for time-delay systems with discontinuous right-hand side and possible limitations on the amplitude of the actuator.

ISS control Lyapunov functions. In our presentation of ISS-CLFs, we follow [60] and [94]. In particular, Theorem 5.9 is due to [60, Theorem 3]. Analogously to the notion of ISS-CLF, one can also introduce the notion of iISS-CLF and show that the feedback (5.17) constructed in Theorem 5.9 integral input-to-state stabilizes the control system (5.11) if V is an iISS-CLF.

Among the first works on ISS-CLFs, we mention [24, 50, 96, 108], where a notion of an ISS control Lyapunov function has been introduced, and it was shown that the existence of such a function with a certain degree of smoothness gives rise to explicit formulas for input-to-state stabilizing control laws. Furthermore, input-to-state stabilizing control laws have nice properties associated with inverse optimality [25, 50].

A notion of iISS control Lyapunov function has been introduced in [59], followed by [30, 55, 60], etc. In these papers, the "universal" construction of state-feedbacks has been proposed that make the closed-loop systems integrally ISS, even if they fail to be input-to-state stabilizable. Input-to-output stable CLFs have been analyzed in [21].

ISS control Lyapunov functions are the basis for the general framework for the modular adaptive design of stabilizers for nonlinear control systems, [49]. Further applications of (i)ISS-CLFs include (i)ISS disturbance attenuation with bounded controls [55] (for systems without disturbances corresponding results have been obtained in [62]), strong integral input-to-state stabilization via saturated controls [11] (see also here an application to spacecraft velocity control using saturated actuators), construction of nonlinear ISS-stabilizing discontinuous feedbacks $u := K(x) + d$ for globally asymptotically controllable control-affine systems (5.87) [66], etc.

Some further extensions of the ISS-CLF concept are discussed in [19]. The theory of ISS-CLFs extends in a natural way the classical theory of control Lyapunov functions, see [8, 92, 94] and many references therein.

Backstepping is one of the most popular methods for the design of stabilizing feedback controllers for nonlinear systems. It emerged at the same time as the ISS concept [17, 39, 40], and soon it was realized that backstepping can be used to design feedback laws that are robust against acting disturbances in the ISS sense [49, 59]. Our brief discussion of this method follows that of [60], in particular, Theorem 5.10 is due to [60, Section 7].

The emergence of adaptive, robust, and observer-based backstepping was described in [42]. We also refer to [41] for a brief review of the history of the early development of this method. Backstepping empowers the control designer with the possibility to incorporate the uncertainties and unknown parameters. Still, its applicability is limited to a class of lower triangular systems that motivated the emergence of *forwarding* techniques introduced in [69, 101, 102], which allow treating of the feedforward systems, see also [82] for a tutorial introduction to the method.

The axial compressor model from Sect. 5.5, is the three-state Galerkin approximation of a nonlinear PDE model derived in [71]. It was studied in numerous works. The bifurcations arising in the uncontrolled system were analyzed in [70]. Full-state feedback law for the global asymptotic stabilization of this system was proposed in [49, Section 2.4]. A controller that globally asymptotically stabilizes the system using only the information about ϕ and ψ was proposed in [48]. In [32, Section 12.7] a controller was proposed that achieves semiglobal asymptotic stabilization of the axial compressor model, using the measurement of ψ only. In [5, Section 4] a globally asymptotically stabilizing controller was proposed that uses measurements of ψ solely. To restore the information about the variable ϕ, in [5] a so-called *reduced-order observer* was proposed.

Our treatment in Sect. 5.5 is motivated by [5]. The key idea behind it is the use of the synergy of the **gain assignment technique** and the small-gain approach. This method was proposed for the stabilization of coupled nonlinear systems in [37]. It was used for robust control of uncertain nonlinear systems [36], for the stabilization of networks of switched systems [18], etc. For a survey on this technique, a reader may consult [34].

Nonlinear detectability and IOSS. For linear time-invariant systems, the notion of observability was introduced by Kalman in [38], and the concept of detectability was introduced by Wonham in [109]. Afterward, the concept of observability was extended to bilinear systems [12], analytic systems [98], and nonlinear systems [29].

The ISS paradigm has changed not only the way how the robust stability is understood but also paved the way for the new formulation of nonlinear detectability concepts. For systems without inputs and disturbances, the concept of *output-to-state stability (OSS)* was proposed as a dual notion of ISS in [97]. For linear systems without inputs, OSS is equivalent to detectability. In [97], it was shown that for nonlinear systems, OSS is equivalent to the existence of a properly defined OSS Lyapunov function.

For nonlinear systems with inputs and outputs

$$\dot{x}(t) = f(x(t), u(t)), \tag{5.88a}$$

$$y(t) = h(x(t), u(t)), \tag{5.88b}$$

the concept of input/output-to-state stability was introduced in [97].

Definition 5.28 A control system with outputs (5.88) is called *input/output to state stable (IOSS)*, if there exist $\beta \in \mathcal{KL}$ and $\gamma_1, \gamma_2 \in \mathcal{K}$ such that for all $x \in \mathbb{R}^n$, $u \in \mathcal{U}$ and all $t \in [0, t_m(x, u))$ the following holds

$$|\phi(t, x, u)| \leq \beta(|x|, t) + \gamma_1(\text{ess sup}_{s\in[0,t]}|u(s)|) + \gamma_2(\text{ess sup}_{s\in[0,t]}|y(s, x, u)|). \tag{5.89}$$

In [97], it was shown that IOSS is sufficient for a nonlinear system to be "zero-detectable". This means that the state remains in a bounded neighborhood of the origin, provided that the inputs and outputs are bounded. In [43] it was shown that IOSS is equivalent to the existence of an IOSS Lyapunov function. Superposition results for the IOSS property have been obtained in [4].

To obtain a proper extension not only of zero-detectability but also of the detectability concept, in [97] the concept of incremental input/output-to-state stability was introduced, and it was shown that incremental IOSS is a necessary condition for a robust observer, as introduced in Definition 5.15. Proposition 5.18 is a variation of this result for systems that are subject to unknown disturbances. The notion of incremental IOSS is widely used for the optimization-based state estimation [83]. The concept of incremental uniform input/output-to-state stability, which is an extension of the incremental input/output-to-state stability property, was used in [2] for discrete-time systems. In [2], it was also argued that Proposition 5.18 could be extended to more general *state estimators* that are not necessarily full-state observers (examples of such estimators include extended Kalman filter (EKF), full information estimation (FIE), moving horizon estimation (MHE), etc.) Relations between incremental IOSS and incremental dissipativity were studied in [85].

Proposition 5.16 was shown in [97, Lemma 21].

Observers for nonlinear systems. For more details on observer theory for nonlinear systems, an interested reader is referred to [1, 3, 33, 68, 90]. In Sect. 5.10, we follow in our analysis [5].

5.11.2 *Obstructions on the Way to Stabilization*

Although input-to-state stabilization of disturbed nonlinear systems

$$\dot{x}(t) = f(x(t), u(t), d(t)) \tag{5.90}$$

is already a challenging problem, in real-world applications of control theory many additional complications appear that we summarized in Table 5.1. In Sect. 5.6, with the help of the ISS theory, we developed an event-based controller, helping to overcome the challenge that the signals in digital controllers are transmitted only at some discrete moments. Here we argue that the ISS framework can systematically handle many further challenges in nonlinear control. Due to space restrictions, we cannot cover these topics in detail. Thus we merely give a short overview of available results.

Table 5.1 Various challenges, appearing in real-world applications of control theory and the approaches how to overcome them

Problem	Remedy
Signals are transmitted at discrete moments of time	Sampled-data control, Event-based control
The full state cannot be measured	Observer/estimator design
Limitations of the transmission capacity of the communication network	Theory of networked control systems
Delays during the transmission	Delay compensation technique
Unmodeled dynamics	Robust control, ISS
Measurement errors, actuator errors	Robust observers/controllers, ISS
Quantization of the sensor and input signals	Quantized controllers
Boundedness of the physical input	Controllers with saturation
Uncertainties of parameters	Adaptive control

Sample-data control. In Sect. 5.6, we have introduced event-based sample-and-hold controllers with state-dependent triggering times. The other methodology for the design of digital controllers is a periodic sample-data control, where it is assumed that the sampling times $(t_k)_{k=1}^{\infty}$ are given by $t_k := kT$ for some time $T > 0$ called the *sampling period*.

As in the section on event-based control, we consider piecewise constant *sample-and-hold* controllers of the form

$$u(t) = u(kT) =: \tilde{u}(k), \quad t \in [kT, (k+1)T), \tag{5.91}$$

for a certain $\tilde{u} : \mathbb{Z}_+ \to \mathbb{R}^m$.

In [104], it was shown that for nonlinear systems, an input-to-state stabilizing continuous-time control law is an input-to-state stabilizing sampled-data control law, provided the sampling is sufficiently fast.

Substituting this signal into (5.90) and introducing a new variable

$$\tilde{x}(k) := x(kT),$$

we derive the corresponding discrete-time system, called the *exact discrete-time model* of (5.90):

$$\begin{aligned}\tilde{x}(k+1) &= \tilde{x}(k) + \int_{kT}^{(k+1)T} f(x(s), \tilde{u}(k), d(s))ds \\ &=: F_T(x(k), u(k), d_T(k)),\end{aligned}$$

where $d_T(k) := d_T(s + kT)$, $s \in [0, T)$.

As it is very hard to compute an exact discrete-time model, one uses various approximations of the exact model. In [74] a framework was proposed, giving sufficient conditions, under which a discrete-time ("sample-data") controller that input-to-state stabilizes an *approximate discrete-time model* of a nonlinear plant with disturbances would also input-to-state stabilize (in an appropriate sense) the exact discrete-time plant model. We refer to [76] for an overview of the early results in the robust sample-data control.

Event-triggered control. In periodic sample-data control, the controller was updated periodically, and the period was chosen small enough to guarantee the desired property in the worst-case scenario. This indicates that periodic sample-data control is not necessarily cost-efficient for the systems with the information constraints. This, in part, motivates the research in event-triggered control, where the controller is updated at some time instants that are dependent on the state of the system. This helps to reduce the usage of communication and computation resources and, at the same time, to achieve the needed properties of the closed-loop system [9]. Some advantages of event-triggered control over the sample-data control were illustrated in [65, Sect. 1.1].

The results presented in Sect. 5.6 are due to [99] that was a milestone in the development of the nonlinear event-based control theory. In [99], Theorem 5.13 was shown for a more general (and more realistic) situation when the update of the control law is delayed by some positive time $\delta \geq 0$ representing the time required to read the state from the sensors, compute the control value and update the actuators. For linear systems, a computational estimate for the time τ in Theorem 5.13 was derived in [99] as well.

Theorem 5.13 tells us that the event-based controller does not have a Zeno behavior if γ and α^{-1} are Lipschitz continuous. In [81], this condition was relaxed within the framework of hybrid systems [26]. The universal formula for the construction of event-triggered stabilizing control laws for input-affine systems has been proposed in [67].

Event-triggered control of linear systems remains an active area of research, see, e.g., [87] and references therein. ISS small-gain theorems can be used to solve the event-based stabilization problems for large-scale systems [20, 63]. For a comprehensive overview of robust event-triggered control, we refer to [65].

Networked control systems. In networked control systems (NCS), several dynamical systems are controlled over the communication network [111]. While such architecture allows organizing a flexible, efficient, and low-cost communication between systems, the constraints in the transmission capacity of the communication network (given in terms of bits or packets per second that can be transported via the network) can significantly reduce the performance of the network.

A classical strategy for control of NCS is the emulation method. Firstly, one designs the controller for the system by ignoring the communication constraints. Then, the controller is implemented via the network, and it is proved that the designed controller stabilizes the system also for the NCS, provided the network satisfies some conditions. Typically, it is assumed that the *maximum allowable transmission interval*

(MATI) is small enough, and when scheduling occurs, the protocol has to satisfy a certain stability property.

One of the first papers on the stability of nonlinear NCS was [106], where the so-called *try-once-discard (TOD) dynamic protocol* was introduced. Input-to-state stability analysis of NCS has been initiated in [77], where a large class of Lyapunov UGAS protocols was introduced that encompasses token ring and TOD protocols as well as several of their modifications. In [14], improved estimates for the MATI guaranteeing the stability of NCS under communication constraints have been proposed. Integral ISS of NCS has been studied in [78]. Modern treatment of the stability analysis and control of NCS uses the formalism of hybrid systems theory [26] and goes beyond the scope of this book.

Delay compensation. Besides communication constraints, the delays in the transmission of the signals constitute another challenge that motivated the development of the *delay compensation* techniques. Since the publication of the paper on *Smith predictor* [91], this topic was of major importance in the control community [13, 46, 47]. In particular, in [88] an observer-based method was proposed for input-to-state stabilization of networked control systems with large uncertain delays.

Quantization. In many applications of control technology, such as power grids and intelligent transportation systems, the signals are digitalized before being transmitted via communication channels. Hence, the controller system does not obtain the state $x(t)$ as an input, but the quantized state $q(x(t))$, where $q : \mathbb{R}^n \to Q$ is a function, called *quantizer* and Q is some finite set.

The origins of the quantized control can be traced back to the 1960s, see, e.g., [52]. ISS methodology is especially powerful for the study of robust stabilization via quantized controllers, as the errors due to quantizations can be treated as disturbance inputs in the ISS framework. Among the first results in this direction are the papers [56–58], where it was proved that if a system can be input-to-state stabilized with the quantization error playing the role of an input, then one can design a quantization strategy to globally asymptotically stabilize the system. Quantized stabilization of high-order nonlinear systems in output-feedback form has been studied with the help of ISS small-gain theorems in [64]. For a more detailed overview of the results in this direction, we refer to the survey [35].

Actuator and sensor dynamics. In general, before the output of the controlled system reaches a controller or an observer, it should be measured by some sensor, which may have its own dynamics. For example, it may involve time delays, or its dynamics may be governed by the heat or wave equations [44, 45]. Before the control signal reaches the system under control, it passes another system, called an actuator, which again may have its own complicated dynamics [44, 45]. The general structure of the closed-loop control system is depicted in Fig. 5.5. The problem of designing a control that would take into account the dynamics of actuator and sensor is called the problem of *compensation of actuator and sensor dynamics*.

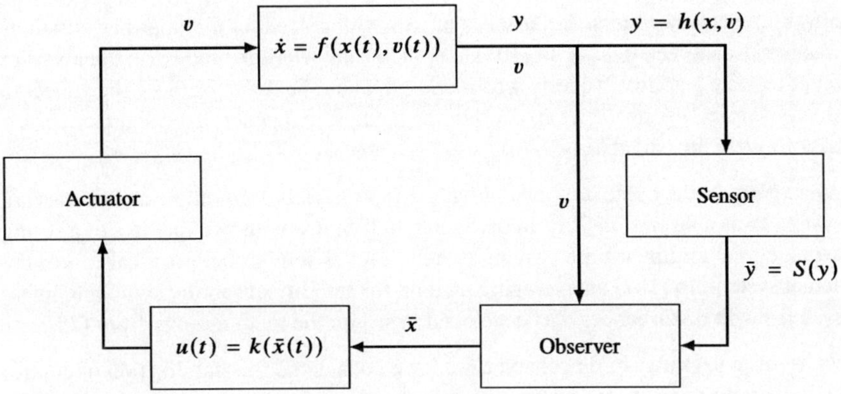

Fig. 5.5 Stabilization problem with actuator disturbances

5.11.3 *Further Control Techniques & ISS*

Model predictive control and IOSS. Model predictive control (MPC), initiated in the late 1970s, has become one of the most influential paradigms in the industrial applications of the control theory and was intensively studied by the systems theoretic community [28, 84].

In MPC, an optimal control task on a very long time horizon is resolved into a sequence of optimal control problems on shorter, overlapping horizons. These problems are much easier to solve numerically, and there are efficient algorithms applicable even for complex problems that enable to overcome the curse of dimensionality, which is typical for the optimal control problems on the infinite horizons [28, 84].

Strict dissipativity with a supply function given by the shifted cost function is one of the properties that are decisive for the proper functioning of the MPC. Strict dissipativity implies the so-called *turnpike property* that ensures a certain similarity between optimal trajectories on varying time horizons [27, 73, 105]. For optimal control problems with nonnegative stage cost, on bounded sets existence of an IOSS-like Lyapunov function implies strict dissipativity for cost functions minimizing a function of the norm of the output [31, Corollary 1].

MPC can also be used as a method for the input-to-state stabilization of nonlinear systems. See, e.g., [53, 61] for some results of this kind in a discrete-time setting.

Lurie systems are control systems of the form

$$\dot{x} = Ax + Bu, \quad y = Cx, \quad u := -\phi(y) + d,$$

obtained from linear control systems by means of a nonlinear output feedback with actuator disturbances. ISS of such systems has been investigated in [6].

Singular perturbations. Robustness of ISS with respect to singular perturbations was firstly analyzed in [15]. In [103], a unified framework for robustness analysis of input-to-state stability property was proposed, and robustness of ISS with respect to slowly varying parameters, rapidly varying signals, and generalized singular perturbations were investigated.

Averaging. In the context of averaging, the ISS property for nonlinear time-varying systems was analyzed in [75]. In particular, in [75] it was shown that ISS of a strong average of a nonlinear time-varying system implies semiglobal practical ISS of the actual system. In [107] an averaging method for stability of rapidly switching linear systems with disturbances was developed based on the methodology from [75].

Extremum seeking. In this chapter, we have considered the stabilization of control systems with respect to a known equilibrium. However, in many applications, the system should be stabilized with respect to an extremum of a reference-to-output equilibrium map.

Consider the single-input-single-output (SISO) control system

$$\dot{x} = f(x, u), \tag{5.92a}$$

$$y = h(x), \tag{5.92b}$$

where $f : \mathbb{R}^n \times \mathbb{R} \to \mathbb{R}^n$ and $h : \mathbb{R}^n \to \mathbb{R}$ are continuously differentiable maps.

Assume that there is a unique x^* such that $y^* = h(x^*)$ is the extremum of the map h. Frequently, the output map h and hence also x^* are not known precisely. The uncertainty of the reference-to-output map requires not only solving the stabilization problem but also implementing an adaptation procedure to find the set point that maximizes the output.

Consider a family of parametrized feedback controllers of the form

$$u(x) := \alpha(x, \theta), \tag{5.93}$$

where $\theta \in \mathbb{R}$ is a scalar tuning parameter that gives us an additional degree of freedom for our controller.

Substituting (5.93) into (5.92a), we obtain a closed-loop system

$$\dot{x} = f(x, \alpha(x, \theta)). \tag{5.94}$$

We impose several assumptions:

Assumption 5.29 There is a continuously differentiable function $l : \mathbb{R} \to \mathbb{R}^n$ such that

$$f(x, \alpha(x, \theta)) = 0 \quad \Leftrightarrow \quad x = l(\theta). \tag{5.95}$$

Assumption 5.29 shows that for each value of a parameter θ, there is a unique equilibrium of (5.94) given by $l(\theta)$.

Assumption 5.30 For each $\theta \in \mathbb{R}$ the corresponding equilibrium $x = l(\theta)$ is GAS, uniformly in θ.

In other words, Assumption 5.30 tells us that for any θ the controller $u : x \mapsto \alpha(x, \theta)$ GAS-stabilizes the corresponding equilibrium of (5.92). Now the question remains on how to choose θ to stabilize the system with respect to the optimal equilibrium. To make this possible, we make one more assumption.

Assumption 5.31 There exists a unique $\theta^* \in \mathbb{R}$ that maximizes the map $Q := h \circ l : \mathbb{R} \to \mathbb{R}$. Furthermore, it holds that

$$Q'(\theta^*) = 0, \quad Q''(\theta^*) < 0, \tag{5.96}$$

and there exists $\alpha_Q \in \mathcal{K}_\infty$ such that

$$Q'(\theta^* + r)r \leq -\alpha_Q(|r|), \quad r \in \mathbb{R}. \tag{5.97}$$

Now the aim is to design the mechanism that results in convergence to the optimal θ. In [100] the following scheme has been proposed:

$$\dot{x}(t) = f\big(x(t), \alpha(x(t), \hat{\theta}(t) + a\sin(\omega t))\big), \tag{5.98a}$$

$$\dot{\hat{\theta}}(t) = kh(x(t))a\sin(\omega t), \tag{5.98b}$$

where k, a, ω are tuning parameters.

Let us introduce new coordinates

$$\tilde{x} := x - x^*, \qquad \tilde{\theta} := \hat{\theta} - \theta^*.$$

In these coordinates the system (5.98) takes the form

$$\dot{\tilde{x}}(t) = f\Big(\tilde{x}(t) + x^*, \alpha\big(\tilde{x}(t) + x^*, \tilde{\theta}(t) + \theta^* + a\sin(\omega t)\big)\Big), \tag{5.99a}$$

$$\dot{\tilde{\theta}}(t) = kh(\tilde{x}(t) + x^*)a\sin(\omega t). \tag{5.99b}$$

It was shown in [100, Theorem 1] that under Assumptions 5.29, 5.30, 5.31, the output of the system can be steered arbitrarily close to the extremum value y^* from an arbitrarily large set of initial conditions by a proper choice of the parameters k, a, ω in the controller (5.99).

In [100, Propositions 1, 2] it was also shown that under Assumptions 5.29, 5.30, 5.31, and after a proper rescaling the subsystems of (5.99) are ISS with respect to the inputs from the other subsystems with the corresponding ISS gains. If the composition of these gains is less than identity, then the small-gain theorem ensures the stability of (5.99) with some sort of uniformity with respect to the choice of the parameters k, a, ω.

The origins of extremum seeking can be traced back to the paper [54]. It was actively developed in 1950–1960s, see, e.g., [72, 79]. A revival of the interest in extremum seeking was due to the paper [51], where the first proof of local asymptotic stability of an extremum seeking feedback scheme for a rather general class of nonlinear (possibly nonaffine in control and open-loop unstable) dynamic systems has been obtained. Semiglobal results have been achieved in [100], and in the above description of the extremum seeking methodology, we follow this paper.

Nowadays, extremum seeking is a prolific research direction with numerous applications to combustion processes, antilock braking systems, bioreactors, control of flight formation, solar cell and radio telescope antenna adjustment to maximize the received signal, etc., [7, 10, 86].

5.12 Exercises

Exercise 5.1 Theorem 5.4 does not always produce the simplest controllers. Check that

$$u(t) := \bar{k}_2(x(t)) = -\frac{2x(t) + x^3(t)}{1 + x^2(t)}$$

also input-to-state stabilizes the system (5.4).

Exercise 5.2 Formulate and prove a result, similar to Theorem 5.4 for the more general input-affine system

$$\dot{x}(t) = g_0(x) + \sum_{i=1}^{m} g_i(x)u_i,$$

where $g_i : \mathbb{R}^n \to \mathbb{R}^n$, $i = 0, \ldots, m$ are Lipschitz continuous, and $u_i(t) \in \mathbb{R}$, for all $i = 1, \ldots, m$.

Exercise 5.3 (Limitations of continuous feedback) Consider the scalar system

$$\dot{x} = g(u) := \begin{cases} 1 - u, & u \geq 1, \\ 0, & u \in [-1, 1], \\ -1 - u, & u \leq -1. \end{cases}$$

Show the following:

(i) Even though g is globally Lipschitz continuous, the control system is not stabilizable by means of a continuous feedback.
(ii) This system is globally stabilizable by a discontinuous feedback

$$u(x) := \begin{cases} x+1, & x > 0, \\ x-1, & x < 0. \end{cases}$$

Exercise 5.4 It is known that any physical input is ultimately limited in magnitude. The following example shows that the saturation of the controller strongly influences the global stability properties of the closed-loop system.

Consider a scalar system

$$\dot{x} = x^2 u + x^3 d. \tag{5.100}$$

Define

$$\text{sat}(v) : = \begin{cases} v & \text{if } |v| \le 1, \\ \frac{1}{|v|} v & \text{if } |v| \ge 1. \end{cases} \tag{5.101}$$

Construct a feedback $u = k(x)$, such that:

(i) (5.100) is ISS with the feedback $u = k(x)$.
(ii) (5.100) is 0-GAS with the feedback $u = \text{sat}(k(x))$.
(iii) (5.100) is not strongly iISS with the feedback $u = \text{sat}(k(x))$.

Exercise 5.5 Find an observer for the system

$$\dot{x}(t) = \begin{bmatrix} 0 & 1 \\ 2 & 3 \end{bmatrix} x(t) + \begin{bmatrix} 0 \\ 1 \end{bmatrix} u(t),$$
$$y(t) = \begin{bmatrix} 1 & 1 \end{bmatrix} x(t).$$

and stabilize it by means of a dynamic feedback.

References

1. Alessandri A (2004) Observer design for nonlinear systems by using input-to-state stability. In: Proceedings of 43rd IEEE Conference on Decision and Control, pp 3892–3897
2. Allan DA, Rawlings J, Teel AR (2021) Nonlinear detectability and incremental input/output-to-state stability. SIAM J Control and Optim 59(4):3017–3039
3. Andrieu V, Praly L, Astolfi A (2009) High gain observers with updated gain and homogeneous correction terms. Automatica 45(2):422–428
4. Angeli D, Ingalls B, Sontag E, Wang Y (2004) Uniform global asymptotic stability of differential inclusions. J Dyn Control Syst 10(3):391–412
5. Arcak M, Kokotović P (2001) Nonlinear observers: a circle criterion design and robustness analysis. Automatica 37(12):1923–1930
6. Arcak M, Teel A (2002) Input-to-state stability for a class of Lurie systems. Automatica 38(11):1945–1949
7. Ariyur KB, Krstic M (2003) Real-time optimization by extremum-seeking control. Wiley

8. Artstein Z (1983) Stabilization with relaxed controls. Nonlinear Anal Theor Methods Appl 7(11):1163–1173
9. Aström KJ (2008) Event based control. Analysis and design of nonlinear control systems. Springer, Berlin, pp 127–147
10. Åström KJ, Wittenmark B (2008) Adaptive control. Dover
11. Azouit R, Chaillet A, Chitour Y, Greco L (2016) Strong iISS for a class of systems under saturated feedback. Automatica 71(272–280)
12. Brockett RW (1972) System theory on group manifolds and coset spaces. SIAM J Control 10(2):265–284
13. Cai X, Meng L, Lin C, Zhang W, Liu L, Liu Y (2017) Input-to-state stabilizing a general nonlinear system with time-varying input delay and disturbances. IFAC-PapersOnLine 50(1):7169–7174
14. Carnevale D, Teel AR, Nesic D (2007) A Lyapunov proof of an improved maximum allowable transfer interval for networked control systems. IEEE Trans Autom Control 52(5):892–897
15. Christofides PD, Teel AR (1996) Singular perturbations and input-to-state stability. IEEE Trans Autom Control 41(11):1645–1650
16. Coron JM (2007) Control and nonlinearity. Am Math Soc
17. Coron JM, Praly L (1991) Adding an integrator for the stabilization problem. Syst Control Lett 17(2):89–104
18. Dashkovskiy S, Pavlichkov S (2012) Global uniform input-to-state stabilization of large-scale interconnections of MIMO generalized triangular form switched systems. Math Control Signals Syst 24(1):135–168
19. Dashkovskiy SN, Efimov DV, Sontag ED (2011) Input to state stability and allied system properties. Autom Remote Control 72(8):1579–1614
20. De Persis C, Sailer R, Wirth F (2013) Parsimonious event-triggered distributed control: a Zeno free approach. Automatica 49(7):2116–2124
21. Efimov D (2002) Universal formula for output asymptotic stabilization. IFAC Proc Volumes 35(1):239–244
22. Fah NCS (1999) Input-to-state stability with respect to measurement disturbances for one-dimensional systems. ESAIM Control Optim Calc Var 4:99–121
23. Freeman R (1995) Global internal stabilizability does not imply global external stabilizability for small sensor disturbances. IEEE Trans Autom Control 40(12):2119–2122
24. Freeman RA, Kokotovic P (1996) Inverse optimality in robust stabilization. SIAM J Control Optim 34(4):1365–1391
25. Freeman RA, Kokotovic PV (2008) Robust nonlinear control design. Birkhäuser, Boston
26. Goebel R, Sanfelice R, Teel AR (2012) Hybrid dynamical systems: modeling, stability, and robustness. Princeton University Press
27. Grüne L, Müller MA (2016) On the relation between strict dissipativity and turnpike properties. Syst Control Lett 90:45–53
28. Grüne L, Pannek J (2017) Nonlinear model predictive control. Springer International Publishing, Berlin
29. Hermann R, Krener A (1977) Nonlinear controllability and observability. IEEE Trans Autom Control 22(5):728–740
30. Hespanha JP, Liberzon D, Morse AS (2002) Supervision of integral-input-to-state stabilizing controllers. Automatica 38(8):1327–1335
31. Höger M, Grüne L (2019) On the relation between detectability and strict dissipativity for nonlinear discrete time systems. IEEE Control Syst Lett 3(2):458–462
32. Isidori A (1999) Nonlinear control systems II. Springer, Berlin
33. Jiang Z-P, Arcak M (2001) Robust global stabilization with ignored input dynamics: an input-to-state stability (ISS) small-gain approach. IEEE Trans Autom Control 46(9):1411–1415
34. Jiang Z-P, Liu T (2018) Small-gain theory for stability and control of dynamical networks: A survey. Annual Reviews in Control 46(58–79)
35. Jiang Z-P, Liu T-F (2013) Quantized nonlinear control-a survey. Acta Autom Sinica 39(11):1820–1830

36. Jiang Z-P, Mareels I, Hill D (1999) Robust control of uncertain nonlinear systems via measurement feedback. IEEE Trans Autom Control 44(4):807–812
37. Jiang Z-P, Teel AR, Praly L (1994) Small-gain theorem for ISS systems and applications. Math Control Signals Syst 7(2):95–120
38. Kalman RE (1960) On the general theory of control systems. In: Proceedings of First International Conference on Automatic Control, pp 481–492
39. Kanellakopoulos I, Kokotovic P, Morse A (1992) A toolkit for nonlinear feedback design. Syst Control Lett 18(2):83–92
40. Kanellakopoulos I, Kokotovic PV, Morse AS (1991) Systematic design of adaptive controllers for feedback linearizable systems. IEEE Trans Autom Control 36(11):1241–1253
41. Kokotović P, Arcak M (2001) Constructive nonlinear control: a historical perspective. Automatica 37(5):637–662
42. Kokotovic PV (1992) The joy of feedback: nonlinear and adaptive. IEEE Control Syst Mag 12(3):7–17
43. Krichman M, Sontag ED, Wang Y (2001) Input-output-to-state stability. SIAM J Control Optim 39(6):1874–1928
44. Krstic M (2009) Compensating a string PDE in the actuation or sensing path of an unstable ODE. IEEE Trans Autom Control 54(6):1362–1368
45. Krstic M (2009) Compensating actuator and sensor dynamics governed by diffusionPDEs. Syst Control Lett 58(5):372–377
46. Krstic M (2009) Delay compensation for nonlinear, adaptive, and PDE systems. Springer, Berlin
47. Krstic M (2010) Input delay compensation for forward complete and strict-feedforward nonlinear systems. IEEE Trans Autom Control 55(2):287–303
48. Krstic M, Fontaine D, Kokotovic P, Paduano J (1998) Useful nonlinearities and global bifurcation control of jet engine stall and surge. IEEE Trans Autom Control 43:1739–1745
49. Krstic M, Kanellakopoulos I, Kokotovic PV (1995) Nonlinear and adaptive control design. Wiley
50. Krstić M, Li Z-H (1998) Inverse optimal design of input-to-state stabilizing nonlinear controllers. IEEE Trans Autom Control 43(3):336–350
51. Krstić M, Wang H-H (2000) Stability of extremum seeking feedback for general nonlinear dynamic systems. Automatica 36(4):595–601
52. Larson R (1967) Optimum quantization in dynamic systems. IEEE Trans Autom Control 12(2):162–168
53. Lazar M, De La Peña DM, Heemels W, Alamo T (2008) On input-to-state stability of min-max nonlinear model predictive control. Syst Control Lett 57(1):39–48
54. Leblanc M (1922) Sur l'electrification des chemins de fer au moyen de courants alternatifs de frequence elevee. Revue générale de l'électricité 12(8):275–277
55. Liberzon D (1999) ISS and integral-ISS disturbance attenuation with bounded controls. In Proceedings of 38th IEEE Conference on Decision and Control, pp 2501–2506
56. Liberzon D (2003) Hybrid feedback stabilization of systems with quantized signals. Automatica 39(9):1543–1554
57. Liberzon D (2003) On stabilization of linear systems with limited information. IEEE Trans Autom Control 48(2):304–307
58. Liberzon D, Hespanha JP (2005) Stabilization of nonlinear systems with limited information feedback. IEEE Trans Autom Control 50(6):910–915
59. Liberzon D, Sontag ED, Wang Y (1999) On integral-input-to-state stabilization. In Proceedings of 1999 American Control Conference, pp 1598–1602
60. Liberzon D, Sontag ED, Wang Y (2002) Universal construction of feedback laws achieving ISS and integral-ISS disturbance attenuation. Syst Control Lett 46(2):111–127
61. Limon D, Alamo T, Raimondo D, De La Peña DM, Bravo J, Ferramosca A, Camacho E (2009) Input-to-state stability: a unifying framework for robust model predictive control. Nonlinear model predictive control. Springer, Berlin, pp 1–26

62. Lin Y, Sontag ED (1991) A universal formula for stabilization with bounded controls. Syst Control Lett 16(6):393–397
63. Liu T, Jiang Z-P (2015) A small-gain approach to robust event-triggered control of nonlinear systems. IEEE Trans Autom Control 60(8):2072–2085
64. Liu T, Jiang Z-P, Hill DJ (2012) Small-gain based output-feedback controller design for a class of nonlinear systems with actuator dynamic quantization. IEEE Trans Autom Control 57(5):1326–1332
65. Liu T, Zhang P, Jiang Z-P (2020) Robust event-triggered control of nonlinear systems. Springer Singapore
66. Malisoff M, Rifford L, Sontag E (2004) Global asymptotic controllability implies input-to-state stabilization. SIAM J Control Optim 42(6):2221–2238
67. Marchand N, Durand S, Castellanos JFG (2012) A general formula for event-based stabilization of nonlinear systems. IEEE Trans Autom Control 58(5):1332–1337
68. Mazenc F, Bernard O (2014) ISS interval observers for nonlinear systems transformed into triangular systems. Int J Robust Nonlinear Control 24(7):1241–1261
69. Mazenc F, Praly L (1996) Adding integrations, saturated controls, and stabilization for feedforward systems. IEEE Trans Autom Control 41(11):1559–1578
70. McCaughan F (1990) Bifurcation analysis of axial flow compressor stability. SIAM J Appl Math 50(5):1232–1253
71. Moore FK, Greitzer EM (1986) A theory of post-stall transients in axial compression systems: Part I-development of equations. J Eng Gas Turbines Power 108(1):68–76
72. Morosanov I (1957) Method of extremum control. Autom Remote Control 18:1077–1092
73. Müller MA, Grüne L, Allgöwer F (2015) On the role of dissipativity in economic model predictive control. IFAC-PapersOnLine 48(23):110–116
74. Nesic D, Laila DS (2002) A note on input-to-state stabilization for nonlinear sampled-data systems. IEEE Trans Autom Control 47(7):1153–1158
75. Nešić D, Teel AR (2001) Input-to-state stability for nonlinear time-varying systems via averaging. Math Control Signals Syst 14(3):257–280
76. Nesić D, Teel (2001) Sampled-data control of nonlinear systems: an overview of recent results. Perspect Robust Control 221–239
77. Nešić D, Teel AR (2004) Input-to-state stability of networked control systems. Automatica 40(12):2121–2128
78. Noroozi N, Geiselhart R, Mousavi SH, Postoyan R, Wirth FR (2019) Integral input-to-state stability of networked control systems. IEEE Trans Autom Control 65(3):1203–1210
79. Ostrovskii II (1957) Extremum regulation. Autom Remote. Control 18(900–907)
80. Pepe P, Ito H (2012) On saturation, discontinuities, and delays, in iISS and ISS feedback control redesign. IEEE Trans Autom Control 57(5):1125–1140
81. Postoyan R, Tabuada P, Nešić D, Anta A (2015) A framework for the event-triggered stabilization of nonlinear systems. IEEE Trans Autom Control 60(4):982–996
82. Praly L (2001) An introduction to forwarding. Control of complex systems. Springer, pp 77–99
83. Rawlings JB, Mayne DQ (2009) Model predictive control: Theory and design. Nob Hill Publishing
84. Rawlings JB, Mayne DQ, Diehl M (2017) Model predictive control: Theory, computation, and design. Nob Hill Publishing Madison, WI
85. Santoso H, Hioe D, Bao J, Lee PL (2012) Operability analysis of nonlinear processes based on incremental dissipativity. J Process Control 22(1):156–166
86. Scheinker A, Krstić M (2017) Model-free stabilization by extremum seeking. Springer, Berlin
87. Selivanov A, Fridman E (2016) Distributed event-triggered control of diffusion semilinear PDEs. Automatica 68(344–351)
88. Selivanov A, Fridman E (2016) Observer-based input-to-state stabilization of networked control systems with large uncertain delays. Automatica 74(63–70)
89. Sepulchre R, Janković M, Kokotović PV (1997) Constructive Nonlinear Control. Springer, Berlin

90. Shim H, Liberzon D (2016) Nonlinear observers robust to measurement disturbances in an ISS sense. IEEE Trans Autom Control 61(1):48–61
91. Smith OJ (1959) A controller to overcome dead time. ISA Journal 6(28–33)
92. Sontag ED (1983) A Lyapunov-like characterization of asymptotic controllability. SIAM J Control Optim 21(3):462–471
93. Sontag ED (1989) Smooth stabilization implies coprime factorization. IEEE Trans Autom Control 34(4):435–443
94. Sontag ED (1989) A 'universal' construction of Artstein's theorem on nonlinear stabilization. Syst Control Lett 13(2):117–123
95. Sontag ED (1998) Mathematical control theory: Deterministic finite-dimensional systems, 2nd edn. Springer, New York
96. Sontag ED, Wang Y (1995) On characterizations of input-to-state stability with respect to compact sets. In: Proceedings of IFAC non-linear control systems design symposium (NOLCOS 1995), pp 226–231
97. Sontag ED, Wang Y (1997) Output-to-state stability and detectability of nonlinear systems. Syst Control Lett 29(5):279–290
98. Sussmann HJ (1976) Existence and uniqueness of minimal realizations of nonlinear systems. Math Syst Theor 10(1):263–284
99. Tabuada P (2007) Event-triggered real-time scheduling of stabilizing control tasks. IEEE Trans Autom Control 52(9):1680–1685
100. Tan Y, Nešić D, Mareels I (2006) On non-local stability properties of extremum seeking control. Automatica 42(6):889–903
101. Teel A (1992) Using saturation to stabilize a class of single-input partially linear composite systems. IFAC Proc Volumes 25(13):379–384
102. Teel AR (1996) A nonlinear small gain theorem for the analysis of control systems with saturation. IEEE Trans Autom Control 41(9):1256–1270
103. Teel AR, Moreau L, Nesic D (2003) A unified framework for input-to-state stability in systems with two time scales. IEEE Trans Autom Control 48(9):1526–1544
104. Teel AR, Nesic D, Kokotovic PV (1998) A note on input-to-state stability of sampled-data nonlinear systems. In: Proceedings of 37th IEEE Conference on Decision and Control, pp 2473–2478
105. Trélat E, Zuazua E (2015) The turnpike property in finite-dimensional nonlinear optimal control. J Differ Equ 258(1):81–114
106. Walsh GC, Ye H, Bushnell LG (2002) Stability analysis of networked control systems. IEEE Trans Control Syst Technol 10(3):438–446
107. Wang W, Nesic D (2010) Input-to-state stability and averaging of linear fast switching systems. IEEE Trans Autom Control 55(5):1274–1279
108. Wang Y (1996) A converse Lyapunov theorem with applications to ISS-disturbance attenuation. In: Proceedings of 13th IFAC World Congress, pp 79–84
109. Wonham W (1967) On pole assignment in multi-input controllable linear systems. IEEE Trans Autom Control 12(6):660–665
110. Zabczyk J (2008) Mathematical control theory. Birkhäuser, Boston
111. Zhang W, Branicky MS, Phillips SM (2001) Stability of networked control systems. IEEE Control Syst Mag 21(1):84–99

Chapter 6
Input-to-State Stability of Infinite Networks

Stability analysis and control of ODE systems remain the backbone of much of systems theory. However, the increasing complexity of large-scale systems makes it hard to apply existing tools for controller synthesis for such systems. For instance, in large-scale vehicle platooning, classical distributed/decentralized control designs may result in nonuniformity in the convergence rate of solutions; i.e., as the number of participating subsystems goes to infinity, the resulting network becomes unstable.

This section develops the basics of the infinite-dimensional ISS theory with the stress on the study of infinite networks with ODE components. Furthermore, we show that a large portion of this theory (in particular, direct and converse ISS Lyapunov theorems, ISS superposition theorems) can be extended to much more broad classes of infinite-dimensional systems, including retarded and partial differential equations as well as switched systems.

Besides its importance in applications, infinite ODE systems illustrate very nicely many challenges arising in the stability theory of infinite-dimensional systems, such as nonuniform convergence rates, dependence of the stability properties on the choice of the norm in the state space, impossibility of a naive extension of superposition theorems for regular ODE systems to the infinite-dimensional case, and distinction between forward completeness and boundedness of reachability sets. We demonstrate these phenomena by means of examples.

As this book is on nonlinear ODE systems, we almost wholly avoid using operator and semigroup theory and do not consider the input-to-state stability of systems governed by partial differential equations (PDEs).

6.1 General Control Systems

We start with a general definition of a control system.

Definition 6.1 Consider a triple $\Sigma = (X, \mathcal{U}, \phi)$ consisting of the following:

(i) A normed vector space $(X, \|\cdot\|_X)$, called the *state space*.

A. Mironchenko, *Input-to-State Stability*, Communications and Control Engineering,
https://doi.org/10.1007/978-3-031-14674-9_6

(ii) A vector space U of input values and a normed vector space of inputs $(\mathcal{U}, \|\cdot\|_{\mathcal{U}})$, where $\mathcal{U}$ is a linear subspace of the space of all maps from $\mathbb{R}_+$ to U. We assume that the following axioms hold:

- *The axiom of shift invariance:* For all $u \in \mathcal{U}$ and all $\tau \geq 0$, the time-shifted function $u(\cdot + \tau)$ belongs to $\mathcal{U}$ with $\|u\|_{\mathcal{U}} \geq \|u(\cdot + \tau)\|_{\mathcal{U}}$.
- *The axiom of concatenation:* for all $u_1, u_2 \in \mathcal{U}$ and for all $t > 0$ the concatenation of u_1 and u_2 at time t, defined by

$$u_1 \underset{t}{\diamond} u_2(\tau) := \begin{cases} u_1(\tau), & \forall \tau \in [0, t], \\ u_2(\tau - t), & \forall \tau \in (t, \infty) \end{cases} \tag{6.1}$$

belongs to $\mathcal{U}$.

(iii) A map $\phi : D_\phi \to X$, $D_\phi \subseteq \mathbb{R}_+ \times X \times \mathcal{U}$, called transition map, so that for all $(x, u) \in X \times \mathcal{U}$ it holds that $D_\phi \cap (\mathbb{R}_+ \times \{(x, u)\}) = [0, t_m) \times \{(x, u)\}$, for a certain $t_m = t_m(x, u) \in (0, +\infty]$. The corresponding interval $[0, t_m)$ is called the *maximal domain of definition* of the mapping $t \mapsto \phi(t, x, u)$, which we call a trajectory of the system.

The triple Σ is called a (control) system if it satisfies the following axioms:

(Σ1) The identity property: For all $(x, u) \in X \times \mathcal{U}$, it holds that $\phi(0, x, u) = x$.
(Σ2) Causality: For all $(t, x, u) \in D_\phi$ and $\tilde{u} \in \mathcal{U}$ such that $u(s) = \tilde{u}(s)$ for all $s \in [0, t]$, it holds that $[0, t] \times \{(x, \tilde{u})\} \subset D_\phi$ and $\phi(t, x, u) = \phi(t, x, \tilde{u})$.
(Σ3) Continuity: For each $(x, u) \in X \times \mathcal{U}$, the trajectory $t \mapsto \phi(t, x, u)$ is continuous on its maximal domain of definition.
(Σ4) The cocycle property: For all $x \in X$, $u \in \mathcal{U}$ and $t, h \geq 0$ so that $[0, t + h] \times \{(x, u)\} \subset D_\phi$, we have $\phi(h, \phi(t, x, u), u(t + \cdot)) = \phi(t + h, x, u)$.

This class of systems encompasses control systems generated by evolution partial differential equations (PDEs), abstract differential equations in Banach spaces with regular right-hand sides, time-delay systems, ordinary differential equations (ODEs), switched systems, as well as important classes of well-posed coupled systems consisting of an arbitrary number of finite- and infinite-dimensional components, with both in-domain and boundary couplings.

Definition 6.2 We say that a system Σ satisfies the *boundedness-implies-continuation (BIC) property* if for each $(x, u) \in X \times \mathcal{U}$, there exists $t_m \in (0, +\infty]$, called a *maximal existence time*, such that $[0, t_m) \times \{(x, u)\} \subset D_\phi$ and for all $t \geq t_m$, it holds that $(t, x, u) \notin D_\phi$. In addition, if $t_m < +\infty$, then for every $M > 0$, there exists $t \in [0, t_m)$ with $\|\phi(t, x, u)\|_X > M$.

As in Chap. 2 we will always assume that

Assumption 6.3 For control systems that we consider in this chapter, we assume that BIC property always holds.

6.2 Infinite Networks of ODE Systems

Although some of the results will be developed for a general class of control systems introduced in Definition 6.1, our main attention will be focused on infinite networks of ODE systems. Using $\mathbb{N}$ as the index set (by default), the ith subsystem is given by

$$\Sigma_i : \dot{x}_i = f_i(x_i, x_{\neq i}, u_i). \tag{6.2}$$

We suppose that:

Assumption 6.4 The family $(\Sigma_i)_{i\in\mathbb{N}}$ comes together with sequences $(N_i)_{i\in\mathbb{N}}$, $(m_i)_{i\in\mathbb{N}}$ of positive integers and *finite* index sets $I_i \subset \mathbb{N}\backslash\{i\}$, $i \in \mathbb{N}$, so that the following hold:

(i) The state vector x_i of Σ_i is an element of $\mathbb{R}^{N_i}$.
(ii) The vector $x_{\neq i}$ is composed of the state vectors x_j, $j \in I_i$. The order of these vectors plays no particular role (as the index set $\mathbb{N}$ does not), so we do not specify it.
(iii) The external input vector u_i is an element of $\mathbb{R}^{m_i}$.
(iv) The right-hand side is a continuous function $f_i : \mathbb{R}^{N_i} \times \mathbb{R}^{N_{\neq i}} \times \mathbb{R}^{m_i} \to \mathbb{R}^{N_i}$, where $N_{\neq i} := \sum_{j\in I_i} N_j$.
(v) Unique local solutions of the ODE (6.2) (in the sense of Carathéodory) exist for all initial states $x_{i0} \in \mathbb{R}^{N_i}$ and all locally essentially bounded functions $x_{\neq i}(\cdot)$ and $u_i(\cdot)$ (which are regarded as time-dependent inputs). We denote the corresponding solution by $\phi_i(t, x_{i0}, x_{\neq i}, u_i)$.

In the ODE (6.2), we consider $x_{\neq i}(\cdot)$ as an *internal input* and $u_i(\cdot)$ as an *external input* (which may be a disturbance or a control input). The interpretation is that the subsystem Σ_i is affected by finitely many neighbors, indexed by I_i, and its external input.

To define the interconnection of the subsystems Σ_i, we consider the state vector $x = (x_i)_{i\in\mathbb{N}}$, the input vector $u = (u_i)_{i\in\mathbb{N}}$, and the right-hand side

$$f(x, u) := (f_i(x_i, x_{\neq i}, u_i))_{i\in\mathbb{N}} \in \prod_{i\in\mathbb{N}} \mathbb{R}^{N_i}.$$

The interconnection is then formally written as

$$\Sigma : \quad \dot{x} = f(x, u). \tag{6.3}$$

To handle this infinite-dimensional ODE properly, we introduce the state space X of Σ as a Banach space of sequences $x = (x_i)_{i\in\mathbb{N}}$ with $x_i \in \mathbb{R}^{N_i}$. The most natural choice is an ℓ^p-space. To define such a space, we first fix a norm on each $\mathbb{R}^{N_i}$ (that might not only depend on the dimension N_i but also on the index i). For brevity, we omit the index in our notation and simply write $|\cdot|$ for each of these norms. Then, for every $p \in [1, \infty)$, we put

$$\ell^p(\mathbb{N}, (N_i)) := \Big\{x = (x_i)_{i\in\mathbb{N}} : x_i \in \mathbb{R}^{N_i},\ \sum_{i\in\mathbb{N}} |x_i|^p < \infty\Big\}.$$

Additionally, we introduce

$$\ell^\infty(\mathbb{N}, (N_i)) := \Big\{x = (x_i)_{i\in\mathbb{N}} : x_i \in \mathbb{R}^{N_i},\ \sup_{i\in\mathbb{N}} |x_i| < \infty\Big\}.$$

The following proposition is proved with standard arguments, see, e.g., [10]. Hence, we omit the details.

Proposition 6.5 *The following statements hold:*

(i) For each choice of norms on $\mathbb{R}^{N_i}$, $i \in \mathbb{N}$, and each $p \in [1, \infty]$, the associated space $\big(\ell^p(\mathbb{N}, (N_i)), |\cdot|_p\big)$ is a Banach space, where for $p \in [1, +\infty)$ we define $|x|_p := \big(\sum_{i\in\mathbb{N}} |x_i|^p\big)^{1/p}$ and $|x|_\infty := \sup_{i\in\mathbb{N}} |x_i|$.

(ii) For each $1 \le p < \infty$, the Banach space $\ell^p(\mathbb{N}, (N_i))$ is separable.

(iii) For each pair (p, q) with $1 \le p < q \le \infty$, the space $\ell^p(\mathbb{N}, (N_i))$ is continuously embedded into $\ell^q(\mathbb{N}, (N_i))$.

As the state space of the system Σ, we consider $X := \ell^p(\mathbb{N}, (N_i))$ for a fixed $p \in [1, \infty]$. The corresponding norm we denote by $\|\cdot\|_X$. Similarly, for a fixed $q \in [1, \infty]$, we consider the *external input space* $U := \ell^q(\mathbb{N}, (m_i))$, where we fix norms on $\mathbb{R}^{m_i}$ that we simply denote by $|\cdot|$ again. The space of admissible *external input functions* is defined by[1]

$$\mathcal{U} := \big\{u : \mathbb{R}_+ \to U : u \text{ is strongly measurable and essentially bounded}\big\}. \tag{6.4}$$

In view of Pettis' theorem, a function is strongly measurable if and only if it is weakly measurable and almost separably valued function. Since $\ell^q(\mathbb{N}, (m_i))$ is separable for all finite q, strong measurability reduces to weak measurability in all of these cases. For a definition of the Bochner integral, we refer the interested readers to [2, Section 1.1] and [14, Chap. III, Sect. 3.7].

Definition 6.6 For fixed $u \in \mathcal{U}$ and $x^0 \in X$, a function $\lambda : J \to X$, where $J \subset \mathbb{R}$ is an interval of the form $[0, T)$ with $0 < T \le \infty$, is called a *solution* of the Cauchy problem

$$\dot{x} = f(x, u), \quad x(0) = x^0.$$

[1] We are using the letter u both for elements of U and $\mathcal{U}$. Since it should become clear from the context if we refer to an input value or an input function, this should not lead to confusion.

provided that $s \mapsto f(\lambda(s), u(s))$ is a locally integrable X-valued function (in the Bochner integral sense) and

$$\lambda(t) = x^0 + \int_0^t f(\lambda(s), u(s))\, ds \quad \forall\, t \in J.$$

Any solution λ is a locally absolutely continuous function, see [2, Proposition 1.2.2].

Definition 6.7 We say that the system Σ is well posed if for every initial value $x^0 \in X$ and every external input $u \in \mathcal{U}$, a unique maximal solution, which we denote by $\phi(\cdot, x^0, u) : [0, t_m(x^0, u)) \to X$ exists, where $0 < t_m(x^0, u) \leq \infty$.

Denoting by $\pi_i : X \to \mathbb{R}^{N_i}$ the canonical projection onto the i-th component (which is a bounded linear operator) and writing $u(t) = (u_1(t), u_2(t), \ldots)$, we obtain

$$\begin{aligned}\pi_i\phi(t, x^0, u) &= x_i^0 + \int_0^t \pi_i f(\phi(s, x^0, u), u(s))ds \\ &= x_i^0 + \int_0^t f_i(\pi_i\phi(s, x^0, u), (\pi_j\phi(s, x^0, u))_{j\in I_i}, u_i(s))ds,\end{aligned}$$

which implies that $t \mapsto \pi_i\phi(t, x^0, u)$ solves the ODE $\dot{x}_i = f_i(x_i, x_{\neq i}, u_i)$ for the internal input $x_{\neq i}(\cdot) := (\pi_j\phi(\cdot, x^0, u))_{j\in I_i}$ and the external input $u_i(\cdot)$. Hence,

$$\pi_i\phi(t, x^0, u) = \phi_i(t, x_i^0, x_{\neq i}, u_i) \quad \forall\, t \in [0, t_m(x^0, u)).$$

Sufficient conditions for the existence and uniqueness of solutions (and forward completeness) can be obtained from the general Carathéodory theory of differential equations on Banach spaces, see [3] as a general reference for systems with bounded infinitesimal generators.

Assumption 6.8 Let the system Σ with state space $X = \ell^p(\mathbb{N}, (N_i))$ and external input space $U = \ell^q(\mathbb{N}, (m_i))$ with given $p, q \in [1, +\infty]$ satisfy the following:

(i) $f : X \times U \to X$ is Lipschitz continuous on bounded subsets of X, i.e., $\forall C > 0\ \exists L_f(C) > 0$, such that $\forall x, y : \|x\|_X \leq C,\ \|y\|_X \leq C, \forall v \in U$: $\|v\|_U \leq C$, it holds that

$$\|f(y, v) - f(x, v)\|_X \leq L_f(C)\|y - x\|_X. \tag{6.5}$$

(ii) $f(x, \cdot)$ is continuous for all $x \in X$.

The following theorem provides a set of conditions that is sufficient for well posedness and BIC property

Theorem 6.9 *Assume that the system Σ with state space $X = \ell^p(\mathbb{N}, (N_i))$ and external input space $U = \ell^q(\mathbb{N}, (m_i))$ for arbitrary $p, q \in [1, +\infty]$ satisfy Assumption 6.8.*

Then Σ is well posed, and BIC property holds.

Proof (Sketch) Well posedness follows from [17]. Furthermore, for any $R > 0$ there is $\tau_R > 0$ such that for all $x \in B_R$ and $u \in B_{R,\mathcal{U}}$ the solutions $\phi(\cdot, x, u)$ exist at least on $[0, \tau_R]$, which easily implies the BIC property, see [5, Theorem 4.3.4] for the related result for systems without inputs.

Analogously to the finite-dimensional case (see Proposition 1.17), one can show that

Proposition 6.10 *Well-posed systems (6.3) are control systems in the sense of Definition 6.1.*

Example 6.11 Assume that the subsystems Σ_i are linear:

$$\Sigma_i : \quad \dot{x}_i = A_i x_i + \tilde{A}_i x_{\neq i} + B_i u_i,$$

with matrices $A_i \in \mathbb{R}^{N_i \times N_i}$, $\tilde{A}_i \in \mathbb{R}^{N_i \times N_{\neq i}}$, and $B_i \in \mathbb{R}^{N_i \times m_i}$. Given a choice of norms on $\mathbb{R}^{N_i}$, on $\mathbb{R}^{N_{\neq i}} = \prod_{j \in I_i} \mathbb{R}^{N_j}$, consider the product norm

$$|x_{\neq i}| := \sum_{j \in I_i} |x_j|. \tag{6.6}$$

We make the following assumptions:

- The operator norms (with respect to the chosen vector norms on $\mathbb{R}^{N_i}$, $\mathbb{R}^{N_{\neq i}}$, and $\mathbb{R}^{m_i}$) of the linear operators A_i, $\tilde{A}_i$, and B_i are uniformly bounded over $i \in \mathbb{N}$.
- $1 \leq p = q < \infty$.
- There exists $m \in \mathbb{N}$ such that $I_i \subset [i - m, i + m] \cap \mathbb{N}$ for all $i \in \mathbb{N}$.

Now, we show that Assumptions 6.8 needed for the invocation of Theorem 6.9 is satisfied. Take $x \in X = \ell^p(\mathbb{N}, (N_i))$ and $u \in U = \ell^p(\mathbb{N}, (m_i))$.

By using the inequality $(a + b)^p \leq 2^{p-1}(a^p + b^p)$ (which holds for all $a, b \geq 0$ due to the convexity of $a \mapsto a^p$) repeatedly, we obtain

$$\begin{aligned}\sum_{i=1}^{\infty} |A_i x_i + \tilde{A}_i x_{\neq i} + B_i u_i|^p &\leq C_1 \sum_{i=1}^{\infty} |x_i|^p + C_2 \sum_{i=1}^{\infty} \sum_{j \in I_i} |x_j|^p + C_3 \sum_{i=1}^{\infty} |u_i|^p \\ &\leq C_1 \|x\|_X^p + C_2(2m+1)\|x\|_X^p + C_3 \|u\|_U^p < \infty,\end{aligned}$$

where $C_1, C_2, C_3 > 0$ are appropriately chosen constants. This shows that right-hand side of our system is well defined as a map from $X \times U$ to X.

To verify Lipschitz continuity, observe that for any $x^1, x^2 \in X$, and $u \in U$ we have

$$\begin{aligned}\|f(x^1,u)-f(x^2,u)\|_X^p &= \sum_{i=1}^{\infty}\big|A_i(x_i^1-x_i^2)+\tilde{A}_i(x_{\neq i}^1-x_{\neq i}^2)\big|^p\\ &\le C_1\sum_{i=1}^{\infty}|x_i^1-x_i^2|^p + C_2\sum_{i=1}^{\infty}|x_{\neq i}^1-x_{\neq i}^2|^p\\ &\le C\cdot\|x^1-x^2\|_X^p, \end{aligned} \tag{6.7}$$

for some constant $C > 0$.

Continuity of f with respect to the second argument can be seen from

$$\|f(x,u^1)-f(x,u^2)\|_X^p \le C_3\sum_{i=1}^{\infty}|u_i^1-u_i^2|^p = C_3\|u^1-u^2\|_U^p,$$

By assumption, u is essentially bounded as a function from $\mathbb{R}_+$ into $\ell^p(\mathbb{N},(m_i))$, which implies that $\|u(\cdot)\|_U$ is locally integrable. □

Remark 6.12 Well posedness with respect to $X := \ell^\infty$ does not imply well posedness with respect to $X := \ell^p$ for any $p < \infty$. Consider an infinite system

$$\dot{x}_i = u(t), \quad i \in \mathbb{N}.$$

For any initial condition and any input $u(t) \equiv 1$, the solution belongs to ℓ^∞, but does not belong to ℓ^p with any $p < \infty$.

Example 6.13 To achieve well posedness with respect to $X := \ell^p$, we have assumed in Theorem 6.9 that f is well defined as a map from $X \times U$ to X and is Lipschitz continuous on bounded balls (in the corresponding norm). Both these conditions are not necessary for well posedness in X, as the following example shows.

Consider the following nonlinear system, consisting of a countable number of the same components.

$$\dot{x}_i = -x_i^{1/3}, \quad i \in \mathbb{N}. \tag{6.8}$$

In this case, the right-hand side is not Lipschitz continuous (even for individual modes), and thus the standard general results are not applicable to obtain the well posedness. Moreover, the right-hand side $f(x) := (-x_1^{1/3}, -x_2^{1/3}, \dots)$ maps ℓ^p into ℓ^{3p}, and thus f is not even defined as a map from $\ell^p \to \ell^p$.

Nevertheless, f is defined as a map from ℓ^∞ to ℓ^∞, and (although it is still not Lipschitz continuous), the system is well posed in the ℓ^∞ norm and one can show that $|x_i(t)| \le |x_i(0)|$, for all $i \in \mathbb{N}$ and all $t \ge 0$, which (after some straightforward argumentation that we omit due to space limitations) implies that the system (6.8) is well posed, forward complete, and globally stable with respect to ℓ^p for any $p \in [1, \infty]$.

Remark 6.14 Example 6.13 indicates that the following method can be used to show well posedness of the system (6.3) with respect to the state space ℓ^p, $p \in [1, +\infty)$. First, one can consider the system (6.3) as the system defined over ℓ^∞ (which is the largest set of all ℓ^p scale). Then, using Lipschitz continuity, one can obtain the local existence and uniqueness of the solutions in the ℓ^∞ space. Then one could use further properties of the system to ensure that the solution actually lies in ℓ^p space and is continuous in ℓ^p-norm, provided that the initial condition also lies in ℓ^p and possibly some further conditions are fulfilled.

6.3 Input-to-State Stability

For a normed vector space W and any $r > 0$ denote $B_{r,W} := \{u \in W : \|u\|_W < r\}$ (the open ball of radius r around 0 in W). If W is a state space X we write simply B_r instead of $B_{r,X}$.

The ISS-like properties can be naturally extended to general control systems.

Definition 6.15 System $\Sigma = (X, \mathcal{U}, \phi)$ is called:

(i) Forward complete (FC), if $D_\phi = \mathbb{R}_+ \times X \times \mathcal{U}$, that is, for every $(x, u) \in X \times \mathcal{U}$ and for all $t \geq 0$ the value $\phi(t, x, u) \in X$ is well defined.

(ii) (Uniformly) Input-to-state stable (ISS), if it is forward complete and there exist $\beta \in \mathcal{KL}$ and $\gamma \in \mathcal{K}_\infty$ such that for all $x \in X$, $u \in \mathcal{U}$ and $t \geq 0$ it holds that

$$\|\phi(t, x, u)\|_X \leq \beta(\|x\|_X, t) + \gamma(\|u\|_{\mathcal{U}}). \tag{6.9}$$

(iii) (Uniformly) Locally input-to-state stable (LISS), if there exist $\beta \in \mathcal{KL}$, $\gamma \in \mathcal{K}_\infty$ and $r > 0$ such that the inequality (6.9) holds for all $x \in \overline{B_r}$, $u \in \overline{B_{r,\mathcal{U}}}$ and $t \geq 0$.

(iv) *Uniformly globally asymptotically stable at zero (0-UGAS)*, if there exists $\beta \in \mathcal{KL}$ such that for all $x \in X$ and $t \geq 0$ it holds that

$$\|\phi(t, x, 0)\|_X \leq \beta(\|x\|_X, t). \tag{6.10}$$

If the system does not have any inputs, i.e., if $\mathcal{U} = \{0\}$, we call the 0-UGAS system just UGAS.

As we assume that for our control systems the BIC property holds, the solutions for which the ISS / LISS / 0-UGAS estimates are valid are global, i.e., exist on $\mathbb{R}_+$.

The next result is an infinite-dimensional counterpart of Proposition 2.7.

Proposition 6.16 *Let $\Sigma = (X, \mathcal{U}, \phi)$ be a control system. If (2.1) is ISS, then (2.1) has CIUCS property, that is, for each $u \in \mathcal{U}$ such that $\lim_{t\to\infty} \|u(t + \cdot)\|_{\mathcal{U}} = 0$, and for any $r > 0$, it holds that*

$$\lim_{t\to\infty} \sup_{\|x\|_X \leq r} \|\phi(t, x, u)\|_X = 0.$$

Proof The proof is analogous to that of Proposition 2.7 and is omitted.

As we assume BIC property, Proposition 2.8 can be extended to the general control systems, in case that $\mathcal{U} := PC_b(\mathbb{R}_+, U)$.

Recall that for nonlinear ODEs, the 0-GAS notion is equivalent to 0-UGAS due to Theorem B.37. For infinite-dimensional systems, even for linear infinite ODEs in the Hilbert space ℓ^2, the situation is quite different. Namely, global attractivity does not imply uniform global asymptotic stability.

Example 6.17 Consider a linear infinite ODE system

$$\dot{x}_k(t) = -\frac{1}{k}x_k(t) + u_k(t), \quad k = 1, \dots, \infty. \tag{6.11}$$

Denote $x := (x_k)_{k=1}^{\infty}$, $x_k \in \mathbb{R}$ and assume that the state space of (6.11) is a Hilbert space ℓ^2. Let also $U := \ell^2$ and $\mathcal{U} := PC_b(\mathbb{R}_+, U)$.

We would like to show that

(i) (6.11) is 0-GAS.
(ii) (6.11) is not 0-UGAS.
(iii) Even vanishing at infinity inputs of an arbitrarily small norm may result in divergent trajectories of (6.11).

We proceed with our analysis step by step.

(i) Let $u := 0$. The unique (classical, continuously differentiable) solution of the system (6.11) can be written as

$$\phi(t, x(0)) = (e^{-\frac{1}{k}t}x_k(0))_{k=1}^{\infty}.$$

As $\|\phi(t, x(0))\|_{\ell^2} \le \|x(0))\|_{\ell^2}$ for all $x(0) \in \ell^2$, the system (6.11) is globally stable.

Let us show global attractivity of (6.11) with $u \equiv 0$.

For any $x(0) \in \ell^2$ and for any $\varepsilon > 0$ there exists $N > 0$: $\sqrt{\sum_{k=N+1}^{\infty} |x_k(0)|^2} < \varepsilon$. Since $|x_k(t)| \le |x_k(0)|$ for any $t \ge 0$, it holds that $\sqrt{\sum_{k=N+1}^{\infty} |x_k(t)|^2} < \varepsilon$ for any $t \ge 0$.

Moreover,

$$\sqrt{\sum_{k=1}^{N} |x_k(t)|^2} \le \sqrt{\sum_{k=1}^{N} e^{-\frac{2}{N}t}|x_k(0)|^2} \le \sqrt{\sum_{k=1}^{N} |x_k(0)|^2 e^{-\frac{1}{N}t}},$$

and thus there exists $t^* > 0$: $\sqrt{\sum_{k=1}^{N} |x_k(t)|^2} \le \varepsilon$ for $t \ge t^*$.

As $\sqrt{a+b} \le \sqrt{a} + \sqrt{b}$ for all $a, b \ge 0$, for any $\varepsilon > 0$ there is a time $t^* = t^*(\varepsilon)$ so that for all $t \ge t^*$ it holds that

$$\sqrt{\sum_{k=1}^{\infty}|x_k(t)|^2} \le \sqrt{\sum_{k=1}^{N}|x_k(t)|^2} + \sqrt{\sum_{k=N+1}^{\infty}|x_k(t)|^2} \le 2\varepsilon.$$

This shows that the system the system (6.11) with $u \equiv 0$ is globally attractive.

Hence, (6.11) is 0-GAS.

(ii) We continue to assume that $u := 0$. Next we show that (6.11) with a state space ℓ^2 is not 0-UGAS. Pick a sequence of states $\hat{x}_i := (\underbrace{0, \ldots, 0}_{i-1}, 1, 0, \ldots, 0, \ldots)$, for all $i \in \mathbb{N}$. Clearly, $\|\hat{x}_i\|_{\ell^2} = 1$ for all $i \in \mathbb{N}$. The solution of (6.11), corresponding to initial condition $\hat{x}_i$, equals to $\phi(t, \hat{x}_i) = e^{-\frac{1}{i}t}\hat{x}_i$.

Assume that $\exists \beta \in \mathcal{KL}$ so that

$$t \ge 0,\ x_0 \in \ell^2 \quad \Rightarrow \quad \|\phi(t, x_0)\|_{\ell^2} \le \beta(\|x_0\|_{\ell^2}, t). \tag{6.12}$$

In particular, this holds for initial conditions $\hat{x}_i$, for all i. This means that

$$t \ge 0,\ i \in \mathbb{N} \quad \Rightarrow \quad \|\phi(t, \hat{x}_i)\|_{\ell^2} \le \beta(1, t).$$

Due to properties of $\mathcal{KL}$ functions, there exist t^* s.t. $\beta(1, t^*) = \frac{1}{2}$. The above inequality implies that

$$e^{-\frac{1}{i}t^*} \le \frac{1}{2} \quad \forall i = 1, \ldots, \infty.$$

Clearly this is false. Hence (6.11) is not 0-UGAS.

(iii) Consider the times $t_1 := 1$, $t_k := t_{k-1} + k$, for $k \ge 2$, and the input

$$u_k(t) := \begin{cases} \frac{2}{\sqrt{k}}, & \text{if } t \in [t_{k-1}, t_k), \\ 0, & \text{otherwise.} \end{cases}$$

Clearly, the input $u = (u_k)$ decays to 0 in ℓ_2-norm. Take $x(0) = 0$.

At the same time, for any $k \in \mathbb{N}$ and for $t \in [0, t_k - t_{k-1}] = [0, k]$, we have

$$\begin{aligned} x_k(t + t_{k-1}) &= x_k(t_{k-1}) + \frac{2}{\sqrt{k}} \int_0^t e^{-\frac{1}{k}(t-s)} ds \\ &= x_k(t_{k-1}) + \frac{2}{\sqrt{k}} e^{-\frac{1}{k}t} \int_0^t e^{\frac{1}{k}s} ds \\ &= x_k(t_{k-1}) + \frac{2}{\sqrt{k}} e^{-\frac{1}{k}t} k(e^{\frac{1}{k}t} - 1) \\ &= x_k(t_{k-1}) + 2\sqrt{k}(1 - e^{-\frac{1}{k}t}). \end{aligned}$$

Noting that $u_k(s) = 0$ for $s < t_{k-1}$, we see that $x_k(t_{k-1}) = 0$. Taking $t = k$, we see from the computations above that

$$x_k(t_k) = x_k(t_{k-1}) + 2\sqrt{k}(1 - e^{-1}) = 2\sqrt{k}(1 - e^{-1}).$$

Thus $\|x(t_k)\|_{\ell_2} \geq |x_k(t_k)| = 2\sqrt{k}(1 - e^{-1})$, and this easily implies that $\|x(t)\|_{\ell_2} \to \infty$, as $t \to \infty$.

For ODE systems, the ISS and 0-UGAS properties are invariant with respect to the choice of the norm in the state space $\mathbb{R}^n$, see Exercise 2.2, which is due to the fact that all the norms in $\mathbb{R}^n$ are equivalent. As in infinite-dimensional spaces the norms are not equivalent, it is natural to expect that ISS and 0-UGAS of infinite-dimensional systems depend on the choice of a norm. The next example confirms this, even if the system is well posed with respect to both norms.

Example 6.18 Consider the following nonlinear system, consisting of a countable number of same components.

$$\dot{x}_i = -x_i^3, \quad i \in \mathbb{N}. \tag{6.13}$$

We are going to show that:

(i) (6.13) is well posed for any state space ℓ^p, $p \in [1, +\infty]$.
(ii) (6.13) is 0-UGAS with respect to ℓ^∞.
(iii) (6.13) is not 0-UGAS with respect to ℓ^p, $p \in [1, +\infty)$.

(i) Local well posedness of this system can be shown by checking the Lipschitz condition. Let $p \in [1, +\infty)$. We have for all $x, y \in \ell^p$

$$\begin{aligned}
\|f(x) - f(y)\|_{\ell^p}^p &= \sum_{i=1}^{\infty} |x_i^3 - y_i^3|^p \\
&= \sum_{i=1}^{\infty} |x_i - y_i|^p |x_i^2 + x_i y_i + y_i^2|^p \\
&\leq \sum_{i=1}^{\infty} |x_i - y_i|^p |2x_i^2 + 2y_i^2|^p \\
&\leq \sum_{i=1}^{\infty} |x_i - y_i|^p 2^p |2\max\{x_i^2, y_i^2\}|^p \\
&= \sum_{i=1}^{\infty} |x_i - y_i|^p 4^p \max\{\|x\|_{\ell^p}^{2p}, \|y\|_{\ell^p}^{2p}\},
\end{aligned}$$

and thus

$$\|f(x) - f(y)\|_{\ell^p} \leq 4\max\{\|x\|_{\ell^p}^2, \|y\|_{\ell^p}^2\}\|x_i - y_i\|_{\ell^p}.$$

This shows that f is Lipschitz continuous on bounded balls, and thus the system (6.13) has for each initial condition $x(0) \in \ell^p$ the unique mild solution, defined on a certain interval.

Global existence one can show as follows. Pick any $p \in [1, +\infty)$. As $\frac{d}{dt}|x_i|^p = p|x_i|^{p-1}\mathrm{sgn}\,(x_i)(-x_i^3) = -p|x_i|^{p+2} \leq 0$, we have that for each $x \in \ell^p$ it holds that

$$\frac{d}{dt}\|x(t)\|_{\ell^p} \leq 0,$$

and thus

$$\|x(t)\|_{\ell^p} \leq \|x(0)\|_{\ell^p}, \quad t \in [0, t^*), \tag{6.14}$$

where $[0, t^*)$ is the maximal interval of existence for the solution corresponding to the initial condition $x(0)$.

By Theorem 6.9, our system satisfies the BIC property. As its solutions are bounded due to (6.14), they can be prolonged on a larger interval.

Hence, if $t^* < \infty$, then $[0, t^*)$ cannot be a maximal interval of existence of solutions. Thus, $t^* = \infty$, and the system is forward complete. In view of (6.14) it is globally stable.

Above argumentation can be performed also for $p = +\infty$.

(ii) Let $p = +\infty$. Pick any $x = (x_1, x_2, \ldots) \in \ell^\infty$. As all subsystems are identical and UGAS, there is $\beta \in \mathcal{KL}$ so that for all $i \in \mathbb{N}$ it holds that

$$|x_i(t)| \leq \beta(|x_i(0)|, t).$$

This implies that

$$\|x(t)\|_{\ell^\infty} = \sup_{i\in\mathbb{N}} |x_i(t)| \leq \sup_{i\in\mathbb{N}} \beta(|x_i(0)|, t) \leq \beta(\|x(0)\|_{\ell^\infty}, t), \tag{6.15}$$

and thus (6.13) is UGAS in ℓ^∞-norm.

(iii) Let us show that (6.13) is not UGAS in ℓ^p-norm for any $p \in [1, +\infty)$. Assume that (6.13) is UGAS in ℓ^p-norm for a certain $p \in [1, +\infty)$. Then there is $\beta \in \mathcal{KL}$ so that for all $x(0) \in \ell^p$ and all $t \geq 0$ it holds that

$$\|x(t)\|_{\ell^p} \leq \beta(\|x(0)\|_{\ell^p}, t). \tag{6.16}$$

In particular, this implies that there is a time $T > 0$ so that for each $x(0) \in \ell^p$ with $\|x(0)\|_{\ell^p} \leq 1$ and all $t \geq T$ it holds that

$$\|x(t)\|_{\ell^p} \leq \frac{1}{2}.$$

Let us show that this condition is violated for our system (6.13).

To see this, consider the i-th subsystem of (6.13). Its solution is given (implicitly) by

$$\frac{1}{x_i^2(t)} - \frac{1}{x_i^2(0)} = 2t. \tag{6.17}$$

Let us compute time within which the state of the i-th subsystem will become twice smaller. For this time that we call $t_{1/2}(x_i(0))$ it holds that

$$\frac{1}{\frac{1}{4}x_i^2(0)} - \frac{1}{x_i^2(0)} = 2t_{1/2}(x_i(0)), \tag{6.18}$$

and thus

$$t_{1/2}(x_i(0)) = \frac{3}{2x_i^2(0)}. \tag{6.19}$$

In particular, the smaller the initial state, the more time it is needed to reduce the norm of the state twice.

Now consider a sequence of initial states $(x_{<k>}) \subset \ell^p$, where

$$x_{<k>}(0) := \Big(\frac{1}{k^{1/p}}, \ldots, \frac{1}{k^{1/p}}, 0, \ldots\Big),$$

where only the first k entries of $x_{<k>}(0)$ are nonzero. It is easy to see that $\|x_{<k>}(0)\|_{\ell^p} = 1$ for all $k \in \mathbb{N}$ and all $p \in [1, \infty)$.

We have that $t_{1/2}(\frac{1}{k^{1/p}}) = \frac{3}{2}k^{2/p}$, and

$$\|x_{<k>}(\frac{3}{2}k^{2/p})\|_{\ell^p} = \|\frac{1}{2}x_{<k>}(0)\|_{\ell^p} = \frac{1}{2}\|x_{<k>}(0)\|_{\ell^p} = \frac{1}{2}.$$

Hence, the larger is k, the longer it takes to twice reduce the ℓ^p-norm of the state. This violates the condition (6.16), and thus (6.13) is not UGAS with respect to ℓ^p-norm for $p \in [1, +\infty)$.

6.4 ISS Superposition Theorems

In this section, we present a generalization of the ISS superposition theorem to general infinite-dimensional systems. First, we recall some basic notions previously introduced for finite-dimensional systems.

Definition 6.19 Consider a forward complete control system $\Sigma = (X, \mathcal{U}, \phi)$.

(i) We call $0 \in X$ an *equilibrium point (of the undisturbed system)* if $\phi(t, 0, 0) = 0$ for all $t \geq 0$.

(ii) We say that Σ has the *continuity at the equilibrium point (CEP) property*, if 0 is an equilibrium and for every $\varepsilon > 0$ and for any $h > 0$ there exists a $\delta = \delta(\varepsilon, h) > 0$, so that

$$t \in [0, h], \quad \|x\|_X \le \delta, \quad \|u\|_{\mathcal{U}} \le \delta \quad \Rightarrow \quad \|\phi(t, x, u)\|_X \le \varepsilon. \quad (6.20)$$

(iii) We say that Σ has *bounded reachability sets (BRS)* if for any $C > 0$ and any $\tau > 0$, it holds that

$$\sup\big\{\|\phi(t, x, u)\|_X : \|x\|_X \le C, \ \|u\|_{\mathcal{U}} \le C, \ t \in [0, \tau]\big\} < \infty.$$

(iv) System Σ is called *uniformly locally stable (ULS)*, if there exist $\sigma \in \mathcal{K}_\infty$, $\gamma \in \mathcal{K}_\infty \cup \{0\}$ and $r > 0$ such that for all $x \in \overline{B_r}$ and all $u \in \overline{B_{r,\mathcal{U}}}$:

$$\|\phi(t, x, u)\|_X \le \sigma(\|x\|_X) + \gamma(\|u\|_{\mathcal{U}}) \quad \forall t \ge 0. \quad (6.21)$$

(v) We say that $\Sigma = (X, \mathcal{U}, \phi)$ has the *uniform limit property on bounded sets (bULIM)*, if there exists $\gamma \in \mathcal{K}_\infty \cup \{0\}$ so that for every $\varepsilon > 0$ and for every $r > 0$ there exists a $\tau = \tau(\varepsilon, r)$ such that

$$\|x\|_X \le r, \quad \|u\|_{\mathcal{U}} \le r \quad \Rightarrow \quad \exists t \le \tau : \ \|\phi(t, x, u)\|_X \le \varepsilon + \gamma(\|u\|_{\mathcal{U}}). \quad (6.22)$$

(vi) We say that Σ has the *uniform limit property (ULIM)*, if there exists $\gamma \in \mathcal{K}_\infty \cup \{0\}$ so that for every $\varepsilon > 0$ and for every $r > 0$ there exists a $\tau = \tau(\varepsilon, r)$ such that for all x with $\|x\|_X \le r$ and all $u \in \mathcal{U}$ there is a $t \le \tau$ such that

$$\|\phi(t, x, u)\|_X \le \varepsilon + \gamma(\|u\|_{\mathcal{U}}). \quad (6.23)$$

We define the attractivity properties for systems with inputs.

Definition 6.20 System $\Sigma = (X, \mathcal{U}, \phi)$ has the:

(i) *Asymptotic gain (AG) property*, if there is a $\gamma \in \mathcal{K}_\infty \cup \{0\}$ such that for all $\varepsilon > 0$, for all $x \in X$ and for all $u \in \mathcal{U}$ there exists a $\tau = \tau(\varepsilon, x, u) < \infty$ such that

$$t \ge \tau \quad \Rightarrow \quad \|\phi(t, x, u)\|_X \le \varepsilon + \gamma(\|u\|_{\mathcal{U}}). \quad (6.24)$$

(ii) *Strong asymptotic gain (sAG) property*, if there is a $\gamma \in \mathcal{K}_\infty \cup \{0\}$ such that for all $x \in X$ and for all $\varepsilon > 0$ there exists a $\tau = \tau(\varepsilon, x) < \infty$ such that for all $u \in \mathcal{U}$

$$t \ge \tau \quad \Rightarrow \quad \|\phi(t, x, u)\|_X \le \varepsilon + \gamma(\|u\|_{\mathcal{U}}). \quad (6.25)$$

(iii) *Uniform asymptotic gain (UAG) property*, if there exists a $\gamma \in \mathcal{K}_\infty \cup \{0\}$ such that for all $\varepsilon, r > 0$ there is a $\tau = \tau(\varepsilon, r) < \infty$ such that for all $u \in \mathcal{U}$ and all $x \in B_r$

$$t \ge \tau \quad \Rightarrow \quad \|\phi(t, x, u)\|_X \le \varepsilon + \gamma(\|u\|_{\mathcal{U}}). \quad (6.26)$$

(iv) *Bounded input uniform asymptotic gain (bUAG) property*, if there exists a $\gamma \in \mathcal{K}_\infty \cup \{0\}$ such that for all $\varepsilon, r > 0$ there is a $\tau = \tau(\varepsilon, r) < \infty$ such that for all $u \in \mathcal{U}$: $\|u\|_{\mathcal{U}} \le r$ and all $x \in B_r$, the implication (6.24) holds.

The ISS superposition Theorem 2.51, which we have shown for ODEs, holds also for the abstract control systems that we consider in this section, without significant differences in the proof.

Theorem 6.21 (ISS superposition theorem) *Let $\Sigma = (X, \mathcal{U}, \phi)$ be a forward complete control system. The following statements are equivalent:*

(i) Σ is ISS.
(ii) Σ is bUAG, CEP, and has BRS.
(iii) Σ is bULIM, ULS, and has BRS.

For ODE systems, we have also shown an ISS superposition theorem for Lipschitz-regular ODE systems (Theorem 2.52) that characterizes ISS in terms of significantly weaker properties. In contrast, even for a special case of infinite ODE systems that are Lipschitz continuous on bounded balls of $X \times U$, we do not get significant improvements in the ISS superposition theorem. Even with this additional regularity, forward completeness remains much weaker than BRS, and LIM property remains much weaker than bULIM property.

The following examples illustrate such limitations.

Example 6.22 **(FC $\wedge$ 0-GAS $\wedge$ 0-UAS $\not\Rightarrow$ BRS)** Let us show that for *nonlinear* infinite ODE systems 0-GAS does not even imply BRS of the undisturbed system. Consider the infinite-dimensional system $\mathcal{S}^1$ defined by

$$\mathcal{S}^1 : \begin{cases} \mathcal{S}^1_k : \begin{cases} \dot{x}_k = -x_k + x_k^2 y_k - \frac{1}{k^2} x_k^3, \\ \dot{y}_k = -y_k. \end{cases} \\ k = 1, 2, \dots, \end{cases} \tag{6.27}$$

with the state space

$$X := \ell^2 = \left\{ (z_k)_{k=1}^\infty : \sum_{k=1}^\infty |z_k|^2 < \infty, \quad z_k = (x_k, y_k) \in \mathbb{R}^2 \right\}. \tag{6.28}$$

We show that $\mathcal{S}^1$ is forward complete, 0-GAS and 0-UAS but does not have bounded reachability sets.

Before we give a detailed proof of these facts, let us give an informal explanation of this phenomenon. If we formally place 0 into the definition of $\mathcal{S}^1_k$ instead of the term $-\frac{1}{k^2} x_k^3$, then the state of $\mathcal{S}^1_k$ (for each k) will exhibit a finite escape time, provided $y_k(0)$ is chosen large enough. The term $-\frac{1}{k^2} x_k^3$ prevents the solutions of $\mathcal{S}^1_k$ from growing to infinity: The solution then looks like a spike, which is then stopped by the damping $-\frac{1}{k^2} x_k^3$, and converges to 0 since $y_k(t) \to 0$ as $t \to \infty$. However, the larger is k, the higher will be the spikes, and hence there is no uniform bound for the solution of $\mathcal{S}^1$ starting from a bounded ball. Now we proceed to the rigorous proof.

First we argue that $\mathcal{S}^1$ is 0-UAS. Indeed, for $r < 1$ the Lyapunov function $V(z) = \|z\|^2_{\ell^2} = \sum_{k=1}^{\infty}(x_k^2 + y_k^2)$ satisfies for all z_k with $|x_k| \le r$ and $|y_k| \le r$, $k \in \mathbb{N}$, the estimate

$$\begin{aligned}\dot{V}(z) &= 2\sum_{k=1}^{\infty}(-x_k^2 + x_k^3 y_k - \frac{1}{k^2}x_k^4 - y_k^2)\\ &\le 2\sum_{k=1}^{\infty}(-x_k^2 + |x_k|^3|y_k| - \frac{1}{k^2}x_k^4 - y_k^2)\\ &\le 2\sum_{k=1}^{\infty}((r^2-1)x_k^2 - y_k^2)\\ &\le 2(r^2-1)V(z).\end{aligned} \tag{6.29}$$

We see that V is an exponential local Lyapunov function for the system (6.27) and thus (6.27) is locally uniformly asymptotically stable. Indeed, it is not hard to show that the domain of attraction contains $\{z \in \ell^2 : |x_k| < r, |y_k| < r \ \forall k \in \mathbb{N}\}$.

To show forward completeness and global attractivity of $\mathcal{S}^1$, we first point out that every $\mathcal{S}_k^1$ is 0-GAS (and hence 0-UGAS, since $\mathcal{S}_k^1$ is finite-dimensional, see Theorem B.37). This follows from the fact that any $\mathcal{S}_k^1$ is a cascade interconnection of an ISS x_k-system (with y_k as an input) and a globally asymptotically stable y_k-system, see Theorem 3.22.

Furthermore, for any $z(0) \in \ell^2$ there exists a finite $N > 0$ such that $|z_k(0)| \le \frac{1}{2}$ for all $k \ge N$. Decompose the norm of $z(t)$ as follows

$$\|z(t)\|_{\ell^2} = \sum_{k=1}^{N-1}|z_k(t)|^2 + \sum_{k=N}^{\infty}|z_k(t)|^2.$$

According to the previous arguments, $\sum_{k=1}^{N-1}|z_k(t)|^2 \to 0$ as $t \to 0$ since all $\mathcal{S}_k^1$ are 0-UGAS for $k = 1, \ldots, N-1$.

Since $|z_k(0)| \le \frac{1}{2}$ for all $k \ge N$, we can apply the computations as in (6.29) in order to obtain (for $r := \frac{1}{2}$) that

$$\frac{d}{dt}\Big(\sum_{k=N}^{\infty}|z_k(t)|^2\Big) \le -\frac{3}{2}\sum_{k=N}^{\infty}|z_k(t)|^2.$$

Hence $\sum_{k=N}^{\infty}|z_k(t)|^2$ decays monotonically and exponentially to 0 as $t \to \infty$. Overall, $\|z(t)\|_{\ell^2} \to 0$ as $t \to \infty$, which shows that $\mathcal{S}^1$ is forward complete, 0-GAS and 0-UAS.

Finally, we show that finite-time reachability sets of $\mathcal{S}^1$ are not bounded. To prove this, it is enough to show that there exists an $r > 0$ and $\tau > 0$ so that for any $M > 0$ there exist $z \in \ell^2$ and $t \in [0, \tau]$ so that $\|z\|_{\ell^2} = r$ and $\|\phi(t, z, 0)\|_{\ell^2} > M$.

Let us consider $\mathcal{S}^1_k$. For $y_k \geq 1$ and for $x_k \in [0, k]$ it holds that

$$\dot{x}_k \geq -2x_k + x_k^2. \tag{6.30}$$

Pick an initial state $x_k(0) = c > 0$ (which is independent of k) so that the solution of $\dot{x}_k = -2x_k + x_k^2$ blows up to infinity in time $t^* = 1$. Now pick $y_k(0) = e$ (Euler's constant) for all $k = 1, 2, \ldots$. For this initial condition we obtain $y_k(t) = e^{1-t} \geq 1$ for $t \in [0, 1]$. Consequently, for $z_k(0) = (c, e)^T$ there exists a time $\tau_k \in (0, 1)$ such that $x_k(\tau_k) = k$ for the solution of $\mathcal{S}^1_k$.

Finally, consider an initial state $z(0)$ for $\mathcal{S}^1$, where $z_k(0) = (c, e)^T$ and $z_j(0) = (0, 0)^T$ for $j \neq k$. For this initial state we have that $\|z(t)\|_{\ell^2} = |z_k(t)|$ and

$$\sup_{t\geq 0} \|z(t)\|_{\ell^2} = \sup_{t\geq 0} |z_k(t)| \geq |x_k(\tau_k)| = k.$$

As $k \in \mathbb{N}$ was arbitrary, the system $\mathcal{S}^1$ does not have BRS.

Example 6.23 **(0-UGAS $\wedge$ sAG $\wedge$ AG with zero gain $\wedge$ UGS with zero gain $\not\Rightarrow$ LISS.)** Consider a system Σ with the state space

$$X = \ell^1 := \left\{(x_k)_{k=1}^{\infty} : \sum_{k=1}^{\infty} |x_k| < \infty\right\}$$

and with the input space $\mathcal{U} := PC_b(\mathbb{R}_+, \mathbb{R})$.

Let the dynamics of the k-th mode of Σ for all $k \in \mathbb{N}$ be given by

$$\dot{x}_k(t) = -\frac{1}{1 + k|u(t)|} x_k(t). \tag{6.31}$$

We use the notation ϕ_k for the flow map of the k-th mode of (6.31). Then for $x = (x_k)$ we have $\phi(t, x, u) = (\phi_k(t, x_k, u))_{k=1}^{\infty}$ (we indicate here that the dynamics of different modes are independent on each other).

Clearly, Σ is 0-UGAS, since for $u \equiv 0$ its dynamics are given by $\dot{x} = -x$. At the same time the inequality $\|\phi(t, x, u)\|_X \leq \|x\|_X$ holds for all $t \geq 0$, $x \in X$ and $u \in \mathcal{U}$, and thus Σ is UGS with zero gain.

Next, we show step by step that:

(i) Σ satisfies Assumption 6.8 and is CEP (and thus belongs to the class of systems that we consider).
(ii) Σ is AG with zero gain.
(iii) Σ is sAG with arbitrarily small (but necessarily nonzero) linear gain. In other words, one must pay for uniformity, but this payment can be made arbitrarily small.
(iv) Finally, Σ is not LISS.

(i) The CEP property is automatically satisfied since, as we mentioned above, the system Σ is UGS with zero gain. Next, we show that Assumption 6.8 also holds.

Denote for all $x \in X$, $v \in U$ the rhs of (6.31) as follows:

$$f(x, v) := (f_k(x, v)), \qquad f_k(x, v) = -\frac{1}{1+k|v|}x_k, \quad k \in \mathbb{N}.$$

Note that f is globally Lipschitz, since for any $x, y \in X$ and any $v \in U$

$$\begin{aligned}\|f(x, v) - f(y, v)\|_X = \sum_{k=1}^{\infty} \frac{1}{1+k|v|}|x_k - y_k| &\le \sum_{k=1}^{\infty} |x_k - y_k| \\ &= \|x - y\|_X.\end{aligned}$$

Now pick any $x \in X$ and let us show that $f(x, \cdot)$ is continuous at any $v \neq 0$. Consider

$$\begin{aligned}&\|f(x, v) - f(x, v_1)\|_X \\ &\quad = \sum_{k=1}^{\infty} \Big|\frac{1}{1+k|v_1|} - \frac{1}{1+k|v|}\Big||x_k| \\ &\quad = \sum_{k=1}^{\infty} \frac{k\big||v_1| - |v|\big|}{(1+k|v_1|)(1+k|v|)}|x_k| \\ &\quad \le |v_1 - v| \sup_{k\in\mathbb{N}} \frac{k}{(1+k|v_1|)(1+k|v|)} \sum_{k=1}^{\infty} |x_k|.\end{aligned}$$

Since $v \neq 0$ and we consider v_1 that are close to v we assume that $v_1 \neq 0$ as well. Then

$$\begin{aligned}&\sup_{k\in\mathbb{N}} \frac{k}{(1+k|v_1|)(1+k|v|)} \le \\ &\le \sup_{k\in\mathbb{N}} \frac{k}{1+k(|v_1|+|v|)} = \lim_{k\to\infty} \frac{k}{1+k(|v_1|+|v|)} = \frac{1}{|v_1|+|v|}.\end{aligned}$$

And overall we get that

$$\|f(x, v) - f(x, v_1)\|_X \le \frac{|v_1 - v|}{|v_1| + |v|}\|x\|_X.$$

Let $|v_1 - v| \le \delta$ for some $\delta \in (0, \frac{|v|}{2})$. Then $|v_1| \ge |v|/2$ and

$$\|f(x, v) - f(x, v_1)\|_X \le \frac{\delta}{3/2|v|}\|x\|_X.$$

Now for any $\varepsilon > 0$ pick $\delta < \frac{|v|}{2}$ so that $\frac{\delta}{3/2|v|}\|x\|_X < \varepsilon$ (which is always possible since $v \neq 0$). This shows continuity of $f(x, \cdot)$ for $v \neq 0$. Note that the choice of δ depends on $\|x\|_X$ but does not depend on x itself.

Next, we show that $f(x, \cdot)$ is continuous at zero as well, but it is no more uniform with respect to the first argument of f. Pick any $x \in X$ again. For any $\varepsilon > 0$ there exists $N = N(\varepsilon, x) > 0$ so that $\sum_{k=N+1}^{\infty} |x_k| < \frac{\varepsilon}{2}$. We have for this x:

$$\begin{aligned}
\|f(x,0) - f(x,v_1)\|_X &= \sum_{k=1}^{\infty} \frac{k|v_1|}{1+k|v_1|}|x_k| \\
&= \sum_{k=1}^{N} \frac{k|v_1|}{1+k|v_1|}|x_k| + \sum_{k=N+1}^{\infty} \frac{k|v_1|}{1+k|v_1|}|x_k| \\
&\leq \sum_{k=1}^{N} \frac{N|v_1|}{1+N|v_1|}|x_k| + \sum_{k=N+1}^{\infty} |x_k| \\
&< \frac{N|v_1|}{1+N|v_1|}\|x\|_X + \frac{\varepsilon}{2} \\
&\leq N|v_1|\|x\|_X + \frac{\varepsilon}{2}.
\end{aligned}$$

Pick $\delta := \frac{\varepsilon}{2\|x\|_X N}$. Then for all $v_1 \in U\colon |v_1| \leq \delta$ it follows that $\|f(x,0) - f(x,v_1)\|_X < \varepsilon$. This shows continuity of $f(x, \cdot)$ at zero for any given $x \in X$. However, note that δ depends on x (since N does) and not only on $\|x\|_X$. Overall, we have shown that Assumption 6.8 is fulfilled for Σ.

(ii) *Σ is AG with zero gain.* For any $x = (x_k) \in X$, for any $\varepsilon > 0$ there exists $N = N(x, \varepsilon) \in \mathbb{N}$ so that $\sum_{k=N+1}^{\infty} |x_k| < \frac{\varepsilon}{2}$.

The norm of the state of Σ at time t can be estimated as follows:

$$\begin{aligned}
\|\phi(t,x,u)\|_X &= \sum_{k=1}^{N} |\phi_k(t,x_k,u)| + \sum_{k=N+1}^{\infty} |\phi_k(t,x_k,u)| \\
&\leq \sum_{k=1}^{N} |\phi_k(t,x_k,u)| + \sum_{k=N+1}^{\infty} |\phi_k(0,x_k,u)| \\
&\leq \sum_{k=1}^{N} |\phi_k(t,x_k,u)| + \frac{\varepsilon}{2}.
\end{aligned}$$

Now we estimate the state of the kth mode of our system for all $k = 1, \ldots, N$:

$$\begin{aligned}|\phi_k(t, x_k, u)| &= e^{-\int_0^t \frac{1}{1+k|u(s)|}ds}|x_k| \\ &\leq e^{-\int_0^t \frac{1}{1+k\|u\|_{\mathcal{U}}}ds}|x_k| \\ &= e^{-\frac{1}{1+k\|u\|_{\mathcal{U}}}t}|x_k| \\ &\leq e^{-\frac{1}{1+N\|u\|_{\mathcal{U}}}t}|x_k|,\end{aligned} \tag{6.32}$$

which holds for any $u \in \mathcal{U}$ and any $x_k \in \mathbb{R}$. Using this estimate we proceed to

$$\|\phi(t, x, u)\|_X \leq \sum_{k=1}^{N} e^{-\frac{1}{1+N\|u\|_{\mathcal{U}}}t}|x_k| + \frac{\varepsilon}{2}.$$

Clearly, for any $u \in \mathcal{U}$ there exists $\tau_a = \tau_a(x, \varepsilon, u)$ so that $\sum_{k=1}^{N} e^{-\frac{1}{1+N\|u\|_{\mathcal{U}}}t}|x_k| \leq \frac{\varepsilon}{2}$ for $t \geq \tau_a$.

Overall, we see that for any $x \in X$, any $t \geq 0$, and all $u \in \mathcal{U}$ there exists $\tau_a = \tau_a(x, \varepsilon, u)$, so that for all $t \geq \tau_a$ it holds that $\|\phi(t, x, u)\|_X \leq \varepsilon$. This shows that Σ is AG with $\gamma \equiv 0$. However, the time τ_a depends on u, and thus the above argument doesn't tell us whether the system is strongly AG.
(iii) Next, we show that Σ is sAG, but we should pay for this by adding a linear gain. However, this linear gain can be made arbitrarily small.

In (6.32), we have obtained the following estimate for the state of the k-th mode of Σ:

$$|\phi_k(t, x_k, u)| \leq e^{-\frac{1}{1+k\|u\|_{\mathcal{U}}}t}|x_k|.$$

This expression can be further estimated as:

$$|\phi_k(t, x_k, u)| \leq e^{-\frac{1}{1+k\max\{\|u\|_{\mathcal{U}}, \frac{2^k}{r}|x_k|\}}t} \cdot \max\{|x_k|, \frac{r}{2^k}\|u\|_{\mathcal{U}}\}. \tag{6.33}$$

For $|x_k| \geq \frac{r}{2^k}\|u\|_{\mathcal{U}}$, we obtain

$$|\phi_k(t, x_k, u)| \leq e^{-\frac{1}{1+k\frac{2^k}{r}|x_k|}t}|x_k|. \tag{6.34}$$

For $|x_k| \leq \frac{r}{2^k}\|u\|_{\mathcal{U}}$, the inequality (6.33) implies

$$|\phi_k(t, x_k, u)| \leq e^{-\frac{1}{1+k\|u\|_{\mathcal{U}}}t}\frac{r}{2^k}\|u\|_{\mathcal{U}} \leq \frac{r}{2^k}\|u\|_{\mathcal{U}}. \tag{6.35}$$

Overall, for any $x_k \in \mathbb{R}$ and any $u \in \mathcal{U}$ we obtain from (6.34) and (6.35):

$$|\phi_k(t, x_k, u)| \leq e^{-\frac{1}{1+k\frac{2^k}{r}|x_k|}t}|x_k| + \frac{r}{2^k}\|u\|_{\mathcal{U}}. \tag{6.36}$$

Having this estimate for the state of the k-th mode, we proceed to the estimate for the whole state of Σ:

$$\begin{aligned}\|\phi(t,x,u)\|_X &= \sum_{k=1}^{\infty} |\phi_k(t,x_k,u)| \\ &\leq \sum_{k=1}^{\infty} e^{-\frac{1}{1+k\frac{2^k}{r}|x_k|}t}|x_k| + \sum_{k=1}^{\infty}\frac{r}{2^k}\|u\|_{\mathcal{U}} \\ &= \sum_{k=1}^{\infty} e^{-\frac{1}{1+k\frac{2^k}{r}|x_k|}t}|x_k| + r\|u\|_{\mathcal{U}}. \end{aligned} \tag{6.37}$$

This estimate is true for all $t \geq 0$, all $x \in X$, all $u \in \mathcal{U}$ and for any $r > 0$.

Now we apply the trick used above in the proof that the system is AG. For any $x \in X$, for any $\varepsilon > 0$ there exists $N = N(x,\varepsilon) \in \mathbb{N}$ so that $\sum_{k=N+1}^{\infty}|x_k| < \frac{\varepsilon}{2}$. Using this fact we continue estimates from (6.37):

$$\begin{aligned}\|\phi(t,x,u)\|_X &\leq \sum_{k=1}^{N} e^{-\frac{1}{1+k\frac{2^k}{r}|x_k|}t}|x_k| \\ &\quad + \sum_{k=N+1}^{\infty} e^{-\frac{1}{1+k\frac{2^k}{r}|x_k|}t}|x_k| + r\|u\|_{\mathcal{U}}. \\ &\leq \sum_{k=1}^{N} e^{-\frac{1}{1+k\frac{2^k}{r}|x_k|}t}|x_k| + \frac{\varepsilon}{2} + r\|u\|_{\mathcal{U}}. \end{aligned}$$

Now for the above ε and x we can find sufficiently large time $\tau_a = \tau_a(\varepsilon, x)$ (clearly, τ_a depends also on N, but N itself depends only on x and ε), so that for all $u \in \mathcal{U}$ and all $t \geq \tau_a$

$$\sum_{k=1}^{N} e^{-\frac{1}{1+k\frac{2^k}{r}|x_k|}t}|x_k| \leq \frac{\varepsilon}{2}.$$

Overall, we obtain that for any $r > 0$, for all $\varepsilon > 0$ and for any $x \in X$ there exists $\tau_a = \tau_a(t,x)$:

$$\|\phi(t,x,u)\|_X \leq \varepsilon + r\|u\|_{\mathcal{U}},$$

for all $u \in \mathcal{U}$ and all $t \geq \tau_a$. This shows that Σ satisfies the strong AG property with the gain $\gamma(s) = rs$.

To finish the proof of (iii), we show that Σ is not sAG for the gain $\gamma \equiv 0$. But this is clear, as already any individual subsystem of Σ is not sAG with zero gain, see Example 2.46.

(iv) Next, we show that Σ is not LISS. Assume that Σ is LISS and hence there exist $r > 0$, $\beta \in \mathcal{KL}$ and $\gamma \in \mathcal{K}_\infty$ so that the inequality (6.9) holds for all $x \in X$: $\|x\|_X \leq r$ and for all $u \in \mathcal{U}$: $\|u\|_\mathcal{U} \leq r$.

Pick a constant input $u(t) = \varepsilon$ for all $t \geq 0$, where $\varepsilon > 0$ is chosen so that $\max\{\varepsilon, 3\gamma(\varepsilon)\} \leq r$. This is always possible since $\gamma \in \mathcal{K}_\infty$. Denote $c := \gamma(\varepsilon)$.

LISS property of Σ implies that for all $x \in X$ with the norm $\|x\|_X = 3c \leq r$ and for all $t \geq 0$ it holds that

$$\|\phi(t, x, u(\cdot))\|_X \leq \beta(3c, t) + c. \tag{6.38}$$

Since $\beta \in \mathcal{KL}$, there exists t^* (which depends on c, but does not depend on x) so that $\beta(3c, t^*) = c^2$ and thus for all $x \in X$ with $\|x\|_X = 3c$ it must hold

$$\|\phi(t^*, x, u(\cdot))\|_X \leq 2c. \tag{6.39}$$

Now consider the initial states of the form $x_{<k>} = 3c \cdot e_k$, where e_k is the k-th standard basis vector of $X = \ell^1$. Certainly, $\|x_{<k>}\|_X = 3c$ for all $k \in \mathbb{N}$. Then

$$\|\phi(t^*, x_{<k>}, u(\cdot))\|_X = |\phi_k(t^*, 3c, u(\cdot))| = e^{-\frac{1}{1+k\varepsilon}t^*}3c.$$

Now, for any $t^* > 0$ there exists k so that $e^{-\frac{1}{1+k\varepsilon}t^*}3c > 2c$, which contradicts to (6.39). This shows that Σ is not LISS.

Example 6.24 **(0-UGAS $\wedge$ sAG $\wedge$ LISS $\wedge$ UGS $\nRightarrow$ ISS)** In the previous example, we have shown that a system that is 0-UGAS, sAG, AG with zero gain, and UGS with zero gain does not have to be LISS. Next, we modify this example to show that *if the system, in addition to the above list of properties, is LISS, this still does not guarantee ISS.*

Consider a system Σ with the state space $X := \ell^1$ and input space $\mathcal{U} := PC_b(\mathbb{R}_+, \mathbb{R})$. Let also the dynamics of the k-th mode of Σ be given by

$$\dot{x}_k(t) = -\frac{1}{1 + |u(t)|^k} x_k(t). \tag{6.40}$$

We continue to use the notation $\phi_k(t, x_k, u)$ for the state of the k-th mode (6.40).

As in Example 6.23, one can prove that this system satisfies Assumption 6.8, is CEP, 0-UGAS, UGS with zero gain, AG with zero gain, and sAG with a nonzero gain. We skip this proof since it is completely analogous. Moreover, for all u: $\|u\|_\mathcal{U} \leq 1$ and for all $x \in X$ it holds that

$$\|\phi(t, x, u)\|_X \leq e^{-\frac{1}{2}t}\|x\|_X, \quad t \geq 0. \tag{6.41}$$

Thus, Σ is LISS with zero gain and with $r = 1$.

---

[2] Note that any $\beta \in \mathcal{KL}$ satisfying the LISS estimate automatically satisfies $\beta(r, 0) \geq r$ for all $r > 0$ (consider $t = 0$ and $u \equiv 0$ in LISS estimate).

The proof that Σ is not ISS goes along the lines of Example 6.23, with the change that the norm of the considered inputs should be larger than 1.

6.5 ISS Lyapunov Functions

As in the ODE case, the concept of the ISS Lyapunov function plays a central role in the ISS analysis.

Definition 6.25 A continuous function $V : D \to \mathbb{R}_+$, $D \subset X$, $0 \in \text{int}\,(D)$ is called a *(coercive) LISS Lyapunov function in a dissipative form* for a system $\Sigma = (X, \mathcal{U}, \phi)$, if there are $\psi_1, \psi_2, \alpha \in \mathcal{K}_\infty$, $\sigma \in \mathcal{K}$ and $r > 0$, such that $B_r \subset D$,

$$\psi_1(\|x\|_X) \le V(x) \le \psi_2(\|x\|_X) \quad \forall x \in D, \tag{6.42}$$

and the Lie derivative of V along the trajectories of Σ satisfies

$$\dot{V}_u(x) \le -\alpha(\|x\|_X) + \sigma(\|u\|_{\mathcal{U}}) \tag{6.43}$$

for all $x \in B_r$ and $u \in B_{r,\mathcal{U}}$.

If $D = X$, and $r = +\infty$, then V is called a *(coercive) ISS Lyapunov function* for Σ.

Importance of ISS Lyapunov functions is due to the following basic result:

Theorem 6.26 (Direct Lyapunov theorem, *[7, Theorem 1], [20, Proposition 1]*) *Let $\Sigma = (X, \mathcal{U}, \phi)$ be a control system satisfying the BIC property (but not necessarily forward complete), and let $x \equiv 0$ be its equilibrium point. If Σ possesses a coercive (L)ISS Lyapunov function (in either dissipative or implication form), then it is (L)ISS.*

Proof In particular, thanks to the generalized comparison principle (Proposition A.35), the proof of this result looks almost the same as the proof of the finite-dimensional Lyapunov theorem (Theorem 2.12). Hence the proof is omitted. □

We will be interested also in the following strengthening of the ISS concept:

Definition 6.27 The system $\Sigma = (X, \mathcal{U}, \phi)$ is called *exponentially input-to-state stable (eISS)* if it is forward complete and there are constants $a, M > 0$ and $\gamma \in \mathcal{K}$ such that for any initial state $x \in X$ and any $u \in \mathcal{U}$ the corresponding solution satisfies

$$\|\phi(t, x, u)\|_X \le Me^{-at}\|x\|_X + \gamma(\|u\|_{\mathcal{U}}) \quad \forall \quad t \ge 0.$$

Exponential input-to-state stability is implied by the existence of a power-bounded exponential ISS Lyapunov function that we define in a dissipative form as follows.

Definition 6.28 A continuous function $V : X \to \mathbb{R}_+$ is called a *(power-bounded) eISS Lyapunov function* for the system $\Sigma = (X, \mathcal{U}, \phi)$ if there exist constants $\underline{\omega}, \overline{\omega}, b, \kappa > 0$ and $\gamma \in \mathcal{K}_\infty$ such that

$$\underline{\omega}\|x\|_X^b \leq V(x) \leq \overline{\omega}\|x\|_X^b, \tag{6.44a}$$

$$\dot{V}_u(x) \leq -\kappa V(x) + \gamma(\|u\|_{\mathcal{U}}), \tag{6.44b}$$

hold for all $x \in X$ and $u \in \mathcal{U}$.

Proposition 6.29 *If there exists a power-bounded eISS Lyapunov function for a control system $\Sigma = (X, \mathcal{U}, \phi)$, then Σ is eISS.*

Proof The proof is a variation of the arguments given in Theorem 2.12, and thus only some of the steps are provided. Let V be a power-bounded eISS Lyapunov function for $\Sigma = (X, \mathcal{U}, \phi)$ as in Definition 6.28 with corresponding constants $\underline{\omega}, \overline{\omega}, b, \kappa > 0$ and function $\gamma \in \mathcal{K}_\infty$.

Pick any $\varepsilon \in (0, \kappa)$ and define $\chi(r) := \frac{1}{\kappa - \varepsilon}\gamma(r)$. For all $x \in X$ and $u \in \mathcal{U}$, we obtain

$$V(x) \geq \chi(\|u\|_{\mathcal{U}}) \quad \Rightarrow \quad \dot{V}_u(x) \leq -\varepsilon V(x).$$

By arguing similar to Theorem 2.12, we obtain the following inequality for all $x \in X$, $u \in \mathcal{U}$, and $t \geq 0$:

$$V(\phi(t, x, u)) \leq e^{-\varepsilon t}V(x) + \chi(\|u\|_{\mathcal{U}}).$$

In view of (6.44a), one gets

$$\underline{\omega}\|\phi(t, x, u)\|_X^b \leq e^{-\varepsilon t}\overline{\omega}\|x\|_X^b + \chi(\|u\|_{\mathcal{U}}),$$

which implies that

$$\begin{aligned}\|\phi(t, x, u)\|_X &\leq \Big(e^{-\varepsilon t}\frac{\overline{\omega}}{\underline{\omega}}\|x\|_X^b + \chi(\|u\|_{\mathcal{U}})\Big)^{\frac{1}{b}} \\ &\leq \Big(2\frac{\overline{\omega}}{\underline{\omega}}\Big)^{\frac{1}{b}} e^{-\frac{\varepsilon}{b}t}\|x\|_X + \big(2\chi(\|u\|_{\mathcal{U}})\big)^{\frac{1}{b}},\end{aligned}$$

showing the exponential ISS property of Σ. In the last estimate, we have used the inequality $\gamma(a + b) \leq \gamma(2a) + \gamma(2b)$, which holds for all $\gamma \in \mathcal{K}$ (so in particular for $\gamma(r) = r^{1/b}$), since $a + b \leq 2\max\{a, b\}$. □

We close the section with a converse Lyapunov result.

Theorem 6.30 (Converse ISS Lyapunov theorem, [23, Theorem 5]) *Let $f : X \times U \to X$ be locally Lipschitz continuous from the Banach space $(X \times U, \|\cdot\|_X + \|\cdot\|_U)$ to the space X.*

Then (7.1) is ISS if and only if there is a locally Lipschitz continuous coercive ISS Lyapunov function for (7.1) in implication form.

Proof The strategy of the proof of this result resembles the proof of the smooth converse ISS Lyapunov theorem (Theorem 2.63). Again, one can construct an auxiliary control system with the disturbances and use the converse UGAS Lyapunov theorem for such systems. The difference is that now we cannot rely on the smooth converse UGAS Lyapunov theorem (Theorem B.32), but instead we can rely on the infinite-dimensional version of Theorem B.31, ensuring the existence of a *Lipschitz continuous* strict Lyapunov function for the modified system (see [15, Section 3.4] for this extension). With this technique, we obtain a Lipschitz continuous ISS Lyapunov function in implication form, as claimed. We refer to [23] for the full proof of this result. □

Remark 6.31 Note that in Theorem 6.30 we do not claim the existence of an ISS Lyapunov function in a dissipative form for an ISS system. Whether the existence of an ISS Lyapunov function in implication form is equivalent to the existence of an ISS Lyapunov function in a dissipative form remains an open problem for infinite-dimensional systems; see [21] for an overview.

6.6 Small-Gain Theorem for Infinite Networks

In this section, we develop small-gain theorems for input-to-state stability of interconnections of countably many ODE subsystems (6.2), depending on certain stability properties of subsystems. There is a number of such results, extending the developments of Chapter 3, see Section 7.1 for an overview. In Chapter 3, we have used the implication form of the ISS property for subsystems. Here we show an ISS small-gain theorem for the case if the ISS property is given in the dissipative form.

6.6.1 The Gain Operator and Its Properties

We assume that each subsystem Σ_i given by (6.2) is eISS and there are eISS Lyapunov functions with linear gains for all Σ_i. As in all previous Lyapunov-based small-gain theorems, we would like to know how each j-th subsystem influences each i-th subsystem. The following assumption provides us with the needed information.

Assumption 6.32 For each $i \in \mathbb{N}$, there are a continuous function $V_i : \mathbb{R}^{N_i} \to \mathbb{R}_+$ and $p, q \in [1, \infty)$ such that the following holds:

- There are $\underline{\alpha}_i, \overline{\alpha}_i > 0$ such that

$$\underline{\alpha}_i |x_i|^p \le V_i(x_i) \le \overline{\alpha}_i |x_i|^p, \quad x_i \in \mathbb{R}^{N_i}. \tag{6.45}$$

- There are $\lambda_i, \gamma_{ij}, \gamma_{iu} > 0$ such that for each $x_i \in \mathbb{R}^{N_i}$, $u_i \in L^\infty(\mathbb{R}_+, \mathbb{R}^{m_i})$, each internal input $x_{\neq i} \in C(\mathbb{R}_+, \mathbb{R}^{N_{\neq i}})$, and for almost all t in the maximal interval of existence of $\phi_i(t) := \phi_i(t, x_i, x_{\neq i}, u_i)$ it holds that

$$D^+(V_i \circ \phi_i)(t) \leq -\lambda_i V_i(\phi_i(t)) + \sum_{j \in I_i} \gamma_{ij} V_j(x_j(t)) + \gamma_{iu}|u_i(t)|^q. \tag{6.46}$$

Here the components of $x_{\neq i}$ are denoted by $x_j(\cdot)$.

- For all t in the maximal interval of the existence of ϕ_i we have

$$D_+(V_i \circ \phi_i)(t) < \infty.$$

We furthermore assume that the following uniformity conditions hold for the constants introduced above.

Assumption 6.33 (a) There are $\underline{\alpha}, \overline{\alpha} > 0$ such that

$$\underline{\alpha} \leq \underline{\alpha}_i \leq \overline{\alpha}_i \leq \overline{\alpha}, \quad i \in \mathbb{N}. \tag{6.47}$$

(b) There is $\underline{\lambda} > 0$ such that

$$\underline{\lambda} \leq \lambda_i, \quad i \in \mathbb{N}. \tag{6.48}$$

(c) There is $\overline{\gamma}_u > 0$ such that

$$\gamma_{iu} \leq \overline{\gamma}_u, \quad i \in \mathbb{N}. \tag{6.49}$$

Note that in the above assumptions, we have dealt with only the properties of subsystems Σ_i of the network Σ. We still need to define the state space for the interconnection as well as the space of input values for the system Σ. The inequalities (6.45) and (6.46) encourage the following assumption:

Assumption 6.34 The system Σ with state space $X = \ell^p(\mathbb{N}, (N_i))$ and external input space $U = \ell^q(\mathbb{N}, (m_i))$ is well posed.

Remark 6.35 The state and input spaces for the network Σ have been defined based on the values of the parameters p, q from Assumption 6.32, which may look artificial as in the applications, the choice of the state space depends on the physical meaning of the variables x_i. In particular, if x_i denotes a mass of i-th subsystem, and we are interested in the dynamics of the total mass of a network Σ, then the state space of the network could be chosen as $X = \ell^1(\mathbb{N}, (N_i))$. Alternatively, suppose the meaning of x_i is a velocity, and we are interested in the dynamics of the total kinetic energy of the network Σ. In that case, a reasonable choice of the state space is $X = \ell^2(\mathbb{N}, (N_i))$. Thus, to meet the needs of applications, the ISS Lyapunov functions V_i should be constructed to ensure some specific values of p, q.

The decay rates (λ_i) and the internal gains (γ_{ij}) induce the following infinite nonnegative matrices

$$\Lambda := \text{diag}(\lambda_1, \lambda_2, \lambda_3, \ldots), \quad \Gamma := (\gamma_{ij})_{i,j \in \mathbb{N}},$$

where we set $\gamma_{ij} := 0$ whenever $j \notin I_i$. A key role will be played by the following infinite matrix

$$\Psi := \Lambda^{-1}\Gamma = (\psi_{ij})_{i,j\in\mathbb{N}}, \quad \psi_{ij} = \frac{\gamma_{ij}}{\lambda_i}. \tag{6.50}$$

Under appropriate boundedness assumptions, the matrix Ψ induces a linear operator on ℓ^1 by

$$(\Psi x)_i = \sum_{j=1}^{\infty} \psi_{ij} x_j \quad \forall \ i \in \mathbb{N}.$$

We call $\Psi : \ell^1 \to \ell^1$ the *gain operator* corresponding to the decay rates λ_i and coefficients γ_{ij}.

To ensure that Γ is a bounded operator from ℓ^1 to ℓ^1, we require

Assumption 6.36 The matrix $\Gamma = (\gamma_{ij})$ satisfies

$$\|\Gamma\|_{1,1} = \sup_{j\in\mathbb{N}} \sum_{i=1}^{\infty} \gamma_{ij} < \infty. \tag{6.51}$$

Here the double index on the left-hand side indicates that we consider the operator norm induced by the ℓ^1-norm both on the domain and codomain of Γ.

Remark 6.37 Assumption 6.36 implies that there is $\overline{\gamma} > 0$ such that $0 < \gamma_{ij} \le \overline{\gamma}$ for all $i \in \mathbb{N}$ and $j \in I_i$.

Under Assumptions 6.36 and 6.33(b), the gain operator Ψ is bounded.

Lemma 6.38 *Suppose that Assumptions 6.36 and 6.33(b) hold. Then $\Psi : \ell^1 \to \ell^1$, defined by* (6.50), *is a bounded operator.*

Proof It holds that

$$\|\Psi\|_{1,1} = \sup_{j\in\mathbb{N}} \sum_{i=1}^{\infty} \psi_{ij} = \sup_{j\in\mathbb{N}} \sum_{i=1}^{\infty} \frac{\gamma_{ij}}{\lambda_i} \le \frac{1}{\underline{\lambda}} \sup_{j\in\mathbb{N}} \sum_{i=1}^{\infty} \gamma_{ij} < \infty,$$

which is equivalent to boundedness of Ψ. □

The following lemma provides a sufficient condition for (6.51).

Lemma 6.39 *If there exists $m \in \mathbb{N}$ so that $I_i \subset [i-m, i+m] \cap \mathbb{N}$ for all $i \in \mathbb{N}$ and $\gamma_{ij} \le \overline{\gamma}$ for all $i, j \in \mathbb{N}$ with a constant $\overline{\gamma} > 0$, then* (6.51) *holds.*

Proof We have the following estimates:

$$\begin{aligned} \sup_{j\in\mathbb{N}} \sum_{i=1}^{\infty} \psi_{ij} &= \sup_{j\in\mathbb{N}} \sum_{i=1}^{\infty} \frac{\gamma_{ij}}{\lambda_i} = \sup_{j\in\mathbb{N}} \sum_{|i-j|\le m} \frac{\gamma_{ij}}{\lambda_i} \\ &\le \sup_{j\in\mathbb{N}} \sum_{|i-j|\le m} \frac{\overline{\gamma}}{\underline{\lambda}} \le (2m+1)\frac{\overline{\gamma}}{\underline{\lambda}} < \infty. \end{aligned}$$

6.6.2 Positive Operators and Their Spectra

This section recalls several basic facts about positive operators on ordered Banach spaces.

Let X be a real Banach space with norm $\|\cdot\|_X$ and $T : X \to X$ be a bounded linear operator. Recall that the *complexification of* X is the complex Banach space $X_\mathbb{C} = \{x + iy : x, y \in X\}$ equipped with the norm

$$\|x + iy\|_{X_\mathbb{C}} := \sup_{t\in[0,2\pi]} \|(\cos t)x + (\sin t)y\|_X.$$

The *complexification of* T is the bounded operator $T_\mathbb{C} : X_\mathbb{C} \to X_\mathbb{C}$ given by

$$T_\mathbb{C}(x + iy) := Tx + iTy.$$

The resolvent of T is the function $R(\lambda, T) := (\lambda I - T_\mathbb{C})^{-1}$, defined for all those $\lambda \in \mathbb{C}$ such that the inverse exists and is a bounded operator. The *resolvent set* of T is given by

$$\rho(T) = \{\lambda \in \mathbb{C} : R(\lambda, T) \text{ exists and is bounded}\}.$$

The *spectrum* is the complement of the resolvent set, that is $\sigma(T) = \mathbb{C}\backslash\rho(T)$. It is well known that the spectrum of a bounded operator is a nonempty compact set. The *spectral radius* of T is defined as

$$r(T) := \max\{|\lambda| : \lambda \in \sigma(T)\},$$

which is finite as $\sigma(T)$ is compact and nonempty. The spectral radius of a bounded operator can be computed by Gelfand's formula [10]:

$$r(T) = \lim_{n\to\infty} \|T^n\|^{1/n} = \inf_{n\in\mathbb{N}} \|T^n\|^{1/n}. \tag{6.52}$$

Let X' be the dual space of X, that is, the Banach space of all bounded linear functionals $x^* : X \to \mathbb{R}$, equipped with the operator norm $\|x^*\|_{X'} = \sup_{\|x\|_X=1} |x^*(x)|$. The *adjoint operator* of T is defined as

$$(T'x^*)(x) := x^*(Tx), \quad x^* \in X', \quad x \in X.$$

It is well known that $\sigma(T') = \sigma(T)$ and $\|T'\| = \|T\|$.

A nonempty subset $X^+ \subset X$ is called a *cone* if it is closed, convex and the following holds:

- If $x \in X^+$ and $\lambda \geq 0$, then $\lambda x \in X^+$.
- If $x, -x \in X^+$, then $x = 0$.

For instance, the convexity of X^+ together with the first of the above properties implies that for any $x, y \in X^+$ and $\lambda, \mu \geq 0$, we have also $\lambda x + \mu y \in X^+$.

The choice of a cone in X induces the partial order $\geq$ given for all $x, y \in X$ by

$$x \geq y \quad \Leftrightarrow \quad x - y \in X^+.$$

The pair (X, X^+) is called an *ordered Banach space*.

In an ordered Banach space (X, X^+), we call a bounded linear operator $T : X \to X$ *positive* if $T(X^+) \subset X^+$. To denote this fact, we write $T > 0$ if $T \neq 0$. The positivity of an operator T can also be expressed by the implication

$$x \geq y \quad \Rightarrow \quad Tx \geq Ty, \quad x, y \in X.$$

If $\operatorname{int}(X^+) \neq \emptyset$, we write $x \ll y$ if $y - x \in \operatorname{int}(X^+)$.

The following lemma will be useful in the sequel.

Lemma 6.40 *Let $T : X \to X$ be a positive bounded linear operator on an ordered Banach space (X, X^+). Further assume that* $\operatorname{int}(X^+) \neq \emptyset$. *Then*

$$r(T) \geq \inf\{\lambda \in \mathbb{R} : \exists x \in \operatorname{int}(X^+)\, s.t.\, Tx \ll \lambda x\}. \tag{6.53}$$

Proof Take any $y \in \operatorname{int} X^+$, any $\lambda > r(T)$, and define $x := (\lambda I - T)^{-1} y$. Using the Neumann series representation of the resolvent operator and using the positivity of T, we have

$$x = \sum_{k=0}^{\infty} \frac{T^k}{\lambda^{k+1}} y = \frac{1}{\lambda} y + \sum_{k=1}^{\infty} \frac{T^k}{\lambda^{k+1}} y \geq \frac{1}{\lambda} y.$$

Here we exploit that $r(T/\lambda) < 1$, which ensures the convergence of the series. As $y \in \operatorname{int} X^+$, this shows that $x \in \operatorname{int} X^+$. Moreover,

$$\begin{aligned} Tx = T(\lambda I - T)^{-1} y &= (T - \lambda I + \lambda I)(\lambda I - T)^{-1} y \\ &= -y + \lambda(\lambda I - T)^{-1} y = -y + \lambda x \ll \lambda x. \end{aligned}$$

Thus, $(r(T), +\infty) \subset \{\lambda \in \mathbb{R} : \exists x \in \operatorname{int}(X^+)\, s.t.\, Tx \leq \lambda x\}$. □

6.6.3 Spectral Radius of the Gain Operator

The adjoint operator of Ψ acts on ℓ^∞ (which is canonically identified with the dual space $(\ell^1)^*$) and can be described by the transpose $\Theta := \Psi^T$, $\Theta = (\theta_{ij}) = (\psi_{ji})$ as

$$(\Theta x)_i = \sum_{j=1}^{\infty} \theta_{ij} x_j = \sum_{j=1}^{\infty} \frac{\gamma_{ji}}{\lambda_j} x_j \quad \forall x \in \ell^\infty.$$

On the Banach space ℓ^∞, we define the cone

$$X^+ := \left\{(x_i)_{i\in\mathbb{N}} \in \ell^\infty : x_i \geq 0 \ \forall i \in \mathbb{N}\right\},$$

and observe that the interior of X^+ is nonempty and is given by

$$\text{int}\ X^+ = \left\{x \in \ell^\infty : \exists \underline{x} > 0\, s.t.\, x_i \geq \underline{x} \ \forall i \in \mathbb{N}\right\}.$$

Clearly, Θ maps the cone X^+ into itself, hence, Θ is a positive operator with respect to this cone. The partial order on ℓ^∞, induced by X^+, is given by

$$x \leq y \quad \Leftrightarrow \quad x_i \leq y_i \quad \forall i \in \mathbb{N}.$$

Using Lemma 6.40, we show next, that the small-gain condition $r(\Psi) < 1$ ensures existence of a certain subeigenvector of the operator $-\Lambda + \Gamma$, that will be useful for the proof of the small-gain theorem for infinite networks.

Lemma 6.41 *Let the sequence (λ_i) be uniformly bounded from above and let $r(\Theta) < 1$. Then the following assertions hold:*

(i) There is a vector $\mu = (\mu_i)_{i\in\mathbb{N}} \in \text{int}\,(X^+)$ and a constant $\lambda_\infty > 0$ such that

$$\frac{[\mu^T(-\Lambda + \Gamma)]_i}{\mu_i} \leq -\lambda_\infty \quad forall \quad i \in \mathbb{N}.$$

(ii) For every $\rho > 0$, one can choose the vector μ and the constant λ_∞ in the statement (i) in a way that

$$\lambda_\infty \geq (1 - r(\Theta))\underline{\lambda} - \rho.$$

Proof By Lemma 6.40 and the assumption that $r(\Theta) < 1$, there exist $\eta \in \text{int}\,(X^+)$ and $\tilde{r} \in (0, 1)$ with $\Theta\eta \leq \tilde{r}\eta$, which is equivalent to $\eta^T\Lambda^{-1}\Gamma \leq \tilde{r}\eta^T$. Defining $\mu^T := \eta^T\Lambda^{-1}$, we can rewrite the previous inequality in a form $\mu^T(-\tilde{r}\Lambda + \Gamma) \leq 0$. This is equivalent to

$$\mu^T(-\Lambda + \Gamma) \leq \mu^T(-(1 - \tilde{r})\Lambda).$$

Hence, we have that

$$\left[\mu^T(-\Lambda + \Gamma)\right]_i \leq -(1 - \tilde{r})\lambda_i\mu_i \leq -(1 - \tilde{r})\underline{\lambda}\,\mu_i$$

with $\underline{\lambda}$ given by (6.48). To show that $\mu \in \text{int}\,(X^+)$, we use the assumption that there is an upper bound $\overline{\lambda}$ for the constants λ_i. It guarantees that the components of μ satisfy

$$\mu_i = \lambda_i^{-1}\eta_i \geq \overline{\lambda}^{-1}\eta_i.$$

Thus, statement (i) is proved.

Statement (ii) is implied by the fact that $\tilde{r}$ can be chosen arbitrarily close to $r(\Theta)$. □

6.6.4 Small-Gain Theorem for Infinite Networks

Now we prove the main result of the section, that guarantees that the network Σ is eISS provided that the spectral radius of the gain operator satisfies $r(\Psi) < 1$. Our proof is constructive as we give an explicit expression for the ISS Lyapunov function for the interconnection.

Theorem 6.42 *Consider the infinite network Σ, consisting of subsystems $(\Sigma_i)_{i\in\mathbb{N}}$, with fixed $p, q \in [1, \infty)$. Let the following hold:*

(i) Σ is well posed as a system with state space $X = \ell^p(\mathbb{N}, (N_i))$, space of input values $U = \ell^q(\mathbb{N}, (m_i))$, and the external input space $\mathcal{U}$, as defined in (6.4).
(ii) Each Σ_i admits a continuous eISS Lyapunov function V_i such that Assumptions 6.32 and 6.33 hold.
(iii) The operator $\Gamma : \ell^1 \to \ell^1$ is bounded, that is, Assumption 6.36 holds.
(iv) The spectral radius of Ψ satisfies $r(\Psi) < 1$.

Then Σ is ISS and admits an eISS Lyapunov function of the form

$$V(x) = \sum_{i=1}^{\infty} \mu_i V_i(x_i), \quad V : X \to \mathbb{R}_+ \tag{6.54}$$

for some $\mu = (\mu_i)_{i\in\mathbb{N}} \in \ell^\infty$ satisfying $\underline{\mu} \le \mu_i \le \overline{\mu}$ with some constants $\underline{\mu}, \overline{\mu} > 0$. More precisely, the function V has the following properties:

(a) V is continuous.
(b) There exists a $\lambda_\infty > 0$ so that for all $x^0 \in X$ and $u \in \mathcal{U}$

$$\dot{V}_u(x^0) \le -\lambda_\infty V(x^0) + \overline{\mu}\,\overline{\gamma}_u \|u\|^q_{\mathcal{U}}.$$

(c) For all $x \in X$ the following sandwich bounds hold

$$\underline{\mu}\,\underline{\alpha}\|x\|^p_X \le V(x) \le \overline{\mu}\,\overline{\alpha}\|x\|^p_X. \tag{6.55}$$

Proof First, we show the theorem for the case that there exists a constant $\overline{\lambda} > 0$ with

$$\lambda_i \le \overline{\lambda} \quad \forall \quad i \in \mathbb{N}. \tag{6.56}$$

Inequality (6.56) means the decay rates of the eISS Lyapunov functions for all subsystems are uniformly bounded. Afterward, we treat the general case.

The proof proceeds in five steps.

Step 1. Definition and coercivity of V. Note that the spectral radii of the operators $\Psi = \Lambda^{-1}\Gamma : \ell^1 \to \ell^1$ and $\Theta = \Psi^T : \ell^\infty \to \ell^\infty$ are equal, since the second operator is the adjoint of the first. Hence, Lemma 6.41 yields a positive vector $\mu = (\mu_i)_{i\in\mathbb{N}} \in \ell^\infty$ whose entries are uniformly bounded from below and above, and yields a constant $\lambda_\infty > 0$ such that

$$\frac{[\mu^T(-\Lambda + \Gamma)]_i}{\mu_i} \leq -\lambda_\infty \quad \forall\, i \in \mathbb{N}. \tag{6.57}$$

To see that V constructed as in (6.54) is well defined, note that for all $x \in X$ we have

$$0 \leq V(x) \leq \sum_{i=1}^{\infty} \mu_i \overline{\alpha}_i |x_i|^p \leq \overline{\alpha}|\mu|_\infty \|x\|_X^p < \infty.$$

This shows also the upper bound for (6.55). The lower bound for (6.55) can be obtained similarly, and thus the inequality (6.44a) holds for V with $b = p$.

Step 2. Continuity of V. Fix a point $x = (x_i) \in X$ and certain $\varepsilon > 0$. Pick $\delta_0, \varepsilon' > 0$ such that

$$\overline{\mu}\,\overline{\alpha} 2^{p-1}(\delta_0^p + \varepsilon') \leq \frac{\varepsilon}{4} \quad \text{and} \quad \varepsilon' \leq \frac{\varepsilon}{4\overline{\alpha}\,\overline{\mu}}.$$

Afterward, choose $N \in \mathbb{N}$ in a way that $\sum_{i=N+1}^{\infty} |x_i|^p \leq \varepsilon'$. Finally, choose $\delta \in (0, \delta_0]$ such that for every $y_i \in \mathbb{R}^{N_i}$ the following holds:

$$|x_i - y_i| < \delta \quad \Rightarrow \quad |V_i(x_i) - V_i(y_i)| < \frac{\varepsilon}{2N\overline{\mu}}, \quad i = 1, \ldots, N.$$

Here we have used the continuity of V_i at x_i. Now let $y = (y_i)_{i\in\mathbb{N}} \in X$ be so that $\|x - y\|_X < \delta$. This implies, in particular, that $|x_i - y_i| < \delta$ for $i = 1, \ldots, N$. Then

$$\begin{aligned} |V(x) - V(y)| &\leq \sum_{i=1}^{\infty} \mu_i |V_i(x_i) - V_i(y_i)| \\ &\leq \frac{\varepsilon}{2} + \overline{\mu} \sum_{i=N+1}^{\infty} |V_i(x_i) - V_i(y_i)|. \end{aligned}$$

The remainder sum we estimate as follows

$$\begin{aligned} \sum_{i=N+1}^{\infty} |V_i(x_i) - V_i(y_i)| &\leq \overline{\alpha} \sum_{i=N+1}^{\infty} |x_i|^p + \overline{\alpha} \sum_{i=N+1}^{\infty} |y_i|^p \\ &\leq \overline{\alpha}\varepsilon' + \overline{\alpha} \sum_{i=N+1}^{\infty} 2^{p-1} \left(|y_i - x_i|^p + |x_i|^p\right) \\ &\leq \overline{\alpha}\varepsilon' + \overline{\alpha} 2^{p-1} \left(\delta_0^p + \varepsilon'\right) \leq \overline{\mu}^{-1} \frac{\varepsilon}{2}. \end{aligned}$$

Here we have used the elementary inequality

$$(a+b)^p \leq 2^{p-1}(a^p + b^p), \quad a, b \geq 0, \; p \geq 1,$$

which follows from the convexity of the map $a \mapsto a^p$. This ensures that $|V(x) - V(y)| < \varepsilon$, and thus V is continuous at x.

Step 3. Estimate of the orbital derivative. Pick an initial state $x^0 \in X$ and an external input $u \in \mathcal{U}$. We write

$$\phi(t) = \phi(t, x^0, u), \quad \phi_i(t) = \pi_i \phi(t), \quad x_{\neq i}(t) = (\pi_j \phi(t))_{j \in I_i},$$

where π_i denotes the projection to the i-th component. Then for any $t > 0$ from the domain of definition of ϕ, it holds that

$$\frac{1}{t}\big(V(\phi(t)) - V(x^0)\big) = \frac{1}{t}\sum_{i=1}^{\infty} \mu_i \big[V_i(\phi_i(t)) - V_i(\phi_i(0))\big].$$

As inequalities (6.46) hold for almost all positive times, the function in the right-hand side of (6.46) is Lebesgue integrable, and as $D_+(V_i \circ \phi_i)(t) < \infty$ for all t, we proceed employing the generalized fundamental theorem of calculus (see [13, Theorem 9 and p. 42, Remark 5.c] or [24, Theorem 7.3, p. 204]) to

$$\frac{1}{t}\big(V(\phi(t)) - V(x^0)\big) \leq \frac{1}{t}\sum_{i=1}^{\infty} \int_0^t \mu_i \Big[-\lambda_i V_i(\phi_i(s)) + \sum_{j \in I_i} \gamma_{ij} V_j(\phi_j(s)) + \gamma_{iu}|u_i(s)|^q\Big] ds.$$

Next we use the Fubini–Tonelli theorem to interchange the infinite sum and the integral (interpreting the sum as an integral associated with the counting measure on $\mathbb{N}$). For this, it suffices to prove that the following integral is finite:

$$\int_0^t \sum_{i=1}^{\infty} \Big|\mu_i \Big[-\lambda_i V_i(\phi_i(s)) + \sum_{j \in I_i} \gamma_{ij} V_j(\phi_j(s)) + \gamma_{iu}|u_i(s)|^q\Big]\Big| ds.$$

Using (6.45), (6.47), (6.49), and (6.56), we upperbound the inner term by

$$\overline{\mu}\Big[\overline{\lambda}\overline{\alpha}|\phi_i(s)|^p + \sum_{j \in I_i} \gamma_{ij}\overline{\alpha}|\phi_j(s)|^p + \overline{\gamma}_u|u_i(s)|^q\Big].$$

Summing all three terms over i, we obtain

$$\overline{\lambda}\overline{\alpha}\sum_{i=1}^{\infty}|\phi_i(s)|^p \le c_1\|\phi(s)\|_X^p,$$

$$\sum_{i=1}^{\infty}\sum_{j\in I_i}\gamma_{ij}\overline{\alpha}|\phi_j(s)|^p \le c_2\sum_{j=1}^{\infty}|\phi_j(s)|^p\sum_{i=1}^{\infty}\gamma_{ij} \le c_3\|\phi(s)\|_X^p,$$

$$\overline{\gamma}_u\sum_{i=1}^{\infty}|u_i(s)|^q = c_4\|u(s)\|_U^q,$$

for some constants $c_1, c_2, c_3, c_4 > 0$. In the inequality for the middle term, we used the boundedness of Γ. Hence,

$$\int_0^t\sum_{i=1}^{\infty}\Big|\mu_i\Big[-\lambda_i V_i(\phi_i(s)) + \sum_{j\in I_i}\gamma_{ij}V_j(\phi_j(s)) + \gamma_{iu}|u_i(s)|^q\Big]\Big|ds$$

$$\le c\int_0^t\left(\|\phi(s)\|_X^p + \|u(s)\|_U^q\right)ds < \infty,$$

for some constant $c > 0$, where we use the essential boundedness of the integrand in the last term ($s \mapsto \|\phi(s)\|_X^p$ is continuous and $s \mapsto |u(s)|_q^q$ is essentially bounded).

Exploiting the notation

$$V_{\text{vec}}(\phi(s)) := (V_1(\phi_1(s)), V_2(\phi_2(s)), \ldots)^T$$

and invoking the Fubini–Tonelli theorem, we yield

$$\begin{aligned}
&\frac{1}{t}(V(\phi(t)) - V(x^0))\\
&\quad\le \frac{1}{t}\int_0^t\sum_{i=1}^{\infty}\mu_i\Big[-\lambda_i V_i(\phi_i(s)) + \sum_{j\in I_i}\gamma_{ij}V_j(\phi_j(s)) + \gamma_{iu}|u_i(s)|^q\Big]ds\\
&\quad= \frac{1}{t}\int_0^t\Big[\mu^T(-\Lambda + \Gamma)V_{\text{vec}}(\phi(s)) + \sum_{i=1}^{\infty}\mu_i\gamma_{iu}|u_i(s)|^q\Big]ds\\
&\quad\le \frac{1}{t}\int_0^t\Big[-\lambda_\infty V(\phi(s)) + \overline{\mu}\,\overline{\gamma}_u|u|_{q,\infty}^q\Big]ds\\
&\quad= \frac{1}{t}\int_0^t -\lambda_\infty V(\phi(s))\,ds + \overline{\mu}\,\overline{\gamma}_u|u|_{q,\infty}^q.
\end{aligned}$$

Here we used (6.57) to show the second inequality above. As $s \mapsto V(\phi(s))$ is continuous, we obtain

$$\begin{aligned}\dot{V}_u(x^0) &= \overline{\lim_{t\to 0+}} \frac{1}{t}\big(V(\phi(t)) - V(x^0)\big)\\ &\le -\lambda_\infty V(x^0) + \overline{\mu}\,\overline{\gamma}_u |u|^q_{q,\infty}.\end{aligned}$$

Consequently, (6.44b) holds for V with $\kappa = \lambda_\infty$ and $\gamma(r) = \overline{\mu}\,\overline{\gamma}_u r^q$.

Step 4. Proof of eISS. We have shown that properties (a)–(c) are satisfied for V. Thus, V is an eISS Lyapunov function for Σ and Σ is eISS by Proposition 6.29. Thus, for uniformly upper-bounded sequence (λ_i) the theorem is proved.

Step 5. Unbounded decay rates λ_i. Let (6.56) do not hold for any $\overline{\lambda}$. Pick any $h > 0$ and define the reduced decay rates $\lambda_i^h := \min\{\lambda_i, h\}$. Thus, $\lambda_i^h \le h$ for all $i \in \mathbb{N}$, which allows us to invoke the previous analysis.

Indeed, as inequalities (6.46) hold with λ_i, they hold also with λ_i^h. Define the modified operators Λ^h, Ψ^h by

$$\Lambda^h := \operatorname{diag}(\lambda_1^h, \lambda_2^h, \ldots), \qquad \Psi^h := (\Lambda^h)^{-1}\Gamma.$$

Considering $(\Lambda^h)^{-1}$ as an operator from ℓ^1 to ℓ^1, one can see that $(\Lambda^h)^{-1} \to \Lambda^{-1}$ as $h \to \infty$. Since we assume that Γ is a bounded operator, it also holds that $\Psi^h \to \Psi$ as $h \to \infty$.

By the small-gain condition, we have $r(\Psi) < 1$. As the spectral radius is upper semicontinuous on the space of bounded operators on a Banach space (see, e.g., [1, Theorem 1.1(i)]), it holds that $r(\Psi^h) < 1$ for h large enough. As the coefficients λ_i^h are uniformly bounded, by feeding the operator Ψ^h to Lemma 6.41, we obtain a vector $\mu = \mu(h)$ and a coefficient $\lambda_\infty = \lambda_\infty(h)$, so that (by the first four steps of this proof) (6.54) is an eISS Lyapunov function for Σ with the decay rate λ_∞. □

Remark 6.43 Theorem 6.42 extends the corresponding small-gain theorem for finite networks, shown in [6, Proposition 3.3]. The result in [6] relies on [6, Lemma 3.1], which is a consequence of the Perron–Frobenius theorem. However, existing infinite-dimensional versions of the Perron–Frobenius theorem, including the Krein–Rutman theorem [18], are *not* applicable to our setting as they require at least quasi-compactness of the gain operator, which is a pretty strong assumption. Lemma 6.41 helps to overcome this obstacle.

Remark 6.44 The parameters μ and λ_∞, exploited for the construction of an eISS Lyapunov function V, are not uniquely determined and depend, e.g., on the disturbance S_ε, the constant ρ (in Lemma 6.41) and (in case of unbounded λ_i) on the parameter h, introduced in Step 5 of the proof of Theorem 6.42.

Remark 6.45 Often eISS Lyapunov functions V_i for subsystems (6.2) are assumed to be continuously differentiable. In this case, the dissipative conditions for ISS Lyapunov functions V_i can be formulated in a computationally simpler style, namely

$$\nabla V_i(x_i) \cdot f_i(x_i, x_{\neq i}, u_i) \leq -\lambda_i V_i(x_i) + \sum_{j \in I_i} \gamma_{ij} V_j(x_j) + \gamma_{iu} |u_i|^q. \tag{6.58}$$

These conditions have to be valid for all $x_i \in \mathbb{R}^{N_i}$, $u_i \in \mathbb{R}^{m_i}$, and $x_{\neq i} \in \mathbb{R}^{N_{\neq i}}$. The expression on the left-hand side of inequality (6.58) represents a formula for the computation of the orbital derivative of V_i under the assumption that V_i is smooth enough.

The proof of the corresponding small-gain theorem goes along the same lines of the proof of Theorem 6.42.

Remark 6.46 The spectral condition $r(\Psi) < 1$ is equivalent to the exponential stability of the discrete-time system induced by the operator Ψ. This result, as well as a number of criteria for this condition that help to verify it in practice, has been shown in [12]. In Proposition C.11 we show such criteria for linear finite-dimensional monotone discrete-time systems.

6.6.5 *Necessity of the Required Assumptions and Tightness of the Small-Gain Result*

The spectral small-gain condition $r(\Psi) < 1$ cannot be removed or relaxed. This is already well known for planar systems (see also Section 3.6 for the tightness analysis of the small-gain condition).

The inequalities (6.45) and (6.47) are necessary for the overall eISS Lyapunov function V to be well defined and coercive. Having relaxed the lower bound in (6.45) or (6.47), we might still be able to prove ISS (though not eISS) of the interconnection by using results on noncoercive ISS Lyapunov functions, cf. [21, Theorem 2.18], but additional assumptions on boundedness of reachable sets might be necessary.

Without the condition (6.48) we do not have a uniform decay rate for the solutions of the subsystems, which prevents us from getting already asymptotic stability for the interconnection, even if the system is linear and all internal and external gains are zero.

To see this, recall the infinite network from Example 6.17:

$$\dot{x}_i = -\frac{1}{i} x_i + u, \quad i \in \mathbb{N},$$

with the input space $\mathcal{U} := L^\infty(\mathbb{R}_+, \mathbb{R})$ and the state space $X = \ell^p$ for any $p \in [1, \infty]$. One can construct eISS Lyapunov functions for subsystems as $V_i(z) = z^2$ for all $i \in \mathbb{N}$ and $z \in \mathbb{R}$. With this choice of Lyapunov functions, all the assumptions that we impose for the small-gain theorem will be satisfied, except for (6.48). At the same time, the network is not even exponentially stable in the absence of inputs. Furthermore, inputs of arbitrarily small magnitude may lead to unboundedness of trajectories, as mentioned in Example 6.17.

The condition (6.51) is essential for the validity of the small-gain theorem, as shown by the following simple example

$$\dot{x}_i = -x_i + iu, \quad i \in \mathbb{N},$$

where we choose $X = \ell^p$ for any $p \in [1, \infty]$. Taking $V_i(z) = z^2$ for all $i \in \mathbb{N}$ and $z \in \mathbb{R}$, after some simple computations, one can see that all other assumptions of the small-gain theorem are fulfilled but the overall system is not ISS.

6.7 Examples

In this section[3], we apply our results to three examples: linear spatially invariant systems, nonlinear spatially invariant systems with a nonlinearity satisfying the sector condition, and a road traffic model. In all cases, we construct eISS Lyapunov functions with linear gains for all subsystems and then apply our small-gain result to construct an exponential ISS Lyapunov function for the overall network.

6.7.1 A Linear Spatially Invariant System

Consider an infinite network of systems Σ_i, given by

$$\begin{aligned}\Sigma_i : \dot{x}_i &= -b_{ii}x_i + b_{i(i-1)}x_{i-1} + b_{i(i+1)}x_{i+1} \\ &=: f_i(x_i, x_{i-1}, x_{i+1}),\end{aligned}$$

where $x_i \in \mathbb{R}$, $b_{ii} > 0$, $b_{i(i-1)}, b_{i(i+1)} \in \mathbb{R}$ for each $i \in \mathbb{N}$, and $b_{10} = 0$. We consider the standard Euclidean norm on $\mathbb{R}$ for each $i \in \mathbb{N}$ and assume that there is a constant $\overline{b} > 0$ so that

$$\max\{b_{ii}, |b_{i(i-1)}|, |b_{i(i+1)}|\} \leq \overline{b} \quad forall \quad i \in \mathbb{N}.$$

From Example 6.11, it immediately follows that the composite system is well posed with $p = 2$.

ISS Lyapunov functions for subsystems. For each subsystem Σ_i, we choose the eISS Lyapunov function candidate $V_i(x_i) = \frac{1}{2}x_i^2$ satisfying (6.45) and (6.47). Using Young's inequality, one can simply verify (6.46) as follows.

---

[3] This section is © 2021 IEEE reprinted (with minor changes), with permission, from "Kawan, A. Mironchenko, A. Swikir, N. Noroozi, M. Zamani. A Lyapunov-based small-gain theorem for infinite networks. IEEE Transactions on Automatic Control, 66(12):5830–5844, 2021".

$$\begin{aligned}\nabla V_i(x_i)\cdot f_i(x_i,x_{i-1},x_{i+1}) &= x_i(-b_{ii}x_i + b_{i(i-1)}x_{i-1} + b_{i(i+1)}x_{i+1})\\ &\leq -(b_{ii}-\varepsilon_i-\delta_i)x_i^2 + \frac{b_{i(i-1)}^2}{4\varepsilon_i}x_{i-1}^2 + \frac{b_{i(i+1)}^2}{4\delta_i}x_{i+1}^2\\ &= -2(b_{ii}-\varepsilon_i-\delta_i)V_i(x_i) + \frac{b_{i(i-1)}^2}{2\varepsilon_i}V_{i-1}(x_{i-1})\\ &\quad + \frac{b_{i(i+1)}^2}{2\delta_i}V_{i+1}(x_{i+1}),\end{aligned}$$

for appropriate choices of $\varepsilon_i, \delta_i > 0$. Hence, we can choose

$$\lambda_i := 2(b_{ii}-\varepsilon_i-\delta_i), \quad \gamma_{i(i-1)} := \frac{b_{i(i-1)}^2}{2\varepsilon_i}, \quad \gamma_{i(i+1)} := \frac{b_{i(i+1)}^2}{2\delta_i}$$

and assume that ε_i, δ_i are such that (6.48) and (6.51) are satisfied.

Small-gain analysis of the network. It follows that the infinite matrix Ψ has the form

$$\Psi = \Lambda^{-1}\Gamma = \begin{pmatrix} 0 & \psi_{12} & 0 & 0 & 0 & 0 & \dots \\ \psi_{21} & 0 & \psi_{23} & 0 & 0 & 0 & \dots \\ 0 & \psi_{32} & 0 & \psi_{34} & 0 & 0 & \dots \\ 0 & 0 & \psi_{43} & 0 & \psi_{45} & 0 & \dots \\ \vdots & \ddots & \ddots & \ddots & \ddots & \ddots & \ddots \end{pmatrix}, \tag{6.59}$$

where $\psi_{ij} = \gamma_{ij}/\lambda_i$. We estimate the spectral radius $r(\Psi)$ by the operator norm $\|\Psi\|$ as

$$r(\Psi) \leq \|\Psi\| = \sup_{j\in\mathbb{N}} \sum_{i=1}^{\infty} \psi_{ij} \leq 2\frac{\overline{\gamma}}{\underline{\lambda}}.$$

Altogether, the following set of sufficient conditions guarantee that the interconnection defined above is eISS.

- $\max\{b_{ii}, |b_{i(i-1)}|, |b_{i(i+1)}|\} \leq \overline{b}$ for all $i \in \mathbb{N}$ with a constant $\overline{b} > 0$ (for well posedness).
- The constants $\varepsilon_i, \delta_i > 0$ are chosen such that

– Assumptions (ii) and (iii) in Theorem 6.42 hold with

$$0 < \underline{\lambda} \leq 2(b_{ii}-\varepsilon_i-\delta_i), \quad \frac{b_{i(i-1)}^2}{2\varepsilon_i} + \frac{b_{i(i+1)}^2}{2\delta_i} \leq \overline{\gamma} < \infty, \quad i \in \mathbb{N},$$

with certain constants $\underline{\lambda}, \overline{\gamma}$;

– the small-gain condition $r(\Psi) < 1$ holds, for which it suffices to have $\frac{b_{i(i-1)}^2}{2\varepsilon_i(b_{ii}-\varepsilon_i-\delta_i)} < 1$ and $\frac{b_{i(i+1)}^2}{2\delta_i(b_{ii}-\varepsilon_i-\delta_i)} < 1$ for all $i \in \mathbb{N}$.

Remark 6.47 As we argued in Remark 6.35, the choice of the "right" Lyapunov functions depends on the physical sense of the variables. Thus, quadratic Lyapunov functions and the corresponding state space $X = \ell^2(\mathbb{N}, (N_i))$ may not be physically appropriate for some applications. However, there are other natural options for Lyapunov functions for the subsystems Σ_i, for example, $W_i(x_i) := |x_i|$, which would lead to other values of the gains and to another expression for the small-gain condition.

6.7.2 A Nonlinear Multidimensional Spatially Invariant System

Here, we analyze a class of nonlinear control systems that appears in many applications, including analysis of neural networks, design of optimization algorithms, Lur'e systems, and so on (see [11] and the references therein).

Consider an infinite network whose subsystems are described by

$$\Sigma_i : \quad \dot{x}_i = A_i x_i + E_i \varphi_i(G_i x_i) + B_i u_i + D_i x_{\neq i},$$

where $A_i \in \mathbb{R}^{N_i \times N_i}$, $E_i \in \mathbb{R}^{N_i}$, $G_i^T \in \mathbb{R}^{N_i}$, $B_i \in \mathbb{R}^{N_i \times m_i}$, $D_i \in \mathbb{R}^{N_i \times N_{\neq i}}$ with $N_{\neq i} = \sum_{j \in I_i} N_j$ and $I_1 = \{i+1\}$, $I_i = \{i-1, i+1\}$ for all $i \geq 2$.

Assumptions. We consider the standard Euclidean norm on each $\mathbb{R}^{N_i}$, $\mathbb{R}^{m_i}$, and $\mathbb{R}^{N_{\neq i}}$, and assume that A_i, E_i, G_i, B_i, D_i are uniformly bounded for all $i \in \mathbb{N}$. That is, $\|A_i\| \leq a$, $\|E_i\| \leq e$, $\|G_i\| \leq g$, $\|B_i\| \leq b$, $\|D_i\| \leq d$, for certain $a, e, g, b, d > 0$ and all i. Additionally, we assume that the nonlinear functions $\varphi_i : \mathbb{R} \to \mathbb{R}$ satisfy

$$\big(\varphi_i(G_i x_i) - r_i G_i x_i\big)\big(\varphi_i(G_i x_i) - l_i G_i x_i\big) \leq 0, \tag{6.60}$$

for all $x_i \in \mathbb{R}^{N_i}$, with $r_i > l_i$, $l_i, r_i \in \mathbb{R}$.

Moreover, we assume that the nonlinear functions $\varphi_i : \mathbb{R} \to \mathbb{R}$ have some regularity properties such that the interconnected system Σ with state space $X := \ell^2(\mathbb{N}, (N_i))$ and input space $U := \ell^2(\mathbb{N}, (m_i))$ is well posed.

ISS Lyapunov functions for subsystems. Now let the function V_i be defined as

$$V_i(x_i) := x_i{}^T M_i x_i, \quad x_i \in \mathbb{R}^{N_i},$$

where $M_i \in \mathbb{R}^{N_i \times N_i}$ is symmetric and positive definite, $\|M_i\| \leq m$ for all $i \in \mathbb{N}$, and

$$0 < \underline{m} \leq \lambda_{\min}(M_i) \leq \lambda_{\max}(M_i) \leq \overline{m} < \infty,$$

where $\lambda_{\min}(\cdot)$ and $\lambda_{\max}(\cdot)$ denote the smallest and largest eigenvalues, respectively.

Assume that for all $i \in \mathbb{N}$ and $x_i \in \mathbb{R}^{N_i}$ the inequality

$$2x_i^T M_i(A_i x_i + E_i \varphi_i(G_i x_i)) \leq -\kappa_i x_i^T M_i x_i \tag{6.61}$$

holds for some κ_i so that $0 < \underline{\kappa} \leq \kappa_i$ for some $\underline{\kappa}$.

The inequality (6.61) is equivalent to

$$\begin{bmatrix} x_i \\ \varphi_i(G_i x_i) \end{bmatrix}^T \begin{bmatrix} A_i^T M_i + M_i A_i + \kappa_i M_i & M_i E_i \\ E_i^T M_i & \mathbf{0} \end{bmatrix} \begin{bmatrix} x_i \\ \varphi_i(G_i x_i) \end{bmatrix} \leq 0,$$

for all $i \in \mathbb{N}$ and $x_i \in \mathbb{R}^{N_i}$ and all $\varphi_i(G_i x_i)$ satisfying (6.60), where $\mathbf{0}$ is a zero matrix with appropriate dimensions.

As $G_i x_i \in \mathbb{R}$, we have that $G_i x_i = (G_i x_i)^T = x_i^T G_i^T$ and

$$-r_i G_i x_i \varphi_i(G_i x_i) - l_i \varphi_i(G_i x_i) G_i x_i = -\varphi_i(G_i x_i)\frac{r_i + l_i}{2} G_i x_i - x_i^T \frac{r_i + l_i}{2} G_i^T \varphi_i(G_i x_i).$$

Thus the inequality (6.60) is equivalent to

$$\begin{bmatrix} x_i \\ \varphi_i(G_i x_i) \end{bmatrix}^T \begin{bmatrix} r_i l_i G_i^T G_i & -\frac{r_i + l_i}{2} G_i^T \\ -\frac{r_i + l_i}{2} G_i & 1 \end{bmatrix} \begin{bmatrix} x_i \\ \varphi_i(G_i x_i) \end{bmatrix} \leq 0. \tag{6.62}$$

By using the S-procedure[4] and (6.62), a sufficient condition for the validity of (6.61) is the validity of the following linear matrix inequality

$$\begin{bmatrix} A_i^T M_i + M_i A_i + \kappa_i M_i - \tau_i r_i l_i G_i^T G_i & M_i E_i + \tau_i \frac{r_i + l_i}{2} G_i^T \\ E_i^T M_i + \tau_i \frac{r_i + l_i}{2} G_i & -\tau_i \end{bmatrix} \preceq 0$$

for some $\tau_i \in \mathbb{R}_+$. Here the notation $S \preceq 0$ means that S is a negative semidefinite matrix.

Properties of V_i. Thus

$$\lambda_{\min}(M_i)|x_i|^2 \leq V_i(x_i) \leq \lambda_{\max}(M_i)|x_i|^2, \quad x_i \in \mathbb{R}^{N_i},$$

and

$$\begin{aligned} \nabla V_i(x_i) \cdot f_i(x_i, x_{\neq i}, u_i) &= 2x_i^T M_i(A_i x_i + E_i \varphi_i(G_i x_i) + B_i u_i + D_i x_{\neq i}) \\ &= 2x_i^T M_i\big(A_i x_i + E_i \varphi_i(G_i x_i)\big) + 2x_i^T M_i B_i u_i + 2x_i^T M_i D_i x_{\neq i}. \end{aligned}$$

As M_i is positive definite, there is a unique square root of M_i, that is the matrix that we denote $\sqrt{M_i}$ that is positive definite and satisfies $\sqrt{M_i}\sqrt{M_i} = M_i$, see [4, p. 431]. Using Cauchy–Schwarz and Young inequalities, respectively, we obtain for any $\varepsilon_i > 0$ that

---

[4] The idea of the *S-procedure* is that if we know that $F_1 \leq 0$, then a simple sufficient condition ensuring that $F_0 \leq 0$ is the existence of $\tau > 0$, such that $F_0 \leq \tau F_1$. S-procedure was introduced in [25]. See also [9] for more on S-procedure.

$$
\begin{aligned}
2x_i^T M_i B_i u_i &= 2x_i^T \sqrt{M_i}\sqrt{M_i} B_i u_i \\
&\le 2|\sqrt{M_i}x_i| \cdot |\sqrt{M_i} B_i u_i| \\
&\le 2|\sqrt{M_i}x_i| \cdot \|\sqrt{M_i} B_i\| |u_i| \\
&\le \varepsilon_i |\sqrt{M_i}x_i|^2 + 2\frac{\|\sqrt{M_i} B_i\|^2 |u_i|^2}{2\varepsilon_i} \\
&= \varepsilon_i x_i^T M_i x_i + \frac{\|\sqrt{M_i} B_i\|^2 |u_i|^2}{\varepsilon_i},
\end{aligned}
$$

and analogously,

$$
2x_i^T M_i D_i x_{\neq i} \le \varepsilon_i x_i^T M_i x_i + \frac{\|\sqrt{M_i} D_i\|^2 |x_{\neq i}|^2}{\varepsilon_i}. \tag{6.63}
$$

With these estimates, we have that

$$
\begin{aligned}
&\nabla V_i(x_i) \cdot f_i(x_i, x_{\neq i}, u_i) \\
&\le -(\kappa_i - 2\varepsilon_i) x_i^T M_i x_i + \frac{\|\sqrt{M_i} B_i\|^2 |u_i|^2}{\varepsilon_i} + \frac{\|\sqrt{M_i} D_i\|^2}{\varepsilon_i} \sum_{j \in I_i} |x_j|^2 \\
&\le -(\kappa_i - 2\varepsilon_i) V_i(x_i) + \frac{\|\sqrt{M_i} B_i\|^2}{\varepsilon_i} |u_i|^2 + \frac{\|\sqrt{M_i} D_i\|^2}{\varepsilon_i} \sum_{j \in I_i} \frac{V_j(x_j)}{\lambda_{\min}(M_j)}.
\end{aligned} \tag{6.64}
$$

Hence, the function $V_i(x_i) = x_i^T M_i x_i$ is an eISS Lyapunov function for the subsystem Σ_i satisfying (6.45) and (6.46) with

$$
\begin{aligned}
\underline{\alpha}_i &:= \lambda_{\min}(M_i), \quad \overline{\alpha}_i := \lambda_{\max}(M_i), \quad \lambda_i := \kappa_i - 2\varepsilon_i, \\
\gamma_{ij} &:= \frac{\|\sqrt{M_i} D_i\|^2}{\lambda_{\min}(M_j)\varepsilon_i}, \quad \gamma_{iu} := \frac{\|\sqrt{M_i} B_i\|^2}{\varepsilon_i}.
\end{aligned}
$$

Small-gain analysis. With $\underline{\alpha} := \underline{m}$ and $\overline{\alpha} := \overline{m}$, (6.47) is satisfied. With a uniformity condition on ε_i, say $0 < \underline{\varepsilon} \le \varepsilon_i \le \overline{\varepsilon} < \infty$ so that $\underline{\kappa} - 2\overline{\varepsilon} > 0$, we see that (6.48) also holds with $\underline{\lambda} := \underline{\kappa} - 2\overline{\varepsilon}$. Finally, we have

$$
0 < \gamma_{ij} \le \frac{md^2}{\underline{m}\,\underline{\varepsilon}} =: \overline{\gamma} < \infty,
$$

showing that (6.51) is satisfied by Lemma 6.39, and

$$
\gamma_{iu} \le \frac{mb^2}{\underline{\varepsilon}} =: \overline{\gamma}_u \quad forall \quad i \in \mathbb{N},
$$

which implies (6.49). Clearly, the infinite matrix $\Psi := \Lambda^{-1}\Gamma$, for Λ and Γ as in (6.50), has the same form as the one in (6.59). In that way, with the same arguments as those in Sect. 6.7.1, one can conclude that any choice of the numbers ε_i such that $\frac{md^2}{(\kappa-2\varepsilon)m\,\varepsilon} < \frac{1}{2}$ for all $i \in \mathbb{N}$ leads to $r(\Psi) < 1$.

Hence, by Theorem 6.42, there exists $\mu = (\mu_i)_{i\in\mathbb{N}} \in \ell^\infty$ satisfying $0 < \underline{\mu} \leq \mu_i \leq \overline{\mu} < \infty$ with constants $\underline{\mu}, \overline{\mu}$ such that function $V(x) = \sum_{i=1}^\infty \mu_i x_i^T M_i x_i$ is an eISS Lyapunov function for the interconnected system Σ.

Remark 6.48 In this example, we have used the Euclidean norm for the space $\mathbb{R}^{N_{\neq i}}$ of $x_{\neq i}$-vectors. However, one could utilize in computations also more specific norms, based on our choice of Lyapunov functions V_i.

Consider the following estimate, obtained above:

$$2x_i^T M_i D_i x_{\neq i} \leq \varepsilon_i x_i^T M_i x_i + \frac{|\sqrt{M_i} D_i x_{\neq i}|^2}{\varepsilon_i}. \tag{6.65}$$

As M_i is positive definite, the map $x_i \mapsto \sqrt{V_i(x_i)} = |\sqrt{M_i} x_i|$ is a norm on $\mathbb{R}^{N_i}$. Thus, the map

$$x_{\neq i} \mapsto |||x_{\neq i}||| := \sqrt{V_{i-1}(x_{i-1}) + V_{i+1}(x_{i+1})}$$

is a norm on $\mathbb{R}^{N_{\neq i}}$.

Now we can continue the estimate (6.63) as follows:

$$2x_i^T M_i D_i x_{\neq i} \leq \varepsilon_i x_i^T M_i x_i + \frac{\|\sqrt{M_i} D_i\|^2_{|||\cdot|||,2} |||x_{\neq i}|||^2}{\varepsilon_i},$$

where the double index on the left-hand side indicates that we consider the operator norm induced by the norm $|||\cdot|||$ on the domain and by Euclidean norm on the codomain of the corresponding operator.

Therefore, we have

$$\begin{aligned}\nabla V_i(x_i) \cdot f_i(x_i, x_{\neq i}, u_i) \leq -(\kappa_i - 2\varepsilon_i) x_i^T M_i x_i + \frac{\|\sqrt{M_i} B_i\|^2 |u_i|^2}{\varepsilon_i} \\ + \frac{\|\sqrt{M_i} D_i\|^2_{|||\cdot|||,2}}{\varepsilon_i} (V_{i-1}(x_{i-1}) + V_{i+1}(x_{i+1})).\end{aligned}$$

and again the function $V_i(x_i) = x_i^T M_i x_i$ is an eISS Lyapunov function for the subsystem Σ_i satisfying (6.45) and (6.46) with the internal gains

$$\gamma_{ij} := \frac{\|\sqrt{M_i} D_i\|^2_{|||\cdot|||,2}}{\varepsilon_i}, \quad j \in \{i-1, i+1\},$$

and other parameters same as above.

6.7.3 A Road Traffic Model

In this example, we apply our approach to a variant of the road traffic model from [8]. We consider a traffic network divided into infinitely many cells, indexed by $i \in \mathbb{N}$. Each cell i represents a subsystem Σ_i described by a differential equation of the following form

$$\Sigma_i : \dot{x}_i = -\Big(\frac{v_i}{l_i} + e_i\Big)x_i + D_i\bar{x}_i + B_i u_i, \quad x_i, u_i \in \mathbb{R}, \tag{6.66}$$

with the following structure

- $e_i = 0$, $D_i = c\frac{v_{i+1}}{l_{i+1}}$, $\bar{x}_i = x_{i+1}$, $B_i = 0$ if $i \in S_1 := \{1\}$;
- $e_i = 0$, $D_i = c\frac{v_{i+4}}{l_{i+4}}$, $\bar{x}_i = x_{i+4}$, $B_i = r > 0$ if $i \in S_2 := \{4 + 8k : k \in \mathbb{N} \cup \{0\}\}$;
- $e_i = 0$, $D_i = c\frac{v_{i-4}}{l_{i-4}}$, $\bar{x}_i = x_{i-4}$, $B_i = \frac{r}{2}$ if $i \in S_3 := \{5 + 8k : k \in \mathbb{N} \cup \{0\}\}$;
- $e_i = 0$, $D_i = c(\frac{v_{i-1}}{l_{i-1}}, \frac{v_{i+4}}{l_{i+4}})^T$, $\bar{x}_i = (x_{i-1}, x_{i+4})$, $B_i = 0$ if $i \in S_4 := \{6 + 8k : k \in \mathbb{N} \cup \{0\}\}$;
- $e_i = e \in (0, 1)$, $D_i = c(\frac{v_{i-4}}{l_{i-4}}, \frac{v_{i+1}}{l_{i+1}})^T$, $\bar{x}_i = (x_{i-4}, x_{i+1})$, $B_i = 0$ if $i \in S_5 := \{9 + 8k : k \in \mathbb{N} \cup \{0\}\}$;
- $e_i = 0$, $D_i = c(\frac{v_{i+1}}{l_{i+1}}, \frac{v_{i+4}}{l_{i+4}})^T$, $\bar{x}_i = (x_{i+1}, x_{i+4})$, $B_i = 0$ if $i \in S_6 := \{2 + 8k : k \in \mathbb{N} \cup \{0\}\}$;
- $e_i = 0$, $D_i = c(\frac{v_{i-4}}{l_{i-4}}, \frac{v_{i-1}}{l_{i-1}})^T$, $\bar{x}_i = (x_{i-4}, x_{i-1})$, $B_i = 0$ if $i \in S_7 := \{7 + 8k : k \in \mathbb{N} \cup \{0\}\}$;
- $e_i = 2e$, $D_i = c(\frac{v_{i-1}}{l_{i-1}}, \frac{v_{i+4}}{l_{i+4}})^T$, $\bar{x}_i = (x_{i-1}, x_{i+4})$, $B_i = 0$ if $i \in S_8 := \{8 + 8k : k \in \mathbb{N} \cup \{0\}\}$;
- $e_i = 0$, $D_i = c(\frac{v_{i-4}}{l_{i-4}}, \frac{v_{i+1}}{l_{i+1}})^T$, $\bar{x}_i = (x_{i-4}, x_{i+1})$, $B_i = 0$ if $i \in S_9 := \{11 + 8k : k \in \mathbb{N} \cup \{0\}\}$;
- $e_i = 0$, $D_i = c\frac{v_{i+1}}{l_{i+1}}$, $\bar{x}_i = x_{i+1}$, $B_i = r/2$ if $i \in S_{10} := \{3\}$;

where for all $i \in \mathbb{N}$, $0 < \underline{v} \le v_i \le \overline{v}$, $0 < \underline{l} \le l_i \le \overline{l}$, and $c \in (0, 0.5)$. In (6.66), l_i is the length of a cell in kilometers (km), and v_i is the flow speed of the vehicles in kilometers per hour (km/h). The state of each subsystem Σ_i, i.e., x_i, is the traffic density, given in vehicles per cell, for each cell i of the road. The scalars B_i represent the number of vehicles that can enter the cells through entries controlled by u_i. In particular, $u_i = 1$ means green light and $u_i = 0$ means red light. Moreover, the constants e_i represent the percentage of vehicles that leave the cells using available exits. The overall system and subsystems are illustrated by Fig. 6.1.

Clearly, the interconnected system Σ with state space $X := \ell^2(\mathbb{N}, (n_i))$ and input space $U := \ell^2(\mathbb{N}, (m_i))$ is well posed (cf. Example 6.11). The choice $U := \ell^2(\mathbb{N}, (m_i))$ means that only finitely many traffic lights can be green at each $t \ge 0$.

Furthermore, each subsystem Σ_i admits an eISS Lyapunov function of the form

$$V_i(x_i) = \frac{1}{2}x_i^2, \quad x_i \in \mathbb{R}.$$

The function V_i satisfies (6.45) and (6.46) for all $i \in \mathbb{N}$ with

$$\underline{\alpha}_i = \overline{\alpha}_i = \frac{1}{2}, \quad \lambda_i = 2(\frac{v_i}{l_i} + e_i - 2\varepsilon_i),$$

$$\gamma_{ij} = \frac{\|D_i\|^2}{2\varepsilon_i} \quad \forall j \in I_i, \quad \gamma_{iu} = \frac{B_i^2}{2\varepsilon_i},$$

for an appropriate choice of $0 < \underline{\varepsilon} \le \varepsilon_i \le \overline{\varepsilon}$ such that $0 < \underline{\lambda} := 2\big(\underline{v}/\overline{l} - 2\overline{\varepsilon}\big) \le \lambda_i$. In that way, one can readily observe that

$$0 < \gamma_{ij} \le \frac{(c\overline{v})^2}{\underline{\varepsilon}\,\underline{l}^2} =: \overline{\gamma} < \infty, \quad 0 < \gamma_{iu} \le \frac{r^2}{2\underline{\varepsilon}} =: \overline{\gamma}_u < \infty.$$

Additionally, the infinite matrix $\Psi := \Lambda^{-1}\Gamma = (\psi_{ij})_{i,j\in\mathbb{N}} = (\gamma_{ij}/\lambda_i)_{i,j\in\mathbb{N}}$, for Λ and Γ defined in (6.50), has the following structure.

- $i \in S_1 \Rightarrow (\gamma_{ij} \neq 0 \Leftrightarrow j = i+1)$;
- $i \in S_2 \Rightarrow (\gamma_{ij} \neq 0 \Leftrightarrow j = i+4)$;
- $i \in S_3 \Rightarrow (\gamma_{ij} \neq 0 \Leftrightarrow j = i-4)$;
- $i \in S_4 \Rightarrow (\gamma_{ij} \neq 0 \Leftrightarrow j \in \{i-1, i+4\})$;
- $i \in S_5 \Rightarrow (\gamma_{ij} \neq 0 \Leftrightarrow j \in \{i-4, i+1\})$;
- $i \in S_6 \Rightarrow (\gamma_{ij} \neq 0 \Leftrightarrow j \in \{i+1, i+4\})$;
- $i \in S_7 \Rightarrow (\gamma_{ij} \neq 0 \Leftrightarrow j \in \{i-4, i-1\})$;
- $i \in S_8 \Rightarrow (\gamma_{ij} \neq 0 \Leftrightarrow j \in \{i-1, i+4\})$;
- $i \in S_9 \Rightarrow (\gamma_{ij} \neq 0 \Leftrightarrow j \in \{i-4, i+1\})$;
- $i \in S_{10} \Rightarrow (\gamma_{ij} \neq 0 \Leftrightarrow j \in \{i+1\})$.

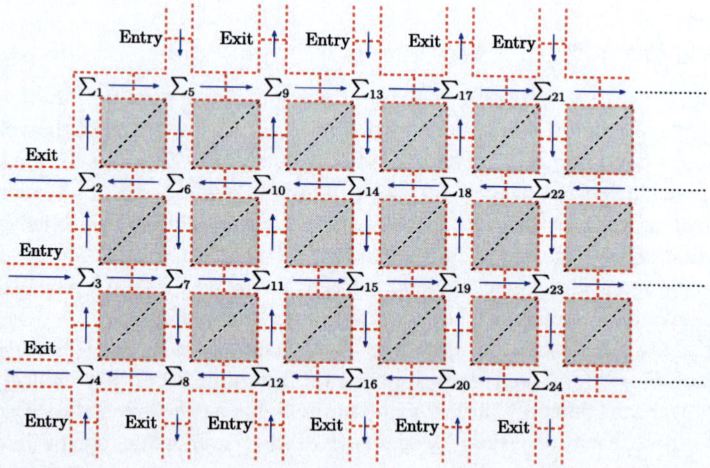

Fig. 6.1 Model of a road traffic network composed of infinitely many subsystems. (This figure is © 2021 IEEE reprinted, with permission, from "Ch. Kawan, A. Mironchenko, A. Swikir, N. Noroozi, M. Zamani. A Lyapunov-based small-gain theorem for infinite networks. IEEE Transactions on Automatic Control, 66(12):5830–5844, 2021")

The spectral radius $r(\Psi)$ can be estimated by

$$r(\Psi) \leq \|\Psi\| = \sup_{j\in\mathbb{N}} \sum_{i=1}^{\infty} \psi_{ij} \leq 2\frac{\overline{\gamma}}{\underline{\lambda}}.$$

Hence, any choice of the constants ε_i such that

$$(2(c\overline{v})^2/\underline{\varepsilon}\,\underline{l}^2)/((\underline{v}/\overline{l}) - 2\overline{\varepsilon}) < 1,$$

for all $i \in \mathbb{N}$ leads to $r(\Psi) < 1$.

Hence, by Theorem 6.42, there exists $\mu = (\mu_i)_{i\in\mathbb{N}} \in \ell^\infty$ satisfying $\underline{\mu} \leq \mu_i \leq \overline{\mu}$ with constants $\underline{\mu}, \overline{\mu} > 0$ such that the function

$$V(x) = \frac{1}{2}\sum_{i=1}^{\infty} \mu_i x_i^2$$

is an eISS Lyapunov function for the interconnected system Σ.

6.8 Concluding Remarks

This chapter is based mostly on the papers [16, 19, 22, 23]. An overview of the available literature, historical remarks, extensions, and discussions a reader may find in Section 7.1. In Sects. 6.6, 6.7 we follow [16] very closely. Lemma 6.40 is due to [12, Proposition 3.10].

6.9 Exercises

Exercise 6.1 Consider a system governed by equations (6.11) with a state space

$$\ell_\infty := \{x = (x_k) : \exists M > 0 : \; x_k \in \mathbb{R} \wedge |x_k| \leq M \; \forall k \in \mathbb{N}\}.$$

What stability properties does this system enjoy?

References

1. Appell J, De Pascale E, Vignoli A (2008) Nonlinear spectral theory. Walter de Gruyter
2. Arendt W, Batty CJ, Hieber M, Neubrander F (2011) Vector-valued Laplace transforms and Cauchy problems. Springer Science & Business Media

3. Aulbach B, Wanner T (1996) Integral manifolds for Carathéodory type differential equations in Banach spaces. In: Six lectures on dynamical systems. World Scientific, River Edge, pp 45–119
4. Bernstein D (2009) Matrix mathematics: theory, facts, and formulas. Princeton University Press
5. Cazenave T, Haraux A (1998) An introduction to semilinear evolution equations. Oxford University Press, New York
6. Dashkovskiy S, Ito H, Wirth F (2011) On a small gain theorem for ISS networks in dissipative Lyapunov form. Eur J Control 17(4):357–365
7. Dashkovskiy S, Mironchenko A (2013) Input-to-state stability of infinite-dimensional control systems. Math Control Signals Syst 25(1):1–35
8. de Wit CC, Ojeda L, Kibangou A (2012) Graph constrained-CTM observer design for the Grenoble south ring. In: Proceedings of 13th IFAC Symposium on Control in Transportation Systems, pp 197–202
9. Derinkuyu K, Pınar MÇ (2006) On the S-procedure and some variants. Math Methods Oper Res 64(1):55–77
10. Dunford N, Schwartz JT (1957) Linear operators. Part I: general theory. New York, Interscience
11. Fetzer M (2017) From classical absolute stability tests towards a comprehensive robustness analysis. PhD thesis, Institute of Mathematical Methods in Engineering, Numerical Analysis and Geometric Modeling, University of Stuttgart
12. Glück J, Mironchenko A (2021) Stability criteria for positive linear discrete-time systems. Positivity 25(5):2029–2059
13. Hagood JW, Thomson BS (2006) Recovering a function from a Dini derivative. Am Math Mon 113(1):34–46
14. Hille E, Phillips RS (1974) Functional analysis and semi-groups. Am Math Soc, Providence, R. I
15. Karafyllis I, Jiang Z-P (2011) Stability and stabilization of nonlinear systems. Springer, London
16. Kawan C, Mironchenko A, Swikir A, Noroozi N, Zamani M (2021) A Lyapunov-based small-gain theorem for infinite networks. IEEE Trans Autom Control 66(12):5830–5844
17. Kawan C, Mironchenko A, Zamani M (2022) A Lyapunov-based ISS small-gain theorem for infinite networks of nonlinear systems. IEEE Trans Autom Control https://arxiv.org/abs/2103.07439
18. Krein MG, Rutman MA (1948) Linear operators leaving invariant a cone in a Banach space. Usp Matematicheskikh Nauk 3(1):3–95
19. Mironchenko A (2016) Local input-to-state stability: characterizations and counterexamples. Syst Control Lett 87(23–28)
20. Mironchenko A, Ito H (2016) Characterizations of integral input-to-state stability for bilinear systems in infinite dimensions. Math Control Relat Fields 6(3):447–466
21. Mironchenko A, Prieur C (2020) Input-to-state stability of infinite-dimensional systems: recent results and open questions. SIAM Rev 62(3):529–614
22. Mironchenko A, Wirth F (2018) Characterizations of input-to-state stability for infinite-dimensional systems. IEEE Trans Autom Control 63(6):1602–1617
23. Mironchenko A, Wirth F (2018) Lyapunov characterization of input-to-state stability for semi-linear control systems over Banach spaces. Syst Control Lett 119:64–70
24. Saks S (1947) Theory of the integral. Courier Corporation
25. Yakubovich VA (1977) S-procedure in nonlinear control theory. Vestnik Leningrad University. Mathematics 4:73–93

Chapter 7
Conclusion and Outlook

In this book, we have developed the ISS theory for systems of ordinary differential equations, covering both fundamental theoretical results and central applications of the ISS theory to nonlinear control and network analysis. Furthermore, we have investigated several properties closely related to ISS, most notably integral ISS, and discussed recent applications of ISS methodology to stability analysis of infinite networks.

Certainly, many important topics remained outside of the scope of this book. This final chapter discusses some of them in a survey-like manner. In Sect. 7.1, we provide a short overview of the input-to-state stability theory of infinite-dimensional systems with a stress on infinite networks. In Sect. 7.2, we briefly mention the main results in the ISS theory of time-varying, discrete-time, impulsive, and hybrid systems. Finally, in Sect. 7.3, we succinctly introduce further ISS-based stability notions and give some references to the corresponding literature.

We hope that together with the knowledge of the ISS theory of ODEs, this will help a reader to get an orientation in the vast and rapidly expanding literature on ISS theory.

7.1 Brief Overview of Infinite-Dimensional ISS Theory

The prominence and richness of the ISS theory of finite-dimensional systems, as well as the importance of modern applications of control theory to the systems governed by partial differential equations (PDEs), delay systems, PDE-ODE cascades, infinite networks, etc., gave an impetus to the development of the input-to-state stability for general infinite-dimensional systems.

In Chap. 6, we have shown several basic results in the infinite-dimensional ISS theory, valid for a broad class of systems introduced in Definition 6.1. In a certain sense, Definition 6.1 is a direct generalization, and a unification of the concepts of strongly continuous nonlinear semigroups (see [130]) with abstract linear control systems considered, e.g., in [143]. Such an abstract definition of a control system has

A. Mironchenko, *Input-to-State Stability*, Communications and Control Engineering,
https://doi.org/10.1007/978-3-031-14674-9_7

been frequently used within the system-theoretic community at least since the 1960s [59, 144]. In a similar spirit, even more general system classes can be considered, containing output maps, a possibility for a solution to jump at certain time instants, etc., see [61, 64].

Furthermore, in Chap. 6, we presented several counterexamples illustrating that finite-dimensional intuition may be wrong in the infinite-dimensional world. Here we put these results into a broader perspective. For a comprehensive and up-to-date survey of the linear and nonlinear infinite-dimensional ISS theory, we refer to [102]. A reader may also consult [69] for an overview of the so-called spectral method and [128] for a survey of results in the linear infinite-dimensional ISS theory.

7.1.1 Fundamental Properties of ISS Systems: General Systems

Direct ISS Lyapunov theorem (Theorem 6.26) has appeared in the context of general control systems (introduced in Definition 6.1) in [28, Theorem 1]. However, similar results have also been used previously in more particular contexts. The ISS superposition theorem (Theorem 6.21) has been shown in [103, Theorem 5] and slightly improved in [95]. Counterexamples considered in Sect. 6.4 are mostly due to [93, 103].

The construction of ISS Lyapunov functions is challenging, especially for systems with boundary inputs. To overcome this challenge, less restrictive noncoercive ISS Lyapunov functions have been proposed in [103], and it was shown that for a general class of control systems (see Definition 6.1) introduced in this chapter, the existence of a noncoercive ISS Lyapunov function for systems with BRS and with a CEP property implies ISS, see [48]. UGAS property for infinite-dimensional nonlinear systems has been analyzed by means of noncoercive Lyapunov functions in [105, 106].

Further progress has been achieved for infinite-dimensional evolution equations of the form

$$\dot{x}(t) = Ax(t) + f(x(t), u(t)), \tag{7.1}$$

where $A : D(A) \subset X \to X$ generates a strongly continuous semigroup (also called C_0-semigroup) $T(\cdot)$ of bounded linear operators, X is a Banach space, U is a normed vector space of input values, $f : X \times U \to X$, and the space of admissible inputs $\mathcal{U} := PC_b(\mathbb{R}_+, U)$ of globally bounded, piecewise continuous functions from $\mathbb{R}_+$ to U that are right-continuous.

This system class includes the infinite networks of ODEs studied in Chap. 6. We refer to excellent books [24, 51, 116] for an overview of the semigroup theory and of the theory of evolution equations.

Equivalence between LISS of (7.1) and uniform asymptotic stability of the undisturbed system (7.1), which extends Theorem 2.37, has been shown in [93]. In the same paper, some problems on the way to the characterization of the ISS property have been revealed. Converse ISS Lyapunov theorems for (7.1) showing equivalence of the ISS property to the existence of an ISS Lyapunov function are proved in [104].

7.1.2 ISS of Linear and Bilinear Boundary Control Systems

Consider a linear infinite-dimensional system

$$\dot{x}(t) = Ax(t) + Bu(t), \tag{7.2}$$

where A is a generator of a strongly continuous semigroup over a Banach space X and $B : U \to X$ is an unbounded input operator. ISS theory of (7.2) including the criterion of ISS in terms of the stability properties of the underlying semigroup and admissibility properties of the input operator has been developed in [49, 50]. Further in [49] the relations between ISS and integral ISS of linear infinite-dimensional systems have been studied. The results on the construction of ISS Lyapunov functions for linear systems with unbounded operators are available only for quite special classes of control systems, see [48].

Alternatively, in [66–68] the ISS analysis of linear parabolic PDEs with Sturm–Liouville operators over a 1-dimensional spatial domain has been performed using the spectral decomposition of the solution, and the approximation of the solution by means of a finite-difference scheme. This method was further developed in the monograph [69], where also the applications of the method to control design have been analyzed.

Integral ISS of infinite-dimensional bilinear systems with distributed input operators has been studied in [97]. ISS of bilinear systems with unbounded input operators has been analyzed in [47], and the results have been applied to the controlled Fokker–Planck equation. Nevertheless, the iISS theory of bilinear systems remains much less developed than the ISS theory of linear systems with admissible input operators.

7.1.3 Lyapunov Methods for ISS Analysis of PDE Systems

The construction of an ISS Lyapunov function is, as a rule, the most efficient and realistic method to prove the input-to-state stability of a nonlinear system. Although direct coercive ISS Lyapunov theorems for infinite-dimensional systems are analogous to finite-dimensional ones, the verification of the dissipation inequality (6.43) even for simple Lyapunov functions requires further tools, such as inequalities for functions from L^p and Sobolev spaces (Friedrichs, Poincare, Agmon, Jensen inequalities, and their relatives), linear matrix inequalities (LMIs), etc. When using the Lyapunov method for designing Lyapunov functions, quadratic Lyapunov candidates are often introduced to prove ISS properties, see in particular [10, 69] where various PDEs are considered.

In [92], the ISS property of a class of semilinear parabolic equations has been studied with the help of ISS Lyapunov functions. In [122, 137], Lyapunov methods are applied for ISS analysis and robust boundary control of hyperbolic systems of the first order. In [64] a general vector Lyapunov small-gain theorem for abstract control systems satisfying a weak semigroup property has been proved. This framework formed the basis for a study of the stability and stabilizability of nonlinear systems in [63].

7.1.4 Small-Gain Theorems for the Stability of Networks

Generalizations of Lyapunov-based small-gain theorems for ODE systems (Theorem 3.28) to networks of n infinite-dimensional systems were provided in [28] in the context of evolution equations in Banach spaces with in-domain couplings but can be generalized to general control systems with an arbitrary type of couplings, see [102, Sect. 6.3]. Lyapunov-based small-gain theorems for interconnections of two integral ISS systems have been developed in [96]. On the basis of [28], ISS small-gain theorems for impulsive infinite-dimensional systems has been developed in [29]. Trajectory-based small-gain theorems have been generalized to feedback interconnections of infinite-dimensional systems in [95].

7.1.5 Infinite Networks

Recent advances in large-scale computing, cheap distributed sensing, and large-scale data management have created the potential for smart applications in which large numbers of dispersed agents need to be regulated for a common objective. For instance, in smart cities, city-wide traffic control based on cheap personal communication, car-to-car communication, and the deployment of numerous sensors can provide a significant step toward safe, energy-efficient, and environmentally friendly traffic concepts. The vision of safe and efficient control of such large, dispersed systems requires tools that are capable of handling an uncertain and time-varying number of participating agents, limited communication, as well as stringent safety specifications.

Standard tools in the literature on stability analysis and control do not scale well to such increasingly common smart networked systems. Networks designed using classic tools may lead to *fragile* systems, where stability and performance indices of the system do depend on the size of the systems in a way that the network tends to instability as the size of the network grows; cf., e.g., [11, 30, 58, 127]. A prolific approach to address this fragility is to overapproximate a finite-but-large network with an *infinite network* consisting of countably many subsystems [6, 7, 9], and analyze this infinite-dimensional approximation.

A prominent place in analysis of infinite networks is occupied by the theory of linear *spatially invariant systems* [7, 8, 11, 23]. In such networks, infinitely many subsystems are coupled via the same pattern. This nice geometrical structure, together with the *linearity* of subsystems, allows developing of powerful criteria for the stability of infinite interconnections. Furthermore, infinite networks naturally appear in vehicle platoon formation [11], ensemble control [84], in representations of the solutions of linear and nonlinear partial differential equations over Hilbert spaces in terms of series expansions with respect to orthonormal or Riesz bases, see, e.g., [83].

Stability analysis of infinite networks was developed in parallel with the ISS theory and was restricted mostly to linear systems. Only very recently, starting with the papers [30, 31], ISS small-gain methods have been applied to infinite networks, in particular with the stress on the development of the theory that would allow the analysis of infinite networks composed of nonlinear systems.

Networks with linear gains. A Lyapunov-based small-gain theorem for exponential input-to-state stability of infinite networks with a linear gain operator, developed in Sect. 6.6, has been shown in [72]. In [113], this result has been extended to networks of systems that are exponentially ISS with respect to closed sets. In this form, it was applied to the stability analysis of infinite time-variant networks and to the design of distributed observers for infinite networks. The main result in [72] deals with infinite networks over ℓ_p-space with a finite p. A quite different Lyapunov-based small-gain theorem for infinite ODE networks with a homogeneous and subadditive gain operator over the state space ℓ_∞ is developed in [101]. Small-gain theorems for infinite networks with linear gains in the *trajectory formulation* have been developed in [99]. In all these papers, the stability of the network is ensured, provided the *spectral small-gain condition* holds, i.e., the spectral radius of the gain operator is less than 1, which is the same as the exponential stability of the discrete-time system, induced by the gain operator. Exhaustive characterizations of the spectral small-gain condition for linear monotone operators in ordered Banach spaces have been derived in [39] (see Proposition C.11 for such criteria for linear monotone operators in $\mathbb{R}^n_+$). Some of these criteria have been extended to a larger class of monotone homogeneous subadditive operators in [101]. The finite-dimensional counterparts of this result are discussed in Sect. C.9.

Lyapunov-based small-gain theorems for networks with nonlinear gains. Small-gain analysis of infinite networks with nonlinear gains was initiated in [30, 31]. In the early work [31], ISS was shown for an infinite network of ISS systems with nonlinear gains, provided that the internal gains capturing the influence of subsystems on each other are all uniformly less than identity, which is a rather conservative condition.

Infinite networks with nonlinear gains and sup-preserving gain operators have been considered in [30]. For such systems, the *robust strong small-gain condition* has been introduced and a method to construct paths of strict decay for the gain operators was proposed, based on the concept of the *strong transitive closure* of the gain operator (see Sect. C.8 of this book for the corresponding technique for finite networks). Based on these results, in [30] a small-gain theorem for infinite networks and the construction of an ISS Lyapunov function for the network were proposed under the assumption that there exists a *linear* path of strict decay for the gain operator Γ. However, the conditions under which such a path exists were not shown in [30].

The first fully nonlinear Lyapunov-based ISS small-gain theorem for networks with nonlinear gains and sup-preserving gain operators has been developed in [73].

In this work, it was shown that *if there exists a (possibly nonlinear) path of strict decay for the operator* Γ *(and some uniformity conditions hold), then the whole network is ISS, and a corresponding ISS Lyapunov function for the network can be constructed*. This result partially extends to the infinite-dimensional setting the nonlinear Lyapunov-based small-gain theorem for finite networks (in maximum formulation) shown in [32], recovers the Lyapunov-based small-gain theorem for infinite networks in [30], and partially recovers the main result in [31], see [73] for the detailed discussion.

Furthermore, the authors study in [73], which properties of the gain operator guarantee the existence of a path of strict decay. By refining the technique for the construction of paths of strict decay introduced in [30], it is shown in [73] that *the robust small-gain condition of the gain operator together with the global attractivity of the dynamical system, induced by the gain operator, guarantee the existence of a path of strict decay for the gain operator*, and thus the existence of an ISS Lyapunov function for the interconnection.

Finally, in [100], an ISS small-gain theorem for systems with linear gains and homogeneous gain operators (including linear and max-linear operators) has been shown, which is firmly based on the nonlinear small-gain theorems in [73]. For homogeneous gain operators, one can introduce the concept of spectral radius via a definition analogous to the well-known Gelfand's formula for the spectral radius r of linear operators. Furthermore, in this case, the existence of a path of strict decay for the gain operator is equivalent to the existence of a point of strict decay, which in turn is equivalent to the spectral small-gain condition.

Trajectory-based small-gain theorems for networks with nonlinear gains. In [99] the trajectory-based small-gain theorems for infinite networks consisting of nonlinear infinite-dimensional systems have been derived. These results show that if all the subsystems are ISS with a uniform $\mathcal{KL}$-transient bound and with the gain operator satisfying the so-called monotone limit property, then the network is ISS. Furthermore, in [99] it was shown that a network consisting of subsystems that are merely uniformly globally stable (UGS) is again UGS provided that the gain operator satisfies a uniform small-gain condition, which is implied by (and probably is weaker than) the so-called monotone limit property.

Distributed observers. The applications of the small-gain theorems for the distributed control of finite and infinite networks are manifold. In particular, there is a vast literature dealing with the distributed estimators, as distributed observers and distributed Kalman filters [75, 114, 115, 142]. Most of the literature deals, however, with finite networks. Based on the small-gain theorem developed in Sect. 6.6, it is possible to develop a framework for the construction of distributed observers for infinite networks, see [112].

Open problems. Nonlinear Lyapunov-based ISS small-gain theorems in [30, 31, 73] have been developed for the case of the maximum formulation of the ISS property for the subsystems and for the sup-preserving gain operators. For semimaximum, summation, and other formulations of the ISS property, further investigations are needed. The solution to this problem for networks with homogenous gains in [100] can be understood as the first step in this direction.

Another *important open problem concerns the relationships between the existence of a path of strict decay and other properties of the gain operator, such as robust strong and uniform small-gain conditions, the monotone limit property, uniform global asymptotic stability of the induced discrete-time system*, etc. This would provide a better understanding of nonlinear ISS small-gain theorems and bring powerful small-gain-type criteria for uniform global asymptotic stability of nonlinear discrete-time systems induced by positive operators. For finite-dimensional nonlinear monotone operators, we study such relationships in Appendix C. For linear infinite-dimensional monotone discrete-time systems, see [39].

7.1.6 ISS of Time-Delay Systems

For time-delay systems, a particular class of infinite-dimensional systems, an extensive ISS theory is available. For this class of systems, two distinct sufficient conditions of Lyapunov-type have been proposed: using ISS Lyapunov–Razumikhin functions [138] and in terms of ISS Lyapunov–Krasovskii functionals [118]. While in [118, 138] time-delay systems with a constant delay have been considered, in [38] the method of Lyapunov–Krasovskii functionals has been applied in combination with the technique of nonlinear matrix inequalities to study ISS of delay systems with a time-varying delay.

In [38, 118], ISS of delay systems was achieved under the assumption that ISS Lyapunov–Krasovskii functionals have a decay rate in terms of the norm of the state or in terms of the Lyapunov–Krasovskii functional itself. The question of whether ISS can be inferred from the existence of an ISS Lyapunov–Krasovskii functional with a *pointwise decay rate* is still an open problem. See [21, 60] for partial solutions in the ISS case, and [20] for the positive answer to this problem for the iISS property.

For converse Lyapunov theorems for input-to-state stability (or, more generally, for input-to-output stability) of time-delay systems in terms of Lyapunov–Krasovskii functionals, see [70].

In [64], a general Lyapunov-based small-gain theorem for abstract systems has been proved, and small-gain results for finite-dimensional and time-delay systems have been provided. General small-gain theorems for abstract systems in a trajectory formulation, shown in [95], are valid in particular for time-delay systems. Small-gain theorems for time-delay systems in a Lyapunov–Krasovskii and Lyapunov–Razumikhin form have been shown in [27], with extensions to the impulsive time-delay systems. To simplify verification of ISS for delay systems, Lyapunov methods are frequently combined with so-called Halanay's inequalities, see [117] and references therein.

We refer to [102, Sect. 7] for a more detailed introduction to ISS of time-delay systems.

7.1.7 Applications

Although the ISS theory for infinite-dimensional systems is a relatively young subject, it has been used in numerous applications. Here we briefly mention some of them. ISS methodology has been applied to stabilize the liquid–solid interface position at the desired position for the one-phase Stefan problem without the heat loss [76–78]. Lyapunov methods have been applied to control of safety factor profile in tokamaks ensuring input-to-state stability against external perturbations [13–16]. Monotonicity methods have been applied to input-to-state stabilization of linear systems with boundary inputs and actuator disturbances in [98]. In [119] the variable structure control approach has been exploited to design discontinuous feedback control laws (with pointwise sensing and actuation) for input-to-state stabilization of linear reaction–diffusion–advection equations w.r.t. actuator disturbances. In [64] vector small-gain theorems for wide classes of systems satisfying the weak semigroup property have been proved and applied to the stabilization of the chemostat in [65]. Finally, let us mention the works dealing with event-triggering control for both hyperbolic PDEs [35–37] and parabolic PDEs [129], where Lyapunov methods have been used for the design of ISS-based triggering controls. In [33], the dynamics of flexible structures with a Kelvin–Voigt damping is studied, and a Lyapunov function approach is employed. The main contribution is the proof of an ISS property of mild solutions, where the external input is defined as the force exerted on the railway track by moving trains, active dampers, or other external forces.

7.2 Input-to-State Stability of Other Classes of Systems

Having given an overview of the infinite-dimensional ISS theory, we proceed to a very brief survey of key results in ISS of finite-dimensional, time-varying, discrete-time, impulsive, and hybrid systems.

7.2.1 Time-Varying Systems

Dynamics of many natural phenomena is described by time-varying ODEs:

$$\dot{x}(t) = f(t, x(t), u(t)), \quad t > t_0, \tag{7.3a}$$
$$x(t_0) = x_0. \tag{7.3b}$$

We assume that for each initial time t_0, each initial condition $x_0 \in \mathbb{R}^n$ and each input $u \in \mathcal{U} := L^\infty(\mathbb{R}, \mathbb{R}^m)$ the corresponding maximal solution of the above system exists on $[t_0, +\infty)$ and is unique. We denote this maximal solution at time $t \geq t_0$ by $\phi(t, t_0, x_0, u)$.

There are several possibilities to extend the notion of ISS from time-invariant systems (2.1) to time-varying systems (7.3). An important role is played by the following one: (7.3) is called *uniformly input-to-state stable (UISS)*, if there exist $\beta \in \mathcal{KL}$ and $\gamma \in \mathcal{K}$, such that the inequality

$$|\phi(t, t_0, x, u)| \le \beta(|x|, t - t_0) + \gamma(\|u\|_\infty) \tag{7.4}$$

holds for all $x \in \mathbb{R}^n$, $t_0 \ge 0$, $t \ge t_0$ and all $u \in \mathcal{U}$.

In this definition, the uniformity indicates that the functions β and γ do not depend on the initial time t_0. This and other related notions have been studied in [34, 71, 89] and other papers. While some results can be easily transferred from the time-invariant to the time-varying case, the other ones do not hold anymore, hence some care is needed. In particular, in [34, p. 3502], the following time-varying scalar system has been considered:

$$\dot{x} = -x + (1 + t)g(u - |x|),$$

where $g(s) = 0$ for $s \le 0$ and $g(s) = s$ for $s \ge 0$.

This system has a Lipschitz continuous right-hand side, and it is ISS with the ISS Lyapunov function in implication form $V : x \mapsto x^2$. At the same time, V is not an ISS Lyapunov function in a dissipative form. As was shown in [43, Proposition 2.9], this system is also not integrally ISS.

7.2.2 Discrete-Time Systems

A large part of the ISS theory for ODEs can be extended to discrete-time systems of the form

$$x(k + 1) = f(x(k), u(k)), \quad k \in \mathbb{N}, \tag{7.5}$$

where $x(k) \in \mathbb{R}^n$ and $u(k) \in \mathbb{R}^m$ for some $m, n \in \mathbb{N}$ and $f : \mathbb{R}^n \times \mathbb{R}^m \to \mathbb{R}^n$. A major simplification arising in discrete-time systems is that for each initial condition $x(0) := x_0 \in \mathbb{R}^n$, and for each input $u \in \mathcal{U} := \ell_\infty(\mathbb{Z}_+, \mathbb{R}^m)$ the corresponding solution of (7.5) exists and is unique globally for any well-defined map f. As usual, we denote this solution at time $k \in \mathbb{Z}_+$ by $\phi(k, x_0, u)$.

We call a discrete-time system (7.5) ISS, if there exist $\beta \in \mathcal{KL}$ and $\gamma \in \mathcal{K}$, such that for all $k \in \mathbb{N}$, $x_0 \in \mathbb{R}^n$ and $u \in \mathcal{U}$ it holds that

$$|\phi(k, x_0, u)| \le \beta(|x_0|, k) + \gamma(\|u\|_\infty). \tag{7.6}$$

Characterizations of ISS for systems (7.5) in terms of ISS Lyapunov functions have been achieved in [56]. Small-gain theorems for general feedback interconnections of two ISS discrete-time systems were established in [56, 80], and their generalization to networks composed of $n \ge 2$ ISS systems has been reported in [91]. Interestingly, for discrete-time systems (7.5) integral ISS is equivalent to 0-GAS, see [1], which is strikingly different from the case of ODE systems, where for forward complete systems iISS is stronger than 0-GAS, see also [133] for related questions.

ISS theory of discrete-time systems plays a central role in model-predictive control of discrete-time systems, see [87, 123].

7.2.3 *Impulsive Systems*

In the modeling of real phenomena, one often has to consider systems that exhibit both continuous and discontinuous behavior. The first systematic treatment of the dynamical systems with impulsive effects is given in the monographs [81, 126].

Let $t_0 \in \mathbb{R}$ be a fixed initial time. Take the space of admissible inputs as $\mathcal{U} := L^\infty([t_0, \infty), \mathbb{R}^m)$ and let $T = (t_i)_{i\in\mathbb{N}}$ be a strictly increasing *sequence of impulse times* without finite accumulation points. Consider a system of the form

$$\dot{x}(t) = f(x(t), u(t)), \quad t \in [t_0, \infty)\backslash T, \tag{7.7a}$$

$$x(t) = g(x^-(t), u^-(t)), \quad t \in T, \tag{7.7b}$$

where $x(t) \in \mathbb{R}^n$, $u \in \mathcal{U}$, and $f, g : \mathbb{R}^n \times \mathbb{R}^m \to \mathbb{R}^n$, where f satisfies standard regularity requirements that ensure local existence and uniqueness of solutions, and for which $x^-(t) = \lim_{s\to t-0} x(s)$ exists for all $t \geq t_0$.

Equation (7.7) together with the sequence of impulse times T define an impulsive system. The first equation of (7.7) describes the continuous dynamics of the system, and the second describes the jumps of the state at impulse times. The maximal solution of (7.7) at time t subject to initial time t_0, initial state x_0, input u, and impulsive sequence T we denote by $\phi_T(t, t_0, x_0, u)$ (as before, we assume that such a solution exists globally and is unique).

Let us introduce the stability properties for system (7.7) that we deal with.

Definition 7.1 For a given sequence T of impulse times we call a system (7.7) *input-to-state stable (ISS)* if there exist $\beta \in \mathcal{KL}$, $\gamma \in \mathcal{K}_\infty$, such that for all $x \in \mathbb{R}^n$, $u \in \mathcal{U}$ and all $t \geq t_0$ it holds that

$$|\phi_T(t, t_0, x, u)| \leq \beta(|x|, t - t_0) + \gamma(\|u\|_\infty). \tag{7.8}$$

System (7.7) is called uniformly input-to-state stable over a given set $\mathcal{S}$ of admissible sequences of impulse times if it is ISS for every sequence in $\mathcal{S}$, with β and γ independent of the choice of the sequence from the class $\mathcal{S}$.

Uniform input-to-state stability of impulsive systems has been investigated in, e.g., [44] (impulsive finite-dimensional systems) and [22, 90, 145] (impulsive time-delay systems). In [27] ISS of interconnected impulsive systems with and without time-delays has been investigated. If both continuous and discontinuous dynamics of the system (taken separately from each other) are ISS, then the resulting dynamics of a finite-dimensional impulsive system are also UISS for all impulse time sequences (it is even strongly UISS, see [44, Theorem 2]).

More challenging is a study of systems for which either continuous or discrete dynamics is not ISS. In this case, input-to-state stability is usually achieved under restrictions on the frequency of impulses, such as *dwell-time* [108], *average dwell-time* [45], *generalized average dwell-time* [29], *reverse average dwell-time* [44], and *nonlinear fixed dwell-time conditions* [29, 126].

In [44], it was proved that impulsive systems possessing an *exponential ISS Lyapunov function* are uniformly ISS over impulse time sequences satisfying the *average dwell-time condition*. In [22], a sufficient condition in terms of exponential Lyapunov–Razumikhin functions is provided, which ensures the uniform ISS of impulsive time-delay systems over impulse time sequences satisfying a fixed dwell-time condition. In these results, only exponential ISS Lyapunov functions have been considered. This restrains the class of systems that can be investigated by Lyapunov methods since, to our knowledge, it is not proved (or disproved) that an exponential ISS Lyapunov function always exists. Even if it does exist, the construction of such a function may be a rather sophisticated task.

In [29], *nonexponential ISS Lyapunov functions* have been introduced and exploited for the ISS analysis by means of suitable *nonlinear fixed dwell-time conditions*, which were used for the analysis of systems without inputs in [126]. In [29], the small-gain theorems for couplings of n impulsive systems have been proposed provided all subsystems have the same time of instability (i.e., either the continuous dynamics of all systems is stable or the discrete dynamics of all systems is stable). Furthermore, the results in [29] have been developed not only for systems (7.7) but also for a more general class of systems, including the differential equations in Banach spaces (7.1).

7.2.4 Hybrid Systems

Hybrid systems generalize the impulsive systems in the sense that the time instants at which the jumps occur depend not only on time but also on the state of the system. A general framework for modeling such phenomena is a hybrid systems theory. There are several different approaches to the study of hybrid systems, see [40, 42, 126, 136, 140].

Input-to-state stability theory for hybrid systems has been developed based on the framework in [40], starting with papers [18, 19]. Superposition theorems for ISS of hybrid systems, as well as characterizations of input-to-state stability in terms of ISS Lyapunov functions, have been developed in [17].

Due to their interactive nature, many hybrid systems can be inherently modeled as feedback interconnections [85, Sect. V]. Significant efforts have been devoted to developing small-gain theorems for interconnected hybrid systems. Trajectory-based small-gain theorems for interconnections of two hybrid systems were reported in

[26, 62, 111], while Lyapunov-based formulations were proposed in [85, 86, 110]. Some of these results were extended to networks composed of $n \geq 2$ ISS hybrid systems in [26].

As in the case of impulsive systems, a more challenging problem is the study of hybrid systems in which either continuous or discrete dynamics is destabilizing (non-ISS). Small-gain results for systems whose subsystems are not necessarily ISS due to unstable continuous or discrete dynamics have been reported for couplings of 2 systems in [85], and in a less conservative form and for couplings of an arbitrary finite number of systems, in [107]. Further results in this direction can be found in [25].

7.3 ISS-Like Stability Notions

Here we provide a brief overview of several further ISS-like notions that play an important role in control theory. Still, our list is by far not exhaustive, as there are further interesting notions, e.g., multiple-measure input stability [74], input-to-state safety [79, 124], etc. Moreover, the notions from the following list can be combined with each other. In particular, one can consider incremental IOS, practical finite-time ISS, etc.

Input-output-to-state stability (IOSS) and its relations to the nonlinear detectability were considered in Sect. 5.11.

7.3.1 *Input-to-Output Stability (IOS)*

Consider an ODE system with outputs of the following form

$$\dot{x}(t) = f(x(t), u(t)), \tag{7.9a}$$

$$y(t) = h(x(t), u(t)), \tag{7.9b}$$

where, as in the previous chapters of the book, we assume that $x(t) \in \mathbb{R}^n$, $u(t) \in \mathbb{R}^m$ for certain $n, m \in \mathbb{N}$ and f is sufficiently regular to ensure existence and uniqueness of the (Carathéodory) solutions of (7.9a). Further let $h : \mathbb{R}^n \times \mathbb{R}^m \to \mathbb{R}^p$ for certain $p \in \mathbb{N}$.

For each initial state x and each input u, we denote by $t_m(x, u)$ the maximal existence time for a solution of (7.9a).

Systems with outputs appear in various contexts in systems and control theory. In particular, in many cases, the state of the system cannot be measured directly, and only certain functions of the state and input are available for control purposes. These functions can be treated as outputs, which gives rise to the system with outputs. On the other hand, in many cases, it is not necessary (or is a too strong assumption) to require the stability of a system with respect to the full set of state variables. This calls for the generalization of the input-to-state stability concept to the systems with outputs, which was introduced in [55] and is given next.

Definition 7.2 System (7.9) is called *input-to-output stable (IOS)*, if there exist $\beta \in \mathcal{KL}$ and $\gamma \in \mathcal{K}$ so that for all initial conditions $x \in \mathbb{R}^n$, all inputs $u \in \mathcal{U}$ and all $t \in [0, t_m(x, u))$ the output of (7.9) satisfies the inequality

$$|y(t, x, u)| \leq \beta(|x|, t) + \gamma(\|u\|_\infty). \tag{7.10}$$

In the special case when the output equals the full state of the system, that is, $h(x, u) = x$, IOS reduces to the classical ISS property. Other choices for the output function can be: tracking error, observer error, drifting error from a targeted set, etc. In this case, IOS represents the robust stability of the controller/observer with respect to the given errors. There are several properties that are closely related to IOS as partial stability [141] and stability with respect to two measures [139].

Lyapunov characterizations of input-to-output stability have been shown in [131]. Superposition theorem for IOS property was obtained in [4]. Trajectory-based small-gain theorems for interconnections of 2 IOS systems have been obtained in [55] and generalized to interconnections of n IOS systems in [57]. Lyapunov-based small-gain theorems for couplings of $n \in \mathbb{N}$ interconnected IOS systems have been reported in [125].

Finally, ISS, IOS, and IOSS are related as follows [132, Sect. 9]:

$$\text{(7.9) is ISS} \quad \Leftrightarrow \quad \text{(7.9) is IOS} \ \wedge \ \text{(7.9) is IOSS}.$$

It is possible to introduce "integral" variants of IOSS and IOS properties by mixing IOSS/IOS with integral ISS features. For these integral variations, it is also possible to introduce the corresponding Lyapunov function concepts, see [5, Sect. VI].

7.3.2 Incremental Input-to-State Stability (Incremental ISS)

Incremental input/output-to-state stability (incremental IOSS) has been introduced in [135], where also its relations to detectability and observer design have been analyzed. Incremental ISS is a special case of incremental IOSS (if the output of the system is equal to zero), namely:

Definition 7.3 System (2.1) is called *incrementally input-to-state stable (incrementally ISS)*, if (2.1) is forward complete and there exist $\beta \in \mathcal{KL}$ and $\gamma \in \mathcal{K}$ such that $\forall x_1, x_2 \in \mathbb{R}^n$, $\forall u_1, u_2 \in \mathcal{U}$, and $\forall t \geq 0$ the following holds

$$|\phi(t, x_1, u_1) - \phi(t, x_2, u_2)| \leq \beta(|x_1 - x_2|, t) + \gamma(\|u_1 - u_2\|_\infty). \tag{7.11}$$

If (7.11) holds with $\gamma = 0$, (2.1) is called *incrementally uniformly globally asymptotically stable (incrementally UGAS)*.

If $f(0,0)=0$, incremental ISS trivially implies ISS, by setting $x_2 := 0$ and $u_2 := 0$.

In [2], incremental UGAS has been characterized in terms of Lyapunov functions. This can be done using the converse Lyapunov theorem for UGAS property from [88] (with respect to closed sets), as incremental UGAS of (2.1) is equivalent to UGAS of the auxiliary system

$$\dot{x}_1 = f(x_1, u), \tag{7.12a}$$
$$\dot{x}_2 = f(x_2, u), \tag{7.12b}$$

with respect to the closed set $\{(x_1, x_2) \in \mathbb{R}^n \times \mathbb{R}^n : x_1 = x_2\}$.

Relations of incremental UGAS to semiglobal versions of this property have been investigated in [2], where it was shown that cascades of incremental ISS systems are again incremental ISS, see [2, Proposition 4.7]. In [3] the notion of incremental iISS has been introduced and characterized in terms of Lyapunov functions. Further, it was shown in [3] that incremental integral ISS is at least as strong as incremental ISS. Incremental ISS motivated the notion of incremental forward completeness, which guarantees the existence of so-called symbolic models [146].

7.3.3 Finite-Time Input-to-State Stability (FTISS)

An input-to-state stable system is called *finite-time input-to-state stable (FTISS)*, if the system is ISS, and β in the estimate

$$|\phi(t,x,u)| \le \beta(|x|,t) + \gamma(\|u\|_\infty), \quad t \ge 0,\ x \in \mathbb{R}^n,\ u \in \mathcal{U}, \tag{7.13}$$

can be chosen so that for each $r > 0$ the function $\beta(r,\cdot)$ monotonically decays to zero in finite time. In particular, for $u \equiv 0$, trajectories of FTISS systems reach the equilibrium in finite time. Finite-time convergence helps us to split in time transients of different interconnected subsystems, which is especially powerful in combination with the small-gain technique.

For ODE systems, the finite-time stability and finite-time ISS are well-studied concepts [46, 109, 120, 121]. Detailed analysis of finite-settling-time behavior of systems with continuous dynamics has been performed in [12]. The concept of finite-time ISS has been introduced in [46] and applied to finite-time control design. Applications of finite-time and fixed-time techniques for control design have been reported in [121]. Nevertheless, converse finite-time ISS Lyapunov theorems are not available even for ODE systems, although Lyapunov characterizations of finite-time stability have been achieved in [109].

7.3.4 Input-to-State Dynamical Stability (ISDS)

ISS estimate depends on the norm of the whole input and for large times it gives the following estimate for the norm of the state: $\lim_{t\to\infty} |\phi(t, x, u)| \le \gamma(\|u\|_\infty)$. At the same time, we know from Proposition 2.7 that if the system is ISS and if $\lim_{t\to\infty} \|u(t + \cdot)\|_\infty = 0$, then also $\lim_{t\to\infty} \sup_{|x|\le r} |\phi(t, x, u)| = 0$ for any $r > 0$. This means that for inputs that are large for small times and then decay to smaller values, the ISS estimate is too rough. To overcome this drawback, the notion of *input-to-state dynamical stability (ISDS)* has been introduced and studied in [41]. ISDS is equivalent to ISS but it provides a less restrictive, although more sophisticated estimate with a so-called memory-fading effect for the norm of the trajectory.

7.3.5 Input-to-State Practical Stability (ISpS)

The system (2.1) is called *input-to-state practically stable (ISpS)*, if it is forward complete and there exist $\beta \in \mathcal{KL}$, $\gamma \in \mathcal{K}_\infty$, and $c > 0$ such that for all $x \in \mathbb{R}^n$, $u \in \mathcal{U}$, and $t \ge 0$ the following holds:

$$|\phi(t, x, u)| \le \beta(|x|, t) + \gamma(\|u\|_\infty) + c. \tag{7.14}$$

Input-to-state practical stability has been introduced in [54, 55] and extends an earlier concept of practical asymptotic stability of dynamical systems [82]. In some works, practical ISS is called ISS with bias [52, 53]. The first characterizations of practical ISS have been developed in [134, Sect. VI]). Characterizations of ISpS for a general class of infinite-dimensional systems have been developed in [94].

References

1. Angeli D (1999) Intrinsic robustness of global asymptotic stability. Syst Control Lett 38(4–5):297–307
2. Angeli D (2002) A Lyapunov approach to incremental stability properties. IEEE Trans Autom Control 47(3):410–421
3. Angeli D (2009) Further results on incremental input-to-state stability. IEEE Trans Autom Control 54(6):1386–1391
4. Angeli D, Ingalls B, Sontag E, Wang Y (2004) Uniform global asymptotic stability of differential inclusions. J Dyn Control Syst 10(3):391–412
5. Angeli D, Sontag ED, Wang Y (2000) A characterization of integral input-to-state stability. IEEE Trans Autom Control 45(6):1082–1097
6. Bamieh B, Jovanovic MR, Mitra P, Patterson S (2012) Coherence in large-scale networks: dimension-dependent limitations of local feedback. IEEE Trans Autom Control 57(9):2235–2249
7. Bamieh B, Paganini F, Dahleh MA (2002) Distributed control of spatially invariant systems. IEEE Trans Autom Control 47(7):1091–1107

8. Bamieh B, Voulgaris PG (2005) A convex characterization of distributed control problems in spatially invariant systems with communication constraints. Syst Control Lett 54(6):575–583
9. Barooah P, Mehta PG, Hespanha JP (2009) Mistuning-based control design to improve closed-loop stability margin of vehicular platoons. IEEE Trans Autom Control 54(9):2100–2113
10. Bastin G, Coron J-M (2016) Stability and boundary stabilization of 1-D hyperbolic systems. Springer
11. Besselink B, Johansson KH (2017) String stability and a delay-based spacing policy for vehicle platoons subject to disturbances. IEEE Trans Autom Control 62(9):4376–4391
12. Bhat SP, Bernstein DS (2000) Finite-time stability of continuous autonomous systems. SIAM J Control Optim 38(3):751–766
13. Bribiesca Argomedo F, Prieur C, Witrant E, Bremond S (2013) A strict control Lyapunov function for a diffusion equation with time-varying distributed coefficients. IEEE Trans Autom Control 58(2):290–303
14. Bribiesca Argomedo F, Witrant E, Prieur C (2012) D^1-input-to-state stability of a time-varying nonhomogeneous diffusive equation subject to boundary disturbances. In: Proceedings of 2012 American Control Conference, pp 2978–2983
15. Bribiesca Argomedo F, Witrant E, Prieur C (2013) Safety factor profile control in a tokamak. Springer
16. Bribiesca Argomedo F, Witrant E, Prieur C, Brémond S, Nouailletas R, Artaud J-F (2013) Lyapunov-based distributed control of the safety-factor profile in a tokamak plasma. Nucl Fusion 53(3):033005–033019
17. Cai C, Teel A (2009) Characterizations of input-to-state stability for hybrid systems. Syst Control Lett 58(1):47–53
18. Cai C, Teel AR, Goebel R (2007) Smooth Lyapunov functions for hybrid systems—part I: existence is equivalent to robustness. IEEE Trans Autom Control 52(7):1264–1277
19. Cai C, Teel AR, Goebel R (2008) Smooth Lyapunov functions for hybrid systems part II: (pre)asymptotically stable compact sets. IEEE Trans Autom Control 53(3):734–748
20. Chaillet A, Goksu G, Pepe P (2021) Lyapunov–Krasovskii characterizations of integral input-to-state stability of delay systems with non-strict dissipation rates. IEEE Trans Autom Control. https://doi.org/10.1109/TAC.2021.3099453
21. Chaillet A, Pepe P, Mason P, Chitour Y (2017) Is a point-wise dissipation rate enough to show ISS for time-delay systems? IFAC-PapersOnLine 50(1):14356–14361
22. Chen W-H, Zheng WX (2009) Input-to-state stability and integral input-to-state stability of nonlinear impulsive systems with delays. Automatica 45(6):1481–1488
23. Curtain R, Iftime OV, Zwart H (2009) System theoretic properties of a class of spatially invariant systems. Automatica 45(7):1619–1627
24. Curtain R, Zwart H (2020) Introduction to infinite-dimensional systems theory: a state-space approach. Springer
25. Dashkovskiy S, Feketa P (2017) Input-to-state stability of impulsive systems and their networks. Nonlinear Anal Hybrid Syst 26:190–200
26. Dashkovskiy S, Kosmykov M (2013) Input-to-state stability of interconnected hybrid systems. Automatica 49(4):1068–1074
27. Dashkovskiy S, Kosmykov M, Mironchenko A, Naujok L (2012) Stability of interconnected impulsive systems with and without time delays, using Lyapunov methods. Nonlinear Anal Hybrid Syst 6(3):899–915
28. Dashkovskiy S, Mironchenko A (2013) Input-to-state stability of infinite-dimensional control systems. Math Control Signals Syst 25(1):1–35
29. Dashkovskiy S, Mironchenko A (2013) Input-to-state stability of nonlinear impulsive systems. SIAM J Control Optim 51(3):1962–1987
30. Dashkovskiy S, Mironchenko A, Schmid J, Wirth F (2019) Stability of infinitely many interconnected systems. IFAC-PapersOnLine 52(16):550–555
31. Dashkovskiy S, Pavlichkov (2020) Stability conditions for infinite networks of nonlinear systems and their application for stabilization. Automatica 112:108643

32. Dashkovskiy S, Rüffer B, Wirth F (2010) Small gain theorems for large scale systems and construction of ISS Lyapunov functions. SIAM J Control Optim 48(6):4089–4118
33. Edalatzadeh MS, Morris KA (2019) Stability and well-posedness of a nonlinear railway track model. IEEE Control Syst Lett 3(1):162–167
34. Edwards HA, Lin Y, Wang Y (2000) On input-to-state stability for time varying nonlinear systems. In: Proceedings of 39th IEEE Conference on Decision and Control, pp 3501–3506
35. Espitia N, Girard A, Marchand N, Prieur C (2016) Event-based control of linear hyperbolic systems of conservation laws. Automatica 70:275–287
36. Espitia N, Girard A, Marchand N, Prieur C (2018) Event-based boundary control of a linear 2×2 hyperbolic system via backstepping approach. IEEE Trans Autom Control 63(8):2686–2693
37. Espitia N, Tanwani A, Tarbouriech S (2017) Stabilization of boundary controlled hyperbolic PDEs via Lyapunov-based event triggered sampling and quantization. In: Proceedings of 56th IEEE Conference on Decision and Control, pp 1266–1271
38. Fridman E, Dambrine M, Yeganefar N (2008) On input-to-state stability of systems with time-delay: a matrix inequalities approach. Automatica 44(9):2364–2369
39. Glück J, Mironchenko A (2021) Stability criteria for positive linear discrete-time systems. Positivity 25(5):2029–2059
40. Goebel R, Sanfelice R, Teel AR (2012) Hybrid dynamical systems: modeling, stability, and robustness. Princeton University Press
41. Grüne L (2002) Input-to-state dynamical stability and its Lyapunov function characterization. IEEE Trans Autom Control 47(9):1499–1504
42. Haddad WM, Chellaboina VS, Nersesov SG (2006) Impulsive and hybrid dynamical systems. Princeton University Press
43. Haimovich H, Mancilla-Aguilar JL (2019) ISS implies iISS even for switched and time-varying systems (if you are careful enough). Automatica 104:154–164
44. Hespanha JP, Liberzon D, Teel AR (2008) Lyapunov conditions for input-to-state stability of impulsive systems. Automatica 44(11):2735–2744
45. Hespanha JP, Morse AS (1999) Stability of switched systems with average dwell-time. In: Proceedings of 38th IEEE Conference on Decision and Control, pp 2655–2660
46. Hong Y, Jiang Z-P, Feng G (2010) Finite-time input-to-state stability and applications to finite-time control design. SIAM J Control Optim 48(7):4395–4418
47. Hosfeld R, Jacob B, Schwenninger F (2022) Integral input-to-state stability of unbounded bilinear control systems. Math Control Signals Syst. https://doi.org/10.1007/s00498-021-00308-9
48. Jacob B, Mironchenko A, Partington JR, Wirth F (2020) Noncoercive Lyapunov functions for input-to-state stability of infinite-dimensional systems. SIAM J Control Optim 58(5):2952–2978
49. Jacob B, Nabiullin R, Partington JR, Schwenninger FL (2018) Infinite-dimensional input-to-state stability and Orlicz spaces. SIAM J Control Optim 56(2):868–889
50. Jacob B, Schwenninger FL, Zwart H (2019) On continuity of solutions for parabolic control systems and input-to-state stability. J Differ Equ 266:6284–6306
51. Jacob B, Zwart HJ (2012) Linear port-Hamiltonian systems on infinite-dimensional spaces. Springer, Basel
52. Jayawardhana B, Logemann H, Ryan EP (2009) Input-to-state stability of differential inclusions with applications to hysteretic and quantized feedback systems. SIAM J Control Optim 48(2):1031–1054
53. Jayawardhana B, Logemann H, Ryan EP (2011) The circle criterion and input-to-state stability. IEEE Control Syst Mag 31(4):32–67
54. Jiang Z-P, Mareels IMY, Wang Y (1996) A Lyapunov formulation of the nonlinear small-gain theorem for interconnected ISS systems. Automatica 32(8):1211–1215
55. Jiang Z-P, Teel AR, Praly L (1994) Small-gain theorem for ISS systems and applications. Math Control Signals Syst 7(2):95–120

56. Jiang Z-P, Wang Y (2001) Input-to-state stability for discrete-time nonlinear systems. Automatica 37(6):857–869
57. Jiang Z-P, Wang Y (2008) A generalization of the nonlinear small-gain theorem for large-scale complex systems. In: Proceedings of 7th World Congress on Intelligent Control and Automation, pp 1188–1193
58. Jovanović MR, Bamieh B (2005) On the ill-posedness of certain vehicular platoon control problems. IEEE Trans Autom Control 50(9):1307–1321
59. Kalman RE, Falb PL, Arbib MA (1969) Topics in mathematical system theory. McGraw-Hill, New York
60. Kankanamalage HG, Lin Y, Wang Y (2017) On Lyapunov–Krasovskii characterizations of input-to-output stability. IFAC-PapersOnLine 50(1):14362–14367
61. Karafyllis I (2007) A system-theoretic framework for a wide class of systems I: applications to numerical analysis. J Math Anal Appl 328(2):876–899
62. Karafyllis I, Jiang Z-P (2007) A small-gain theorem for a wide class of feedback systems with control applications. SIAM J Control Optim 46(4):1483–1517
63. Karafyllis I, Jiang Z-P (2011) Stability and stabilization of nonlinear systems. Springer, London
64. Karafyllis I, Jiang Z-P (2011) A vector small-gain theorem for general non-linear control systems. IMA J Math Control Inf 28:309–344
65. Karafyllis I, Jiang Z-P (2012) A new small-gain theorem with an application to the stabilization of the chemostat. Int J Robust Nonlinear Control 22(14):1602–1630
66. Karafyllis I, Krstic M (2016) ISS with respect to boundary disturbances for 1-D parabolic PDEs. IEEE Trans Autom Control 61(12):3712–3724
67. Karafyllis I, Krstic M (2017) ISS in different norms for 1-D parabolic PDEs with boundary disturbances. SIAM J Control Optim 55(3):1716–1751
68. Karafyllis I, Krstic M (2018) Decay estimates for 1-D parabolic PDEs with boundary disturbances. ESAIM Control Optim Calc Var 24(4):1511–1540
69. Karafyllis I, Krstic M (2019) Input-to-state stability for PDEs. Springer, Cham
70. Karafyllis I, Pepe P, Jiang Z-P (2008) Input-to-output stability for systems described by retarded functional differential equations. Eur J Control 14(6):539–555
71. Karafyllis I, Tsinias J (2004) Nonuniform in time input-to-state stability and the small-gain theorem. IEEE Trans Autom Control 49(2):196–216
72. Kawan C, Mironchenko A, Swikir A, Noroozi N, Zamani M (2021) A Lyapunov-based small-gain theorem for infinite networks. IEEE Trans Autom Control 66(12):5830–5844
73. Kawan C, Mironchenko A, Zamani M (2022) A Lyapunov-based ISS small-gain theorem for infinite networks of nonlinear systems. IEEE Trans Autom Control. https://arxiv.org/abs/2103.07439
74. Kellett CM, Dower PM (2012) A generalization of input-to-state stability. In: Proceedings of 51st IEEE Conference on Decision and Control, pp 2970–2975
75. Kim T, Shim H, Cho DD (2016) Distributed Luenberger observer design. In: Proceedings of 55th IEEE Conference on Decision and Control, pp 6928–6933
76. Koga S, Bresch-Pietri D, Krstic M (2020) Delay compensated control of the Stefan problem and robustness to delay mismatch. Int J Robust Nonlinear Control 30(6):2304–2334
77. Koga S, Diagne M, Krstic M (2018) Control and state estimation of the one-phase Stefan problem via backstepping design. IEEE Trans Autom Control 64(2):510–525
78. Koga S, Karafyllis I, Krstic M (2018) Input-to-state stability for the control of Stefan problem with respect to heat loss at the interface. In: Proceedings of 2018 American Control Conference, pp 1740–1745
79. Kolathaya S, Ames AD (2018) Input-to-state safety with control barrier functions. IEEE Control Syst Lett 3(1):108–113
80. Laila DS, Nešić D (2003) Discrete-time Lyapunov-based small-gain theorem for parameterized interconnected ISS systems. IEEE Trans Autom Control 48(10):1783–1788
81. Lakshmikantham V, Bainov DD, Simeonov PS (1989) Theory of impulsive differential equations. World Scientific Publishing Co., Inc., Teaneck, NJ

82. Lakshmikantham V, Leela S, Martynyuk AA (1990) Practical stability of nonlinear systems. World Scientific
83. Lhachemi H, Shorten R (2019) ISS property with respect to boundary disturbances for a class of Riesz-spectral boundary control systems. Automatica 109:108504
84. Li J-S (2011) Ensemble control of finite-dimensional time-varying linear systems. IEEE Trans Autom Control 56(2):345–357
85. Liberzon D, Nesic D, Teel AR (2014) Lyapunov-based small-gain theorems for hybrid systems. IEEE Trans Autom Control 59(6):1395–1410
86. Liberzon D, Nešić D (2006) Stability analysis of hybrid systems via small-gain theorems. In: Proceedings of 9th international workshop on hybrid systems: computation and control, pp 421–435
87. Limon D, Alamo T, Raimondo D, De La Peña DM, Bravo J, Ferramosca A, Camacho E (2009) Input-to-state stability: a unifying framework for robust model predictive control. In: Nonlinear model predictive control. Springer, pp 1–26
88. Lin Y, Sontag ED, Wang Y (1996) A smooth converse Lyapunov theorem for robust stability. SIAM J Control Optim 34(1):124–160
89. Lin Y, Wang Y, Cheng D (2005) On nonuniform and semi-uniform input-to-state stability for time varying systems. IFAC Proc Vol 38(1):312–317
90. Liu J, Liu X, Xie W-C (2011) Input-to-state stability of impulsive and switching hybrid systems with time-delay. Automatica 47(5):899–908
91. Liu T, Jiang Z-P, Hill DJ (2012) Lyapunov formulation of the ISS cyclic-small-gain theorem for hybrid dynamical networks. Nonlinear Anal Hybrid Syst 6(4):988–1001
92. Mazenc F, Prieur C (2011) Strict Lyapunov functions for semilinear parabolic partial differential equations. Math Control Relat Fields 1(2):231–250
93. Mironchenko A (2016) Local input-to-state stability: characterizations and counterexamples. Syst Control Lett 87:23–28
94. Mironchenko A (2019) Criteria for input-to-state practical stability. IEEE Trans Autom Control 64(1):298–304
95. Mironchenko A (2021) Small gain theorems for general networks of heterogeneous infinite-dimensional systems. SIAM J Control Optim 59(2):1393–1419
96. Mironchenko A, Ito H (2015) Construction of iISS Lyapunov functions for interconnected parabolic systems. In: Proceedings of 2015 European Control Conference, pp 37–42
97. Mironchenko A, Ito H (2016) Characterizations of integral input-to-state stability for bilinear systems in infinite dimensions. Math Control Relat Fields 6(3):447–466
98. Mironchenko A, Karafyllis I, Krstic M (2019) Monotonicity methods for input-to-state stability of nonlinear parabolic PDEs with boundary disturbances. SIAM J Control Optim 57(1):510–532
99. Mironchenko A, Kawan C, Glück J (2021) Nonlinear small-gain theorems for input-to-state stability of infinite interconnections. Math Control Signals Syst 33:573–615
100. Mironchenko A, Noroozi N, Kawan C, Zamani M (2021) ISS small-gain criteria for infinite networks with linear gain functions. Syst Control Lett 157:105051
101. Mironchenko A, Noroozi N, Kawan C, Zamani M (2021) A small-gain approach to ISS of infinite networks with homogeneous gain operators. In: Proceedings of 60th IEEE Conference on Decision and Control, pp 4817–4822
102. Mironchenko A, Prieur C (2020) Input-to-state stability of infinite-dimensional systems: recent results and open questions. SIAM Rev 62(3):529–614
103. Mironchenko A, Wirth F (2018) Characterizations of input-to-state stability for infinite-dimensional systems. IEEE Trans Autom Control 63(6):1602–1617
104. Mironchenko A, Wirth F (2018) Lyapunov characterization of input-to-state stability for semilinear control systems over Banach spaces. Systems & Control Letters 119:64–70
105. Mironchenko A, Wirth F (2019) Existence of non-coercive Lyapunov functions is equivalent to integral uniform global asymptotic stability. Math Control Signals Syst 31(4):1–26
106. Mironchenko A, Wirth F (2019) Non-coercive Lyapunov functions for infinite-dimensional systems. Journal of Differential Equations 105:7038–7072

107. Mironchenko A, Yang G, Liberzon D (2018) Lyapunov small-gain theorems for networks of not necessarily ISS hybrid systems. Automatica 88:10–20
108. Morse AS (1996) Supervisory control of families of linear set-point controllers—part I. Exact matching. IEEE Trans Autom Control 41(10):1413–1431
109. Moulay E, Perruquetti W (2006) Finite time stability and stabilization of a class of continuous systems. J Math Anal Appl 323(2):1430–1443
110. Nesic D, Teel AR (2008) A Lyapunov-based small-gain theorem for hybrid ISS systems. In: Proceedings of 47th IEEE Conference on Decision and Control, pp 3380–3385
111. Nešić D, Liberzon D (2005) A small-gain approach to stability analysis of hybrid systems. In: Proceedings of 44th IEEE Conference on Decision and Control, pp 5409–5414
112. Noroozi N, Mironchenko A, Kawan C, Zamani M (2021) Set stability of infinite networks: ISS small-gain theory and its applications. IFAC-PapersOnLine 54(9):72–77
113. Noroozi N, Mironchenko A, Kawan C, Zamani M (2022) Small-gain theorem for stability, cooperative control, and distributed observation of infinite networks. Eur J Control. https://doi.org/10.1016/j.ejcon.2022.100634
114. Olfati-Saber R (2007) Distributed Kalman filtering for sensor networks. In: Proceedings of 46th IEEE Conference on Decision and Control, New Orleans, pp 5492–5498
115. Olfati-Saber R, Shamma JS (2005) Consensus filters for sensor networks and distributed sensor fusion. In: Proceedings of 44th IEEE Conference on Decision and Control, pp 6698–6703
116. Pazy A (1983) Semigroups of linear operators and applications to partial differential equations. Springer, New York
117. Pepe P (2021) A nonlinear version of Halanay's inequality for the uniform convergence to the origin. Math Control Relat Fields
118. Pepe P, Jiang Z-P (2006) A Lyapunov–Krasovskii methodology for ISS and iISS of time-delay systems. Syst Control Lett 55(12):1006–1014
119. Pisano A, Orlov Y (2017) On the ISS properties of a class of parabolic DPS' with discontinuous control using sampled-in-space sensing and actuation. Automatica 81:447–454
120. Polyakov A (2011) Nonlinear feedback design for fixed-time stabilization of linear control systems. IEEE Trans Autom Control 57(8):2106–2110
121. Polyakov A, Efimov D, Perruquetti W (2015) Finite-time and fixed-time stabilization: implicit Lyapunov function approach. Automatica 51:332–340
122. Prieur C, Mazenc F (2012) ISS-Lyapunov functions for time-varying hyperbolic systems of balance laws. Math Control Signals Syst 24(1–2):111–134
123. Rawlings JB, Mayne DQ, Diehl M (2017) Model predictive control: theory, computation, and design. Nob Hill Publishing, Madison, WI
124. Romdlony MZ, Jayawardhana B (2016) On the new notion of input-to-state safety. In: Proceedings of 55th IEEE Conference on Decision and Control pp 6403–6409
125. Rüffer B (2007) Monotone dynamical systems, graphs, and stability of large-scale interconnected systems. PhD thesis, Fachbereich 3 (Mathematik & Informatik) der Universität Bremen
126. Samoilenko AM, Perestyuk NA (1995) Impulsive differential equations. World Scientific Publishing Co. Inc., River Edge, NJ
127. Sarkar T, Roozbehani M, Dahleh MA (2018) Robustness sensitivities in large networks. In: Emerging applications of control and systems theory. Springer, pp 81–92
128. Schwenninger FL (2020) Input-to-state stability for parabolic boundary control: linear and semilinear systems. In: Control theory of infinite-dimensional systems. Springer, pp 83–116
129. Selivanov A, Fridman E (2016) Distributed event-triggered control of diffusion semilinear PDEs. Automatica 68:344–351
130. Showalter RE (2013) Monotone operators in Banach space and nonlinear partial differential equations. Am Math Soc
131. Sontag E, Wang Y (2000) Lyapunov characterizations of input to output stability. SIAM J Control Optim 39(1):226–249

132. Sontag ED (2008) Input to state stability: basic concepts and results. In: Nonlinear and optimal control theory, chap 3. Springer, Heidelberg, pp 163–220
133. Sontag ED, Krichman M (2003) An example of a GAS system which can be destabilized by an integrable perturbation. IEEE Trans Autom Control 48(6):1046–1049
134. Sontag ED, Wang Y (1996) New characterizations of input-to-state stability. IEEE Trans Autom Control 41(9):1283–1294
135. Sontag ED, Wang Y (1997) Output-to-state stability and detectability of nonlinear systems. Syst Control Lett 29(5):279–290
136. Stamova I (2009) Stability analysis of impulsive functional differential equations. Walter de Gruyter GmbH & Co. KG, Berlin
137. Tanwani A, Prieur C, Tarbouriech S (2017) Disturbance-to-state stabilization and quantized control for linear hyperbolic systems
138. Teel AR (1998) Connections between Razumikhin-type theorems and the ISS nonlinear small-gain theorem. IEEE Trans Autom Control 43(7):960–964
139. Teel AR, Praly L (2000) A smooth Lyapunov function from a class-estimate involving two positive semidefinite functions. ESAIM Control Optim Calc Var 5:313–367
140. van der Schaft A, Schumacher H (2000) An introduction to hybrid dynamical systems. Springer London Ltd., London
141. Vorotnikov VI (2005) Partial stability and control: the state-of-the-art and development prospects. Autom Remote Control 66(4):511–561
142. Wang L, Morse AS (2018) A distributed observer for a time-invariant linear system. IEEE Trans Autom Control 63(7):2123–2130
143. Weiss G (1989) Admissibility of unbounded control operators. SIAM J Control Optim 27(3):527–545
144. Willems JC (1972) Dissipative dynamical systems part I: general theory. Arch Ration Mech Anal 45(5):321–351
145. Yang S, Shi B, Hao S (2013) Input-to-state stability for discrete-time nonlinear impulsive systems with delays. Int J Robust Nonlinear Control 23(4):400–418
146. Zamani M, Pola G, Mazo M, Tabuada P (2011) Symbolic models for nonlinear control systems without stability assumptions. IEEE Trans Autom Control 57(7):1804–1809

Appendix A
Comparison Functions and Comparison Principles

In this appendix, we introduce the classes of comparison functions, which are very useful for developing the stability theory, and derive several essential properties of these functions. Next, we define the Dini derivatives and introduce several comparison principles based on the formalism of comparison functions. Finally, we collect important facts from the analysis that will be frequently used in this book.

A.1 Comparison Functions

In this section, we study properties of the following classes of comparison functions:

$$\begin{aligned}
\mathcal{P} &:= \{\gamma \in C(\mathbb{R}_+) : \gamma(0) = 0 \text{ and } \gamma(r) > 0 \text{ for } r > 0\},\\
\mathcal{K} &:= \{\gamma \in \mathcal{P} : \gamma \text{ is strictly increasing}\},\\
\mathcal{K}_\infty &:= \{\gamma \in \mathcal{K} : \gamma \text{ is unbounded}\},\\
\mathcal{L} &:= \{\gamma \in C(\mathbb{R}_+) : \gamma \text{ is strictly decreasing with } \lim_{t\to\infty} \gamma(t) = 0\},\\
\mathcal{KL} &:= \{\beta \in C(\mathbb{R}_+ \times \mathbb{R}_+, \mathbb{R}_+) : \beta(\cdot, t) \in \mathcal{K}\ \forall t \geq 0,\ \beta(r, \cdot) \in \mathcal{L}\ \forall r > 0\}.
\end{aligned}$$

Functions of class $\mathcal{P}$ are also called *positive definite functions*. Functions of class $\mathcal{K}$ will be frequently called also class $\mathcal{K}$ functions or simply $\mathcal{K}$-functions. The same convention we use with other classes of comparison functions.

We will see later that the stability properties of the system (1.1) can be restated in terms of comparison functions, and proofs of many results in the stability theory can be significantly simplified by using standard manipulations with comparison functions.

A. Mironchenko, *Input-to-State Stability*, Communications and Control Engineering,
https://doi.org/10.1007/978-3-031-14674-9

A.1.1 Elementary Properties of Comparison Functions

This section gives a short account of the basic properties of comparison functions that will be used frequently in this book.

Here we denote by $\circ$ a composition of the maps, that is, $f \circ g(\cdot) := f(g(\cdot))$.

Proposition A.1 (Elementary properties of $\mathcal{K}$- and $\mathcal{L}$-functions) *The following properties hold:*

(i) *For all $f, g \in \mathcal{K}$ it follows that $f \circ g \in \mathcal{K}$.*
(ii) *For any $f \in \mathcal{K}_\infty$ there exists f^{-1} that also belongs to $\mathcal{K}_\infty$.*
(iii) *For any $f \in \mathcal{K}$, $g \in \mathcal{L}$ it holds that $f \circ g \in \mathcal{L}$ and $g \circ f \in \mathcal{L}$.*
(iv) *$(\mathcal{K}_\infty, \circ)$ is a group.*
(v) *For any $\gamma_1, \gamma_2 \in \mathcal{K}$ it follows that $\gamma_+ : s \mapsto \max\{\gamma_1(s), \gamma_2(s)\}$ and $\gamma_- : s \mapsto \min\{\gamma_1(s), \gamma_2(s)\}$ also belong to class $\mathcal{K}$.*
(vi) *For any $\beta_1, \beta_2 \in \mathcal{KL}$ it follows that $\beta_+ : (s, t) \mapsto \max\{\beta_1(s, t), \beta_2(s, t)\}$ and $\beta_- : (s, t) \mapsto \min\{\beta_1(s, t), \beta_2(s, t)\}$ also belong to class $\mathcal{KL}$.*

Proof (i) Since $f, g \in C(\mathbb{R}_+)$, then $f \circ g \in C(\mathbb{R}_+)$. In view of $f(0) = g(0) = 0$ it holds also $f(g(0)) = 0$. The monotonicity of $f \circ g$ is clear.
(ii) Since $f \in \mathcal{K}_\infty$, it is a bijection on $\mathbb{R}_+$ and there exists an inverse unbounded map $f^{-1} \in C(\mathbb{R}_+, \mathbb{R}_+)$. Clearly, $f^{-1}(0) = 0$. For any $r_1 < r_2$ we have $f\big(f^{-1}(r_1)\big) < f\big(f^{-1}(r_2)\big)$. Since f is monotonically increasing, it holds $f^{-1}(r_1) < f^{-1}(r_2)$.
(iii)–(iv) Clear.
(v) Let $\gamma_1, \gamma_2 \in \mathcal{K}$. Clearly, γ_- is continuous, $\gamma_-(0) = 0$. Let $s_1, s_2 \in \mathbb{R}_+$: $s_2 > s_1$. Then $\gamma_-(s_2) = \min\{\gamma_1(s_2), \gamma_2(s_2)\} > \min\{\gamma_1(s_1), \gamma_2(s_1)\} = \gamma_-(s_1)$. Hence $\gamma_- \in \mathcal{K}$. Analogously, $\gamma_+ \in \mathcal{K}$.
(vi) Let $\beta_1, \beta_2 \in \mathcal{KL}$. Define $\beta_- : (s, t) \mapsto \min\{\beta_1(s, t), \beta_2(s, t)\}$. Next, $\beta_-(\cdot, t) \in \mathcal{K}$ due to item (v). Now let $t_1, t_2 \geq 0$ be so that $t_2 > t_1$. Then for any $s \in \mathbb{R}_+$ $\beta_-(s, t_2) = \min\{\beta_1(s, t_2), \beta_2(s, t_2)\} < \min\{\beta_1(s, t_1), \beta_2(s, t_1)\} = \beta_-(s, t_1)$. Clearly, β_- is continuous and convergent to 0 w.r.t. the second argument. Hence, $\beta_1 \in \mathcal{KL}$. □

Remark A.2 If $f \in \mathcal{K}$, then f^{-1} is defined on $\mathrm{Im}(f) \subset \mathbb{R}_+$. We will slightly abuse the notation by setting $f(r) := +\infty$, for $r \in \mathbb{R}_+ \backslash \mathrm{Im}(f)$.

Elementary but useful result on $\mathcal{K}$-functions is

Proposition A.3 (Weak triangle inequality) *For any $\gamma \in \mathcal{K}$ and any $\sigma \in \mathcal{K}_\infty$ it holds for all $a, b \geq 0$ that:*

$$\gamma(a+b) \leq \max\left\{\gamma(a+\sigma(a)), \gamma(b+\sigma^{-1}(b))\right\}. \tag{A.1}$$

Proof Pick any $\sigma \in \mathcal{K}_\infty$. Then either $b \leq \sigma(a)$ or $a \leq \sigma^{-1}(b)$ holds. This immediately implies (A.1). □

In particular, setting $\sigma :=$ id (identity operator on $\mathbb{R}_+$) in (A.1) we obtain for any $\gamma \in \mathcal{K}$ a simple inequality

$$\gamma(a+b) \leq \max\{\gamma(2a), \gamma(2b)\}.$$

Another useful inequality is

Proposition A.4 ($\mathcal{K}_\infty$-inequality) *For all $a, b \geq 0$, all $\alpha \in \mathcal{K}_\infty$, and all $r > 0$ it holds that*

$$ab \leq r\alpha(a) + b\alpha^{-1}\left(\tfrac{b}{r}\right). \tag{A.2}$$

Proof Pick any $\alpha \in \mathcal{K}_\infty$. Then either $b \leq r\alpha(a)$ or $b \geq r\alpha(a)$. The latter inequality can be equivalently written as $a \leq \alpha^{-1}(\frac{b}{r})$. This shows that for any $a, b \in \mathbb{R}_+$ the inequality (A.2) holds. □

Lemma A.5 *For any $\eta \in \mathcal{K}_\infty$ with $\mathrm{id} - \eta \in \mathcal{K}_\infty$ it holds that:*

(i) There is $\rho \in \mathcal{K}_\infty$ such that $(\mathrm{id} - \eta)^{-1} = \mathrm{id} + \rho$.
(ii) There are $\eta_1, \eta_2 \in \mathcal{K}_\infty$ such that $\mathrm{id} - \eta_1, \mathrm{id} - \eta_2 \in \mathcal{K}_\infty$ and $\mathrm{id} - \eta = (\mathrm{id} - \eta_1) \circ (\mathrm{id} - \eta_2)$.

Proof (i) Define $\rho := \eta \circ (\mathrm{id} - \eta)^{-1}$. As ρ is a composition of $\mathcal{K}_\infty$-functions, $\rho \in \mathcal{K}_\infty$. It holds that $(\mathrm{id} + \rho) \circ (\mathrm{id} - \eta) = \mathrm{id} - \eta + \eta \circ (\mathrm{id} - \eta)^{-1} \circ (\mathrm{id} - \eta) = \mathrm{id} - \eta + \eta = \mathrm{id}$, and thus $\mathrm{id} + \rho = (\mathrm{id} - \eta)^{-1}$.
(ii) Choose $\eta_2 := \frac{1}{2}\eta$ and $\eta_1 := \frac{1}{2}\eta \circ (\mathrm{id} - \eta_2)^{-1}$. Clearly, $\mathrm{id} - \eta_2 \in \mathcal{K}_\infty$. Direct calculation shows that $\mathrm{id} - \eta = (\mathrm{id} - \eta_1) \circ (\mathrm{id} - \eta_2)$. Multiplying this from right on $(\mathrm{id} - \eta_2)^{-1}$, we see that $\mathrm{id} - \eta_1 \in \mathcal{K}_\infty$. □

Analogously, the following holds:

Lemma A.6 *For any $\eta \in \mathcal{K}_\infty$ it holds that:*

(i) There is $\rho \in \mathcal{K}_\infty$ such that $(\mathrm{id} + \eta)^{-1} = \mathrm{id} - \rho$.
(ii) There are $\eta_1, \eta_2 \in \mathcal{K}_\infty$ such that $\mathrm{id} + \eta = (\mathrm{id} + \eta_1) \circ (\mathrm{id} + \eta_2)$.

Proof (i) Define $\rho := \eta \circ (\mathrm{id} + \eta)^{-1}$. As ρ is a composition of $\mathcal{K}_\infty$-functions, $\rho \in \mathcal{K}_\infty$. It holds that $(\mathrm{id} - \rho) \circ (\mathrm{id} + \eta) = \mathrm{id} + \eta - \eta \circ (\mathrm{id} + \eta)^{-1} \circ (\mathrm{id} + \eta) = \mathrm{id} + \eta - \eta = \mathrm{id}$, and thus $\mathrm{id} - \rho = (\mathrm{id} + \eta)^{-1}$.
(ii) Choose $\eta_2 := \frac{1}{2}\eta$ and $\eta_1 := \frac{1}{2}\eta \circ (\mathrm{id} + \eta_2)^{-1}$. Direct calculation shows the claim. □

A.1.2 Sontag's $\mathcal{KL}$-Lemma

One of the most useful results concerning comparison functions is a so-called Sontag's $\mathcal{KL}$-Lemma. In particular, it is beneficial for the proof of converse Lyapunov theorems.

Proposition A.7 (Sontag's $\mathcal{KL}$-Lemma) *For each $\beta \in \mathcal{KL}$ and any $\lambda > 0$ there exist $\alpha_1, \alpha_2 \in \mathcal{K}_\infty$:*

$$\alpha_1(\beta(s,t)) \le \alpha_2(s)e^{-\lambda t}, \quad t, s \in \mathbb{R}_+. \tag{A.3}$$

Proof First pick $\rho \in \mathcal{K}_\infty$ and $\theta \in \mathcal{L}$ so that

$$\beta(\rho(t), t) \le \theta(t) \quad \forall t \ge 0. \tag{A.4}$$

Step 1. Let us show that such functions exist.
First, it is clear that for all $r > 0$ and any $q \in (0, \beta(r, 0)]$ there exists a unique $t \ge 0$ so that $\beta(r, t) = q$. Pick a monotonically decreasing sequence $(q_i)_{i=1}^\infty$: $q_i \to 0$, $i \to \infty$ and a monotonically increasing sequence $(r_i)_{i=1}^\infty$: $r_i \to \infty$ together with a corresponding sequence $(\tilde{t}_i)_{i=1}^\infty$ so that for all $i = 1, \ldots, \infty$ we have

$$\beta(r_i, \tilde{t}_i) = q_i.$$

Pick now any sequence $(t_i)_{i=1}^\infty$: t_i is strictly increasing, $t_i \to \infty$ as $i \to \infty$ and $t_i \ge \tilde{t}_i$ for all $i = 1, \ldots, \infty$.

Define also $t_0 := 0$, $r_0 := r_1/2$ and $q_0 > \max\{q_1, \beta(r_0, 0)\}$. Since $\beta \in \mathcal{KL}$, we have

$$\beta(r_i, t_i) \le \beta(r_i, \tilde{t}_i) = q_i.$$

Now define θ so that $\theta(t_i) = q_{i-1}$ for all $i = 1, \ldots, \infty$ and extend θ to the whole $\mathbb{R}_+$ in an arbitrary way so that $\theta \in \mathcal{L}$.

Also define $\rho(t_i) = r_{i-1}$ for all $i = 1, \ldots, \infty$, and extend ρ to the whole $\mathbb{R}_+$ in an arbitrary way so that $\rho \in \mathcal{K}_\infty$.

By construction we have for any $i = 0, 1, \ldots, \infty$ and for all $t \in [t_i, t_{i+1}]$ it holds that

$$\beta(\rho(t), t) \le \beta(\rho(t_{i+1}), t_i) = \beta(r_i, t_i) \le q_i = \theta(t_{i+1}) < \theta(t).$$

This shows that (A.4) holds with these ρ, θ.

Step 2. Denote the inverse of θ as θ^{-1}, which is defined, continuous and strictly decreasing on the interval $(0, \theta(0)]$. Pick any $\lambda > 0$. Then the function $s \mapsto e^{-2\lambda\theta^{-1}(s)}$ is defined for $s \in (0, \theta(0)]$ and is continuous and strictly increasing on its domain of definition. Since $\theta^{-1}(s) \to +\infty$ as $s \to +0$, $e^{-2\lambda\theta^{-1}(s)} \to 0$ as $s \to +0$. Setting $e^{-2\lambda\theta^{-1}(0)} := 0$, we extend the map $s \mapsto e^{-2\lambda\theta^{-1}(s)}$ by continuity to $[0, \theta(0)]$.

Obviously, there exist $\alpha_1 \in \mathcal{K}_\infty$ so that for all $s \in [0, \theta(0)]$ it holds that

$$\alpha_1(s) \le e^{-2\lambda\theta^{-1}(s)}. \tag{A.5}$$

Substituting $s := \theta(t)$ in this inequality and using (A.4) we derive that

$$\alpha_1(\beta(\rho(t), t))e^{2\lambda t} \le \alpha_1(\theta(t))e^{2\lambda t} \le 1 \quad \forall t \ge 0. \tag{A.6}$$

Using (A.6) and the fact that $\beta \in \mathcal{KL}$, we obtain that for $s, t \in \mathbb{R}_+ : \ 0 < s \le \rho(t)$

$$\begin{aligned}\alpha_1(\beta(s, t))e^{\lambda t} &= \sqrt{\alpha_1(\beta(s, 0))}\frac{\sqrt{\alpha_1(\beta(s, t))}}{\sqrt{\alpha_1(\beta(s, 0))}}\sqrt{\alpha_1(\beta(s, t))e^{2\lambda t}} \\ &\le \sqrt{\alpha_1(\beta(s, 0))}\sqrt{\alpha_1(\beta(\rho(t), t))e^{2\lambda t}} \\ &\le \sqrt{\alpha_1(\beta(s, 0))}.\end{aligned} \tag{A.7}$$

On the other hand, for $s \ge \rho(t)$ it holds that

$$\alpha_1(\beta(s, t))e^{\lambda t} \le \alpha_1(\beta(s, 0))e^{\lambda\rho^{-1}(s)}. \tag{A.8}$$

Pick any $\alpha_2 \in \mathcal{K}_\infty$: for all $s \ge 0$

$$\alpha_2(s) \ge \max\{\alpha_1(\beta(s, 0))e^{\lambda\rho^{-1}(s)}, \sqrt{\alpha_1(\beta(s, 0))}\}.$$

This implies that (A.3) holds for all $s, t \ge 0$. □

The significance of $\mathcal{KL}$-lemma is due to the fact that each $\mathcal{KL}$-function of two arguments (t, r) can be upperbounded after a proper scaling by another $\mathcal{KL}$-function, where the dependence on r and t is separated.

We state also another useful variation of Proposition A.7:

Corollary A.8 *For each $\beta \in \mathcal{KL}$ and any $\lambda > 0$ there exist $\tilde{\alpha}_1, \tilde{\alpha}_2 \in \mathcal{K}_\infty$:*

$$\beta(s, t) \le \tilde{\alpha}_1(\tilde{\alpha}_2(s)e^{-\lambda t}) \quad \forall s, t \ge 0. \tag{A.9}$$

A.1.3 *Positive Definite Functions*

In this section, we characterize positive definiteness in terms of comparison functions.

Definition A.9 Let X, Y be metric spaces.

(i) A map $f : X \to Y$ is called *proper* if the preimage of any compact subset of Y is compact in X.
(ii) We say that a *sequence* $(x_i)_{i\ge1}$ *escapes to infinity* if for every compact set $S \subset X$ only finitely many points of $(x_i)_{i\ge1}$ are contained in S.

Properness can be characterized as follows:

Proposition A.10 *A continuous map $f : X \to Y$ is proper $\Leftrightarrow$ for every sequence of points $(x_i)_{i\geq 1} \subset X$ escaping to infinity the sequence $(f(x_i))_{i\geq 1}$ also escapes to infinity.*

Proof "$\Rightarrow$". Let f be proper and pick any sequence $(x_i)_{i\geq 1} \subset X$ that escapes to infinity. Assume that $(f(x_i))_{i\geq 1}$ doesn't escape to infinity. Then there exists a compact set $S \subset Y$ so that the cardinality of $\{i : f(x_i) \in S\}$ is infinite. But since f is proper, $f^{-1}(S)$ is compact, and since (x_i) escapes to infinity, there is only a finite number of x_i in $f^{-1}(S)$, a contradiction to the infinite cardinality of $\{i : f(x_i) \in S\}$.

"$\Leftarrow$". Assume that for each sequence $(x_i)_{i\geq 1}$ escaping to infinity, the corresponding sequence $(f(x_i))_{i\geq 1}$ also escapes to infinity. Let f be not proper. Then there exists $S \subset Y$, which is compact and so that $f^{-1}(S)$ is not compact. Assume that $f^{-1}(S)$ is a bounded set. Pick $y \in \partial f^{-1}(S) \backslash f^{-1}(S)$ (which is possible since we assumed that $f^{-1}(S)$ is not compact) and pick a sequence $(y_i) \subset f^{-1}(S)$ so that $y_i \to y$ as $i \to \infty$. Since f is a continuous function, $f(y_i) \to f(y)$ as $i \to \infty$. But $f(y_i) \in S$ for all i and thus due to compactness of S we have that $f(y) \in S$, and hence $y \in f^{-1}(S)$, a contradiction.

Let ρ be a metric in X. Assume that $f^{-1}(S)$ is not bounded and consider a sequence $(x_i)_{i\geq 1} \subset f^{-1}(S)$ so that $\rho(0, x_i)$ monotonically increases to infinity as $i \to \infty$. Due to construction $(x_i)_{i\geq 1}$ escapes to infinity, and thus $(f(x_i))_{i\geq 1}$ also escapes to infinity. However, $f(x_i) \in S$ for all $i \in \mathbb{N}$, and we come to the contradiction. This shows the claim. □

As a corollary of Proposition A.10, we obtain the criterion of properness for $V \in C(\mathbb{R}^n, \mathbb{R}_+)$.

Corollary A.11 *Let $V \in C(\mathbb{R}^n, \mathbb{R}_+)$ be given. The following assertions are equivalent:*

(i) V is proper.
(ii) V is radially unbounded, *that is, $V(x) \to \infty$ as $|x| \to \infty$.*
(iii) There is $\psi \in \mathcal{K}_\infty$ such that $V(x) \geq \psi(|x|)$ for all $x \in \mathbb{R}^n$.

Proof (i) $\Leftrightarrow$ (ii) Follows by Proposition A.10.
(iii) $\Rightarrow$ (ii) Clear.
(ii) $\Rightarrow$ (iii) The argument is similar to the proof of Proposition A.14 and thus omitted. □

We extend the definition of positive definite functions to functions defined on $\mathbb{R}^n$.

Definition A.12 $V \in C(\mathbb{R}^n, \mathbb{R}_+)$ is called *positive definite* if $V(0) = 0$ and $V(x) > 0$ for any $x \neq 0$.

Positive definiteness can be characterized in terms of comparison functions.

Proposition A.13 *$\rho \in C(\mathbb{R}^n, \mathbb{R}_+)$ is positive definite if and only if $\rho(0) = 0$ and there are $\omega \in \mathcal{K}_\infty$ and $\sigma \in \mathcal{L}$ such that*

$$\rho(x) \geq \omega(|x|)\sigma(|x|), \quad x \in \mathbb{R}^n. \tag{A.10}$$

Proof "$\Leftarrow$". Clear.

"$\Rightarrow$". Assume that $\rho(x) \to 0$ as $|x| \to \infty$. Otherwise, we can consider any $\bar{\rho}$: $\mathbb{R}^n \to \mathbb{R}_+$ such that $\bar{\rho}(x) \leq \rho(x)$ for all $x \in \mathbb{R}^n$ and $\bar{\rho}(x) \to 0$ as $|x| \to \infty$.

As $\rho(x) \to 0$ as $|x| \to \infty$, ρ has a positive global maximum attained at some $x^* \in \mathbb{R}^n$. We define the positive, and nondecreasing function

$$\tilde{\omega}(s) := \min_{\min\{s,|x^*|\} \leq |y| \leq |x^*|} \frac{\rho(y)}{\rho(x^*)}, \quad s \geq 0. \tag{A.11}$$

We also define the positive and nonincreasing function

$$\tilde{\sigma}(s) := \min_{|x^*| \leq |y| \leq \max\{s,|x^*|\}} \frac{\rho(y)}{\rho(x^*)}, \quad s \geq 0. \tag{A.12}$$

Continuity of $\tilde{\omega}$, $\tilde{\sigma}$ can be shown using ideas from Lemma A.27.

For $|x| \leq |x^*|$, it holds that

$$\tilde{\omega}(|x|) = \min_{|x| \leq |y| \leq |x^*|} \frac{\rho(y)}{\rho(x^*)} \leq \frac{\rho(x)}{\rho(x^*)},$$

and analogously for $|x| \geq |x^*|$, we have

$$\tilde{\sigma}(|x|) = \min_{|x^*| \leq |y| \leq |x|} \frac{\rho(y)}{\rho(x^*)} \leq \frac{\rho(x)}{\rho(x^*)}.$$

As $\tilde{\omega}(s) \leq 1$ and $\tilde{\sigma}(s) \leq 1$ for all $s \geq 0$, we see that

$$\tilde{\omega}(|x|)\tilde{\sigma}(|x|) \leq \frac{\rho(x)}{\rho(x^*)}. \tag{A.13}$$

Define now

$$\omega(s) := \rho(x^*)\tilde{\omega}(s)s, \quad \sigma(s) := \tilde{\sigma}(s)\frac{1}{1+s}, \quad s \geq 0.$$

Clearly, $\omega \in \mathcal{K}_\infty$ and $\sigma \in \mathcal{L}$. Exploiting (A.13), we have for all $x \in \mathbb{R}^n$:

$$\omega(|x|)\sigma(|x|) = \rho(x^*)\tilde{\omega}(|x|)\tilde{\sigma}(|x|)\frac{|x|}{1+|x|} \leq \rho(x).$$

This finishes the proof. □

We close the section with the criterion of positive definiteness and properness.

Proposition A.14 *$V \in C(\mathbb{R}^n, \mathbb{R}_+)$ is positive definite and proper $\Leftrightarrow$ there are $\psi_1, \psi_2 \in \mathcal{K}_\infty$ such that*

$$\psi_1(|x|) \leq V(x) \leq \psi_2(|x|), \quad x \in \mathbb{R}^n. \tag{A.14}$$

Proof Trivially, (A.14) implies that V is positive definite and proper. Let us prove the converse implication.

Define $\psi_1(r) := \inf_{|x|\geq r} V(x)$ and $\psi_2(r) := \sup_{|x|\leq r} V(x)$. Then (A.14) holds by construction. By Lemma A.27, ψ_1, ψ_2 are nondecreasing continuous functions, with $\psi_1(0) = \psi_2(0) = 0$. Furthermore, since V is proper, by Corollary A.11 it holds that $\lim_{r\to\infty} \psi_1(r) = \lim_{r\to\infty} \psi_2(r) = +\infty$. Positive definiteness of V implies that both ψ_1 and ψ_2 are also positive definite.

Define $\tilde{\psi}_1(r) := \psi_1(r)(1 - e^{-r})$ and $\tilde{\psi}_2(r) := \psi_2(r) + r$. Then $\tilde{\psi}_1, \tilde{\psi}_2 \in \mathcal{K}_\infty$ and (A.14) holds with $\tilde{\psi}_1, \tilde{\psi}_2$ instead of ψ_1, ψ_2. □

A.1.4 Approximations, Upper and Lower Bounds

It is ultimately helpful to have conditions that ensure the possibility to estimate a given function $z : \mathbb{R}_+ \to \mathbb{R}_+$ from above/below by a $\mathcal{K}$-function. In addition, smooth approximations of $\mathcal{K}$-functions are often needed. In this section, we provide desirable conditions of this kind.

We will need the following technical result:

Proposition A.15 *For any $z \in C(\mathbb{R}_+, \mathbb{R})$ and any $\varepsilon > 0$ there is a partition $(R_k)_{k\in\mathbb{Z}} \subset (0, +\infty)$ satisfying the following properties:*

(i) $R_k \to 0$ *as* $k \to -\infty$.
(ii) $R_k \to +\infty$ *as* $k \to +\infty$.
(iii) $R_k < R_{k+1}$ *for all* $k \in \mathbb{Z}$.
(iv) $(R_k)_{k\in\mathbb{Z}}$ *has no finite accumulation points, other than zero.*
(v) $\sup_{s\in[R_k,R_{k+1}]} z(s) - \inf_{s\in[R_k,R_{k+1}]} z(s) < \varepsilon$, *for all* $k \in \mathbb{Z}$.

Proof It is easy to see that a partition satisfying the properties (i)–(iv) always exists (and can be chosen independently on z). Pick any of such partitions and denote it $(S_k)_{k\in\mathbb{Z}}$. Since $z \in C(\mathbb{R}_+, \mathbb{R})$, by Heine–Cantor's theorem z is uniformly continuous on $[S_k, S_{k+1}]$ for any $k \in \mathbb{Z}$. Thus, there exists a partition of $[S_k, S_{k+1}]$ into finite number of intervals $S_k = S_{k1}, \dots, S_{km(k)} = S_{k+1}$ so that $S_{ki} < S_{kj}$ for $i < j$ and $\sup_{s\in[S_{ki},S_{ki+1}]} z(s) - \inf_{s\in[S_{ki},S_{ki+1}]} z(s) < \varepsilon$, for each $i = 1, \dots, m(k) - 1$.

Inserting in the infinite vector $(S_k)_{k\in\mathbb{Z}}$ between each S_k and S_{k+1} the elements $S_{k2}, \dots, S_{km(k)-1}$, we obtain a new infinite vector, which we denote $(R_k)_{k\in\mathbb{Z}}$. Clearly, the partition $(R_k)_{k\in\mathbb{Z}}$ satisfies all the properties in the claim of the lemma. □

The next result is very helpful

Proposition A.16 *Let $z : \mathbb{R}_+ \to \mathbb{R}_+$ be nondecreasing and continuous at zero function satisfying $z(0) = 0$ and $z(r) > 0$ for $r > 0$. Then:*

(i) there exist $z_1, z_2 \in \mathcal{K} \bigcap C^\infty\big((0, +\infty)\big)$ *so that*

$$z_1(r) \leq z(r) \leq z_2(r) \quad r \geq 0. \tag{A.15}$$

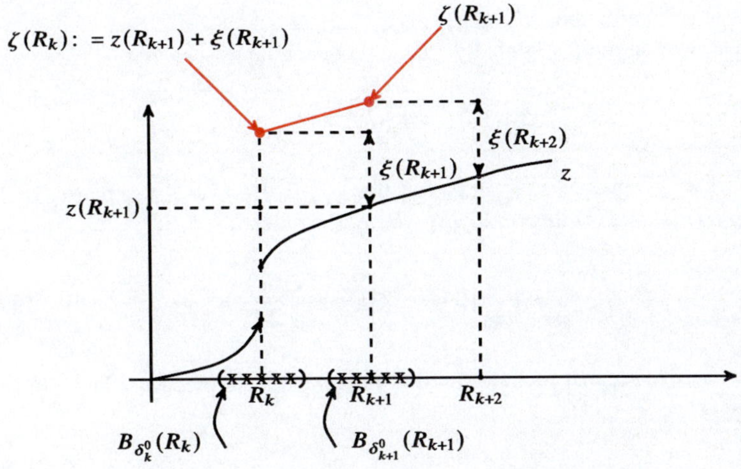

Fig. A.1 Illustration to the proof of Proposition A.16

(ii) *if in addition* $\lim_{r\to\infty} z(r) = \infty$, *then* z_1, z_2 *can be chosen from class* $\mathcal{K}_\infty$.
(iii) *if* $z \in C(\mathbb{R}_+, \mathbb{R}_+)$, *then for every* $\varepsilon > 0$ *the functions* z_1, z_2 *can be chosen so that* $\sup_{s\in\mathbb{R}_+} |z_2(s) - z(s)| < \varepsilon$ *and* $\sup_{s\in\mathbb{R}_+} |z_1(s) - z(s)| < \varepsilon$.

Proof We will prove the above properties for an upper bound z_2. The proof for z_1 is entirely analogous.

Pick a partition $(R_k)_{k\in\mathbb{Z}} \subset (0, +\infty)$ satisfying the properties (i)–(iv) of Proposition A.15. Pick any $\varepsilon > 0$ and any $\xi \in C^\infty(\mathbb{R}_+, \mathbb{R}_+) \bigcap \mathcal{K}$ so that $\lim_{r\to\infty} \xi(r) = \varepsilon$.

Define now a function (see Fig. A.1 for an illustration):

$$\zeta(R_k) := z(R_{k+1}) + \xi(R_{k+1}), \quad k \in \mathbb{Z}. \tag{A.16}$$

Since z is nondecreasing and ξ is strictly increasing, we obtain that $\zeta(R_k)$ is strictly increasing with k. Since $\xi \in \mathcal{K}$, $z(0) = 0$, and $R_k \to 0$ as $k \to -\infty$, it holds that $\zeta(R_k) \to 0$ as $k \to -\infty$.

Enlarge the domain of definition of ζ using linear splines: define for any $k \in \mathbb{Z}$ and any $s \in [R_k, R_{k+1}]$

$$\zeta(s) := \zeta(R_k) + \frac{s - R_k}{R_{k+1} - R_k}\big(\zeta(R_{k+1}) - \zeta(R_k)\big). \tag{A.17}$$

Clearly, ζ is continuous and increasing on $(0, +\infty)$. Finally, define $\zeta(0) := 0 = \lim_{s\to+0} \zeta(s)$. Thus, $\zeta \in \mathcal{K}$ and ζ is C^∞ everywhere except of $\bigcup_{k\in\mathbb{N}} R_k \cup \{0\}$.

For any $k \in \mathbb{Z}$ and for any $s \in [R_k, R_{k+1}]$ it holds that

$$\zeta(s) - z(s) \geq \zeta(R_k) - z(s) = z(R_{k+1}) - z(s) + \xi(R_{k+1}) \geq \xi(R_{k+1}) > 0,$$

and thus ζ upperbounds z on $\mathbb{R}_+$.

Next, we "smooth" ζ globally. Pick a sequence $(\delta_k^0)_{k\in\mathbb{Z}}$ so that

$$B_{\delta_k^0}(R_k)\bigcap B_{\delta_j^0}(R_j)=\emptyset,\quad k\neq j.$$

For each $k\in\mathbb{Z}$ pick $\delta_k<\delta_k^0$ and a function $\zeta_k: B_{\delta_k}(R_k)\to\mathbb{R}_+$ so that ζ_k is C^∞ on $B_{\delta_k}(R_k)$, $\zeta_k(s)=\zeta(s)$ for $s\in B_{\delta_k}(R_k)\backslash B_{\delta_k/2}(R_k)$ and

$$\sup_{s\in B_{\delta_k/2}(R_k)}|\zeta_k(s)-\zeta(s)|<\frac{\xi(R_k)}{2}<\frac{\varepsilon}{2}. \tag{A.18}$$

It is not hard to see that such functions always exist (ζ is a piecewise linear function). Define

$$z_2(s):=\begin{cases}\zeta_k(s), & \text{if } s\in B_{\delta_k}(R_k) \text{ for some } k\in\mathbb{Z},\\ \zeta(s), & \text{otherwise.}\end{cases}$$

By construction, $z_2\in C^\infty((0,+\infty),\mathbb{R}_+)$ and for any $k\in\mathbb{Z}$ and for any $s\in[R_k,R_{k+1}]$ it holds that

$$z_2(s)-z(s)=z_2(s)-\zeta(s)+\zeta(s)-z(s)\geq\xi(R_{k+1})-\frac{\xi(R_k)}{2}\geq\frac{\xi(R_k)}{2}>0.$$

Hence $z_2(s)-z(s)\geq 0$ for all $s\geq 0$, and the right estimate in (A.15) is shown. The left estimate can be shown similarly.
(ii) Follows by construction.
(iii) Let $z\in C(\mathbb{R}_+,\mathbb{R}_+)$. Pick any $\varepsilon>0$ and the corresponding partition $(R_k)_{k\in\mathbb{Z}}\subset(0,+\infty)$, satisfying axioms (i)–(v) of Proposition A.15. Construct for this partition the functions ζ and z_2 as above.

For any $k\in\mathbb{Z}$ it holds that

$$\sup_{s\in[R_k,R_{k+1}]}|z_2(s)-z(s)|\leq\sup_{s\in[R_k,R_{k+1}]}|z_2(s)-\zeta(s)|+\sup_{s\in[R_k,R_{k+1}]}|\zeta(s)-z(s)|.$$

Let us consider both terms independently. From (A.18) we obtain

$$\sup_{s\in[R_k,R_{k+1}]}|z_2(s)-\zeta(s)|=\max\Big\{\sup_{s\in[R_k,R_k+\delta_k/2]}|\zeta_k(s)-\zeta(s)|,$$
$$\sup_{s\in[R_{k+1}-\delta_{k+1}/2,R_{k+1}]}|\zeta_{k+1}(s)-\zeta(s)|\Big\}\leq\frac{\varepsilon}{2}$$

and according to the definition of ζ and in view of item (v) of Proposition A.15 (which holds since $z\in C(\mathbb{R}_+,\mathbb{R}_+)$) we obtain

$$\sup_{s\in[R_k,R_{k+1}]} |\zeta(s) - z(s)| \le \sup_{s\in[R_k,R_{k+1}]} \big(\zeta(R_{k+1}) - z(s)\big)$$
$$= \sup_{s\in[R_k,R_{k+1}]} \big(z(R_{k+2}) + \xi(R_{k+2}) - z(s)\big)$$
$$\le \xi(R_{k+2}) + \big(z(R_{k+2}) - z(R_{k+1})\big)$$
$$+ \sup_{s\in[R_k,R_{k+1}]} \big(z(R_{k+1}) - z(s)\big) < 3\varepsilon.$$

Overall,

$$\sup_{s\in\mathbb{R}_+} |z_2(s) - z(s)| < 3.5\varepsilon,$$

and item (iii) is shown. □

Next we prove a partial counterpart of Proposition A.16 for estimates of functions $\psi : \mathbb{R}_+ \times \mathbb{R}_+ \to \mathbb{R}_+$ by a $\mathcal{KL}$-function.

Proposition A.17 *Let $\psi : \mathbb{R}_+ \times \mathbb{R}_+ \to \mathbb{R}_+$ be any nondecreasing and continuous at 0 w.r.t. the first argument map that is nonincreasing w.r.t. the second argument and so that $\lim_{t\to\infty} \psi(r, t) = 0$ for any $r \ge 0$. Let also $\psi(0, t) = 0$ for any $t \ge 0$. Then there exists $\beta \in \mathcal{KL}$:*

$$\psi(r, t) \le \beta(r, t) \quad \forall r, t \ge 0.$$

Proof Pick a partition $R := (R_k)_{k\in\mathbb{Z}} \subset (0, +\infty)$ satisfying the properties (i)–(iv) of Proposition A.15 and another partition $\tau := (\tau_m)_{m\in\mathbb{N}} \subset [0, +\infty)$ so that $\tau_0 = 0$, (τ_m) is increasing, has no finite accumulation points and $\tau_m \to \infty$ as $m \to \infty$. Cartesian product $R \times \tau$ defines a mesh over $\mathbb{R}_+ \times \mathbb{R}_+$. Pick also any $\omega \in \mathcal{KL}$.

For each $k \in \mathbb{Z}$ and $m \in \mathbb{N}$, $m \ne 0$ define

$$\beta(R_k, \tau_m) := \psi(R_{k+1}, \tau_{m-1}) + \omega(R_{k+1}, \tau_{m-1}).$$

For $m = 0$ define

$$\beta(R_k, 0) = \beta(R_k, \tau_0) := 2\psi(R_{k+1}, 0) + \omega(R_{k+1}, 0).$$

For each $k \in \mathbb{Z}$ and $m \in \mathbb{N}$ define β within triangles (R_{k+1}, τ_m), (R_k, τ_{m+1}), (R_k, τ_m) and (R_{k+1}, τ_m), (R_{k+1}, τ_{m+1}), (R_k, τ_{m+1}) by a plane that intersects the values of β at the vertices (such plane is unique). This defines the values of β in $(0, +\infty) \times [0, +\infty) = \bigcup_{k\in\mathbb{Z}} \bigcup_{m\in\mathbb{N}} [R_k, R_{k+1}] \times [\tau_k, \tau_{k+1}]$.

Defining $\beta(0, t) := 0$ we see that β is defined over $\mathbb{R}_+ \times \mathbb{R}_+$, strictly increasing w.r.t. the first argument, strictly decreasing w.r.t. the second argument. Since $\lim_{t\to\infty} \psi(r, t) = 0$ for any $r \ge 0$, it holds also that $\lim_{t\to\infty} \beta(r, t) = 0$ for any $r \ge 0$. Hence β is continuous and thus $\beta \in \mathcal{KL}$.

It remains to show that β estimates ψ from above. To see this, pick any $k \in \mathbb{Z}$ and $m \in \mathbb{N}$. For every $(r, s) \in [R_k, R_{k+1}] \times [\tau_m, \tau_{m+1}]$ we have:

$$\begin{aligned}\beta(r,s)-\psi(r,s) &\geq \beta(R_k,\tau_{m+1})-\psi(R_{k+1},\tau_m)\\ &= \psi(R_{k+1},\tau_m)+\omega(R_{k+1},\tau_m)-\psi(R_{k+1},\tau_m)\\ &= \omega(R_{k+1},\tau_m).\end{aligned}$$

Hence, $\beta(r,s) \geq \psi(r,s)$ for all $r, s \geq 0$. □

The following result can be used, e.g., for the proof of the converse Lyapunov Theorem B.31:

Lemma A.18 *The following statements hold:*

(i) For any $\alpha \in \mathcal{P}$ and for any $L > 0$ there is $\rho \in \mathcal{P}$ so that $\rho(s) \leq \alpha(s)$ for all $s \in \mathbb{R}_+$ and ρ is globally Lipschitz with a Lipschitz constant L. One of such maps ρ is given by

$$\rho(r) := \inf_{y\geq 0}\big\{\alpha(y)+L|y-r|\big\}. \tag{A.19}$$

(ii) If in (i) $\alpha \in \mathcal{K}$, then ρ given by (A.19) is a $\mathcal{K}$-function.
(iii) If in (i) $\alpha \in \mathcal{K}_\infty$, then ρ given by (A.19) is a $\mathcal{K}_\infty$-function.

Proof (i) Consider ρ given by (A.19). Note that for any $r > 0$ it holds that $\alpha(y) + L|y-r| \to \infty$ as $y \to \infty$. Thus there is $r^* > 0$ such that $\rho(r) = \inf_{y\in[0,r^*]}\big\{\alpha(y) + L|y-r|\big\}$, and as α is continuous, there is $y^* = y^*(r)$ such that $\rho(r) = \alpha(y^*) + L|y^* - r|$.

Clearly, $0 \leq \rho(r) \leq \alpha(r)$ for all $r \geq 0$. Assume that $\rho(r) = 0$ for some $r \geq 0$. By above argument, $\rho(r) = \alpha(r)$, and as $\alpha(r) = 0$ if and only if $r = 0$, it follows that $\rho(0) = 0$ and $\rho(r) > 0$ for $r > 0$.

Next, for any $r_1, r_2 \geq 0$ we have by the triangle inequality

$$\begin{aligned}\rho(r_1)-\rho(r_2) &= \inf_{y\geq 0}\big\{\alpha(y)+L|y-r_1|\big\}-\inf_{y\geq 0}\big\{\alpha(y)+L|y-r_2|\big\}\\ &\leq \inf_{y\geq 0}\big\{\alpha(y)+L|y-r_2|+L|r_2-r_1|\big\}-\inf_{y\geq 0}\big\{\alpha(y)+L|y-r_2|\big\}\\ &= L|r_2-r_1|.\end{aligned}$$

Similarly, using the triangle inequality for the second term, we obtain that

$$\rho(r_1)-\rho(r_2) \geq -L|r_2-r_1|,$$

and thus ρ is globally Lipschitz with a Lipschitz constant L and is of class $\mathcal{P}$.
(ii) Let $\alpha \in \mathcal{K}$. Pick any $r_1, r_2 \geq 0$ with $r_1 > r_2$ and let $y_1 \geq 0$ be so that $\rho(r_1) = \alpha(y_1) + |y_1 - r_1|$. Consider the expression

$$\rho(r_1)-\rho(r_2) = \alpha(y_1)+L|y_1-r_1|-\inf_{y\geq 0}\{\alpha(y)+L|y-r_2|\}. \tag{A.20}$$

If $y_1 \geq r_2$, then $\rho(r_1) - \rho(r_2) \geq \alpha(y_1) + L|y_1 - r_1| - \alpha(r_2) > 0$, as α is increasing.

If $y_1 < r_2$, then

$$\rho(r_1) - \rho(r_2) \geq \alpha(y_1) + L|y_1 - r_1| - \big(\alpha(y_1) + L|y_1 - r_2|\big) > 0.$$

(iii) Let $\alpha \in \mathcal{K}_\infty$. Assume to the contrary that ρ is bounded: $\rho(r) \leq M$ for all r. Then for every r there is r' with $\alpha(r') + L|r - r'| \leq 2M$. Looking at the second term, we see that $r' \to \infty$ as $r \to \infty$. But then $\alpha(r') \to \infty$, a contradiction. □

As a corollary of Lemma A.18, we obtain

Lemma A.19 *For any $\alpha \in \mathcal{K}_\infty$ there is $\eta \in \mathcal{K}_\infty$ such that $\eta(r) \leq \alpha(r)$ for all $r \geq 0$, and* $\mathrm{id} - \eta \in \mathcal{K}_\infty$.

Proof Take any $L \in (0, 1)$ and construct $\rho \in \mathcal{K}_\infty$ globally Lipschitz with a constant L as in Lemma A.18.

Clearly, $(\mathrm{id} - \rho)(0) = 0$, and $\mathrm{id} - \rho$ is continuous. For $r, s \geq 0$ with $r > s$ we have

$$\begin{aligned} r - \rho(r) - (s - \rho(s)) &= r - s - (\rho(r) - \rho(s)) \geq r - s - L(r - s) \\ &= (1 - L)(r - s) > 0, \end{aligned}$$

and thus $\mathrm{id} - \rho$ is increasing. Furthermore, $r - \rho(r) \geq (1 - L)r \to \infty$ as $r \to \infty$ and thus $\mathrm{id} - \rho \in \mathcal{K}_\infty$. □

We proceed with a useful result on functions of 2 arguments that are $\mathcal{K}$-functions in both arguments.

Lemma A.20 *Let $\gamma \in C(\mathbb{R}_+ \times \mathbb{R}_+, \mathbb{R}_+)$ be such that $\gamma(r, \cdot) \in \mathcal{K}$ for each $r > 0$ and $\gamma(\cdot, s) \in \mathcal{K}$ for each $s > 0$. Then there is $\sigma \in \mathcal{K}_\infty$, such that*

$$\gamma(r, s) \leq \sigma(r)\sigma(s), \quad r, s \geq 0. \tag{A.21}$$

Proof See [2, Corollary IV.5]. □

We close this section with a result for upper bounds on families of $\mathcal{K}_\infty$-functions.

Lemma A.21 *Let $(\gamma_M)_{M\in\mathbb{N}}$ be a family of class $\mathcal{K}_\infty$ functions. Then, there exists $\sigma \in \mathcal{K}_\infty$ such that*

$$\gamma_M(r) \leq \sigma(M)\sigma(r), \quad M \in \mathbb{N},\ r \geq 0. \tag{A.22}$$

Proof See [1, Lemma 3.9]. □

A.2 Marginal Functions

In this book, we often deal with the functions defined as suprema or infima of continuous functions over certain regions. Such functions are occasionally called marginal functions. In this section, we prove basic results concerning such functions. We start with

Proposition A.22 *Let (S_X, ρ_X) and (S_Y, ρ_Y) be metric spaces with the metrics ρ_X and ρ_Y, respectively, $X \subset S_X$ be a compact set, and $Y \subset S_Y$ be locally compact (that is, each point of Y has a compact neighborhood). Assume that $f : X \times Y \to \mathbb{R}$ is continuous on $X \times Y$. Then function $g : y \mapsto \max_{x \in X} f(x, y)$ is continuous on Y.*

Proof For a given $y \in Y$ denote by $X_y := \arg\max_{x \in X} f(x, y)$ the set of values $x \in X$ on which a function $f(\cdot, y)$ takes its maximum. The set X_y is nonempty as X is compact.

Step 1. At first, we prove an auxiliary statement, namely: for all $y_0 \in Y$ and all $\omega > 0$ there exists $\delta > 0$, such that for all $y \in Y$, the following implication holds:

$$\rho_Y(y, y_0) < \delta \quad \Rightarrow \quad X_y \subset B_\omega(X_{y_0}) = \bigcup_{x \in X_{y_0}} B_\omega(x).$$

Assume that this statement does not hold. Then there exist $y_0 \in Y$, $\omega_0 > 0$ and sequences $(\delta_n) \subset \mathbb{R}$, $(y_n) \subset Y$, and $(x_n) \subset X$, such that

$$\lim_{n \to \infty} \delta_n = 0, \qquad \rho_Y(y_n, y_0) < \delta_n, \qquad x_n \in X_{y_n} \backslash B_{\omega_0}(X_{y_0}).$$

By construction $\lim_{n \to \infty} y_n = y_0$. As X is compact, from the sequence $(x_n)_{n=1}^\infty$ one can find a convergent subsequence $(x_{n_k})_{k=1}^\infty$. Let $\lim_{k \to \infty} (x_{n_k}, y_{n_k}) = (x^*, y_0) \in X \times Y$.

If $x^* \in X_{y_0}$, then some elements of $(x_{n_k})_{k=1}^\infty$ belong to $B_{\omega_0}(x^*) \subset B_{\omega_0}(X_{y_0})$, and we have a contradiction.

Let $x^* \notin X_{y_0}$. Then $f(x^*, y_0) < f(x_0, y_0)$ for some $x_0 \in X_{y_0}$, and therefore disjoint balls $B_s((x^*, y_0)) \subset X \times Y$ and $B_s((x_0, y_0)) \subset X \times Y$ for some $s > 0$ exist, such that $\forall (x', y') \in B_s((x^*, y_0))$, $\forall (x, y) \in B_s((x_0, y_0))$ it holds $f(x', y') < f(x, y)$. But for all $s > 0$ infinitely many elements of the sequence (x_{n_k}, y_{n_k}) belong to $B_s((x^*, y_0))$. Let one of them be $(x_{n_{k1}}, y_{n_{k1}})$. We have that function $f(\cdot, y_{n_{k1}})$ does not possess maximum at $x_{n_{k1}}$, and therefore $x_{n_{k1}} \notin X_{y_{n_{k1}}}$, and we come to a contradiction. Our statement is proven.

Step 2. Now we can prove the claim of the lemma. Take any $y_0 \in Y$. As Y is a locally compact space, there is a compact neighborhood S of y_0. As X is compact, $X \times S$ is compact. Since f is a continuous function, f is uniformly continuous on $X \times S$ according to Heine-Cantor's theorem. Pick any $\varepsilon > 0$. The uniform continuity ensures that

$$\exists \omega > 0 : \forall x_1, x_0 \in X : \ \rho_X(x_1, x_0) < \omega, \ \forall y \in S : \ \rho_Y(y, y_0) < \omega$$
$$\Rightarrow \ |f(x_1, y) - f(x_0, y_0)| < \varepsilon.$$

By the claim proved above, for this ω there is $\delta \in (0, \omega)$, such that for all $y \in Y$ with $\rho_Y(y, y_0) < \delta$ there exist $x_1 \in X_y$ and $x_0 \in X_{y_0}$ satisfying $\rho_X(x_1, x_0) < \omega$. Finally, we have

$$\begin{aligned} |g(y) - g(y_0)| &= |\max_{x\in X} f(x, y) - \max_{x\in X} f(x, y_0)| \\ &= |f(x_1, y) - f(x_0, y_0)| \\ &\leq \varepsilon. \end{aligned}$$

Thus, g is continuous at y_0. Since y_0 was chosen arbitrarily, g is continuous in all Y. □

We have the following corollary:

Corollary A.23 *Let $g : \mathbb{R}^n \to \mathbb{R}^p$, $n, p \in \mathbb{N}$ be a continuous function. Then $v, w : \mathbb{R}_+ \to \mathbb{R}_+$ defined by*

$$v(r) := \sup_{|x|=r} |g(x)|, \quad w(r) := \inf_{|x|=r} |g(x)|,$$

are continuous functions.

Proof Switching from Cartesian to spherical coordinates in n dimensions, we can write that $g(x) = \tilde{g}(r, \theta)$, where θ is a $n-1$-dimensional vector of angle coordinates on the hypersphere. Since we have made a continuous change of coordinates, $\tilde{g} : \mathbb{R}_+ \times [0, 2\pi]^{n-1} \to \mathbb{R}^p$ is continuous on its domain of definition (we include the angle 2π in order to have a compact domain for angles). Now

$$v(r) := \sup_{|x|=r} |g(x)| = \sup_{\theta\in[0,2\pi]^{n-1}} |\tilde{g}(r, \theta)|.$$

Applying Proposition A.22 to $(r, \theta) \mapsto |g(r, \theta)|$, we see that v is a continuous function.

Analogous treatment shows continuity of w. □

Lemma A.24 *Let $g : \mathbb{R}_+ \to \mathbb{R}^p$, $p \in \mathbb{N}$ be a continuous function. Then $\psi_1, \psi_2 : \mathbb{R}_+ \to \mathbb{R}_+$ defined for $r \geq 0$ by $\psi_1(r) := \inf_{x\geq r} |g(x)|$ and $\psi_2(r) := \sup_{x\leq r} |g(x)|$ are continuous and nondecreasing functions.*

Proof Clearly, ψ_1, ψ_2 are well-defined and nondecreasing. Pick any $r_1 \geq 0$. Continuity of g at point r_1 implies that for any $\varepsilon > 0$ there is a $\delta > 0$ s.t.

$$|x - r_1| \leq \delta \ \Rightarrow \ |g(x) - g(r_1)| < \varepsilon.$$

For this δ pick any $r \in B_\delta(r_1)$. Assume that $r > r_1$. Then we have

$$\begin{aligned}
\psi_2(r) &= \sup_{x \le r} |g(x)| \\
&= \max\Big\{ \sup_{x \le r_1} |g(x)|,\ \sup_{r_1 \le x \le r} |g(x)| \Big\} \\
&\le \max\Big\{ \sup_{x \le r_1} |g(x)|,\ \sup_{r_1 \le x \le r} \big(|g(x) - g(r_1)| + |g(r_1)| \big) \Big\} \\
&\le \max\Big\{ \sup_{x \le r_1} |g(x)|, \varepsilon + |g(r_1)| \Big\} \\
&\le \psi_2(r_1) + \varepsilon,
\end{aligned}$$

which shows that $\psi_2(r) - \psi_2(r_1) \le \varepsilon$. The case $r < r_1$ can be treated by exchange of r and r_1 in the above computations. Overall, ψ_2 is continuous. Similar argument shows continuity of ψ_1. □

Now we can state a corollary of Proposition A.22 and Lemma A.24:

Corollary A.25 *Let $g : \mathbb{R}_+ \times \mathbb{R}_+ \to \mathbb{R}$ be a continuous function. Then $\psi_1, \psi_2 : \mathbb{R}_+ \times \mathbb{R}_+ \to \mathbb{R}_+$ defined by*

$$\psi_1(r, t) := \inf_{x \ge r} |g(x, t)|, \quad \psi_2(r, t) := \sup_{x \le r} |g(x, t)|$$

are continuous and nondecreasing w.r.t. the first argument functions. If in addition $g(x, t) \ge 0$ for all $x, t \in \mathbb{R}_+ \times \mathbb{R}_+$ and $g(r, \cdot) \in \mathcal{L}$ for each $r > 0$, then also $\psi_2(r, \cdot) \in \mathcal{L}$ for all $r > 0$.

Proof It is clear that ψ_1, ψ_2 are nondecreasing w.r.t. the first argument.

Pick any $r_1, t_1 \ge 0$. Continuity of g at point (r_1, t_1) implies that for any $\varepsilon > 0$ there is a $\delta > 0$ s.t.

$$|x - r_1| \le \delta, \quad |t - t_1| \le \delta \quad \Rightarrow \quad |g(x, t) - g(r_1, t_1)| < \varepsilon.$$

For this δ pick any $r \in B_\delta(r_1)$, any $t \in B_\delta(t_1)$, and assume that $r > r_1$. We have:

$$\begin{aligned}
\psi_2(r, t) &= \sup_{x \le r} |g(x, t)| \\
&= \max\Big\{ \sup_{x \le r_1} |g(x, t)|,\ \sup_{r_1 \le x \le r} |g(x, t)| \Big\} \\
&\le \max\Big\{ \sup_{x \le r_1} |g(x, t)|, \\
&\qquad \sup_{r_1 \le x \le r} \big(|g(x, t) - g(r_1, t_1)| + |g(r_1, t_1) - g(r_1, t)| + |g(r_1, t)| \big) \Big\} \\
&\le \max\Big\{ \sup_{x \le r_1} |g(x, t)|, 2\varepsilon + |g(r_1, t)| \Big\} \\
&\le \psi_2(r_1, t) + 2\varepsilon,
\end{aligned}$$

which shows that

$$r \in [r_1, r_1 + \delta), \quad |t - t_1| \leq \delta \quad \Rightarrow \quad \psi_2(r, t) - \psi_2(r_1, t) \leq 2\varepsilon.$$

The case $r < r_1$ can be treated by exchange of r and r_1 in the above computations. Overall,

$$|r - r_1| \leq \delta, \quad |t - t_1| \leq \delta \quad \Rightarrow \quad |\psi_2(r, t) - \psi_2(r_1, t)| \leq 2\varepsilon. \qquad \text{(A.23)}$$

Hence, for any $r \in B_\delta(r_1)$ and any $t \in B_\delta(t_1)$

$$|\psi_2(r, t) - \psi_2(r_1, t_1)| \leq |\psi_2(r, t) - \psi_2(r_1, t)| + |\psi_2(r_1, t) - \psi_2(r_1, t_1)|.$$

As r_1 is fixed, the continuity of $\psi_2(r_1, \cdot)$ is implied by Proposition A.22 (in this case $X = [0, r_1]$). Hence, there is $\delta_2 \in (0, \delta)$, such that for all $t \in B_{\delta_2}(t_1)$ it holds that $|\psi_2(r_1, t) - \psi_2(r_1, t_1)| \leq \varepsilon$.

Combining this with (A.23), we obtain that

$$|r - r_1| \leq \delta_2, \quad |t - t_1| \leq \delta_2 \quad \Rightarrow \quad |\psi_2(r, t) - \psi_2(r_1, t_1)| \leq 3\varepsilon, \qquad \text{(A.24)}$$

which shows continuity of ψ_2. A similar argument shows the continuity of ψ_1.

Now let $g(x, t) \geq 0$ for all x, t and let $g(r, \cdot) \in \mathcal{L}$ for each $r > 0$. Pick any $r \geq 0$ and any $t_1, t_2 \in \mathbb{R}_+$: $t_1 < t_2$. Then for a certain $x_2 \in [0, r]$ it holds that

$$\psi_2(r, t_2) = \sup_{x \leq r} g(x, t_2) = g(x_2, t_2) < g(x_2, t_1) \leq \sup_{x \leq r} g(x, t_1) = \psi_2(r, t_1).$$

This finishes the proof. □

Remark A.26 Note that in the proof of Corollary A.25, the continuity of $\psi_1(\cdot, t)$ for each fixed $t \in \mathbb{R}_+$ follows by Lemma A.24 and continuity of $\psi_1(r, \cdot)$ for any fixed $r \in \mathbb{R}_+$ is implied by Proposition A.22 (in this case $X = [0, r]$, which is compact for a fixed r). However, this does not imply the joint continuity of ψ_1 in both arguments. Hence in the proof of Corollary A.25, we have mimicked the arguments from the proof of Lemma A.24 once again.

Lemma A.27 *Let $g : \mathbb{R}^n \to \mathbb{R}^p$, where $n, p \in \mathbb{N}$, be a continuous function and $g(0) = 0$. Then $\psi_1, \psi_2 : \mathbb{R}_+ \to \mathbb{R}_+$ defined by*

$$\psi_1(r) := \inf_{|x| \geq r} |g(x)|, \quad \psi_2(r) := \sup_{|x| \leq r} |g(x)|,$$

are well-defined, continuous, nondecreasing, and $\psi_1(0) = \psi_2(0) = 0$.

Proof Clearly, ψ_1, ψ_2 are well-defined, nondecreasing and $\psi_1(0) = \psi_2(0) = 0$. Define $v, w : \mathbb{R}_+ \to \mathbb{R}_+$ by $v(r) := \sup_{|x|=r} |g(x)|$ and $w(r) := \inf_{|x|=r} |g(x)|$. These functions are continuous according to Corollary A.23. Now, $\psi_1(r) = \inf_{s \geq r} w(s)$ and

$\psi_2(r) = \sup_{0\le s\le r} \nu(s)$. An application of Lemma A.24 shows continuity of ψ_1 and ψ_2. □

Lemma A.28 *Let $\xi \in \mathcal{P}$ and $\underline{\lim}_{r\to\infty} \xi(r) > 0$. Then there is a $\psi \in \mathcal{K}$: $\psi(r) \le \xi(r)$ for $r \in \mathbb{R}_+$ and $\lim_{r\to\infty} \psi(r) = \underline{\lim}_{r\to\infty} \xi(r)$. In particular, if $\underline{\lim}_{r\to\infty} \xi(r) = \infty$, then $\psi \in \mathcal{K}_\infty$.*

Proof Define $\tilde{\psi}(r) := \inf_{s\ge r} \xi(s)$. By Lemma A.27, $\tilde{\psi}$ is nondecreasing, continuous, $\xi(r) \ge \tilde{\psi}(r)$ for all $r \in \mathbb{R}_+$ and $\tilde{\psi}(0) = 0$. Furthermore, by definition $\lim_{r\to\infty} \psi(r) = \underline{\lim}_{r\to\infty} \xi(r)$.

Now the $\mathcal{K}$-function $\psi(r) := \tilde{\psi}(r)(1 - e^{-r})$, $r \in \mathbb{R}_+$ satisfies all the needed properties. □

A.3 Dini Derivatives

As we know (see, e.g., Theorem 1.2), absolutely continuous functions are differentiable almost everywhere. However, sometimes one has to "differentiate" functions, which are merely continuous. To do this, we exploit Dini derivatives. For a continuous function $y : \mathbb{R} \to \mathbb{R}$, define the right upper Dini derivative and the right lower Dini derivative by

$$D^+y(t) := \overline{\lim_{h\to+0}} \frac{y(t+h) - y(t)}{h} \quad \text{and} \quad D_+y(t) := \underline{\lim_{h\to+0}} \frac{y(t+h) - y(t)}{h},$$

respectively.

We start with simple properties of Dini derivatives

Lemma A.29 *The following properties of Dini derivatives hold:*

(i) For all $f, g \in C(\mathbb{R})$ it holds that $D^+(f + g) \le D^+(f) + D^+(g)$.
(ii) For any $f \in C(\mathbb{R})$ it holds that $D^+(-f) = -D_+(f)$.

Proof Is left to the reader (Exercise A.15). □

We proceed with the following generalization of the chain rule:

Lemma A.30 *Let g be a continuous function on a bounded interval $[a, b]$ and let t be such that $f : g([a, b]) \to \mathbb{R}$ is continuously differentiable at $g(t)$ and $\frac{df}{dr}(r)\Big|_{r=g(t)} > 0$. Then*

$$D^+(f \circ g)(t) = \frac{df}{dr}(r)\Big|_{r=g(t)} D^+g(t).$$

Proof Consider a function $h : t \in \mathbb{R} \to \mathbb{R}$ that is continuous at $t = c$ and satisfies $h(c) > 0$. Let $k : \mathbb{R} \to \mathbb{R}$ be an arbitrary continuous function. Then the following holds (allowing the limits to be equal $\pm\infty$)

$$\overline{\lim_{t\to c}} \big(h(t)k(t)\big) = \lim_{t\to c} h(t) \cdot \overline{\lim_{t\to c}} k(t).$$

Then for f, g, and t as in the statement of the lemma, the following holds

$$\begin{aligned} D^+(f\circ g)(t) &= \overline{\lim_{h\to+0}} \frac{f\circ g(t+h) - f\circ g(t)}{h} \\ &= \overline{\lim_{h\to+0}} \frac{f(g(t+h)-g(t)+g(t)) - f\circ g(t)}{g(t+h)-g(t)} \cdot \frac{g(t+h)-g(t)}{h} \\ &= \lim_{h\to+0} \frac{f(g(t+h)-g(t)+g(t)) - f\circ g(t)}{g(t+h)-g(t)} \overline{\lim_{h\to+0}} \frac{g(t+h)-g(t)}{h} \\ &= \frac{df}{dr}(r)\Big|_{r=g(t)} D^+g(t). \end{aligned}$$

We state also the following version of the fundamental theorem of calculus (which is a special case of [7, Theorem 9], [16, Theorem 7.3, pp. 204–205]), which is useful, in particular, for the proof of direct Lyapunov theorem (Proposition 2.19).

Proposition A.31 *Let $f : [a, b] \to \mathbb{R}$ be continuous and $g : [a, b] \to \mathbb{R}$ be locally Lebesgue integrable and bounded on $[a, b]$. Suppose that*

$$D_+ f(t) \le g(t), \quad t \in [a, b]. \tag{A.25}$$

Then for all $t \in [a, b]$ we have

$$f(t) - f(a) \le \int_a^t g(s)ds. \tag{A.26}$$

Proof Define for $t \in [a, b]$ the function $G(t) := \int_a^t g(s)ds$.

First, assume that g is continuous. Then it follows for all $t \ge 0$ that

$$D_+\left(f(t) - G(t)\right) = (D_+ f(t)) - g(t) \le 0.$$

It follows from [18, Theorem 2.1] that $f + G$ is nonincreasing. As $G(a) = 0$ the claim follows.

Now let g be locally Lebesgue integrable and bounded on $[a, b]$, and pick an arbitrary $t > 0$. As continuous functions are dense in L^1 and since g has a uniform bound on $[0, t]$, there is a sequence of continuous functions (g_i) and $M > 0$ so that $|g_i(r)| \le M$ for all $i \in \mathbb{N}$ and all $r \in [0, t]$ and

$$\int_0^t |g_i(r) - g(r)|dr < \frac{1}{i}, \qquad i \in \mathbb{N}.$$

Without loss of generality, we may choose the functions g_i satisfying $g_i \ge g$ pointwise.

Then for all $i \in \mathbb{N}$ it holds that

$$f(t) - f(a) \leq \int_a^t g_i(s)ds = \int_a^t g(s)ds + \int_a^t g_i(s) - g(s)ds.$$

Taking the limit $i \to \infty$, we obtain that

$$f(t) - f(a) \leq \int_a^t g(s)ds.$$

The proposition is shown. □

We close the section with a lemma on derivatives of monotone functions.

Lemma A.32 *Let $b : \mathbb{R}_+ \to \mathbb{R}$ be a nonincreasing function. Then for each $t \in \mathbb{R}_+$ it holds that*

$$b(t) \geq D^+ \int_0^t b(s)ds \geq D_+ \int_0^t b(s)ds \geq \lim_{h \to +0} b(t+h). \tag{A.27}$$

Proof As b is nonincreasing, it is Riemann integrable on any finite subinterval of $\mathbb{R}_+$, and thus the integrals in the formulation of the result make sense. Pick any $t \geq 0$. By definition of a Dini derivative, it holds that

$$\begin{aligned} D^+ \int_0^t b(s)ds &= \overline{\lim_{h \to +0}} \frac{1}{h} \Big(\int_0^{t+h} b(s)ds - \int_0^t b(s)ds \Big) \\ &= \overline{\lim_{h \to +0}} \frac{1}{h} \int_t^{t+h} b(s)ds \\ &\leq \overline{\lim_{h \to +0}} \frac{1}{h} \int_t^{t+h} b(t)ds = b(t). \end{aligned}$$

On the other hand, we have that

$$D_+ \int_0^t b(s)ds \geq \underline{\lim_{h \to +0}} \frac{1}{h} \int_t^{t+h} b(t+h)ds = \lim_{h \to +0} b(t+h).$$

The inequality $D^+ \int_0^t b(s)ds \geq D_+ \int_0^t b(s)ds$ is clear. □

A.4 Comparison Principles

Comparison principles are a type of results where the properties of certain differential equations or inequalities are derived by relating them to the properties of some auxiliary equations or inequalities.

We start with a well-known basic result.

Proposition A.33 *Let f satisfy Assumption 1.8. Furthermore, let $u \in \mathcal{U}$, $T > 0$ and $x, y \in AC([0, T), \mathbb{R})$ be such that $x(0) \leq y(0)$ and*

$$\dot{x}(t) - f(x(t), u(t)) \leq \dot{y}(t) - f(y(t), u(t)), \quad \text{for a.e. } t \in [0, T). \tag{A.28}$$

Then $x(t) \leq y(t)$ for every $t \in [0, T)$.

Proof Let the statement be false. Then there is $s > 0$ such that $x(s) = y(s)$ and $x(t) > y(t)$ for $t \in (s, s + \varepsilon]$ for some $\varepsilon > 0$. Define $z := x - y$. As x, y are absolutely continuous, they are differentiable almost everywhere, and an easy argument shows that for almost every $t \in [s, s + \varepsilon)$ it holds that

$$\begin{aligned}\dot{z}(t) = \dot{x}(t) - \dot{y}(t) &\leq f(x(t), u(t)) - f(y(t), u(t)) \\ &\leq |f(x(t), u(t)) - f(y(t), u(t))| \\ &\leq L|x(t) - y(t)| = Lz(t),\end{aligned} \tag{A.29}$$

where we have used that $z(t) > 0$, $t \in [s, s + \varepsilon)$. Here L is a Lipschitz constant that is uniform for all $t \in [0, s + \varepsilon]$, as x and y are uniformly bounded over $[s, s + \varepsilon]$, and f is Lipschitz continuous on bounded balls.

The inequality (A.29) together with $z(s) = 0$ leads to $z(t) \leq z(s)e^{L(t-s)} = 0$, for $t \in [s, s + \varepsilon]$, which is a contradiction. □

A variation of the above principle for the Dini derivatives is as follows:

Proposition A.34 *Let Assumption 1.8 hold, let $u \in C(\mathbb{R}_+, \mathbb{R}^m)$, $T > 0$, and let $x, y \in C([0, T), \mathbb{R})$ be such that $x(0) \leq y(0)$ and*

$$D^+x(t) - f(x(t), u(t)) \leq D_+y(t) - f(y(t), u(t)), \quad \text{for all } t \in [0, T). \tag{A.30}$$

Then $x(t) \leq y(t)$ for every $t \in [0, T)$.

Proof Let the statement be false. Then there is $s > 0$ such that $x(s) = y(s)$ and $x(t) > y(t)$ for $t \in (s, s + \varepsilon]$ for some $\varepsilon > 0$. Define $z := x - y$.

For every $t \in [s, s + \varepsilon)$ it holds that

$$\begin{aligned}D^+z(t) = D^+(x(t) - y(t)) \leq D^+x(t) + D^+(-y)(t) &= D^+x(t) - D_+y(t) \\ &\leq f(x(t), u(t)) - f(y(t), u(t)) \\ &\leq |f(x(t), u(t)) - f(y(t), u(t))| \\ &\leq L|x(t) - y(t)| = Lz(t),\end{aligned}$$

where we have used that $z(t) > 0, t \in [s, s+\varepsilon)$. Here L is the Lipschitz constant that is uniform for all $t \in [0, s+\varepsilon]$, as x and y are uniformly bounded over $[s, s+\varepsilon]$, and f is Lipschitz continuous on bounded balls. By Proposition A.31, this implies that

$$z(t) - z(s) \le L \int_s^t z(\tau)d\tau, \quad t \in [s, s+\varepsilon],$$

which by Grönwall's inequality leads to

$$z(t) \le z(s)e^{L(t-s)}, \quad t \in [s, s+\varepsilon].$$

As $z(s) = 0$, we come to a contradiction. □

Next, we would like to prove another comparison principle, ensuring a uniform estimate for solutions to certain differential inequalities. This result can be understood as a nonlinear version of the Grönwall's inequality.

Proposition A.35 (Comparison principle) *For any $\alpha \in \mathcal{P}$ there exists $\beta \in \mathcal{KL}$ so that for any $y \in C(\mathbb{R}_+, \mathbb{R}_+)$ satisfying the differential inequality*

$$D^+y(t) \le -\alpha(y(t)) \quad \forall t > 0, \tag{A.31}$$

it holds that

$$y(t) \le \beta(y(0), t) \quad \forall t \ge 0. \tag{A.32}$$

Proof From (A.31), it follows that

$$\frac{D^+y(t)}{\alpha(y(t))} \le -1 \tag{A.33}$$

for all t so that $y(t) \ne 0$.

Define $\eta : \mathbb{R}_+ \to [-\infty, \infty)$ by

$$\eta(s) := \int_1^s \frac{dr}{\alpha(r)}. \tag{A.34}$$

Without loss of generality, we assume $\lim_{s\to+0} \eta(s) = -\infty$ and $\lim_{s\to+\infty} \eta(s) = +\infty$.[1] If this is not the case, following the idea in [12, Lemma 4.4], we can replace α with a function $s \mapsto \min\{s, \alpha(s)\}$, $s \in \mathbb{R}_+$ to obtain a function of class $\mathcal{P}$, satisfying (A.31) with $\lim_{s\to+0} \eta(s) = -\infty$ and $\lim_{s\to+\infty} \eta(s) = +\infty$. Since α is continuous, η

[1] Property $\lim_{s\to+\infty} \eta(s) = +\infty$ is not necessary since (A.37) is defined for $t \in \mathbb{R}_+$.

is continuously differentiable, and its derivative is positive on its domain of definition. Thus, Lemma A.30 implies

$$\frac{D^+y(t)}{\alpha(y(t))} = D^+\eta(y(t)),$$

and (A.33) can be rewritten as

$$D^+\eta(y(t)) \leq -1. \tag{A.35}$$

Applying Proposition A.31 with $f := \eta(y(\cdot))$, we obtain that

$$\eta(y(t)) - \eta(y(0)) \leq -t. \tag{A.36}$$

Since η is strictly increasing, η is invertible and its inverse η^{-1} is a strictly increasing function, defined over $[-\infty, +\infty)$. An easy manipulation with (A.36) leads to

$$y(t) \leq \eta^{-1}(\eta(y(0)) - t) \tag{A.37}$$

for all $t \in [0, T)$, where $T := \min\{t \in \mathbb{R}_+ : y(t) = 0\}$. Here, let $T = \infty$ if $\{t \in \mathbb{R}_+ : y(t) = 0\} = \emptyset$. If $y(\tau) = 0$ holds for some $\tau \in \mathbb{R}_+$, then $y(t) = 0$ is satisfied for all $t \geq \tau$, due to (A.31) and hence $y(t) \geq 0$. Define $\beta : \mathbb{R}_+ \times \mathbb{R}_+ \to \mathbb{R}_+$ by

$$\beta(r, t) := \begin{cases} \eta^{-1}(\eta(r) - t), & \text{if } r > 0, \\ 0, & \text{if } r = 0. \end{cases}$$

Due to the strict increasing property of η and since $\lim_{s \to +0} \eta(s) = -\infty$, we have $\beta \in \mathcal{KL}$. With this β the inequality (A.32) follows from (A.37) and $y(t) = 0$, for all $t \geq T$. □

We state also a comparison principle for systems with inputs:

Proposition A.36 *For any $\rho \in \mathcal{P}$ there is $\beta \in \mathcal{KL}$ such that for any $\tilde{t} \in (0, +\infty]$, any $y \in C([0, \tilde{t}), \mathbb{R})$ with $y(0) \geq 0$ and any $v \in C([0, \tilde{t}), \mathbb{R}_+)$ if*

$$D^+y(t) \leq -\rho\big(\max\{y(t) + v(t), 0\}\big) \tag{A.38}$$

holds for all $t \in [0, \tilde{t})$, then denoting by v_t the restriction of v to $[0, t]$, we have

$$y(t) \leq \max\big\{\beta(y(0), t), \|v_t\|_\infty\big\}, \quad t \in [0, \tilde{t}). \tag{A.39}$$

Proof By Proposition A.13, for $\rho \in \mathcal{P}$ pick $\omega \in \mathcal{K}_\infty$ and $\sigma \in \mathcal{L}$ such that $\rho(r) \geq \omega(r)\sigma(r)$, $r \geq 0$. With this estimate, we proceed from (A.38) to

$$D^+y(t) \leq -\omega\big(\max\{y(t) + v(t), 0\}\big)\sigma\big(\max\{y(t) + v(t), 0\}\big). \tag{A.40}$$

Define now $t_0 := \min\{t \in [0, \tilde{t}) : y(t) \le \|v_t\|_\infty\}$ and set $t_0 := \tilde{t}$, if $y(t) > \|v_t\|_\infty$ for all $t \in [0, \tilde{t})$.

As $D^+y(t) \le 0$ for $t \ge 0$, y is nonincreasing by Proposition A.31. Since $t \mapsto \|v_t\|_\infty$ is nondecreasing, from the definition of t_0 it follows that

$$y(t) \le \|v_t\|_\infty, \quad t \in [t_0, \tilde{t}). \tag{A.41}$$

For $t \le t_0$ it holds that

$$2y(t) \ge \max\{y(t) + v(t), 0\} \ge y(t) \ge v(t) \ge 0,$$

and we proceed from (A.40) to

$$D^+y(t) \le -\omega(y(t))\sigma(2y(t)), \quad t \in [0, t_0). \tag{A.42}$$

As $r \mapsto \omega(r)\sigma(2r)$ is a positive definite function, we can apply Proposition A.35 to infer existence of $\beta \in \mathcal{KL}$ such that

$$y(t) \le \beta(y(0), t), \quad t \in [0, t_0). \tag{A.43}$$

Combining (A.41) and (A.43), we obtain the claim. □

The final result in this section is

Proposition A.37 *For any $\rho \in \mathcal{P}$ there is $\beta \in \mathcal{KL}$ such that for any $\tilde{t} \in (0, +\infty]$, any $y \in AC([0, \tilde{t}), \mathbb{R}_+)$ and any $v \in L^\infty([0, \tilde{t}), \mathbb{R}_+)$ if*

$$\dot{y}(t) \le -\rho(y(t)) + v(t) \tag{A.44}$$

holds for almost all $t \in [0, \tilde{t})$ such that $y(t) > 0$, then

$$y(t) \le \beta(y(0), t) + \int_0^t 2v(s)ds, \quad t \in [0, \tilde{t}). \tag{A.45}$$

Proof We assume that (A.44) holds for almost all $t \in [0, \tilde{t})$ such that $y(t) > 0$. Let $y(t) = 0$ on a certain interval $[t_1, t_2] \subset [0, \tilde{t})$. Then $\dot{y}(t) = 0$ on (t_1, t_2), and in particular, the estimate (A.44) holds on this interval. This shows that (A.44) holds in fact for almost all $t \in [0, \tilde{t})$.

By Lemma A.18, there is a globally Lipschitz continuous ρ_1, such that $\rho \ge \rho_1$ pointwise. Consider the following initial value problem:

$$\dot{w}(t) = -\rho_1(|w(t)|) + v(t), \quad w(0) = y(0). \tag{A.46}$$

As ρ_1 is globally Lipschitz, there is a unique (locally) absolutely continuous solution w of (A.46).

Since

$$\dot{y}(t) \le -\rho_1(y(t)) + v(t), \quad \text{for a.e. } t \in [0, \tilde{t}) \tag{A.47}$$

using Proposition A.33, we obtain that $0 \le y(t) \le w(t)$ for all $t \in [0, \tilde{t})$.

Define for $t \in [0, \tilde{t})$

$$v_1(t) := \int_0^t v(s)ds, \qquad w_1(t) = w(t) - v_1(t).$$

As w, v_1, and w_1 are absolutely continuous, they are differentiable outside of the sets S_1, S_2, and S_3, respectively, all of which have measure zero. Thus, all of them are differentiable simultaneously for all $t \notin S_1 \cup S_2 \cup S_3$, and for all such t it holds that

$$\dot{w}_1(t) = \dot{w}(t) - \dot{v}_1(t).$$

Thus, for almost all $t \in [0, \tilde{t})$, we have

$$\dot{w}_1(t) = \dot{w}(t) - v(t) = -\rho_1(|w(t)|) = -\rho_1(w(t)),$$

where the last equality holds since $w(t) \ge 0$ for all t.

Thus, for almost all $t \in [0, \tilde{t})$, we obtain

$$\dot{w}_1(t) = -\rho_1(\max\{w_1(t) + v_1(t), 0\}).$$

Applying Proposition A.36, there is $\beta \in \mathcal{KL}$ depending on ρ_1 solely, such that for all $t \in [0, \tilde{t})$ that

$$w_1(t) \le \max\{\beta(w_1(0), t), \|v_{1,t}\|_\infty\}.$$

As $w_1(0) = w(0) = y(0)$, we infer for all $t \in [0, \tilde{t})$ that

$$\begin{aligned} y(t) &\le w(t) = w_1(t) + v_1(t) \\ &\le \beta(y(0), t) + \|v_{1,t}\|_\infty + v_1(t) \\ &= \beta(y(0), t) + 2\int_0^t v(s)ds. \end{aligned}$$

The result is shown. □

A.5 Brouwer's Theorem and Its Corollaries

In this section, we state some fundamental results from topology of $\mathbb{R}^n$.

Definition A.38 For $S \subset \mathbb{R}^n$, we call $x \in S$ a *fixed point* of a map $f : S \to S$ if $f(x) = x$.

We start with the celebrated

Theorem A.39 (Brouwer's fixed point theorem) *Let B be a closed ball in $\mathbb{R}^n$. Then every $f \in C(B)$ has a fixed point.*

One of the most important consequences of Brouwer's theorem is

Theorem A.40 (Brouwer's invariance of domain theorem) *Let S be an open subset of $\mathbb{R}^n$, and let $f : S \to \mathbb{R}^n$ be a continuous injective map. Then $f(S)$ is also open.*

We introduce several further definitions:

Definition A.41 $f : \mathbb{R}^n \to \mathbb{R}^n$ is called an *open map*, if f maps open sets into open sets. If f maps closed sets to closed sets, then f is called a *closed map*.

Definition A.42 A map $T : \mathbb{R}^n \to \mathbb{R}^n$ is called a *homeomorphism* if T is continuous, bijective and has a continuous inverse.

Invariance of domain theorem immediately implies the following important properties of homeomorphisms:

Corollary A.43 *Any homeomorphism $T : \mathbb{R}^n \to \mathbb{R}^n$ is an open and closed map.*

Proof Let $T : \mathbb{R}^n \to \mathbb{R}^n$ be a homeomorphism. Then it is an open map by Theorem A.40.

Now let C be a closed subset of $\mathbb{R}^n$. Denote $S := \mathbb{R}^n \backslash C$. Since S is open, $T(S)$ is again open. But T is a bijection, which means that $\mathbb{R}^n = T(S) \cup T(C)$ and $T(S) \bigcap T(C) = \emptyset$. Thus, $T(C) = \mathbb{R}^n \backslash T(S)$ and hence $T(C)$ is closed. □

Next lemma shows that homeomorphisms on $\mathbb{R}^n$ are positive definite and radially unbounded transformations.

Lemma A.44 *For any homeomorphism $T : \mathbb{R}^n \to \mathbb{R}^n$ satisfying $T(0) = 0$ there exist $\psi_1, \psi_2 \in \mathcal{K}_\infty$:*

$$\psi_1(|x|) \leq |T(x)| \leq \psi_2(|x|) \quad \forall x \in \mathbb{R}^n. \tag{A.48}$$

Proof The precise bounds for the homeomorphism T are given by

$$\psi_1(r) := \inf_{|x| \geq r} |T(x)|, \qquad \psi_2(r) := \max_{|x| \leq r} |T(x)|. \tag{A.49}$$

Clearly, ψ_1, ψ_2 are well-defined. Due to Lemma A.27, ψ_1, ψ_2 are continuous.

Since $T(0) = 0$, and since (by Corollary A.43) T is an open map, for every $r > 0$ there exists $\delta_r > 0$ so that $\overline{B_{\delta_r}} \subset T\big(B_r\big)$. Injectivity of T now implies that $\overline{B_{\delta_r}} \bigcap T\big(\mathbb{R}^n \backslash B_r\big) = \emptyset$, which shows that $\psi_1(r) = \min_{|x| \geq r} |T(x)| > 0$ for any $r > 0$.

Since $T(0) = 0$, $\psi_1(0) = \psi_2(0) = 0$, and T is a bijection, ψ_1 and ψ_2 are positive definite functions. Due to construction, they are also nondecreasing. Let us show that ψ_1, ψ_2 are strictly increasing.

Pick any $r_1, r_2 > 0$: $r_2 > r_1$ and set $r_m := \frac{1}{2}(r_1 + r_2)$. By Corollary A.43 and due to bijectivity of T, it holds that $T(\overline{B_{r_1}}) \subset T(B_{r_m}) \subset T(\overline{B_{r_2}})$, and the inclusion is strict. This implies $\psi_2(r_2) > \psi_2(r_1)$, and hence ψ_2 is strictly increasing. Analogously, ψ_1 is also strictly increasing.

Finally, since T is surjective and maps compact sets into compact sets, $\psi_2(r) \to \infty$ and $\psi_1(r) \to \infty$ as $r \to \infty$. This shows that $\psi_1, \psi_2 \in \mathcal{K}_\infty$. □

A.6 Concluding Remarks

Comparison functions. Classes of comparison functions were introduced by Jose Massera in the 1950s and were widely used by Hahn in his book [8] to characterize stability notions of dynamical systems. The usefulness of these classes of functions in the ISS theory and other contexts made their use standard within the systems theoretic society.

Lemma A.6 is due to [15, Lemmas 1.1.3, 1.1.5]. Proposition A.7 is due to Sontag, see [17, Proposition 7]. The proof that we employed is due to [19, Lemma 3, p. 326]. Proposition A.13 was shown for scalar positive definite functions in [2, Lemma IV.1], and extended to general positive definite functions in [11, Lemma 18]. Proposition A.16 generalizes [5, Lemma 2.5] (estimates from below/above by infinitely differentiable $\mathcal{K}$-functions) and [15, Lemma 1.1.6] (approximation by C^∞ functions staying in ε-neighborhood of a given function) and unifies these two features. Lemma A.18 is due to [10, p. 130]. Lemma A.21 was shown in [1, Lemma 3.9]. Lemma A.22 is well-known. A reader may consult [3, Theorem 1.4.16] for more general results of this kind.

Some further properties of comparison functions can be found in a survey [11], which is a great complement to the material described in this chapter. Some general ways for recasting $\varepsilon - \delta$ definitions into comparison function formalism are analyzed in [14].

Dini derivatives. Dini has introduced the derivatives of continuous functions in [6]. For more on this theory we refer to [4, Sect. 6], [5, 7], [9, Chap. 3].

Comparison principles. Proposition A.33 is rather well-known, though the author does not have a precise reference for it. For absolutely continuous y, Proposition A.35 was proved in [12, Lemma 4.4], Proposition A.36 was shown in [2, Lemma IV.2], and Proposition A.37 was shown in [2, Corollary IV.3]. We have extended it to continuous y by using the Dini derivatives formalism. In the time-delay realm, a notable result, which is closely connected to comparison principles, is a nonlinear Halanay's inequality [13].

A.7 Exercises

Comparison functions

Exercise A.1 *(Another weak triangle inequality)* Show that for any $\gamma \in \mathcal{K}$, any $\rho \in \mathcal{K}_\infty$ satisfying $\rho - \mathrm{id} \in \mathcal{K}_\infty$, and for any $a, b \in \mathbb{R}_+$ it holds that:

$$\gamma(a+b) \leq \gamma \circ \rho(a) + \gamma \circ \rho \circ (\rho - \mathrm{id})^{-1}(b). \tag{A.50}$$

Exercise A.2 *(Quadrille inequality)* Show that for any $x_1, x_2, y_1, y_2 \in \mathbb{R}^n$ it holds that

$$\big||x_1 - x_2| - |y_1 - y_2|\big| \leq |x_1 - y_1| + |x_2 - y_2|. \tag{A.51}$$

Exercise A.3 Prove Lemma A.5.

Exercise A.4 Show that for any $\alpha \in \mathcal{K}$ and any $a, b \geq 0$ it holds that

$$\alpha(a+b) \geq \frac{1}{2}\alpha(a) + \frac{1}{2}\alpha(b). \tag{A.52}$$

Exercise A.5 Prove that $f : \mathbb{R}^n \to \mathbb{R}^m$ is continuous at point $x_0 \in \mathbb{R}^n$ if and only if there is $\gamma \in \mathcal{K}_\infty$ and $r > 0$, such that for all $x \in \mathbb{R}^n$: $|x| \leq r$ it holds that $|f(x + x_0) - f(x_0)| \leq \gamma(|x|)$.

Exercise A.6 We call $f : D \subset \mathbb{R}^n \to \mathbb{R}^m$ $\mathcal{K}$-continuous, if there is $\gamma \in \mathcal{K}$, such that for all $x, y \in D$ it holds that $|f(x) - f(y)| \leq \gamma(|x - y|)$.

Show that if f is $\mathcal{K}$-continuous, then f is uniformly continuous. Show that if D is compact, then $\mathcal{K}$-continuity and uniform continuity are equivalent properties.

Exercise A.7 Let $f : \mathbb{R}^n \to \mathbb{R}^n$ be Lipschitz continuous on bounded balls. Show that there exist $w \in \mathcal{K}$ and $L > 0$ so that

$$|f(x) - f(0)| \leq \big(L + w(|x|)\big)|x|, \quad x \in \mathbb{R}^n.$$

Exercise A.8 Let $\gamma \in \mathcal{K}_\infty$ satisfy the condition $\gamma(s) < s$ for all $s > 0$. Prove or disprove that there is $\alpha \in \mathcal{K}$ so that

$$(\mathrm{id} + \alpha) \circ \gamma(s) < s, \quad s > 0. \tag{A.53}$$

Can we find $\alpha \in \mathcal{P}$ with the same properties?

Exercise A.9 Show that for any $\gamma \in \mathcal{K}$ there are a convex function $\gamma_v \in \mathcal{K} \bigcap C^1(\mathbb{R}_+, \mathbb{R}_+)$ and a concave function $\gamma_c \in \mathcal{K} \bigcap C^1((0, +\infty), \mathbb{R}_+)$ so that

$$\gamma(r) \leq \gamma_c \circ \gamma_v(r), \quad r \geq 0. \tag{A.54}$$

Exercise A.10 Let $f : \mathbb{R}_+ \to \mathbb{R}_+$ be continuous. Show the following:

(i) If there exists $\lim_{t\to\infty} f(t) =: f^* < \infty$, then f is uniformly continuous on $\mathbb{R}_+$ (in particular, any $f \in \mathcal{K}\backslash\mathcal{K}_\infty$ is uniformly continuous on $\mathbb{R}_+$).
(ii) Show that if $\lim_{t\to\infty} f(t)$ doesn't exist, then the statement (i) is not valid any-more in general, even if f is globally bounded.
(iii) Give an example of a function $f : \mathbb{R}_+ \to \mathbb{R}_+$ that is uniformly continuous on $\mathbb{R}_+$ but $\lim_{t\to\infty} f(t)$ doesn't exist.

Exercise A.11 For which $n \in \mathbb{N}$ does the following inequality hold for all $x_i \geq 0$ and all $\alpha \in \mathcal{K}_\infty$

$$\alpha\left(\frac{1}{n}\sum_{i=1}^{n} x_i\right) \leq \frac{2}{n}\sum_{i=1}^{n} \alpha(x_i)?$$

Marginal functions and Dini derivatives

Exercise A.12 Let assumptions of Corollary A.25 hold and let $g(x,t) \geq 0$ for all $x, t \in \mathbb{R}_+$ and $g(r,\cdot) \in \mathcal{L}$ for each $r > 0$. Does it hold that $\psi_1(r,\cdot) \in \mathcal{L}$ for all $r > 0$?

Exercise A.13 Prove that for any $\rho \in \mathcal{P}$ such that $\rho(r) \to \infty$ as $r \to +\infty$, there exists $\alpha \in \mathcal{K}_\infty$ such that

$$\rho(r) \geq \alpha(r), \quad r \geq 0. \tag{A.55}$$

Exercise A.14 Can the condition $\frac{df}{dr}(r)\Big|_{r=g(t)} > 0$ be dropped in Lemma A.30?

Comparison principles

Exercise A.15 Show Lemma A.29.

Exercise A.16 Let $f, g : [0,1] \to \mathbb{R}$ be continuous on $(0,1]$. Does it hold that

$$\overline{\lim_{h\to+0}} f(h)g(h) \leq \overline{\lim_{h\to+0}} f(h) \cdot \overline{\lim_{h\to+0}} g(h)? \tag{A.56}$$

Exercise A.17 Let $a_i : \mathbb{R} \to \mathbb{R}$, $i = 1, \ldots, n$ be continuous. Show that

$$\overline{\lim_{t\to+\infty}} \sum_{i=1}^{n} a_i(t) \leq \sum_{i=1}^{n} \overline{\lim_{t\to+\infty}} a_i(t). \tag{A.57}$$

Show that this inequality may be strict.

Exercise A.18 Consider an infinite sequence $a_i : \mathbb{R} \to \mathbb{R}, i \in \mathbb{N}$ of continuous functions. Does it hold that

$$\overline{\lim_{t\to\infty}} \sum_{i=1}^{\infty} a_i(t) \leq \sum_{i=1}^{\infty} \overline{\lim_{t\to+\infty}}\, a_i(t) \tag{A.58}$$

under assumption that the series on both sides exist?

Exercise A.19 Show that the claim of Proposition A.33 does not hold if f is merely a continuous function.

Brouwer's theorem and its corollaries

Exercise A.20 Prove Brouwer's Theorem A.39 for $n = 1$.

Exercise A.21 Show that without the assumption of injectivity, Theorem A.40 does not hold anymore.

Exercise A.22 Show that $f \in \mathcal{K}_\infty$ if and only if f is a homeomorphism on $\mathbb{R}_+$.

Exercise A.23 *(Reverse of Lemma A.44)* Given any $\alpha \in \mathcal{K}_\infty$ and $n \in \mathbb{N}$, there are homeomorphisms $T_l, T_u : \mathbb{R}^n \to \mathbb{R}^n$ satisfying $T_u(0) = T_l(0) = 0$ so that

$$|T_l(y)| \leq \alpha(|y|) \leq |T_u(y)| \quad \forall y \in \mathbb{R}^n.$$

Hint: consider, e.g., $T_{u,i}(y) = \text{sgn}\,(y_i)\alpha\big(|y_i|\sqrt{n}\big),\ i = 1, \ldots, n.$

Exercise A.24 Let X be a Banach space and let $L : X \to X$ be a linear operator. Show that:

(i) If $X = \mathbb{R}^p$ for some $p \in \mathbb{N}$, then: L is surjective $\Leftrightarrow$ L is injective $\Leftrightarrow$ L is bijective.
(ii) Show that for an arbitrary Banach space X, the property in item (i) does not necessarily hold. Moreover, in general, neither injectivity implies surjectivity, nor does surjectivity implies injectivity.

Exercise A.25 Show that invariance of domain theorem does not hold in infinite-dimensional Banach spaces, even for linear maps. Consider the right shift map $A : \ell_\infty \to \ell_\infty$, defined for $x = (x_1, x_2, \ldots) \in \ell_\infty$ by $Ax = (0, x_1, x_2, \ldots)$. Show that A does not map open sets into open sets.

References

1. Angeli D, Sontag E, Wang Y (2000) Further equivalences and semiglobal versions of integral input to state stability. Dyn Control 10(2):127–149
2. Angeli D, Sontag ED, Wang Y (2000) A characterization of integral input-to-state stability. IEEE Trans Autom Control 45(6):1082–1097
3. Aubin J, Frankowska H (2009) Set-valued analysis. Birkhäuser, Boston

4. Bruckner A, Leonard J (1966) Derivatives. Am Math Mon 73(4):24–56
5. Clarke FH, Ledyaev Y, Stern RJ (1998) Asymptotic stability and smooth Lyapunov functions. J Differ Equ 149(1):69–114
6. Dini U (1878) Fondamenti per la Teorica Delle Funzioni di Variabili Reali. T. Nistri e C.
7. Hagood JW, Thomson BS (2006) Recovering a function from a Dini derivative. Am Math Mon 113(1):34–46
8. Hahn W (1967) Stability of motion. Springer, New York
9. Kannan R, Krueger CK (2012) Advanced analysis: on the real line. Springer Science & Business Media
10. Karafyllis I, Jiang Z-P (2011) Stability and stabilization of nonlinear systems. Springer, London
11. Kellett CM (2014) A compendium of comparison function results. Math Control Signals Syst 26(3):339–374
12. Lin Y, Sontag ED, Wang Y (1996) A smooth converse Lyapunov theorem for robust stability. SIAM J Control Optim 34(1):124–160
13. Pepe P (2022) A nonlinear version of Halanay's inequality for the uniform convergence to the origin. Math Control Rel Fields 12(3):789–811
14. Rawlings J, Risbeck M (2015) On the equivalence between statements with epsilon-delta and K-functions. Technical report
15. Rüffer B (2007) Monotone dynamical systems, graphs, and stability of large-scale interconnected systems. PhD thesis, Fachbereich 3 (Mathematik & Informatik) der Universität Bremen
16. Saks S (1947) Theory of the integral. Courier Corporation
17. Sontag ED (1998) Comments on integral variants of ISS. Syst Control Lett 34(1–2):93–100
18. Szarski J (1965) Differential inequalities. Polish Sci. Publ. PWN, Warszawa, Poland
19. Teel AR, Praly L (2000) A smooth Lyapunov function from a class-estimate involving two positive semidefinite functions. ESAIM Control Optim Calc Var 5:313–367

Appendix B
Stability

Consider the system

$$\dot{x}(t) = f(x(t), d(t)), \quad x(t) \in \mathbb{R}^n, \quad d(t) \in D, \tag{B.1}$$

where $D \subset \mathbb{R}^m$ is the set of values that disturbances may attain and the *disturbance input* d belongs to the set $\mathcal{D} := L^\infty(\mathbb{R}_+, D)$.

In this appendix, we treat time-varying parameters in (B.1) as *disturbances*, as we aim to derive conditions guaranteeing that these parameters carry out an asymptotically negligible impact on the evolution of the system (though they do influence its dynamics on finite-time intervals). As this role is different from the role of the inputs in most of this book, we use the notation d to denote such parameters, which is common in the control literature.

The theory that we develop naturally extends the classical stability theory for dynamical systems

$$\dot{x} = f(x), \quad x(t) \in \mathbb{R}^n.$$

The main aim of this chapter is, however, to provide the basis for the development of the input-to-state stability theory in Chap. 2.

At first, we formalize the notions of the equilibrium point, robust forward completeness, stability, and asymptotic stability and show relationships between these concepts. Next, we characterize these notions in terms of comparison functions introduced in Sect. A.1. Afterward, we define the notion of uniform global asymptotic stability (which can also be called input-to-stability with zero gain) and introduce Lyapunov functions, which generalize in a certain sense the concept of the full energy of a system. This concept plays a fundamental role in stability analysis. This is because uniform global asymptotic stability is equivalent to the existence of a Lyapunov function, which we show in Sects. B.4 and B.5.

A. Mironchenko, *Input-to-State Stability*, Communications and Control Engineering,
https://doi.org/10.1007/978-3-031-14674-9

B.1 Forward Completeness and Stability

Throughout this appendix, we assume that (B.1) satisfies Assumption 1.8 and is forward complete. Hence, the flow map of (B.1) is well defined, and as in Chap. 1, we denote by $\phi(t, x, d)$ the state of (B.1) at the moment $t \in \mathbb{R}_+$ corresponding to an initial condition $x \in \mathbb{R}^n$ and a disturbance $d \in \mathcal{D}$.

Definition B.1 We call $x^* \in \mathbb{R}^n$ an *equilibrium point* of (B.1), if $\phi(t, x^*, d) = x^*$ for all $t \geq 0$ and all $d \in \mathcal{D}$.

Definition B.2 We call $x^* \in \mathbb{R}^n$ a *robust equilibrium point (REP)* of (B.1), if x^* is an equilibrium point of (B.1), and for every $\varepsilon > 0$ and every $h > 0$ there exists $\delta = \delta(\varepsilon, h) > 0$, so that

$$t \in [0, h],\ |x - x^*| \leq \delta,\ d \in \mathcal{D} \quad \Rightarrow \quad |\phi(t, x, d) - x^*| \leq \varepsilon. \tag{B.2}$$

Forward completeness of (B.1) guarantees the existence (and boundedness on finite intervals) of any trajectory, corresponding to any state and any disturbance input. In Definition 1.30, we have introduced the boundedness of reachability sets property, which guarantees the finite-time reachability sets $\phi([0, h], B_R, B_{R,\mathcal{U}})$ are bounded for any $R > 0$ and any $h > 0$. In the context of this chapter, further strengthening of this property is of importance, guaranteeing that the upper bound for the finite-time reachability sets can be chosen uniformly w.r.t. disturbances.

Definition B.3 (B.1) is called *robustly forward complete (RFC)* if for any $C > 0$ and any $h > 0$ it holds that

$$\sup_{|x| \leq C,\ d \in \mathcal{D},\ t \in [0,h]} |\phi(t, x, d)| < \infty.$$

Next, we show that an equilibrium point is necessarily robust, provided that (B.1) is RFC, and f is sufficiently regular.

Definition B.4 We say that the *flow of* (B.1) *is Lipschitz continuous on compact intervals*, if for any $\tau > 0$ and any $R > 0$ there exists $L > 0$ so that

$$x, y \in B_R,\ t \in [0, \tau],\ d \in \mathcal{D} \ \Rightarrow \ |\phi(t, x, d) - \phi(t, y, d)| \leq L|x - y|. \tag{B.3}$$

Lemma B.5 *Let the flow of (B.1) be Lipschitz continuous on compact intervals. If $x^* \in \mathbb{R}^n$ is an equilibrium point of (B.1), then x^* is a robust equilibrium point of (B.1).*

Proof Let x^* be an equilibrium point. Pick any $d \in \mathcal{D}$, $\varepsilon > 0$, $h \geq 0$, $\delta > 0$, and any $x \in \mathbb{R}^n$ with $|x - x^*| \leq \delta$. Due to Lipschitz continuity of ϕ on the compact intervals there exists a function $L : \mathbb{R}_+ \times \mathbb{R}_+ \to \mathbb{R}_+$, increasing on both arguments so that

$$\sup_{t\in[0,h]} |\phi(t,x,d)-x^*| = \sup_{t\in[0,h]} |\phi(t,x,d)-\phi(t,x^*,d)|$$
$$\leq L(h,\delta)|x-x^*|$$
$$\leq L(h,\delta)\delta.$$

Choosing δ so that $L(h,\delta)\delta \leq \varepsilon$ (which is clearly possible) proves that x^* is a robust equilibrium point of (B.1). □

Next, we derive a sufficient condition for Lipschitz continuity of the flow on compact intervals.

Lemma B.6 *Let the following conditions hold:*

(i) (B.1) be robustly forward complete.
(ii) f be Lipschitz continuous uniformly w.r.t. the second argument.

Then (B.1) has a flow that is Lipschitz continuous on compact intervals.

Proof Pick any $C > 0$, any x_1^0, x_2^0: $|x_i^0| \leq C$, $i = 1, 2$ and any $d \in \mathcal{D}$. Let $x_i(t) = \phi(t, x_i^0, d)$, $i = 1, 2$ be the solutions of (B.1) defined for all $t \in \mathbb{R}_+$.

Pick any $\tau > 0$ and set $K(C,\tau) := \sup_{|x|\leq C,\ d\in\mathcal{D},\ t\in[0,\tau]} |\phi(t,x,d)|$, which is finite since (B.1) is RFC. We have for any $t \in [0,\tau]$:

$$|x_1(t)-x_2(t)| \leq |x_1^0-x_2^0| + \int_0^t |f(x_1(r),d(r)) - f(x_2(r),d(r))|dr$$

$$\leq |x_1^0-x_2^0| + L(K(C,\tau)) \int_0^t |x_1(r)-x_2(r)|dr.$$

According to Grönwall's inequality, we obtain

$$|x_1(t)-x_2(t)| \leq |x_1^0-x_2^0| e^{L(K(C,\tau))t},$$

which proves the lemma. □

Summarizing the results of Lemmas B.5 and B.6, we obtain:

Corollary B.7 *Let f be Lipschitz continuous uniformly w.r.t. the second argument. Let further (B.1) be robustly forward complete and let x^* be an equilibrium of (B.1). Then x^* is a robust equilibrium point of (B.1).*

The following example shows that the claim of Corollary B.7 does not necessarily hold if f is not Lipschitz continuous uniformly w.r.t. the second argument.

Example B.8 Let $D = \mathbb{R}$ and $x(t) \in \mathbb{R}$. The following examples illustrate the relations between forward completeness, robust forward completeness, and robustness of the equilibrium point.

(i) (B.1) is RFC, but 0 is not REP of (B.1): $\dot{x}(t) = |d(t)|\big(x(t) - x^3(t)\big)$.
(ii) (B.1) is forward complete, but not RFC and 0 is not REP: $\dot{x}(t) = d(t)x(t)$.
(iii) 0 is REP of (B.1), (B.1) is forward complete, but not RFC:

$$\dot{x}(t) = \frac{1}{|d(t)|+1}x(t) + d(t)\max\{|x(t)| - 1, 0\}.$$

(iv) 0 is REP of (B.1) and (B.1) is RFC: $\dot{x}(t) = \frac{1}{|d(t)|+1}x(t)$.

Next, we introduce important notions of Lyapunov and Lagrange stability.

Definition B.9 An equilibrium x^* of the system (B.1) is called *(locally) Lyapunov stable over $\mathcal{D}$*, if for any $\varepsilon > 0$ there exists $\delta = \delta(\varepsilon)$ so that

$$t \geq 0,\ |x - x^*| \leq \delta,\ d \in \mathcal{D} \quad \Rightarrow \quad |\phi(t, x, d)| \leq \varepsilon. \tag{B.4}$$

If 0 is a locally Lyapunov stable equilibrium, we also say that the system (B.1) is *(locally) Lyapunov stable over $\mathcal{D}$*.

We proceed with global notions.

Definition B.10 The system (B.1) is called

(i) *Lagrange stable (over $\mathcal{D}$)*, if for any $C > 0$

$$\sup_{|x|\leq C,\ d\in\mathcal{D},\ t\geq 0} |\phi(t, x, d)| < \infty.$$

(ii) *Globally Lyapunov stable (over $\mathcal{D}$)*, if (B.1) is locally Lyapunov stable over $\mathcal{D}$ and Lagrange stable over $\mathcal{D}$.

For simplicity of notation, we drop the words "over $\mathcal{D}$" whenever it does not induce any ambiguity. Definitions of stability notions given in the ε-δ language are sometimes awkward to use. Next, we give simple restatements of these properties in terms of comparison functions, introduced in Sect. A.1.

Proposition B.11 *Consider a system (B.1). The following statements are equivalent:*

(i) (B.1) is Lyapunov stable.
(ii) There exist $r > 0$ and $\sigma \in \mathcal{K}_\infty$ so that

$$|x| \leq r,\ t \geq 0,\ d \in \mathcal{D} \quad \Rightarrow \quad |\phi(t, x, d)| \leq \sigma(|x|). \tag{B.5}$$

(iii) 0 is REP of (B.1) and for any $\varepsilon > 0$ there exist $T = T(\varepsilon)$ and $\delta = \delta(\varepsilon)$ so that

$$t \geq T,\ |x| \leq \delta,\ d \in \mathcal{D} \quad \Rightarrow \quad |\phi(t, x, d)| \leq \varepsilon. \tag{B.6}$$

Proof The proof is similar to that of Lemma 2.41 and is omitted. □

In particular, item (iii) of Proposition B.11 shows that robustness of the 0-equilibrium of (B.1) is a "local in time" component of Lyapunov stability property. The next proposition indicates that the RFC property is a "local in time" component of Lagrange stability. The combination of the facts that (B.1) is RFC and 0 is REP of (B.1) is thus a "local in time" version of global stability. This combination is also useful and has several interesting restatements in terms of comparison functions, see, e.g., [11, Proposition 2.15].

Proposition B.12 *Consider a system (B.1). The following statements are equivalent:*

(i) (B.1) is Lagrange stable.
(ii) There exist $c > 0$ and $\sigma \in \mathcal{K}_\infty$ so that

$$x \in \mathbb{R}^n,\ t \geq 0,\ d \in \mathcal{D} \quad \Rightarrow \quad |\phi(t, x, d)| \leq \sigma(|x|) + c.$$

(iii) There exist $c_2 > 0$ and $\sigma_2 \in \mathcal{K}_\infty$ so that

$$x \in \mathbb{R}^n,\ t \geq 0,\ d \in \mathcal{D} \quad \Rightarrow \quad |\phi(t, x, d)| \leq \sigma_2(|x| + c_2).$$

(iv) (B.1) is RFC and for each $r > 0$ there is $T = T(r)$:

$$\sup_{|x| \leq r,\ d \in \mathcal{D},\ t \geq T(r)} |\phi(t, x, d)| < \infty. \tag{B.7}$$

Proof The proof is similar to that of Lemma 2.42 and is omitted. □

Finally, we characterize global Lyapunov stability.

Proposition B.13 *Consider a system (B.1). The following statements are equivalent:*

(i) (B.1) is globally Lyapunov stable.
(ii) There exists $\sigma \in \mathcal{K}_\infty$ so that

$$x \in \mathbb{R}^n,\ t \geq 0,\ d \in \mathcal{D} \quad \Rightarrow \quad |\phi(t, x, d)| \leq \sigma(|x|).$$

B.2 Attractivity and Asymptotic Stability

Having defined the stability notions, we proceed to the attractivity concepts.

Definition B.14 An equilibrium x^* of the system (B.1) is called:

(i) *Locally attractive (ATT)* if there exists $r > 0$ so that

$$|x - x^*| \leq r,\ d \in \mathcal{D} \quad \Rightarrow \quad \lim_{t \to \infty} |\phi(t, x, d) - x^*| = 0.$$

(ii) *Globally attractive (GATT)* if $\lim_{t \to \infty} |\phi(t, x, d) - x^*| = 0$ for all $x \in \mathbb{R}^n$ and all $d \in \mathcal{D}$.

(iii) *Locally asymptotically stable (AS)*, if x^* is Lyapunov stable and locally attractive.
(iv) *Globally asymptotically stable (GAS)*, if x^* is Lyapunov stable and globally attractive.

Remark B.15 When $x^* = 0$, we say (as it is usual in control theory) that *the system* (B.1) *is ATT, GATT, AS, and GAS*, respectively.

Attractivity and stability are conceptually different notions in the sense that neither of them implies another one, even if no disturbances are acting on the system. It is easy to find a globally Lyapunov stable nonattractive system (just consider $\dot{x} = 0$). Next, we show an example of a system that possesses an equilibrium, which attracts all the points of $\mathbb{R}^2\backslash\{0\}$ but is not Lyapunov stable.

Example B.16 Consider a planar system without disturbances, defined in polar coordinates (r, θ) as

$$\dot{r} = r(1 - r), \tag{B.8a}$$

$$\dot{\theta} = \sin^2(\theta/2). \tag{B.8b}$$

This system has two equilibrium points: $(0, 0)$ and $(1, 0)$. It is easy to see that the equilibrium $(0, 0)$ is unstable, and does not attract any trajectory of the system (B.8). At the same time the point $(1, 0)$ attracts the trajectories, starting at any point $x_0 \in \mathbb{R}^2\backslash\{0\}$. However, the point $(1, 0)$ is not stable since there exist points starting arbitrarily close to $(1, 0)$, which go around the unit circle before they converge to $(1, 0)$. We leave a formal proof of global attractivity and instability of the equilibrium $(1, 0)$ as an exercise for the reader. Velocity field and some of trajectories of (B.8) are depicted in Fig. B.1.

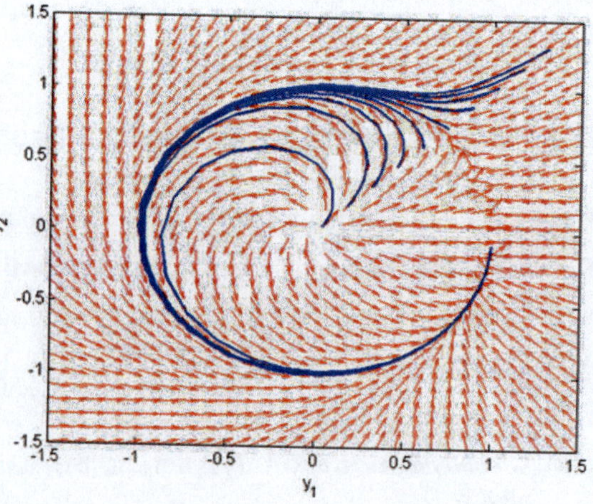

Fig. B.1 Velocity field and some of trajectories of (B.8)

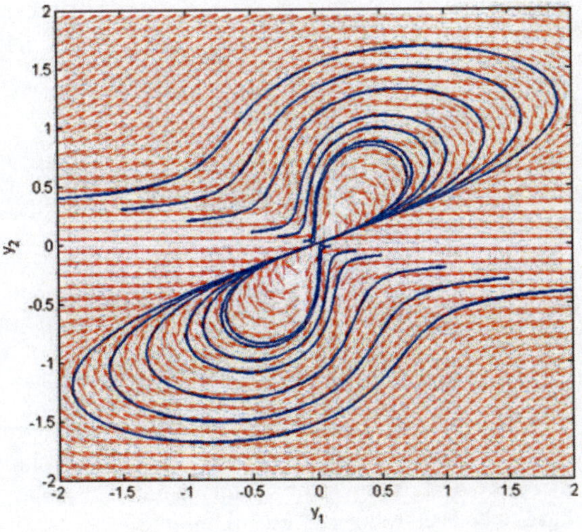

Fig. B.2 Velocity field and some of trajectories of (B.9)

Example B.17 A more involved example of an unstable globally attractive system was proposed by Vinograd [14]:

$$\dot{x} = \frac{x^2(y - x) + y^5}{r^2(1 + r^4)}, \qquad \dot{y} = \frac{y^2(y - 2x)}{r^2(1 + r^4)}, \tag{B.9}$$

where $r^2 = x^2 + y^2$.

The equilibrium $(x, y) = (0, 0)$ is globally attractive but it is not stable. The phase portrait of this system is given by Fig. B.2. For a detailed analysis of this system, a reader may consult [10, p. 120].

Proposition B.11 tells us that Lyapunov stability is a "uniform" notion in the sense that for all initial values with the same norm, we can find a common upper bound for the maximal deviation of trajectories from the origin. In contrast to this, ATT and GATT are not uniform notions. All trajectories of a GATT system converge to the origin, but the rate of convergence to the origin may vary drastically for the initial values with the same norm.

B.3 Uniform Attractivity and Uniform Asymptotic Stability

Though global asymptotic stability ensures convergence of every individual trajectory to the origin, it does not guarantee the existence of a convergence rate that is uniform w.r.t. the disturbances acting on the system. In particular, consider the following example:

Example B.18 Consider the system

$$\dot{x}(t) = -\frac{1}{1+d^2(t)}x(t). \tag{B.10}$$

with $d \in \mathcal{D} := PC_b(\mathbb{R}_+, \mathbb{R})$.

Clearly, $|\phi(t, x, d)| \leq |x|$ for any $t \geq 0$, $x \in \mathbb{R}$, $d \in \mathcal{D}$. However, for $d \equiv const$ the solution of the system (B.10) reads

$$\phi(t, x, d) = e^{-\frac{1}{1+d^2}t}x,$$

and $\frac{1}{1+d^2}$, which can be understood as a convergence rate, tends to 0 as $d \to \infty$.

System (B.10) does not have uniform attraction times but it is globally Lyapunov stable, in particular, it has an upper bound for solutions, which is independent on d. As the next example shows, this is not always the case, and GAS systems do not have to be even robustly forward complete.

Example B.19 Consider the system

$$\dot{x}(t) = d(t)y(t)x(t) - x^3(t), \tag{B.11a}$$
$$\dot{y}(t) = -y(t), \tag{B.11b}$$

with $d \in \mathcal{D} = PC_b(\mathbb{R}_+, \mathbb{R})$. The last equation can be solved explicitly: $y(t) = y(0)e^{-t}$. Thus,

$$\dot{x}(t) = d(t)y(0)e^{-t}x(t) - x^3(t).$$

Now pick $V(x) := x^2$. Then

$$\begin{aligned}\dot{V}(x) = 2x\dot{x} &\leq 2\|d\||y(0)|e^{-t}x^2(t) - 2x^4(t)\\ &= 2\|d\||y(0)|e^{-t}V(x(t)) - 2V^2(x(t)).\end{aligned}$$

Direct analysis shows that $V(x(t)) \to 0$ as $t \to \infty$, and hence $x(t) \to 0$ as $t \to \infty$.

At the same time, the system is not RFC, which is easy to see.

Above examples motivate the following uniform stability notion:

Definition B.20 The system (B.1) is called *uniformly globally asymptotically stable (UGAS)* if (B.1) is forward complete and there exists a $\beta \in \mathcal{KL}$ such that

$$x \in \mathbb{R}^n,\ d \in \mathcal{D},\ t \geq 0 \quad \Rightarrow \quad |\phi(t, x, d)| \leq \beta(|x|, t). \tag{B.12}$$

For characterization of UGAS property we need the following strengthening of global attractivity:

Definition B.21 A control system (B.1) is called *uniformly globally attractive (UGATT)*, if for any $r, \varepsilon > 0$ there exists $\tau = \tau(r, \varepsilon)$ so that

$$|x| \le r,\ d \in \mathcal{D},\ t \ge \tau(r, \varepsilon) \quad \Rightarrow \quad |\phi(t, x, d)| \le \varepsilon. \tag{B.13}$$

Using the terminology of Chap. 2, one can call the UGAS property of (B.1) the ISS property with zero gain and the UGATT property the UAG property with zero gain.

The definition shows that UGAS systems are GAS, robustly forward complete, and uniformly globally attractive. The following theorem shows that the last two properties, together with the robustness of the trivial equilibrium, already ensure UGAS. This result is a counterpart of the ISS superposition Theorem 2.51.

Theorem B.22 (UGAS superposition principle) *(B.1) is UGAS if and only if the following three properties hold:*

(i) (B.1) is RFC,
(ii) 0 is a robust equilibrium point for (B.1),
(iii) (B.1) is uniformly globally attractive.

Proof "⇒". UGAS implies that

$$x \in \mathbb{R}^n,\ d \in \mathcal{D},\ t \ge 0 \quad \Rightarrow \quad |\phi(t, x, d)| \le \beta(|x|, 0),$$

which shows global stability, which in turn implies that 0 is a robust equilibrium of (B.1) and that (B.1) is RFC.

Take arbitrary $\varepsilon, \delta > 0$. Define $\tau_a = \tau_a(\varepsilon, \delta)$ as a solution of an equation $\beta(\delta, \tau_a) = \varepsilon$ (if it exists, then it is unique, because of monotonicity of β w.r.t. the second argument, if it does not exist, we put $\tau_a(\varepsilon, \delta) = 0$). Then for all $t \ge \tau_a$, all $x \in \mathbb{R}^n$: $|x| \le \delta$ and all $d \in \mathcal{D}$

$$|\phi(t, x, u)| \le \beta(|x|, t) \le \beta(|x|, \tau_a) \le \varepsilon,$$

which shows UGATT of (B.1).

"⇐". Assume that (B.1) is RFC and 0 is its robust equilibrium point. Since (B.1) is UGATT, the properties in item (iii) in Proposition B.11 and item (iv) in Proposition B.12 hold and thus by these Propositions, (B.1) is Lyapunov and Lagrange stable, and thus globally Lyapunov stable. Hence by Proposition B.13 there exists $\sigma \in \mathcal{K}_\infty$ so that

$$t \ge 0,\ |x| \le \delta,\ d \in \mathcal{D} \quad \Rightarrow \quad |\phi(t, x, d)| \le \sigma(\delta). \tag{B.14}$$

Define $\varepsilon_n := \frac{1}{2^n}\sigma(\delta)$, for $n \in \mathbb{Z}_+$. Uniform global attractivity ensures that there exists a sequence of times $\tau_n := \tau(\varepsilon_n, \delta)$ that we assume without loss of generality to be strictly increasing, such that

$$t \ge \tau_n,\ |x| \le \delta,\ d \in \mathcal{D} \quad \Rightarrow \quad |\phi(t, x, d)| \le \varepsilon_n.$$

Fig. B.3 Construction of the function $\omega(\delta, \cdot)$

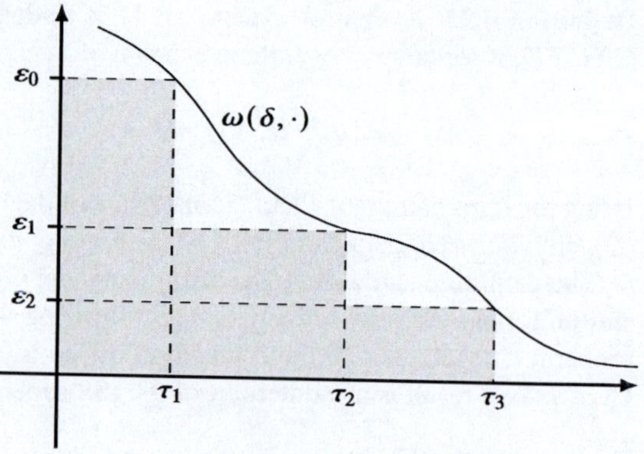

From (B.14) we see that we may set $\tau_0 := 0$. Define $\omega(\delta, \tau_n) := \varepsilon_{n-1}$, for $n \in \mathbb{N}$, and $\omega(\delta, 0) := 2\varepsilon_0 = 2\sigma(\delta)$.

Now extend the function $\omega(\delta, \cdot)$ for $t \in \mathbb{R}_+\backslash\{\tau_n, n \in \mathbb{N}\}$ so that $\omega(\delta, \cdot) \in \mathcal{L}$ (see Fig. B.3). All such functions satisfy the estimate (B.12), because for all $t \in (\tau_n, \tau_{n+1})$ it holds that $|\phi(t, x, d)| \leq \varepsilon_n < \omega(\delta, t)$. Doing this for all $\delta \in \mathbb{R}_+$ we obtain function ω.

Let us check that $\psi : (r, t) \mapsto \sup_{0\leq s\leq r} \omega(s, t) \geq \omega(r, t)$ satisfies all assumptions of Proposition A.17.

Pick $t_1, t_2 \geq 0$ so that $t_1 < t_2$. Fix any $r > 0$. Then for some $r_2 \in [0, r]$ we have

$$\psi(r, t_2) = \omega(r_2, t_2) < \omega(r_2, t_1) \leq \psi(r, t_1).$$

Hence, ψ is decreasing w.r.t. the second argument.

Next, $\psi(0, \cdot) = 0$ by construction. Moreover, for any $t > 0$ it holds that

$$\psi(r, t) \leq \psi(r, 0) = \sup_{0\leq s\leq r} \omega(s, 0) = \omega(r, 0) = 2\sigma(r) \to 0, \quad r \to +0.$$

Finally, ψ is nondecreasing w.r.t. the first argument. Then according to Proposition A.17 there is a $\beta \in \mathcal{KL}$: $\beta(r, t) \geq \psi(r, t) \geq \omega(r, t)$. □

The following example shows that the assumptions of the robustness of an equilibrium and robust forward completeness cannot be dropped in Theorem B.22.

Example B.23 Let $\mathcal{D} := L^\infty(\mathbb{R}_+, \mathbb{R})$, $x(t), y(t) \in \mathbb{R}$ and consider

$$\dot{x}(t) = d(t)x(t)y(t) - x^3(t) - x^{1/3}(t), \tag{B.15a}$$
$$\dot{y}(t) = -y^3(t) - y^{1/3}(t). \tag{B.15b}$$

We are going to show that (B.15) is UGATT, forward complete, has 0 as a sole equilibrium point but (B.15) is not RFC and 0 is a nonrobust equilibrium. This shows that such systems are quite frequent, at least in the presence of disturbances.

According to Example 1.24, we know that there is a time $t_c > 0$ so that the solutions of the y-equation subject to any initial condition converge to 0 in time less than t_c. This immediately implies that also the state of the whole system (B.15) converges to 0 at the time not larger than $2t_c$, for any initial condition and any disturbance from $\mathcal{D}$. This shows the UGATT of (B.15). At the same time, (B.15) is not RFC, and 0 is not a robust equilibrium of (B.15), which can be shown by choice of a sufficiently large constant disturbance d.

B.4 Lyapunov Functions

Usually, it is hard to verify UGAS or global Lyapunov stability of nonlinear systems (B.1) directly. The restatements of the UGAS property that we have proved in Sect. B.3 are important for the theory, but they are hard to use in practice. In this section, we develop Lyapunov theory giving a powerful method to prove UGAS of nonlinear systems, with or without disturbances.

Definition B.24 A continuous function $V : \mathbb{R}^n \to \mathbb{R}_+$ is called a *(strict) Lyapunov function* for the system (B.1), if there are $\psi_1, \psi_2 \in \mathcal{K}_\infty$ and $\alpha \in \mathcal{P}$ so that

$$\psi_1(|x|) \leq V(x) \leq \psi_2(|x|) \quad x \in \mathbb{R}^n, \tag{B.16}$$

and

$$\dot{V}_d(x) \leq -\alpha(V(x)), \quad x \in \mathbb{R}^n,\ d \in \mathcal{D}, \tag{B.17}$$

where the Lie derivative of V is defined as follows:

$$\dot{V}_d(x) := \overline{\lim_{t \to +0}} \frac{1}{t}\big(V(\phi(t, x, d)) - V(x)\big). \tag{B.18}$$

Instead of (B.17) one can require merely

$$\sup_{d \in \mathcal{D}} \dot{V}_d(x) < 0, \quad x \in \mathbb{R}^n. \tag{B.19}$$

It is easy to see that this condition implies the existence of $\alpha \in \mathcal{P}$, such that (B.17) holds.

By Lemma 2.18, if V is continuously differentiable, and d is continuous, then the Lie derivative of V can be computed as

$$\dot{V}_d(x) = \nabla V(x) \cdot f(x, d(0)). \tag{B.20}$$

Definition B.25 A continuous function $V : \mathbb{R}^n \to \mathbb{R}_+$ is called a *nonstrict Lyapunov function* for the system (B.1), if $\exists \psi_1, \psi_2 \in \mathcal{K}_\infty$ so that for all $x \in \mathbb{R}^n$ the estimate (B.16) holds and for all $x \in \mathbb{R}^n$ and all $d \in \mathcal{D}$ we have

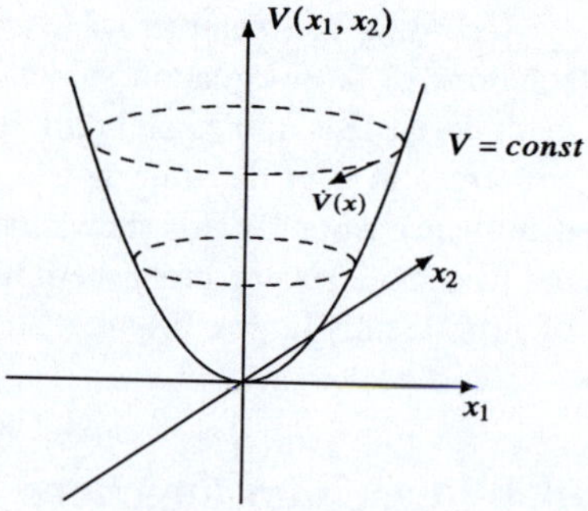

Fig. B.4 Level sets of a typical Lyapunov function V (dashed line)

$$\dot{V}_d(x) \leq 0. \tag{B.21}$$

Lyapunov functions are, in some sense, a generalization of the energy of physical systems. In fact, for many mechanical systems, the full energy is a Lyapunov function. In Fig. B.4, we show a graph of a typical Lyapunov function with several *level sets* (sets of the form $\{x \in \mathbb{R}^n : V(x) = c\}$ for a certain $c \geq 0$). Since the derivative of a strict Lyapunov function is negative, the state of the system always moves to lower energy levels with a certain uniform speed of convergence that does not depend on the disturbance d and the norm of the state. In fact, every such trajectory asymptotically approaches the equilibrium, at which the system's energy equals zero. Using comparison principle (Proposition A.35), we can make this argument precise.

Theorem B.26 *If there exists a strict Lyapunov function for (B.1), then (B.1) is UGAS.*

Proof Pick any $x \neq 0$ and any $d \in \mathcal{D}$. Since V is a Lyapunov function we have

$$\dot{V}_d(x) \leq -\alpha(V(x)).$$

From the comparison principle (Proposition A.35) with $y(t) := V(x(t))$ it follows that $\exists \beta \in \mathcal{KL}$, such that

$$V(\phi(t, x, d)) \leq \beta(V(x), t), \quad t \in \mathbb{R}_+,\ x \in \mathbb{R}^n,\ d \in \mathcal{D}.$$

The inequality (B.16) implies that for certain $\psi_1, \psi_2 \in \mathcal{K}_\infty$ the following estimate holds

$$\psi_1(|\phi(t, x, d)|) \leq V(\phi(t, x, d)) \leq \beta(V(x), t) \leq \beta(\psi_2(|x|), t)$$

for any $t \in \mathbb{R}_+$, $x \in \mathbb{R}^n$, $d \in \mathcal{D}$, which immediately implies for the same t, x, d that

$$|\phi(t, x, d)| \leq \psi_1^{-1} \circ \beta(\psi_2(|x|), t) := \tilde{\beta}(|x|, t).$$

Clearly, $\tilde{\beta} \in \mathcal{KL}$. This shows that (B.1) is UGAS. □

Next we apply the developed machinery to stability analysis of two classical systems: to Lorenz equations and to the nonlinear pendulum.

Example B.27 *(Lorenz system)* Consider the famous *Lorenz system*

$$\dot{x} = \sigma(y - x), \tag{B.22a}$$
$$\dot{y} = rx - y - xz, \tag{B.22b}$$
$$\dot{z} = xy - bz, \tag{B.22c}$$

where $\sigma, r, b > 0$.

This system has been proposed by Lorenz as an extensively simplified model of atmospheric convection. It is obtained from a complicated system of nonlinear partial differential equations proposed by Rayleigh by considering the dynamics only of three modes of these PDEs and neglecting all the other ones. The parameters σ and r are called *Prandtl and Rayleigh numbers*. System (B.22) is famous for being the first discovered system, possessing for certain values of parameters a so-called "strange attractor". Our aim in this example is a modest one: we will show by employing the Lyapunov method that for $r < 1$, the origin is a UGAS equilibrium (and thus, a chaotic behavior does not occur).

To this end, consider a Lyapunov function candidate

$$V(x, y, z) := \frac{1}{\sigma}x^2 + y^2 + z^2, \quad x, y, z \in \mathbb{R}.$$

Lie derivative of V w.r.t. the system (B.22) equals

$$\begin{aligned}\frac{d}{dt}V(x, y, z) &= \frac{2}{\sigma}x\dot{x} + 2y\dot{y} + 2z\dot{z}\\ &= 2x(y - x) + 2y(rx - y - xz) + 2z(xy - bz)\\ &= -2x^2 + 2(1 + r)xy - 2y^2 - 2bz^2.\end{aligned}$$

For $r < 1$, the last expression is negative for all x, y, z: $x^2 + y^2 + z^2 \neq 0$. This shows that V is a strict Lyapunov function for (B.22) and thus, the origin is a UGAS equilibrium.

For $r = 1$, we see that $\frac{d}{dt}V(x, y, z) = -2(x - y)^2 - 2bz^2 \leq 0$ for all $x, y, z \in \mathbb{R}$, and $\frac{d}{dt}V(x, y, z) = 0$ iff $x = y$ and $z = 0$. This shows by means of Exercise B.15 global stability for $r = 1$. Employing the Barabashin–Krasovskiy–LaSalle principle, it is possible to show that for $r = 1$, the system is UGAS.

Example B.28 *(Nonlinear pendulum)* The equation of motion of a *mathematical pendulum* (see Fig. B.5) is given by the Newton's law:

$$\ddot{x} = -\frac{g}{L}\sin x. \tag{B.23}$$

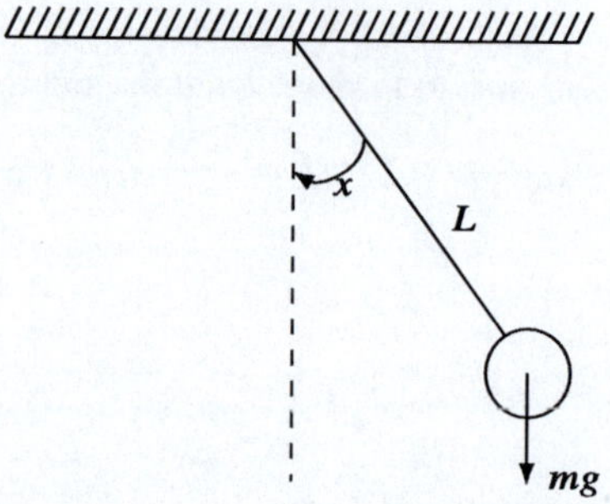

Fig. B.5 Mathematical pendulum

Here x is the angle as depicted in Fig. B.5, g is the gravitational acceleration on Earth, and L is the length of the string. We assume that $x \in (-\pi, \pi]$. Let us rewrite (B.23) as a system of 2 equations of the first order:

$$\dot{x} = \frac{g}{L}y,$$
$$\dot{y} = -\sin x. \tag{B.24}$$

There are 2 equilibrium points (for $x \in (-\pi, \pi]$) for the nonlinear pendulum: $(x, y) = (0, 0)$ and $(x, y) = (\pi, 0)$.

We are going to prove that the equilibrium $(0, 0)$ is Lyapunov stable, but it is not locally asymptotically stable (see Figs. B.6 and B.7).

Let us pick the following Lyapunov function candidate for (B.24):

$$V(x, y) := \frac{1}{2}y^2 + \frac{L}{g}(1 - \cos x). \tag{B.25}$$

It is easy to see that $V(0, 0) = 0$, and $V(x, y) > 0$ unless $x = y = 0$. Lie derivative of V w.r.t (B.23) equals:

$$\frac{d}{dt}V(x, y) = y\dot{y} + \frac{L}{g}\dot{x} \cdot \sin x = 0.$$

This means that the energy of a nonlinear pendulum is conserved, and thus the system (B.24) is Lyapunov stable. However, it is not locally asymptotically stable.

B.5 Converse Lyapunov Theorem

We have shown in Sect. B.4 that the existence of a strict Lyapunov function is sufficient for UGAS of (B.1). To start the search for such a Lyapunov function, it would be highly desirable to know in advance that it does exist for all UGAS systems

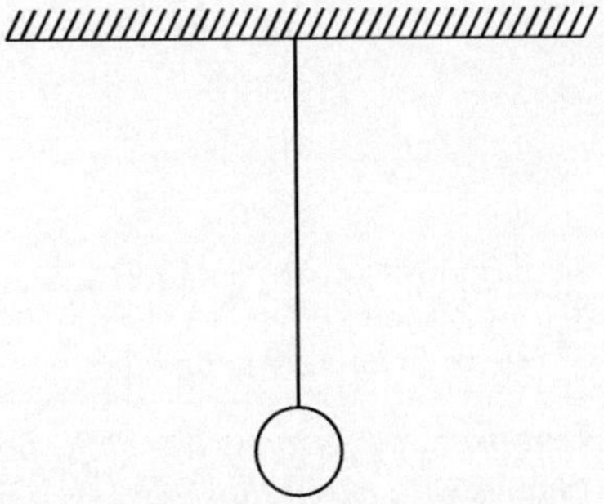

Fig. B.6 Equilibrium (0, 0)

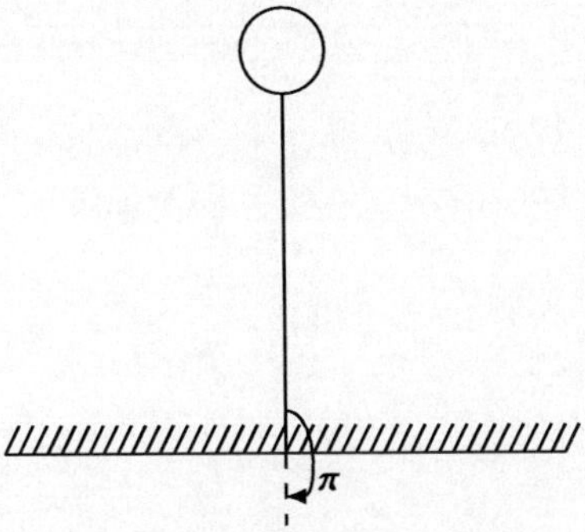

Fig. B.7 Equilibrium $(\pi, 0)$

of the form (B.1). For many reasons, it is desirable to have a Lyapunov function with a certain degree of regularity. In this section, we show that for any UGAS system (B.1) satisfying certain mild conditions, a Lipschitz continuous Lyapunov function necessarily exists.

We start with several simple lemmas.

Lemma B.29 *Let* $f, g : D \to \mathbb{R}_+$ *be any functions for which* $\sup_{d \in D} f(d)$ *is finite. Then*

$$\sup_{d \in D} f(d) - \sup_{d \in D} g(d) \leq \sup_{d \in D} (f(d) - g(d)). \tag{B.26}$$

Proof If $\sup_{d \in D} g(d) = +\infty$, the claim is clear as the supremum in the right-hand side of (B.26) is finite. In the rest of the proof let $\sup_{d \in D} g(d)$ be finite.

Since $\sup_{d \in D} f(d) < \infty$, there exists a sequence $(d_n)_{n=1}^{\infty} \subset D$, so that

$$\sup_{d \in D} f(d) - \sup_{d \in D} g(d) \leq f(d_n) + \frac{1}{2^n} - \sup_{d \in D} g(d),$$

for any $n \in \mathbb{N}$. We obtain

$$\sup_{d\in D} f(d) - \sup_{d\in D} g(d) \le f(d_n) + \frac{1}{2^n} - g(d_n)$$
$$\le \sup_{d\in D}(f(d) - g(d)) + \frac{1}{2^n}.$$

Letting $n \to \infty$, we obtain (B.26). □

We need another simple lemma

Lemma B.30 *For any $k \in \mathbb{N}$ the function $G_k(r) := \max\{r - \frac{1}{k}, 0\}$ is Lipschitz continuous with a unit Lipschitz constant, i.e., for all $r_1, r_2 \ge 0$ it holds that*

$$\big|G_k(r_1) - G_k(r_2)\big| \le |r_1 - r_2|. \tag{B.27}$$

Proof For $r_1, r_2 \in [\frac{1}{k}, \infty)$ and for $r_1, r_2 \in [0, \frac{1}{k}]$ the inequality (B.27) holds trivially. For $r_1 \ge \frac{1}{k}$ and $r_2 < \frac{1}{k}$ we have

$$\Big|\max\Big\{r_1 - \frac{1}{k}, 0\Big\} - \max\Big\{r_2 - \frac{1}{k}, 0\Big\}\Big| = \Big|r_1 - \frac{1}{k}\Big| \le |r_1 - r_2|.$$

The cases $r_2 \ge \frac{1}{k}$ and $r_1 < \frac{1}{k}$ are analogous. □

Next, we prove a Lipschitz continuous converse Lyapunov theorem. We use ideas from [6, Sect. 3.4] that are based on earlier local converse Lyapunov theorems, see, e.g., [15, Theorem 19.3] and on the Sontags' $\mathcal{KL}$-Lemma (Proposition A.7). Lipschitz continuity of the flow of (B.1) will guarantee Lipschitz continuity of the constructed Lyapunov function.

Theorem B.31 (Converse UGAS Lyapunov theorem) *Assume that f is continuous on $\mathbb{R}^n \times D$, and f is Lipschitz continuous uniformly w.r.t. the second argument. If (B.1) is UGAS, then there exists a strict Lyapunov function for (B.1) that is Lipschitz continuous on bounded balls.*

Furthermore, in the latter case, one can always find a strict Lyapunov function with a linear decay rate.

Proof Let (B.1) be UGAS. Then there exists a $\beta \in \mathcal{KL}$, so that for all $x \in \mathbb{R}^n$ and all $t \ge 0$ the estimate (B.12) holds. Due to Sontag's Lemma (Proposition A.7) there exist $\alpha_1, \alpha_2 \in \mathcal{K}_\infty$ so that

$$\alpha_2^{-1}(\beta(s,t)) \le \alpha_1(s)e^{-t}, \quad t, s \in \mathbb{R}_+, \tag{B.28}$$

is satisfied.

Pick any Lipschitz continuous function $\rho \in \mathcal{K}_\infty$ with a unit Lipschitz constant so that: $\rho(r) \le \alpha_2^{-1}(r)$, $r \in \mathbb{R}_+$, which is possible due to Lemma A.18. For any $k \in \mathbb{N}$ define $G_k(r) := \max\{r - \frac{1}{k}, 0\}$ and consider a function

$$V_k^\eta(x) := \sup_{d \in \mathcal{D}} \max_{s \geq 0} e^{\eta s} G_k\big(\rho(|\phi(s, x, d)|)\big), \tag{B.29}$$

for any $\eta \in (0, 1)$ and all $x \in \mathbb{R}^n$. Clearly,

$$\begin{aligned}
0 \leq V_k^\eta(x) &\leq \sup_{d \in \mathcal{D}} \max_{s \geq 0} e^{\eta s} \rho\big(|\phi(s, x, d)|\big) \\
&\leq \sup_{d \in \mathcal{D}} \max_{s \geq 0} e^{\eta s} \alpha_2^{-1}\big(|\phi(s, x, d)|\big) \\
&\leq \max_{s \geq 0} e^{\eta s} \alpha_2^{-1}\big(\beta(|x|, s)\big) \\
&\leq \max_{s \geq 0} e^{\eta s} \alpha_1(|x|) e^{-s} \\
&= \alpha_1(|x|),
\end{aligned}$$

and thus V_k^η is bounded from above.

Next, we show that the Lie derivative of V_k^η is negative along the trajectories of (B.1). Observe that for any $v \in \mathcal{D}$ and for any $x \in \mathbb{R}^n$

$$\begin{aligned}
V_k^\eta(\phi(t, x, v)) &= \sup_{d \in \mathcal{D}} \max_{s \geq 0} e^{\eta s} G_k\big(\rho\big(|\phi(s, \phi(t, x, v), d)|\big)\big) \\
&= e^{-\eta t} \sup_{d \in \mathcal{D}} \max_{s \geq 0} e^{\eta(s+t)} G_k\Big(\rho\Big(\big|\phi(s + t, x, v \underset{t}{\Diamond} d)\big|\Big)\Big),
\end{aligned} \tag{B.30}$$

where the concatenation $v \underset{t}{\Diamond} d$ is defined by (6.1) and belongs to $\mathcal{D}$.

We proceed with

$$V_k^\eta(\phi(t, x, v)) = e^{-\eta t} \sup_{d \in \mathcal{D}} \max_{s \geq t} G_k\big(e^{\eta s} \rho(|\phi(s, x, v \underset{t}{\Diamond} d)|)\big).$$

Since the supremum in the above expression is taken only over a subset of all disturbances, the expression cannot become smaller if we take the supremum over all disturbances $d \in \mathcal{D}$. This leads to

$$\begin{aligned}
V_k^\eta(\phi(t, x, v)) &\leq e^{-\eta t} \sup_{d \in \mathcal{D}} \max_{s \geq t} e^{\eta s} G_k\big(\rho(|\phi(s, x, d)|)\big) \\
&\leq e^{-\eta t} V_k^\eta(x).
\end{aligned} \tag{B.31}$$

Now we are able to compute the Lie derivative of V_k^η for any $x \in \mathbb{R}^n$ and any acting disturbance $v \in \mathcal{D}$:

$$\begin{aligned}
\dot{V}_{k,v}^\eta(x) &:= \overline{\lim_{t \to +0}} \frac{1}{t}\Big(V_k^\eta(\phi(t, x, v)) - V_k^\eta(x)\Big) \\
&\leq \overline{\lim_{t \to +0}} \frac{1}{t}(e^{-\eta t} - 1) V_k^\eta(x) \\
&= -\eta\, V_k^\eta(x).
\end{aligned} \tag{B.32}$$

Let us show that V_k^η are Lipschitz continuous on bounded balls. Since (B.1) is UGAS, for any $d \in \mathcal{D}$, $x \in \mathbb{R}^n$ and $t \geq 0$ it holds that

$$\rho\big(|\phi(t,x,d)|\big) \leq \alpha_2^{-1}\big(\beta(|x|,t)\big) \leq e^{-t}\alpha_1(|x|).$$

Pick any $R > 0$ and $x \in B_R$. Then for any $d \in \mathcal{D}$ and for any $t \geq T(R,k) := \ln(1 + k\alpha_1(R))$ it holds that

$$\rho(|\phi(t,x,d)|) \leq \frac{1}{k},$$

and thus

$$G_k\big(\rho(|\phi(t,x,d)|)\big) = 0, \quad t \geq T(R,k),\ x \in B_R,\ d \in \mathcal{D}.$$

Thus, the maximum in V_k^η is attained on the finite-time interval, i.e., for any $x \in B_R$, the following is true

$$V_k^\eta(x) := \sup_{d\in\mathcal{D}} \max_{s\in[0,T(R,k)]} e^{\eta s} G_k\big(\rho(|\phi(s,x,d)|)\big). \tag{B.33}$$

Pick any $R > 0$ and any $x, y \in B_R$. By Lemma B.29 we have

$$\begin{aligned}\big|V_k^\eta(x) - V_k^\eta(y)\big| &\leq \Big|\sup_{d\in\mathcal{D}} \max_{s\in[0,T(R,k)]} e^{\eta s} G_k\big(\rho(|\phi(s,x,d)|)\big)\\ &\quad - \sup_{d\in\mathcal{D}} \max_{s\in[0,T(R,k)]} e^{\eta s} G_k\big(\rho(|\phi(s,y,d)|)\big)\Big|\\ &\leq \sup_{d\in\mathcal{D}} \max_{s\in[0,T(R,k)]} e^{\eta s}\Big|G_k\big(\rho(|\phi(s,x,d)|)\big)\\ &\quad - G_k\big(\rho(|\phi(s,y,d)|)\big)\Big|.\end{aligned}$$

Due to Lemma B.30, G_k is Lipschitz with a unit Lipschitz constant. This property is shared also by the function ρ. Thus, we can proceed with the estimates to obtain

$$\begin{aligned}\big|V_k^\eta(x) - V_k^\eta(y)\big| &\leq \sup_{d\in\mathcal{D}} \max_{s\in[0,T(R,k)]} e^{\eta s}\big||\phi(s,x,d)| - |\phi(s,y,d)|\big|\\ &\leq \sup_{d\in\mathcal{D}} \max_{s\in[0,T(R,k)]} e^{\eta s}|\phi(s,x,d) - \phi(s,y,d)|.\end{aligned}$$

Since f is Lipschitz continuous uniformly w.r.t. the second argument and (B.1) is RFC (since it is UGAS), Lemma B.6 ensures that the flow ϕ is Lipschitz continuous on compact time intervals. Hence there exists a function $L = L(R,k)$ that we assume to be strictly increasing w.r.t. both arguments, so that

$$\begin{aligned}\big|V_k^\eta(x) - V_k^\eta(y)\big| &\leq \max_{s\in[0,T(R,k)]} e^{\eta s} L(R,k)|x-y|\\ &\leq e^{\eta T(R,k)} L(R,k)|x-y|.\end{aligned}$$

This ensures that V_k^η are Lipschitz continuous on bounded balls. Now define $M(R,k) := e^{\eta T(R,k)} L(R,k)$, which is a strictly increasing function w.r.t. both of its arguments.

Define $W^\eta : \mathbb{R}^n \to \mathbb{R}_+$ by

$$W^\eta(x) := \sum_{k=1}^{\infty} \frac{2^{-k}}{1+M(k,k)} V_k^\eta(x) \quad \forall x \in \mathbb{R}^n. \tag{B.34}$$

Since for all $x \in \mathbb{R}^n$ it holds that $V_k^\eta(x) \le \alpha_1(|x|)$, and

$$V_k^\eta(x) \ge \sup_{d \in \mathcal{D}} G_k\big(\rho(|\phi(0,x,d)|)\big) = G_k(\rho(|x|)),$$

we see that

$$\psi_1(|x|) \le W^\eta(x) \le \alpha_1(|x|), \quad x \in \mathbb{R}^n,$$

where ψ_1 is defined for any $r \ge 0$ as

$$\psi_1(r) := \sum_{k=1}^{\infty} \frac{2^{-k}}{1+M(k,k)} G_k\big(\rho(r)\big). \tag{B.35}$$

Since $\rho \in \mathcal{K}_\infty$, $\psi_1(0) = 0$ and $r \mapsto G_k(\rho(r))$ is continuous and increasing on the interval $\big(\rho^{-1}(\frac{1}{k}), \infty\big)$, ψ_1 is increasing over $\mathbb{R}_+$. For $r \in \big(\rho^{-1}(\frac{1}{N+1}), \rho^{-1}(\frac{1}{N})\big)$ it holds that $\psi_1(r) = \sum_{k=N+1}^{\infty} \frac{2^{-k}}{M(k,k)}(\rho(r) - \frac{1}{k})$, which is a continuous function. From this fact, it is easy to see that ψ_1 is continuous over $\mathbb{R}_+$ and hence $\psi_1 \in \mathcal{K}_\infty$.

Due to the estimate (B.32), we have

$$\dot{W}^\eta(x) \le -\eta W^\eta(x).$$

This shows that W^η is an exponential strict Lyapunov function for (B.1).

Now, pick any $R > 0$ and $x, y \in B_R$. It holds that

$$\begin{aligned}
\big|W^\eta(x) - W^\eta(y)\big| &= \Big|\sum_{k=1}^{\infty} \frac{2^{-k}}{1+M(k,k)} (V_k^\eta(x) - V_k^\eta(y))\Big| \\
&\le \sum_{k=1}^{\infty} \frac{2^{-k} M(R,k)}{1+M(k,k)} |x-y| \\
&\le \left(1 + \sum_{k=1}^{[R]+1} \frac{2^{-k} M(R,k)}{1+M(k,k)}\right) |x-y|.
\end{aligned}$$

This shows that W^η is a Lipschitz continuous (on bounded balls) Lyapunov function for (B.1). □

Using the smoothing technique, a stronger result has been obtained in [8, Theorem 2.9]:

Theorem B.32 *Assume that f is continuous on $\mathbb{R}^n \times D$, and Lipschitz continuous uniformly w.r.t. the second argument. (B.1) is UGAS if and only if there exists an infinitely differentiable Lyapunov function for (B.1), with certain decay rate $\alpha \in \mathcal{K}_\infty$.*

Proof For the proof, please consult [8, proof of Theorem 2.9]. The reasoning in [8, proof of Theorem 2.9] ensures the existence of a Lyapunov function with a certain $\alpha \in \mathcal{P}$. However, arguing as in Proposition 2.17, one can find another Lyapunov function with a $\mathcal{K}_\infty$-decay rate. □

B.6 Systems with Compact Sets of Input Values

We have seen in Sect. B.3 that UGAS is a stronger notion than GAS for the class of systems that we have considered so far. However, in practice, disturbances are often uniformly bounded. That is, the disturbance values belong to the compact subset of $\mathbb{R}^m$. In this section, we show that for such systems, GAS is equivalent to UGAS.

Proposition B.33 *Let Assumption 1.8 hold, (B.1) be forward complete and let the space of input values D be compact. Then (B.1) is robustly forward complete.*

Proof This follows directly from Theorem 1.31. □

Proposition B.34 *Let Assumption 1.8 hold, (B.1) be forward complete and let the space of input values D be compact. If (B.1) is GAS, then (B.1) is UGATT.*

Proof Pick arbitrary $\varepsilon > 0$ and $r > 0$. Since (B.1) is Lyapunov stable, there is a $\delta > 0$ so that

$$t \geq 0, \quad |x| \leq \delta, \quad d \in \mathcal{D} \quad \Rightarrow \quad |\phi(t, x, d)| \leq \varepsilon.$$

Define $\mathcal{S} := D$, $K := \overline{B_{\delta/2}}$, $\Omega := B_\delta$, and $C := B_r$. Hence $\mathcal{D}_{\mathcal{S}} = \mathcal{D}$ and in view of the global attractivity of (B.1), the assumptions of Theorem 1.44 are satisfied. Define the function τ as in (1.35). We have

$$T := \sup\{\tau(x, \Omega, d) : x \in C, d \in \mathcal{D}\} < +\infty.$$

Thus for every $x \in C$ and every $d \in \mathcal{D}$ there exists time $\tau = \tau(x, \Omega, d) \leq T$: $|\phi(\tau, x, d)| \leq \delta$. But, due to Lyapunov stability and the cocycle property, $|\phi(\tau, x, d)| \leq \varepsilon$ holds for every $t \geq \tau$ (and hence for all $t \geq T$). Since T depends on r and ε only, this shows UGATT of (B.1). □

As a corollary, we obtain a criterion of UGAS for the systems with uniformly bounded disturbances.

Theorem B.35 *Let Assumption 1.8 hold, (B.1) be forward complete and let the space of input values D be compact. Then (B.1) is GAS $\Leftrightarrow$ (B.1) is UGAS.*

Proof "$\Leftarrow$". Clear.

"$\Rightarrow$". Assumptions of the theorem and Proposition B.34 imply that (B.1) is UGATT. Lyapunov stability of (B.1) implies that 0 is a robust equilibrium point of (B.1). Finally, forward completeness is equivalent to robust forward completeness by Proposition B.33. Hence (B.1) is UGAS by Theorem B.22. □

Remark B.36 Theorem B.35 remains true if the assumption of forward completeness is dropped. This can be shown using "boundedness-implies-continuation property" (Proposition 1.20). See Proposition 2.8 for more details.

B.7 UGAS of Systems Without Inputs

Classical stability theory studies asymptotic properties of undisturbed systems

$$\dot{x} = f(x), \tag{B.36}$$

where f is a locally Lipschitz continuous function, and $x(t) \in \mathbb{R}^n$. Let us specialize the previously obtained results to this important special case.

Theorem B.37 *Consider a forward complete system (B.36) with a locally Lipschitz f. The following statements are equivalent:*

(i) (B.36) is UGAS.
(ii) (B.36) is GAS.
(iii) (B.36) is UGATT and 0 is an equilibrium point of (B.36).
(iv) (B.36) is LIM and is Lyapunov stable.
(v) (B.36) possesses an infinitely differentiable strict Lyapunov function.

Proof (i) $\Leftrightarrow$ (ii) by Theorem B.35.
(ii) $\Rightarrow$ (iii) Let (B.36) be GAS. Since the input space equals $\{0\}$, Proposition B.34 implies that (B.36) is UGATT.
(iii) $\Rightarrow$ (i) Since f is locally Lipschitz, the solution of (B.36) depends continuously on the initial states according to Theorem 1.40. For a system without disturbances and with $f(0) = 0$, a continuous dependence of a flow on the initial state at 0 is equivalent to the fact that 0 is a robust equilibrium point of (B.36). At the same time, for (B.36) forward completeness is equivalent to RFC by Proposition B.33. Finally, Theorem B.22 shows the implication.
(ii) $\Leftrightarrow$ (iv) This follows by Exercise B.12.
(v) $\Leftrightarrow$ (i) This holds due to Theorem B.32. □

Remark B.38 We have already seen in Example B.8 that GATT does not imply LS because the solutions can move far away before they converge to the origin.

Theorem B.37 shows another property of GATT systems that are not LS: their time of convergence to zero is not uniform for the initial states with the same norm.

The items (i) and (ii) of Theorem B.37 are no more equivalent for infinite-dimensional systems, see Example 6.17.

B.8 Concluding Remarks

The stability theory of undisturbed systems originated from the works of Poincare [12], who was interested in global stability properties of planar systems, and of Lyapunov [9] who investigated Lyapunov stability properties of n-dimensional systems. Another milestone for the early stages of development of the dynamical systems theory was a monograph of Birkhoff [3], which paved the way for the topological and ergodic theory of dynamical systems and made dynamical systems a separate branch of mathematics.

We developed the stability theory of dynamical systems in this section in a way to provide a firm basis for the development of input-to-state stability theory. Therefore many interesting topics in dynamical systems theory have been omitted. For a more broad picture of results and methods of dynamical systems theory for systems without disturbances, you may consult many excellent books, e.g., [1, 10, 13]. A short but very nice survey of the early results in the stability theory one can find in the last sections of each chapter in the book [2].

For a survey of classical converse Lyapunov theorems, a reader may consult [4, Chap. VI] and for a more recent overview [7]. Theorem B.22 is a special case of [6, Theorem 2.2].

For the works related to weak attractivity, see Sect. 2.13.

Example B.16 is taken from [10, p. 119].

B.9 Exercises

Forward completeness and stability

Exercise B.1 Consider a robustly forward complete system (B.1) with the nonlinearity f that is Lipschitz continuous w.r.t. the first argument but not uniformly w.r.t. the second one. Show that it may have a flow that is not Lipschitz continuous on compact intervals.

Exercise B.2 Find sufficient conditions for (B.1) to be continuously dependent on initial values uniformly over $\mathcal{D}$, that is, for each $x \in \mathbb{R}^n$, each $\tau > 0$ and each $\varepsilon > 0$ there is a $\delta = \delta(x, \tau, \varepsilon)$: for all $y \in \mathbb{R}^n$: $|x - y| \leq \delta$ holds

$$\sup_{d \in \mathcal{D},\ t \in [0,\tau]} |\phi(t, x, d) - \phi(t, y, d)| < \varepsilon.$$

Exercise B.3 Prove Propositions B.12 and B.13.

Exercise B.4 Let $A \in \mathbb{R}^{n\times n}$. Show that for linear systems

$$\dot{x} = Ax$$

the notions of Lyapunov stability, Lagrange stability, and global stability coincide.

Attractivity and asymptotic stability

Exercise B.5 Provide a detailed treatment of Example B.16.

Exercise B.6 The system (B.1) is called *globally nonuniformly stable* if for all $x \in \mathbb{R}^n$ and all $d \in \mathcal{D}$ it holds that

$$\sup_{t\geq 0} |\phi(t, x, d)| < \infty.$$

Prove that GATT of (B.1) implies global nonuniform stability of (B.1).

Exercise B.7 Show that (B.1) is Lyapunov stable iff the function $w : x \mapsto \sup_{t\geq 0,\ d\in\mathcal{D}} |\phi(t, x, d)|$ is well-defined in a certain neighborhood of 0 and is continuous at the origin.

Exercise B.8 Let $A \in \mathbb{R}^{n\times n}$. Consider a linear undisturbed system

$$\dot{x} = Ax. \tag{B.37}$$

Prove that for (B.37) local attractivity is equivalent to global asymptotic stability.

Exercise B.9 Prof. Brainstormer has shown the following result:

Theorem. *The equilibrium $x \equiv 0$ is globally attractive for the system $\dot{x} = x^2$, $x(t) \in \mathbb{R}$.*

Proof The set of solutions of $\dot{x} = x^2$ consists of trajectories $x_c(t) = -\frac{1}{t+c}$, $c \in \mathbb{R}$, and of the trivial solution $x \equiv 0$. For any $c \in \mathbb{R}$ it holds that $\lim\limits_{t\to\infty} -\frac{1}{t+c} = 0$. Thus, every solution converges to 0. □

But strangely, $\dot{x}(t) > 0$ whenever $x(t) > 0$. Why?

Exercise B.10 The solution $x \equiv 0$ of the equation $\dot{x}(t) = ax(t)$, $a \in \mathbb{R}$, $x(t) \in \mathbb{R}$ is globally asymptotically stable iff $a < 0$. Now consider equation $\dot{x}(t) = a(t)x(t)$, $x(t) \in \mathbb{R}$ with $a \in C(\mathbb{R}, \mathbb{R})$. Does the condition $a(t) < 0$, $t \geq 0$ guarantee that the solutions of the equation $\dot{x}(t) = a(t)x(t)$ converge to zero?

Exercise B.11 Let f be Lipschitz continuous in x, uniformly w.r.t. d. We know that Lyapunov stability of (B.1) implies that 0 is an equilibrium of (B.1). In this exercise, we study relations between attractivity and the equilibrium concept.

(i) Let (B.1) be RFC. Then $0 \in \mathbb{R}^n$ is an equilibrium if and only if for every $\varepsilon > 0$ there is $y \in \mathbb{R}^n$: $|\phi(t, y, d)| \le \varepsilon$ for all $t \ge 0$ and all $d \in \mathcal{D}$.
(ii) For the system (B.1) without disturbances if $\phi(t, y) \to 0$ as $t \to \infty$ for a certain $y \in \mathbb{R}^n$, then 0 is an equilibrium.

Exercise B.12 (B.1) is *globally weakly attractive* (this property could also be called the limit property with zero gain), if

$$x \in \mathbb{R}^n,\ d \in \mathcal{D} \quad \Rightarrow \quad \inf_{t\ge 0} |\phi(t, x, d)| = 0.$$

That is, the system (B.1) is globally weakly attractive whenever every its trajectory approaches the origin arbitrarily close.

(i) Prove that for (B.1) GAS is equivalent to global weak attractivity combined with Lyapunov stability.
(ii) Show that if (B.1) is globally weakly attractive and locally attractive, then (B.1) is globally attractive.
(iii) Show that (B.1) is globally weakly attractive if and only if

$$x \in \mathbb{R}^n,\ d \in \mathcal{D} \quad \Rightarrow \quad \underline{\lim}_{t\ge 0} |\phi(t, x, d)| = 0. \tag{B.38}$$

Exercise B.13 Obviously, global attractivity implies global weak attractivity. Prove or disprove the converse implication.

Uniform attractivity and uniform asymptotic stability
In view of Theorem B.22, it is interesting to see, which properties does a system have that satisfies the RFC property and is UGATT but fails to satisfy the REP property. In the following exercise, a reader is invited to investigate this question.

Definition B.39 We say that a system Σ is *completely uniformly globally attractive (CUGATT)*, if there are $\beta \in \mathcal{KL}$ and $C > 0$ so that for all $x \in \mathbb{R}^n$, all $t \ge 0$ and all $d \in \mathcal{D}$ it holds that

$$|\phi(t, x, d)| \le \beta(|x| + C, t). \tag{B.39}$$

Exercise B.14 Show that (B.1) is CUGATT iff (B.1) is UGATT and RFC.
Hint: Proposition B.12 and ideas from the proof of Theorem B.22 may be helpful.

Lyapunov functions

Exercise B.15 Show that if there exists a continuous nonstrict Lyapunov function V for (B.1), then (B.1) is globally Lyapunov stable.

Exercise B.16 Let V be a Lyapunov function for (B.1). Show that for any continuously differentiable $\psi \in \mathcal{K}_\infty$ satisfying $\frac{d\psi}{dr}(r) > 0$ for any $r > 0$, the function $W(x) := \psi(V(x))$ is again a Lyapunov function.

Exercise B.17 Show that the equilibrium $(\pi, 0)$ in Example B.28 is unstable.

Exercise B.18 A linear approximation for small angles x of a nonlinear pendulum, investigated in Example B.28, has the form

$$\ddot{x} + a^2 x = 0,$$

for $a > 0$. This approximation is called a *harmonic oscillator*.

Talking below about the stability properties of the harmonic oscillator, we mean the stability properties of the corresponding first-order system in the variables $(x, \dot{x})$.

(i) Prove that the solution $x \equiv 0$ is stable but not asymptotically stable.
(ii) Prove a similar statement for the problem

$$\ddot{x} + \Omega x = 0,$$

$x(t) \in \mathbb{R}$, where $\Omega \in \mathbb{R}^{n \times n}$ is a positive definite matrix and $x \in \mathbb{R}^n$.
(iii) Find a function $k = k(x, \dot{x})$, so that the solution $x \equiv 0$ for

$$\ddot{x} + a^2 x = k(x, \dot{x})$$

is asymptotically stable.

Exercise B.19 Prove that we obtain an equivalent definition of a C^1 (UGAS) Lyapunov function for $\dot{x} = f(x)$ if we require merely $\dot{V}(x) = \nabla V(x) f(x) < 0$ instead of $\dot{V}(x) = \nabla V(x) f(x) < -\alpha(|x|)$ for some $\alpha \in \mathcal{P}$.
Comment: This result shows that the trajectories of UGAS systems corresponding to the initial values with the same norm have a uniform speed of convergence. For general infinite-dimensional systems, the claim of this exercise is false, even for linear systems, see [5, Ex. 8.2, p. 108].

Exercise B.20 Investigate the stability of the equilibrium $x \equiv 0$ for the systems

$$\text{(i)} \begin{cases} \dot{x} = y - x + xy \\ \dot{y} = x - y - x^2 - y^3 \end{cases}, \quad \text{(ii)} \begin{cases} \dot{x} = 2y^3 - x^5 \\ \dot{y} = -x - y^3 + y^5 \end{cases}, \quad \text{(iii)} \begin{cases} \dot{x} = y - 3x - x^3 \\ \dot{y} = 6x - 2y \end{cases}.$$

UGAS of systems without inputs

Exercise B.21 In this exercise, we continue investigations started in Exercises B.3, B.6, and B.7. Let us consider the system (B.36), and define

$$q : x \mapsto \sup_{t \geq 0} |\phi(t, x)|.$$

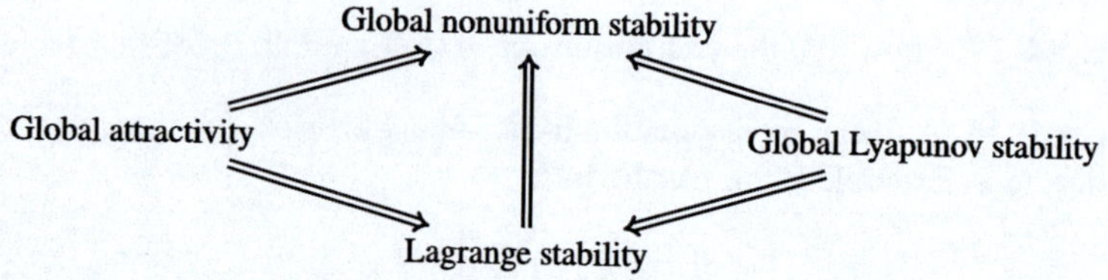

Fig. B.8 Relations between global attractivity, global Lyapunov stability, global nonuniform stability, and other properties

Your objective is to show the implications in Fig. B.8.

(i) Prove that global attractivity implies Lagrange stability.
(ii) It is easy to see that every Lagrange stable system is automatically nonuniformly globally stable. Does the converse hold?
(iii) Prove that if q is continuous over $\mathbb{R}^n$, then (B.36) is globally Lyapunov stable. Show by means of a counterexample that converse does not hold.
(iv) Prove that if q is continuous over $\mathbb{R}^n \setminus \{0\}$, then (B.36) is Lagrange stable. Show by means of a counterexample that converse does not hold.

References

1. Arnold V (1992) Ordinary differential equations. Springer
2. Bhatia NP, Szegö GP (2002) Stability theory of dynamical systems. Springer Science & Business Media
3. Birkhoff G (1927) Dynamical systems. Am Math Soc
4. Hahn W (1967) Stability of motion. Springer, New York
5. Jacob B, Zwart HJ (2012) Linear port-Hamiltonian systems on infinite-dimensional spaces. Springer, Basel
6. Karafyllis I, Jiang Z-P (2011) Stability and stabilization of nonlinear systems. Springer, London
7. Kellett CM (2015) Classical converse theorems in Lyapunov's second method. Discrete Contin Dyn Syst Ser B 20(8):2333–2360
8. Lin Y, Sontag ED, Wang Y (1996) A smooth converse Lyapunov theorem for robust stability. SIAM J Control Optim 34(1):124–160
9. Lyapunov A (1892) Obshchaya Zadacha ob Ustoichivosti Dvizhenia. Kharkov Mathematical Society
10. Meiss JD (2007) Differential dynamical systems. Society for Industrial and Applied Mathematics
11. Mironchenko A, Wirth F (2019) Non-coercive Lyapunov functions for infinite-dimensional systems. J Differ Equ 105:7038–7072
12. Poincaré H (1893) Les Méthodes Nouvelles de la Mécanique Céleste: Méthodes De MM. Newcomb, Glydén, Lindstedt Et Bohlin. Gauthier-Villars it fils
13. Teschl G (2012) Ordinary differential equations and dynamical systems. Am Math Soc

14. Vinograd R (1957) The inadequacy of the method of characteristic exponents for the study of nonlinear differential equations (in Russian). Mat Sb 41(4):431–438
15. Yoshizawa T (1966) Stability theory by Liapunov's second method. Math Soc Jpn

Appendix C
Nonlinear Monotone Discrete-Time Systems

In this chapter, we systematically analyze the properties of linear and nonlinear monotone operators as well as of discrete-time dynamical systems induced by these operators. This chapter is self-contained and can be read independently from the rest of the book. We apply this theory to the stability analysis of networks of ISS systems in Chap. 3.

We start by introducing the main concepts, which are followed by a detailed analysis of discrete-time linear monotone systems in Sect. C.2. In Sect. C.3 we introduce small-gain conditions for monotone operators and study relationships between them. In Sect. C.4, we analyze the properties of decay sets of monotone operators, which is a preparation for characterizing the local and global asymptotic stability of nonlinear monotone systems in Sect. C.5. In Sect. C.6, we introduce the concept of a path of strict decay, which is a key ingredient for the small-gain analysis of networks of ISS systems in Chap. 3.

In the rest of this chapter, we analyze the important special cases of nonlinear monotone operators: the gain operators (Sect. C.7), max-preserving operators (Sect. C.8), and homogeneous operators (Sect. C.9). In all these cases, we prove powerful characterizations of global asymptotic stability in terms of small-gain conditions. In the two latter cases, we also provide efficient constructions of paths of strict decay.

C.1 Basic Notions

We start with a basic concept

Definition C.1 Let X be any set. *A partial order* on X is a binary relation $\leq$ that is

(i) *reflexive*: $x \leq x$ for all $x \in X$,
(ii) *transitive*: $x \leq y$ and $y \leq z$ implies $x \leq z$,
(iii) *antisymmetric*: $x \leq y$ and $y \leq x$ implies $x = y$.

A. Mironchenko, *Input-to-State Stability*, Communications and Control Engineering,
https://doi.org/10.1007/978-3-031-14674-9

Definition C.2 A nonempty set together with a partial order is called a *partially ordered set*.

In any partially ordered set one can define the notions of supremum and infimum as follows:

Definition C.3 Let X be a partially ordered set and let $S \subset X$. If there is $a_0 \in X$ so that $a \le a_0$ for all $a \in S$, then a_0 is called a *majorant* of S. If in addition $a_0 \le b$ for any other majorant b of S, then a_0 is called *supremum* of S. *Infimum* of a sct $S \subset X$ is defined analogously.

Assumption C.4 For two vectors $x, y \in \mathbb{R}^n$ we say that $x \ge y$ if $x_i \ge y_i$ for all $i = 1, \dots, n$. The relation $\ge$ defines a partial order on $\mathbb{R}^n$.

In this chapter and in the whole book, we identify $\mathbb{R}^n$ with the ordered normed vector space $(\mathbb{R}^n, \ge)$ endowed with the Euclidean norm.

It is easy to see that with such a definition, if $x \le y$ for some $x, y \in \mathbb{R}^n_+$, then $|x| \le |y|$, that is, the Euclidean norm is monotone with respect to the partial order $\ge$.

We write $x \ll y$ to express the fact that $x_i < y_i$ for all $i = 1, \dots, n$. Analogously, we write $x \gg y$ if $x_i > y_i$ for all $i = 1, \dots, n$.

For any finite sequence $w_i = (w_{1i}, \dots, w_{ni})^T \in \mathbb{R}^n$, $i = 1, \dots, m$, the supremum exists and is given by $z := \sup_{i=1}^m w_i \in \mathbb{R}^n$ with $z_i := \max\{w_{i1}, \dots, w_{im}\}$.

The *supremum* (or *join*) of two elements $x, y \in \mathbb{R}^n$ we denote by $x \vee y = \sup\{x, y\}$. Define also the *positive part* of a vector $x \in \mathbb{R}^n$ by $x^+ := x \vee 0 = (\max\{x_i, 0\})_{i=1}^n$.

Remark C.5 It is easy to check that for any $x, y, z \in \mathbb{R}^n$ it holds that

$$\sup\{x, \sup\{y, z\}\} = \sup\{x, y, z\}. \tag{C.1}$$

Definition C.6 A nonlinear operator $A : \mathbb{R}^n \to \mathbb{R}^n$ is called

(i) *monotone* if for all $x, y \in \mathbb{R}^n$ such that $x \ge y$, it follows that $A(x) \ge A(y)$,
(ii) *nonnegative* if $A(\mathbb{R}^n_+) \subset \mathbb{R}^n_+$.

A nonlinear monotone operator $A : \mathbb{R}^n_+ \to \mathbb{R}^n_+$ gives rise to the *induced discrete-time system*

$$x(k+1) = A(x(k)), \quad k \in \mathbb{Z}_+. \tag{C.2}$$

The state of this system at time $k \in \mathbb{Z}_+$ corresponding to an initial condition $x \in \mathbb{R}^n$ we denote by $\phi(k, x) \in \mathbb{R}^n$.

We define the following central stability notions for discrete-time systems.

Definition C.7 Discrete-time system (C.2) is called

(i) *Uniformly globally asymptotically stable (UGAS)*, if there is $\beta \in \mathcal{KL}$ so that for each $k \ge 0$ and all $x \in \mathbb{R}^n$ it holds that

$$|\phi(k, x)| \le \beta(k, |x|). \tag{C.3}$$

(ii) *Uniformly globally exponentially stable (UGES)*, if there are $a \in (0, 1)$ and $M > 0$ so that for each $k \geq 0$ and all $x \in \mathbb{R}^n$ it holds that

$$|\phi(k, x)| \leq Ma^k|x|. \tag{C.4}$$

C.2 Linear Monotone Systems

Before we analyze nonlinear monotone systems, let us look at linear monotone systems for which a particularly strong theory is available. As we will see, some of these results can be extended to the nonlinear context as well.

We denote by $r(A)$ the spectral radius of a matrix $A \in \mathbb{R}^{n\times n}$. We need an important Perron–Frobenius theorem for nonnegative matrices [1, Theorem 2.1.1, p. 26].

Theorem C.8 *Let $A \in \mathbb{R}^{n\times n}_+$. Then:*

(i) $r(A)$ is an eigenvalue of A.
(ii) There is $x \in \mathbb{R}^n_+$, $x \neq 0$: $Ax = r(A)x$.

As the coefficients of the characteristic polynomial depend continuously on the entries of its matrix, and since the roots of a polynomial depend continuously on the coefficients of the polynomial, we obtain:

Proposition C.9 *The spectral radius $r(\cdot)$ is a continuous function on the space $\mathbb{R}^{n\times n}$.*

Finally, we will need (see [1, Theorem 2.1.11, p. 28])

Proposition C.10 *Let $A \in \mathbb{R}^{n\times n}_+$. If for some $x \in \operatorname{int}(\mathbb{R}^n_+)$ and $b > 0$ it holds that $Ax \leq bx$, then $r(A) \leq b$.*

The following proposition characterizes the exponential stability of linear discrete-time systems.

Proposition C.11 (Criteria for exponential stability of linear discrete-time systems) *Let $A \in \mathbb{R}^{n\times n}_+$. Then the following statements are equivalent:*

(i) Spectral small-gain condition*:* $r(A) < 1$.
(ii) $A^k \to 0$, for $k \to \infty$.
(iii) (C.2) is UGES.
(iv) (C.2) is UGAS.
(v) There is $P \in \mathbb{R}^{n\times n}_+$ so that $A(I + P)x \not\geq x$ for all $x \in \mathbb{R}^n_+$, $x \neq 0$.
(vi) $Ax \not\geq x$ for all $x \in \mathbb{R}^n_+$, $x \neq 0$.
(vii) $(I - A)^{-1}$ exists and is a nonnegative matrix.
(viii) There is $x \in \operatorname{int}(\mathbb{R}^n_+)$: $Ax \ll x$.
(ix) There is $\lambda \in (0, 1)$ and $x \in \operatorname{int}(\mathbb{R}^n_+)$: $Ax \leq \lambda x$.
(x) There is $\xi \in \mathcal{K}_\infty$ such that for all $w, v \in \mathbb{R}^n_+$ it holds that $(I - A)w \leq v \Rightarrow |w| \leq \xi(|v|)$.

(xi) *There is* $\xi \in \mathcal{K}_\infty$ *such that for all* $w, v \in \mathbb{R}^n_+$ *it holds that* $w \le (Aw) \vee v \Rightarrow |w| \le \xi(|v|)$.

Proof (i) $\Rightarrow$ (ii) Due to Gelfand's formula, it holds that

$$r(A) = \lim_{k\to\infty} \|A^k\|^{1/k}.$$

As $r(A) < 1$, there are certain $k_0 \in \mathbb{Z}_+$ and $a < 1$ such that $\|A^k\|^{1/k} \le a$ for all $k \ge k_0$. Thus, $\|A^k\| \le a^k$ for all $k \ge k_0$.
Since $a < 1$, we have that $\|A^k\| \to 0$ as $k \to \infty$.
(ii) $\Rightarrow$ (iii) By assumption $\|A^k\| \to 0$ as $k \to \infty$, and thus there is $k > 0$ such that $\|A^k\| < 1$.

Now for each $j \in \mathbb{Z}_+$ pick $r \in \mathbb{Z}_+$ and $s \in \mathbb{Z}_+$ with $s < k$ so that $j = kr + s$. Then for any $x \in \mathbb{R}^n_+$

$$\begin{aligned}
|\phi(j, x)| &= |A^j x| = |A^{kr+s}x| \\
&\le \|A^k\|^r \|A^s\| |x| \\
&= \left(\|A^k\|^{\frac{1}{k}}\right)^j \left(\|A^k\|^{\frac{1}{k}}\right)^{-s} \|A^s\| |x| \\
&\le \left(\|A^k\|^{\frac{1}{k}}\right)^j \max_{s\in[0,k-1]} \left(\left(\|A^k\|^{\frac{1}{k}}\right)^{-s} \|A^s\|\right)|x|.
\end{aligned}$$

As $\|A^k\| < 1$, and thus also $\|A^k\|^{\frac{1}{k}} < 1$, we see that the discrete-time system (C.2) is UGES.
(iii) $\Rightarrow$ (iv) Trivial.
(iv) $\Rightarrow$ (i) For all $x \in \mathbb{R}^n$, it holds that

$$|\phi(k, x)| = |A^k x| = |x| \left|A^k\left(\frac{x}{|x|}\right)\right| = |x||\phi(k, \frac{x}{|x|})| \le |x|\beta(1, k).$$

As $\beta \in \mathcal{KL}$, there is $s \in \mathbb{Z}_+$ so that $\beta(1, s) < 1$, and thus

$$\|A^s\| \le \beta(1, s) < 1.$$

Now for any $k \in \mathbb{Z}_+$ there are unique $a, b \in \mathbb{Z}_+$ so that $k = as + b$ and $b < s$. Then

$$\|A^k\| = \|A^{as+b}\| \le \|A^b\|\|A^s\|^a \le \max_{r\in[0,s]} \|A^r\|\big(\beta(1, s)\big)^a.$$

Finally, by Gelfand's formula and since $\frac{a}{k} = \frac{a}{as+b} \ge \frac{a}{as+s}$ we obtain that

$$\begin{aligned}
r(A) &= \lim_{k\to\infty} \|A^k\|^{1/k} \leq \lim_{k\to\infty} \left[\left(\max_{r\in[0,s]} \|A^r\| \right)^{\frac{1}{k}} (\beta(1,s))^{a/k} \right] \\
&\leq \lim_{k\to\infty} \left(\max_{r\in\mathbb{Z}_+\cap[0,s]} \|A^r\| \right)^{\frac{1}{k}} (\beta(1,s))^{\frac{a}{as+s}} \\
&= (\beta(1,s))^{\frac{a}{as+s}} \\
&< 1.
\end{aligned}$$

(i) $\Rightarrow$ (v) Since (i) holds, there is $\varepsilon > 0$ so that $r((1+\varepsilon)A) < 1$ and by the already proved equivalence between (i) and (iii), the system (C.2) with a monotone operator $\tilde{A} := (1+\varepsilon)A = A(I+\varepsilon I)$ instead of A is UGES.

Now, let (v) do not hold with $P = \varepsilon I$. Then there is a certain $x \in \mathbb{R}^n$, $x \neq 0$, so that $\tilde{A}x \geq x$. But then, in view of the monotonicity of $\tilde{A}$, we have that

$$\tilde{A}^n x \geq \tilde{A}^{n-1} x \geq \ldots \geq x,$$

for all $n \in \mathbb{Z}_+$. As the Euclidean norm is monotone with respect to the order in $\mathbb{R}^n$, it holds that

$$|\tilde{A}^n x| \geq |x|,$$

and thus the solution of (C.2) with $\tilde{A}$ operator does not converge to 0 for an initial condition x, which contradicts to UGES of this system.
(v) $\Rightarrow$ (vi) Assume that (vi) does not hold, and thus there is $y \in \mathbb{R}^n_+\backslash\{0\}$ so that $Ay \geq y$. Then for any $P \in \mathbb{R}^{n\times n}_+$ it holds that $(I+P)y = y + Py \geq y$ and as A is an increasing operator it holds that $A(I+P)y \geq Ay \geq y$. This contradicts to (v).
(vi) $\Rightarrow$ (i) Assume that (i) doesn't hold, that is, $r(A) \geq 1$. By Theorem C.8, there is $x \in \mathbb{R}^n_+$, $x \neq 0$ so that $Ax = r(A)x \geq x$, which contradicts to (vi).
(i) $\Rightarrow$ (vii) As $r(A) < 1$, $(I-A)^{-1}$ can be computed via Neumann series:

$$(I-A)^{-1} = \sum_{k=0}^{\infty} A^k.$$

Clearly, $(I-A)^{-1}$ is a nonnegative matrix.
(vii) $\Rightarrow$ (viii) By (vii), we have $(I-A)^{-1}(\mathbb{R}^n_+) \subset \mathbb{R}^n_+$. Applying $I-A$ on both sides yields $\mathbb{R}^n_+ \subset (I-A)\mathbb{R}^n_+$. That is, every $x \in \mathbb{R}^n_+$ can be written as $x = y - Ay$ for some $y \in \mathbb{R}^n_+$. For any $x \in \text{int}\,(\mathbb{R}^n_+)$, it holds for a corresponding y that $Ay \ll y$. As A is nonnegative, $Ay \in \mathbb{R}^n_+$, and hence $y \in \text{int}\,(\mathbb{R}^n_+)$.
(viii) $\Rightarrow$ (ix) Clear.
(ix) $\Rightarrow$ (i) Follows from Proposition C.10.
(vii) $\Rightarrow$ (x) Let $w, v \in \mathbb{R}^n_+$ be such that $(I-A)w \leq v$. Then $w \leq (I-A)^{-1}v$, and by monotonicity of the norm with respect to the order we have that $|w| \leq |(I-A)^{-1}v| \leq |(I-A)^{-1}||v| =: \xi(|v|)$.
(x) $\Rightarrow$ (xi) Let $\xi \in \mathcal{K}_\infty$ be as in (x) and let $w, v \in \mathbb{R}^n_+$. As $w, v \in \mathbb{R}^n_+$, $w \leq (Aw) \vee v$ and since $A(\mathbb{R}^n_+) \subset \mathbb{R}^n_+$, it follows that $w \leq Aw + v$, and hence $|w| \leq \xi(|v|)$.

(xi) $\Rightarrow$ (vi) Let there is $x \in \mathbb{R}^n_+ \backslash \{0\}$: $A(x) \geq x$. Then $x \leq (Ax) \vee 0$, and by (xi) we have that $x = 0$. A contradiction. $\square$

Remark C.12 *(Vector of strict decay I)* One may ask, how to find λ and x as in Proposition C.11 (ix), provided that $r(A) < 1$ holds. If A is *strictly positive*, that is, $A(\mathbb{R}^n_+ \backslash \{0\}) \subset \text{int}\,(\mathbb{R}^n_+)$, a more constructive result can be shown based on Perron–Frobenius theorem for strictly positive matrices [1, Theorem 1.3.26, p. 13]. It ensures that for a strictly positive A we have:

(i) $r(A)$ is a simple eigenvalue of A, greater than the magnitude of any other eigenvalue of A.
(ii) An eigenvector corresponding to $r(A)$ (called *Perron–Frobenius eigenvector*) lies in $\text{int}\,(\mathbb{R}^n_+)$.

Hence, if $r(A) < 1$, then there is $x \in \text{int}\,(\mathbb{R}^n_+)$: $Ax = r(A)x$, which shows the condition in Proposition C.11 (ix).

Remark C.13 *(Vector of strict decay II)* If $A \in \mathbb{R}^{n \times n}_+$, Perron–Frobenius theorem for strictly positive matrices cannot be applied in general, but as we are interested in item Proposition C.11 (ix) merely in a subeigenvector, another constructions are possible, see Exercise C.5.

Although the spectral small-gain condition (i) makes sense only for linear systems, the conditions (v), (vi), (viii), (x), (xi) do make sense for nonlinear systems and operators, and will be important in the sequel.

C.3 Small-Gain Conditions

In Proposition C.11, we characterized the UGAS property of linear discrete-time systems by several equivalent conditions that can also be interpreted for nonlinear systems. Many of such properties play a crucial role in the stability analysis of networks.

We start with a nonlinear counterpart of the property (x) in Proposition C.11.

Definition C.14 Let $A : \mathbb{R}^n_+ \to \mathbb{R}^n_+$ be a nonlinear operator. We say that $\text{id} - A$ has the *monotone bounded invertibility (MBI) property* if there exists $\xi \in \mathcal{K}_\infty$ such that for all $v, w \in \mathbb{R}^n_+$

$$(\text{id} - A)(w) \leq v \quad \Rightarrow \quad |w| \leq \xi(|v|).$$

Recall that for a set $S \subset \mathbb{R}^n$ and for $x \in \mathbb{R}^n$ we denote the *distance from x to S* by $\text{dist}\,(x, S) := \inf_{y \in S} |y - x|$.

We are going to relate the MBI property to the following small-gain conditions.

Definition C.15 We say that a nonlinear operator $A : \mathbb{R}^n_+ \to \mathbb{R}^n_+$ satisfies:

(i) The *small-gain condition* if

$$A(x) \not\geq x \quad \text{for all } x \in \mathbb{R}^n_+ \setminus \{0\}. \tag{C.5}$$

(ii) The *strong small-gain condition* if there exists $\rho \in \mathcal{K}_\infty$ such that applying $\text{id} + \rho$ componentwise we obtain

$$(\text{id} + \rho) \circ A(x) \not\geq x \quad \text{for all } x \in \mathbb{R}^n_+ \setminus \{0\}. \tag{C.6}$$

(iii) The *uniform small-gain condition* if there is an $\eta \in \mathcal{K}_\infty$ such that

$$\text{dist}\,(A(x) - x, \mathbb{R}^n_+) \geq \eta(|x|), \quad x \in \mathbb{R}^n_+. \tag{C.7}$$

The following lemma is left as Exercise C.9.

Lemma C.16 *Let $A : \mathbb{R}^n_+ \to \mathbb{R}^n_+$ be a nonlinear operator, and $\rho \in \mathcal{K}_\infty$. Then*

$$(\text{id} + \rho) \circ A(x) \not\geq x \;\; \forall x \in \mathbb{R}^n_+ \setminus \{0\} \quad \Leftrightarrow \quad A \circ (\text{id} + \rho)(x) \not\geq x \;\; \forall x \in \mathbb{R}^n_+ \setminus \{0\}.$$

Now we give a criterion for the strong small-gain condition.

Proposition C.17 *A nonlinear operator $A : \mathbb{R}^n_+ \to \mathbb{R}^n_+$ satisfies the strong small-gain condition if and only if there are $\eta \in \mathcal{K}_\infty$ and an operator $\vec{\eta} : \mathbb{R}^n_+ \to \mathbb{R}^n_+$, defined by*

$$\vec{\eta}(x) := (\eta(x_i))_{i=1}^n \quad \textit{for all } x \in \mathbb{R}^n_+, \tag{C.8}$$

such that

$$A(x) \not\geq x - \vec{\eta}(x) \quad \textit{for all } x \in \mathbb{R}^n_+ \setminus \{0\}. \tag{C.9}$$

Proof "$\Rightarrow$": Let the strong small-gain condition hold with corresponding ρ. Then for any $x = (x_i)_{i=1}^n \in \mathbb{R}^n_+ \setminus \{0\}$ it holds that

$$\exists i \in I : \quad \big[(\text{id} + \rho) \circ \big(A(x)\big)\big]_i = (\text{id} + \rho)([A(x)]_i) < x_i. \tag{C.10}$$

As $\rho \in \mathcal{K}_\infty$, Lemma A.5 ensures existence of $\eta \in \mathcal{K}_\infty$ such that $\text{id} - \eta = (\text{id} + \rho)^{-1} \in \mathcal{K}_\infty$. Thus, (C.10) is equivalent to

$$\exists i \in I : \quad A(x)_i < x_i - \eta(x_i) = \Big[x - \vec{\eta}(x)\Big]_i, \tag{C.11}$$

which is the same as (C.9).

"$\Leftarrow$": Let (C.9) hold with a certain $\eta_1 \in \mathcal{K}_\infty$ and a corresponding $\vec{\eta}_1$. By Lemma A.19, one can choose $\eta \in \mathcal{K}_\infty$, such that $\eta \leq \eta_1$ and $\text{id} - \eta \in \mathcal{K}_\infty$. Then (C.9) holds with this η and a corresponding $\vec{\eta}$, i.e., we have

$$\exists i \in I : \quad A(x)_i < x_i - \eta(x_i) = (\text{id} - \eta)(x_i).$$

As $\eta \in \mathcal{K}_\infty$ satisfies $\text{id} - \eta \in \mathcal{K}_\infty$, by Lemma A.5(i) there is $\rho \in \mathcal{K}_\infty$ such that $(\text{id} - \eta)^{-1} = \text{id} + \rho$, and thus (C.10) holds, which shows that A satisfies the strong small-gain condition.
□

Denote by $\mathbb{1} := (1, 1, \dots, 1) \in \mathbb{R}^n$ the vector in $\mathbb{R}^n$ whose all components equal to 1.

Now we characterize the uniform small-gain condition:

Proposition C.18 (Criteria for the uniform small-gain condition) *The following statements are equivalent:*

(i) $\text{id} - A$ *has the MBI property.*
(ii) A *satisfies the uniform small-gain condition.*
(iii) There is an $\eta \in \mathcal{K}_\infty$ *such that*

$$A(x) \not\geq x - \eta(|x|)\mathbb{1} \quad \textit{for all } x \in \mathbb{R}^n_+ \setminus \{0\}. \tag{C.12}$$

Furthermore, all the above conditions imply the strong small-gain condition.

Proof Denote the negative part of a vector $x \in \mathbb{R}^n$ by $x^- := -(-x)^+$ and note that

$$\text{dist}\,(x, \mathbb{R}^n_+) = |x^-| \quad \text{for all } x \in \mathbb{R}^n.$$

(i) ⇒ (ii) Fix $x \in \mathbb{R}^n_+$ and observe that

$$(\text{id} - A)(x) = -(A(x) - x) \leq -(A(x) - x)^-.$$

As $-(A(x) - x)^- \in \mathbb{R}^n_+$, the MBI property implies

$$\text{dist}\,(A(x) - x, \mathbb{R}^n_+) = | - (A(x) - x)^-| \geq \xi^{-1}(|x|),$$

which proves (ii) with $\eta := \xi^{-1}$.
(ii) ⇒ (iii) Pick any $x \in \mathbb{R}^n_+ \setminus \{0\}$ and observe that

$$\begin{aligned}\eta(|x|) &\leq \text{dist}\,(A(x) - x, \mathbb{R}^n_+) = |(A(x) - x)^-| \\ &= |(x - A(x))^+| \leq |\max_{i=1}^n (x - A(x))_i \mathbb{1}| = \sqrt{n} \max_{i=1}^n (x - A(x))_i.\end{aligned}$$

This inequality implies the existence of an index i (depending on x) such that

$$\frac{1}{\sqrt{n}}\eta(|x|) \leq (x - A(x))_i.$$

This yields $x - A(x) \not\leq \frac{1}{\sqrt{n}}\eta(|x|)\mathbb{1}$, which proves (iii) with $\frac{1}{\sqrt{n}}\eta$ instead of η.
(iii) ⇒ (i) By assumption, for each $x \in \mathbb{R}^n_+ \setminus \{0\}$ there is i such that $(x - A(x))_i > \eta(|x|)$. Then $\text{dist}\,(A(x) - x, \mathbb{R}^n_+) = |(x - A(x))^+| \geq \max_{i=1}^n (x - A(x))_i > \eta(|x|)$.

Now let $(\mathrm{id} - A)(x) \leq w$ for some $w \in \mathbb{R}^n_+$. Then $A(x) - x \geq -w$ and hence $\eta(|x|) < \mathrm{dist}\,(-w, \mathbb{R}^n_+) = |(-w)^-| = |-w| = |w|$, and hence $|x| \leq \eta^{-1}(|w|)$, which shows (i).

Finally, let $\eta \in \mathcal{K}_\infty$ be as in (iii). Then for all $x \in \mathbb{R}^n_+$ it holds that $\eta(|x|)\mathbb{1} \geq \vec{\eta}(x)$, where $\vec{\eta}$ is defined by (C.8). Hence $A(x) \ngeq x - \vec{\eta}(x)$ for all $x \in \mathbb{R}^n_+ \setminus \{0\}$, and by Proposition C.17, A satisfies the strong small-gain condition. □

Now we relate the MBI property to other properties that have been equivalent to MBI for linear systems (which follows from Proposition C.11), but are no more equivalent for nonlinear systems.

Proposition C.19 *Let $A : \mathbb{R}^n_+ \to \mathbb{R}^n_+$ be a nonlinear operator. Consider the following statements:*

(i) $\mathrm{id} - A$ *has the MBI property.*
(ii) There is $\xi \in \mathcal{K}_\infty$ such that for all $w, v \in \mathbb{R}^n_+$ it holds that $w \leq A(w) \vee v \Rightarrow |w| \leq \xi(|v|)$.
(iii) $A(x) \ngeq x$ *for all* $x \in \mathbb{R}^n_+\backslash\{0\}$.

The following implications hold:

(i) ⇒ (ii) ⇒ (iii).

Proof (i) ⇒ (ii) Let $\xi \in \mathcal{K}_\infty$ be as in the definition of the MBI property, and let $w, v \in \mathbb{R}^n_+$ be such that $w \leq A(w) \vee v$. As $A(\mathbb{R}^n_+) \subset \mathbb{R}^n_+$, it follows that $w \leq A(w) + v$, and by the MBI property of $\mathrm{id} - A$ we have that $|w| \leq \xi(|v|)$.
(ii) ⇒ (iii) Let there is $x \in \mathbb{R}^n_+\backslash\{0\}$: $A(x) \geq x$. Then $x \leq A(x) \vee 0$, and by (ii) we have that $x = 0$. A contradiction. □

The next lemma relates the MBI property to the existence of a monotone inverse of $\mathrm{id} - A$:

Lemma C.20 *Let $A : \mathbb{R}^n_+ \to \mathbb{R}^n_+$ be a monotone operator, $A(0) = 0$, and $\mathrm{id} - A$ be a homeomorphism with a monotone inverse $(\mathrm{id} - A)^{-1}$. Then $\mathrm{id} - A$ satisfies the MBI property.*

Proof Pick any $w, v \in \mathbb{R}^n_+$ such that $(\mathrm{id} - A)(w) \leq v$ holds. As $(\mathrm{id} - A)^{-1}$ is monotone, this implies that $w \leq (\mathrm{id} - A)^{-1}(v)$, and by monotonicity of the norm with respect to the order, this implies that

$$|w| \leq |(\mathrm{id} - A)^{-1}(v)| \leq \xi(|v|),$$

where $\xi(r) := \sup_{|x| \leq r} |(\mathrm{id} - A)^{-1}(x)|$.

As $A(0) = 0$, then $(\mathrm{id} - A)(0) = 0$, and thus also $(\mathrm{id} - A)^{-1}(0) = 0$. Since $(\mathrm{id} - A)^{-1}$ is continuous, all conditions of Lemma A.27 are fulfilled with $g = (\mathrm{id} - A)^{-1}$, and thus Lemma A.27 implies that ξ is continuous, nondecreasing and $\xi(0) = 0$. Hence, it can be upperbounded by a $\mathcal{K}_\infty$-function. □

C.4 Sets of Decay

In this section, we relate the small-gain condition to the properties of decay sets of discrete-time systems.

Definition C.21 For any $x \in \mathbb{R}^n_+$, define the *orbit* of $x \in \mathbb{R}^n_+$ by $O(x) := \{\phi(k, x) : k \in \mathbb{Z}_+\}$. By $\omega(x)$ we denote the *ω-limit set* of x, that is, $\omega(x) := \cap_{k \in \mathbb{Z}_+} \{\phi(t, x) : t \in \mathbb{Z}_+,\ t \geq k\} = \cap_{k \in \mathbb{Z}_+} O(\phi(k, x))$.

Definition C.22 A set $S \subset \mathbb{R}^n_+$ is called:

(i) *Unbounded*, if for each $r > 0$ there is $x \in S$: $|x| > r$.
(ii) *Cofinal* (or *jointly unbounded*), if for any $x \in \mathbb{R}^n_+$ there is $y \in S$ such that $x \leq y$.

Clearly, cofinality is a stronger property than unboundedness.

We denote by $\mathcal{M}(\mathbb{R}^n_+)$ the set of all continuous monotone operators, mapping $\mathbb{R}^n_+$ to $\mathbb{R}^n_+$. An important role will be played by the following sets of decay:

Definition C.23 Define the *set of nonstrict decay* $\Psi(A)$, the *set of strict decay* $\Omega(A)$, and the *set of strict decay on i-th coordinate* $\Omega_i(A)$ by

$$\Psi(A) := \{x \in \mathbb{R}^n_+ : A(x) \leq x\}, \tag{C.13a}$$

$$\Omega_i(A) := \{x \in \mathbb{R}^n_+ : (A(x))_i < x_i\}, \tag{C.13b}$$

$$\Omega(A) := \cap_{i=1}^n \Omega_i(A) = \{x \in \mathbb{R}^n_+ : A(x) \ll x\}. \tag{C.13c}$$

Proposition C.24 *Let $A \in \mathcal{M}(\mathbb{R}^n_+)$ with $A(0) = 0$. Then $\Psi(A)$ is a closed and invariant set, containing* 0.

Proof Clear. □

Let us introduce some notation. Let $(e_i)_{i=1}^n$ be the standard basis of $\mathbb{R}^n$. For $Q \subset \mathbb{R}^n$ we denote by $\mathrm{co}(Q)$ the *convex hull* of all vectors in Q. For any $r > 0$ the set $S_r := \mathrm{co}((re_i)_{i=1}^n) \cap \mathbb{R}^n_+$ we call a *$(n-1)$-simplex*. The vectors re_i, $i = 1, \ldots, n$ are the vertices of the $(n-1)$-simplex S_r. In other words,

$$S_r = \Big\{s \in \mathbb{R}^n_+ : \sum_i s_i = r\Big\} = \{x \in \mathbb{R}^n_+ : \|x\|_1 = r\}. \tag{C.14}$$

A face is a simplex spanned by a subset of the vertices of S_r, i.e., for $I \subset \{1, \ldots, n\}$ we call $\sigma = \mathrm{co}((re_i)_{i \in I})$ a *face of* S_r. By I_σ we denote the indices of the vertices spanning σ.

We need the well-known Knaster–Kuratowski–Mazurkiewicz (KKM) theorem.

Theorem C.25 (Knaster–Kuratowski–Mazurkiewicz, 1929) *Let $r > 0$ and let S_r be a $(n-1)$-simplex. If a family $(R_i)_{i=1,\ldots,n}$ of subsets of S_r is such that all the sets R_i are closed or all are open, and each face σ of S_r is contained in the corresponding union $\cup_{i \in I_\sigma} R_i$, then*

$$\cap_{i=1}^{n} R_i \neq \emptyset. \tag{C.15}$$

It was shown originally in [9]. Our formulation is taken from [14, Theorem 1.5.7]. See also [10], where the authors provide a simple proof of the KKM theorem from Brouwer's fixed point theorem.

With the help of the KKM theorem, we can show the following important property of the sets of strict decay:

Proposition C.26 *Let $A \in \mathcal{M}(\mathbb{R}^n_+)$, $r > 0$, and $A(x) \not\geq x$ for all $x \in S_r$. Then*

$$\Omega(A) \cap S_r \neq \emptyset, \tag{C.16}$$

In particular, if $A(x) \not\geq x$ for all $x \neq 0$, then the set $\Omega(A)$ is unbounded.

Proof Let $r > 0$ and S_r be the corresponding $(n-1)$-simplex. Define $R_i := \Omega_i(A) \cap S_r$. All $\Omega_i(A)$ are open sets in $\mathbb{R}^n_+$, as A is continuous, and thus R_i are also open in the induced topology of S_r.

Let σ be any face of S_r, I_σ be the set of vertices of σ and let $x \in \sigma$. As $A(x) \not\geq x$, there is some i such that $(A(x))_i < x_i$. If $i \notin I_\sigma$, then $(A(x))_i < x_i = 0$, a contradiction. Hence, $i \in I_\sigma$, and thus, $x \in \Omega_i(A) \cap S_r = R_i \subset \cup_{i \in I_\sigma} R_i$. Now, KKM theorem implies (C.16).

If $A(x) \not\geq x$ for all $x \neq 0$, then $\Omega(A) \cap S_r \neq \emptyset$ for all $r > 0$, which implies that the set $\Omega(A)$ is unbounded. □

C.5 Asymptotic Stability of Induced Systems

In this section, we proceed to the stability analysis of nonlinear monotone systems. We need several further stability notions for discrete-time systems.

Definition C.27 Discrete-time system (C.2) is called:

(i) *(Locally) attractive*, if there is $r > 0$ such that for any $x \in B_r$ it holds that $\phi(k, x) \to 0$ as $k \to \infty$.
(ii) *Globally attractive (GATT)*, if for any $x \in \mathbb{R}^n_+$ it holds that $\phi(k, x) \to 0$ as $k \to \infty$.
(iii) *(Locally) weakly attractive*, if there is $r > 0$ such that for any $x \in B_r$ it holds that $\inf_{k \in \mathbb{Z}_+} |\phi(k, x)| = 0$.
(iv) *Globally weakly attractive (GWATT)*, if for any $x \in \mathbb{R}^n_+$ it holds that $\inf_{k \in \mathbb{Z}_+} |\phi(k, x)| = 0$.
(v) *Lyapunov stable*, if for any $\varepsilon > 0$ there is $\delta > 0$ such that for any $x \in B_\delta$ it follows that $|\phi(k, x)| \leq \varepsilon$ for all $k \in \mathbb{Z}_+$.
(vi) *(Locally) asymptotically stable*, if (C.2) is attractive and stable.

Furthermore:

Definition C.28 *We call* (C.2) weakly attractive with an attraction region $R \subset \mathbb{R}^n_+$, if $0 \in \text{int}\,(R)$ (in the induced topology of $\mathbb{R}^n_+$), R is forward invariant (that is $\phi(k, x) \in R$ for all $x \in R$ and $k \in \mathbb{Z}_+$) and for any $x \in R$ it holds that $\inf_{q\in\mathbb{Z}_+} |\phi(q, x)| = 0$.

For any $x \in \mathbb{R}^n_+$ we denote $A^k(x) := A \circ A^{k-1}(x)$ for all $k \in \mathbb{N}$ and $A^0(x) = x$. We start with some basic relations:

Lemma C.29 *Let $A : \mathbb{R}^n_+ \to \mathbb{R}^n_+$ be a monotone operator satisfying $A(x) \not\geq x$ for all $x \in \mathbb{R}^n_+\backslash\{0\}$. Then $A(0) = 0$.*

Proof Suppose $A(0) = x > 0$. Then by monotonicity $A(x) \geq A(0) = x > 0$ contradicting $A(x) \not\geq x$ for $x \neq 0$. This implies $A(0) = 0$. □

Lemma C.30 *Let $A : \mathbb{R}^n_+ \to \mathbb{R}^n_+$ be a monotone operator, and let there exist $y \in \mathbb{R}^n_+\backslash\{0\}$: $\inf_{k\in\mathbb{Z}_+} |A^k(y)| = 0$. Then $A(0) = 0$.*

Proof Take y as in the formulation of the lemma and suppose $A(0) = x > 0$. Then by monotonicity $A(y) \geq A(0) = x > 0, A^2(y) \geq A(x) \geq A(0) = x$, etc. By induction we obtain that $A^k(y) \geq x$, and thus $\inf_{k\geq 1} |A^k(y)| \geq |x| > 0$. As $\inf_{k\geq 0} |A^k(y)| = 0$, then $y = 0$, a contradiction. □

Proposition C.31 *Let $A : \mathbb{R}^n_+ \to \mathbb{R}^n_+$ be a monotone operator. If the origin of* (C.2) *is weakly attractive with an attraction region R, then for all $x \in R\backslash\{0\}$ and all $k \in \mathbb{N}$ it holds that $A^k(x) \not\geq x$.*

Proof Let $x \in R\backslash\{0\}$ satisfy $A^k(x) \geq x$ for some $k \in \mathbb{N}$. By monotonicity of A, we have for any $s = 0, \ldots, k-1$ and for any $p \geq 1$ that

$$A^{kp+s}(x) = A^s(A^{kp}(x)) \geq A^s(x),$$

and hence also

$$|\phi(kp+s, x)| = |A^{kp+s}(x)| \geq |A^s(x)|.$$

Assume that $A^s(x) = 0$ for a certain $s \in \{0, \ldots, k-1\}$. As A is monotone, and (C.2) is weakly attractive, Lemma C.30 guarantees that $A(0) = 0$, and thus, $A^{q+s}(x) = A^q(A^s(x)) = 0$ for all $q \geq 0$. In particular, $A^k(x) = 0 < x$, a contradiction.

Thus,

$$\inf_{q\in\mathbb{Z}_+} |\phi(q, x)| = \inf\{|x|, |A(x)|, \ldots, |A^{k-1}(x)|\} > 0,$$

a contradiction to $x \in R$. □

A partial converse to previous proposition is given by the following lemma

Lemma C.32 *Let $A \in \mathcal{M}(\mathbb{R}^n_+)$ satisfy $A(y) \not\geq y$ for all $y \in \mathbb{R}^n_+\backslash\{0\}$. Then for any $x \in \mathbb{R}^n_+$ with a precompact $O(x)$, it follows that $\lim_{k\to\infty} \phi(k, x) = 0$.*

Proof As $O(x)$ is precompact, $\omega(x)$ is nonempty and compact, see [19, Lemma 6.6]. Hence $z := \sup\big(\omega(x)\big) \in \mathbb{R}^n_+$ is finite and well-defined. By monotonicity of A and invariance of $\omega(x)$ we have that

$$A(z) \geq A(\omega(x)) = \omega(x),$$

and thus $A(z) \geq z$. By assumption, this implies $z = 0$, and thus $\omega(x) = \{0\}$. This implies the claim. □

A local version of the above lemma is

Lemma C.33 *Let for some $x \in \mathbb{R}^n_+$ the set $R := \overline{O(x)}$ be compact. If $A(y) \not\geq y$ for all $y \in R\backslash\{0\}$, then $\lim_{k\to\infty} \phi(k, x) = 0$.*

The following result shows that the local weak attractivity implies Lyapunov stability for monotone systems with a continuous A. This is a very strong property that does not hold for arbitrary nonlinear systems and which does not also hold for infinite-dimensional monotone systems (even if we assume attractivity instead of weak attractivity).

Lemma C.34 *Let $A \in \mathcal{M}(\mathbb{R}^n_+)$. If (C.2) is locally weakly attractive, then (C.2) is stable.*

Proof Let (C.2) be locally weakly attractive with an attraction region R such that $0 \in \operatorname{int}(R)$. By Proposition C.31, $A(x) \not\geq x$ for $x \in R\backslash\{0\}$. Thus, for any $\varepsilon > 0$, there is $r > 0$ such that $S_r \subset B_\varepsilon \cap R$. By Proposition C.26, there is $y \in \Omega(A) \cap S_r \subset \operatorname{int}(\mathbb{R}^n_+)$. Thus, there is $\delta > 0$: for all $x \in B_\delta$ it holds that $x \leq y$. By monotonicity of A it holds that $A(x) \leq A(y) \ll y$. Thus, $|A(x)| \leq \varepsilon$ for all $x \in B_\delta$ and (C.2) is stable. □

Now we can reap the fruits. We start by showing that local asymptotic stability is equivalent to the validity of the small-gain condition in a neighborhood of the origin.

Theorem C.35 (Criteria for local asymptotic stability of monotone systems) *Let $A \in \mathcal{M}(\mathbb{R}^n_+)$. The following statements are equivalent:*

(i) (C.2) is locally asymptotically stable.
(ii) (C.2) is locally attractive.
(iii) (C.2) is locally weakly attractive.
(iv) There is a neighborhood of the origin U such that $A(x) \not\geq x$ for any $x \in U \cap \mathbb{R}^n_+$.

Proof (i) ⇒ (ii) ⇒ (iii) Clear.
(iii) ⇒ (iv) Follows by Proposition C.31.
(iv) ⇒ (i) As $A(x) \not\geq x$ for any $x \in U$, there is $r > 0$ such that $S_r \subset U$. By Proposition C.26, there is $y \in \Omega(A) \cap S_r$. Pick $\delta > 0$ such that for all $x \in B_\delta$ it holds that $x \leq y$. By monotonicity of A, it holds that $A(x) \leq A(y) \ll y$, and thus $A^k(x) \leq y$ for all $k \geq 1$, and $O(x)$ is precompact.

Lemma C.33 shows that B_δ is in the attraction region of (C.2). Hence, (C.2) is attractive and by Lemma C.34, it is stable. Overall, (C.2) is locally asymptotically stable. □

In contrast to the above local result, UGAS of monotone systems is not equivalent to the global small-gain condition, but we have the following result.

Theorem C.36 (Criteria for UGAS of monotone systems) *Let $A \in \mathcal{M}(\mathbb{R}^n_+)$. Consider the following statements:*

(i) (C.2) is UGAS.
(ii) (C.2) is GATT.
(iii) (C.2) is GWATT.
(iv) $A(x) \not\geq x \quad \forall x \in \mathbb{R}^n_+$.
(v) $\Omega(A)$ is unbounded, and furthermore, $A(x) = x \Leftrightarrow x = 0$.

The following holds:

$$(i) \quad \Leftrightarrow \quad (ii) \quad \Leftrightarrow \quad (iii) \quad \Rightarrow \quad (iv) \quad \Rightarrow \quad (v).$$

Furthermore, in general (iv) $\not\Rightarrow$ (iii).

Proof (i) $\Rightarrow$ (ii) $\Rightarrow$ (iii) Clear.
(iii) $\Rightarrow$ (i) By Lemma C.34, GWATT implies Lyapunov stability, and in particular, $A(0) = 0$. Using the cocycle property, it is not hard to show that

$$\text{GWATT} \ \wedge \ \text{Lyapunov stability} \quad \Rightarrow \quad \text{GATT}.$$

Now pick any $r > 0$. Then for any $x \in \mathbb{R}^n_+$: $|x| < r$ it follows that $x \leq r\mathbb{1}$, and by monotonicity of A we obtain that

$$0 = A^k(0) \leq A^k(x) \leq A^k(r\mathbb{1}), \quad k \in \mathbb{Z}_+.$$

Thus, the system (C.2) is uniformly globally attractive, i.e., for any $r > 0$ and any $\varepsilon > 0$ there is $\tau = \tau(\varepsilon, r) > 0$, such that

$$|x| \leq r, \quad k \geq \tau \quad \Rightarrow \quad |A^k(x)| \leq \varepsilon.$$

Now UGAS follows using the arguments as in Theorem B.22.
(iii) $\Rightarrow$ (iv) Holds by Proposition C.31.
(iv) $\Rightarrow$ (v) Holds by Proposition C.26.
Example, showing that the condition $A(x) \not\geq x \ \forall x \in \mathbb{R}^n_+$ in general does not imply GWATT, is left as Exercise C.8. □

As we have seen from Theorem C.36, merely unboundedness of $\Omega(A)$ together with the fact that 0 is the only fixed point of A is not sufficiently strong to imply UGAS of (C.2). At the same time, we have the following:

Lemma C.37 *Let $A \in \mathcal{M}(\mathbb{R}^n_+)$ satisfy $A(0) = 0$ and be without other fixed points. If $\Psi(A)$ is cofinal, then (C.2) is UGAS.*

Proof As $\Psi(A)$ is cofinal, for each $x \in \mathbb{R}^n_+$ there is $y \in \Psi(A)$ such that $x \leq y$. As $y \in \Psi(A)$, we have $Ay \leq y$, and by monotonicity of A we have that $(A^k(y))_{k\in\mathbb{Z}_+}$ is a nonincreasing (w.r.t. order in $\mathbb{R}^n$) sequence. Thus, $(A^k(y))_{k\in\mathbb{Z}_+}$ converges to a certain $z \in \mathbb{R}^n_+$. As A is continuous, $z = \lim_{k\to\infty} A^{k+1}(y) = A(\lim_{k\to\infty} A^k(y)) = A(z)$, which implies $z = 0$ as 0 is the only fixed point of A.

Again by monotonicity we have that $A^k(x) \le A^k(y)$. Since $A^k(y) \to 0$ as $k \to \infty$, then also $A^k(x) \to 0$ as $k \to \infty$. Thus, (C.2) is GATT, and due to Theorem C.36, (C.2) is UGAS. □

C.6 Paths of Strict Decay

Certain paths in the set $\Omega(A)$ play an important role for Lyapunov-based small-gain theory developed in Sect. 3.5.

Definition C.38 Let $A \in \mathcal{M}(\mathbb{R}^n_+)$. A map $\sigma : \mathbb{R}_+ \to \mathbb{R}^n_+$ is called a *path of strict decay with respect to* A, if $\sigma = (\sigma_i)_{i=1}^n$ with $\sigma_i \in \mathcal{K}_\infty$ for all i and there exists $\rho \in \mathcal{K}_\infty$ such that

$$A(\sigma(r)) \le (\mathrm{id} + \rho)^{-1}(\sigma(r)), \quad r \ge 0. \tag{C.17}$$

Proposition C.39 *Let $A \in \mathcal{M}(\mathbb{R}^n_+)$. Consider the following properties:*

(i) There is $\xi \in \mathcal{K}_\infty$, such that $\Omega((\mathrm{id} + \xi) \circ A)$ is cofinal and there is a path of strict decay with respect to $(\mathrm{id} + \xi) \circ A$.
(ii) There is a path of strict decay with respect to A.
(iii) There is $\xi \in \mathcal{K}_\infty$, such that $x(k + 1) = (\mathrm{id} + \xi) \circ A(x(k))$ is UGAS.
(iv) A satisfies the strong small-gain condition.

It holds that (i) ⇔ (ii) ⇒ (iii) ⇒ (iv).

Proof (i) ⇒ (ii) Clear.
(ii) ⇒ (i) Let σ be a path of strict decay, and ρ be as in Definition C.38.

By Lemma A.5, there are $\rho_1, \rho_2 \in \mathcal{K}_\infty$, such that $\mathrm{id} + \rho = (\mathrm{id} + \rho_1) \circ (\mathrm{id} + \rho_2)$, and thus

$$(\mathrm{id} + \rho)^{-1} = (\mathrm{id} + \rho_2)^{-1} \circ (\mathrm{id} + \rho_1)^{-1}.$$

Substituting this into (C.17), we obtain that

$$(\mathrm{id} + \rho_2) \circ A(\sigma(r)) \le (\mathrm{id} + \rho_1)^{-1}(\sigma(r)), \quad r \ge 0. \tag{C.18}$$

Thus, $\sigma(\mathbb{R}_+) \subset \Omega((\mathrm{id} + \rho_2) \circ A)$, and since all $\sigma_i \in \mathcal{K}_\infty$, this shows that $\Omega((\mathrm{id} + \rho_2) \circ A)$ is cofinal.
(ii) ⇒ (iii) By the previous argument, the estimate (C.18) holds. Now assume that there is $x \in \mathbb{R}^n_+$, $x \ne 0$ such that $(\mathrm{id} + \rho_2) \circ A(x) \ge x$. As $\sigma(0) = 0$, and σ is continuous with $\mathcal{K}_\infty$-components, there is the minimal $r^* \ge 0$, such that $x \le \sigma(r^*)$. As r^* is chosen to be minimal, there is i, such that $x_i = \sigma_i(r^*)$.

By monotonicity of $(\mathrm{id} + \rho_2) \circ A$, we have

$$x \le (\mathrm{id} + \rho_2) \circ A(x) \le (\mathrm{id} + \rho_2) \circ A(\sigma(r^*)) \le (\mathrm{id} + \rho_1)^{-1}(\sigma(r^*)),$$

and in particular

$$x_i \le (\mathrm{id} + \rho_1)^{-1}(\sigma_i(r^*)) = (\mathrm{id} + \rho_1)^{-1}(x_i) < x_i,$$

a contradiction. Hence $(\mathrm{id} + \rho_2) \circ A(x) \not\ge x$ for all $x \in \mathbb{R}^n_+ : x \ne 0$.

Substituting $r = 0$ into (C.18), we see that $(\mathrm{id} + \rho_2) \circ A(0) = 0$, and 0 is the only fixed point of the operator $(\mathrm{id} + \rho_2) \circ A$. Since $\Omega((\mathrm{id} + \rho_2) \circ A)$ is cofinal, Lemma C.37 shows UGAS of $x(k+1) = (\mathrm{id} + \rho_2) \circ A(x(k))$.

(iii) ⇒ (iv) This follows by Theorem C.36. □

Showing the existence of paths of strict decay for nonlinear operators is a complex problem. For linear systems, however, we have the following:

Proposition C.40 *Let $A \in \mathcal{M}(\mathbb{R}^n_+)$ be a linear operator (restricted to $\mathbb{R}^n_+$). Then $r(A) < 1$ if and only if there is $x \in \mathrm{int}\,(\mathbb{R}^n_+)$ such that $\sigma : \mathbb{R}_+ \to \mathbb{R}^n_+$ defined by*

$$\sigma(r) := rx, \quad r \in \mathbb{R}_+ \tag{C.19}$$

is a path of strict decay with respect to the operator A with a linear ρ (see Definition C.38 of the path of strict decay).

Proof By Proposition C.11, the condition $r(A) < 1$ is equivalent to existence of $\lambda \in (0, 1)$ and $x \in \mathrm{int}\,(\mathbb{R}^n_+)$ such that $Ax \le \lambda x$. This implies the claim.

Conversely, if σ is a path of strict decay with respect to the operator A with a linear ρ, then setting $r := 1$, we obtain that $A(x) \le \lambda x$, for a certain $\lambda \in (0, 1)$ and $x \in \mathrm{int}\,(\mathbb{R}^n_+)$, which implies $r(A) < 1$ by Proposition C.11. □

C.7 Gain Operators

Till now, we have been dealing with the stability theory of general nonlinear discrete-time systems governed by continuous monotone operators. In Sect. C.2, we have seen that for linear operators, much more can be proved. Next, we consider several further classes of operators for which stronger results are available. In this section, we study so-called *gain operators*, which are of prime importance in the small-gain analysis of large-scale networks, see Sect. 3.5. In Sect. C.8, we consider max-preserving operators, and finally in Sect. C.9, we proceed to homogeneous operators.

The next theorem strengthens previously developed results for the operators, composed from $\mathcal{K}_\infty$-functions via monotone aggregation functions, introduced in Definition 3.23.

Theorem C.41 (Small-gain properties of gain operators) *Consider any* $\Gamma := (\gamma_{ij})_{i,j=1}^n$ *with* $\gamma_{ij} \in \mathcal{K}_\infty \cup \{0\}$ *for all* i, j, *any vector of monotone aggregation functions* $\mu = (\mu_i)_{i=1}^n$, *and a corresponding induced continuous monotone operator* $A := \Gamma_\mu$, *defined by* (3.58).

Then the following statements are equivalent:

(i) id $- A$ *has the MBI property.*
(ii) A *satisfies the uniform small-gain condition.*
(iii) *There is an* $\eta \in \mathcal{K}_\infty$ *such that*

$$A(x) \not\geq x - \eta(|x|)\mathbb{1} \quad \text{for all } x \in \mathbb{R}^n_+ \setminus \{0\}. \tag{C.20}$$

(iv) A *satisfies the strong small-gain condition.*
(v) *There is a path of strict decay with respect to* A.
(vi) *There exists* $\rho \in \mathcal{K}_\infty$, *such that the induced system*

$$x(k+1) = (\text{id} + \rho) \circ A(x(k)), \quad k \in \mathbb{Z}_+ \tag{C.21}$$

is UGAS.

Furthermore, if any of above conditions holds, a path of strict decay with respect to A *can be chosen to be piecewise linear on* $(0, +\infty)$.

Proof Implications (i) $\Leftrightarrow$ (ii) $\Leftrightarrow$ (iii) $\Rightarrow$ (iv) hold due to Proposition C.18.
(iv) $\Rightarrow$ (i) Follows by [15, Theorem 6.1] (originally shown in [4, Lemma 13, p. 102] for $A = \Gamma_{\boxplus}$).
(iv) $\Rightarrow$ (v) See [15, Theorem 5.10].
(v) $\Rightarrow$ (vi) $\Rightarrow$ (iv) Follows by Proposition C.39.

The last claim of piecewise linearity of the path of strict decay was shown in [15, Theorem 5.10]. □

C.8 Max-Preserving Operators

In this section, we specialize the class of nonlinear gain operators further, which allows us to build a rich theory that is beneficial for the formulation of so-called max-type small-gain theorems. Recall that by $x \vee y$ we denote the supremum (or join) of vectors $x, y \in \mathbb{R}^n_+$.

It is not hard to restate the definition of a monotone operator in terms of max-operation:

Proposition C.42 *The operator* $A : \mathbb{R}^n_+ \to \mathbb{R}^n_+$ *is monotone if and only if for all* $x, y \in \mathbb{R}^n_+$ *it holds that*

$$A(x \vee y) \geq A(x) \vee A(y). \tag{C.22}$$

Proof "$\Rightarrow$". For any $x, y \in \mathbb{R}^n_+$ it holds that $x \vee y \geq x$ and $x \vee y \geq y$. As A is monotone, we have that $A(x \vee y) \geq A(x)$ and $A(x \vee y) \geq A(y)$. This shows (C.22).

"$\Leftarrow$". Pick any $x, y \in \mathbb{R}^n_+$ such that $x \geq y$. Then $x \vee y = x$ and thus $A(x) = A(x \vee y) \geq A(x) \vee A(y)$. This inequality holds if and only if $A(x) \geq A(y)$. □

A special class of monotone operators will be important for us

Definition C.43 $A : \mathbb{R}^n_+ \to \mathbb{R}^n_+$ is called *max-preserving (or join-morphism)* if for every $x, y \in \mathbb{R}^n_+$ the following equality holds:

$$A(x \vee y) = A(x) \vee A(y). \tag{C.23}$$

We denote by $\mathrm{MP} = \mathrm{MP}(\mathbb{R}^n_+)$ the set of all max-preserving operators $A : \mathbb{R}^n_+ \to \mathbb{R}^n_+$. For $A, B \in \mathrm{MP}$ denote by $A \circ B$ the composition of maps A and B and by $A \vee B$ the operator defined by

$$(A \vee B)(z) := A(z) \vee B(z), \quad z \in \mathbb{R}^n_+. \tag{C.24}$$

Definition C.44 A *semiring* is a set R endowed with binary operations $+$ and $\circ$, called addition and multiplication, such that:

(i) $(R, +)$ is a commutative monoid with identity element 0:

- $(a + b) + c = a + (b + c)$,
- $0 + a = a + 0 = a$,
- $a + b = b + a$.

(ii) $(R, \circ)$ is a monoid with identity element 1:

- $(a \circ b) \circ c = a \circ (b \circ c)$,
- $1 \circ a = a \circ 1 = a$.

(iii) Multiplication left and right distributes over addition:

- $a \circ (b + c) = (a \circ b) + (a \circ c)$,
- $(a + b) \circ c = (a \circ c) + (b \circ c)$.

(iv) Multiplication by 0 annihilates R:

- $0 \circ a = a \circ 0 = 0$.

The following lemma [16, Lemma 1] reveals the algebraic structure of MP set.

Proposition C.45 *The set* MP *is a* $(\circ, \vee)$*-semiring with identity element* $\mathrm{id} : \mathbb{R}^n_+ \to \mathbb{R}^n_+$ *and neutral element* 0.

Max-preserving operators can be characterized in terms of comparison functions:

Proposition C.46 *$A : \mathbb{R}^n_+ \to \mathbb{R}^n_+$ with $A(x) = (A_1(x), \dots, A_n(x))^T$ is max-preserving if and only if there exist nondecreasing functions $\gamma_{ij} : \mathbb{R}_+ \to \mathbb{R}_+$, $i, j = 1, \dots, n$ with $A_i(x) = \max_{j=1,\dots,n} \gamma_{ij}(x_j)$, for all $x \in \mathbb{R}^n_+$, $i = 1, \dots, n$.*

Proof "$\Rightarrow$". Let $(e_i)_{i=1}^n$ be the canonical basis of $\mathbb{R}^n$. Define $\gamma_{ij}(r) := A_i(re_j)$ for all $i, j = 1, \dots, n$. By Proposition C.42, A is monotone. Hence, all γ_{ij}, $i, j = 1, \dots, n$ are nondecreasing as well.

For any $x \in \mathbb{R}^n_+$ it holds that $x = x_1e_1 + \ldots + x_ne_n$, where x_i is the i-th component of x in the canonical basis of $\mathbb{R}^n$. In particular, it holds that $x = \sup_{i=1}^n x_ie_i$, and since A is max-preserving, $A(x) = \sup_{j=1}^n A(x_je_j)$, and thus

$$A_i(x) = \sup_{j=1}^{n} A_i(x_je_j) = \max_{j=1}^{n}\{\gamma_{ij}(x_j)\}.$$

"$\Leftarrow$". Straightforward. □

Thus, the action of a max-preserving operator A over $x \in \mathbb{R}^n_+$ can be interpreted as a max-algebraic multiplication of the matrix $(\gamma_{ij})_{i,j}$ over x. Alternatively, max-preserving operators can be considered as operators induced by the monotone aggregation function $\mu_\otimes$ (see (3.55)) from nondecreasing functions (γ_{ij}).

Now let $A^n(x) := A \circ A^{n-1}(x)$, for all $n \geq 2$. Define a map $A^\times : \mathbb{R}^n_+ \to \mathbb{R}^n_+$ by

$$A^\times(x) := \sup_{i=0}^{n-1} A^i(x). \tag{C.25}$$

Proposition C.47 *The following holds:*

(i) Let $A \in \mathrm{MP}$. It holds that $A \circ A^\times = A^\times \circ A$.
(ii) Let A be continuous and $A(0) = 0$. Then $A^\times$ is continuous, $A^\times(0) = 0$, and there is $\xi \in \mathcal{K}_\infty$ such that $|A^\times(x)| \leq \xi(|x|)$.

Proof (i) As MP is a semiring by Proposition C.45, it holds for any $x \in \mathbb{R}^n_+$ that

$$\begin{aligned} A \circ A^\times(x) &= A \circ \sup_{k=0}^{n-1} A^k(x) \\ &= \sup_{k=0}^{n-1} A^{k+1}(x) = \sup_{k=0}^{n-1} A^k(A(x)) \\ &= (\sup_{k=0}^{n-1} A^k) \circ A(x) = A^\times \circ A(x). \end{aligned}$$

(ii) Clearly, $A^\times$ is continuous as a composition of continuous maps, and $A^\times(0) = 0$ by definition. Define $\tilde{\xi}(r) := \sup_{z \in \mathbb{R}^n_+ : |z| \leq r} |A^\times(z)|$. By Lemma A.27, $\tilde{\xi}$ is continuous, nondecreasing and $\tilde{\xi}(0) = 0$. Taking any $\xi \in \mathcal{K}_\infty$ such that $\xi(r) \geq \tilde{\xi}(r)$ for all $r \geq 0$, we obtain the claim. □

Motivated by Proposition C.11, we relate various types of small-gain conditions:

Theorem C.48 *Let $A \in \text{MP}$ be continuous. The following statements are equivalent:*

(i) (C.2) *is UGAS.*
(ii) $A(x) \not\geq x$ *for all* $x \in \mathbb{R}^n_+ \backslash \{0\}$.
(iii) (Cyclic small-gain condition) *for any* $p \in \{1, \ldots, n\}$ *and for any sequence* $(k_1, \ldots, k_p) \in \{1, \ldots, n\}$: $k_1 = k_p$ *it holds that*

$$\gamma_{k_1k_2} \circ \gamma_{k_2k_3} \circ \ldots \circ \gamma_{k_{p-1}k_p}(r) < r, \quad r > 0. \tag{C.26}$$

(iv) *the claim (ii) holds and for each* $k \in \mathbb{N}$ *and each* $x \in \mathbb{R}^n_+$ *it holds that* $A^k(x) \leq A^\times(x)$.
(v) *There is* $\xi \in \mathcal{K}_\infty$ *such that for all* $w, v \in \mathbb{R}^n_+$ *it holds that* $w \leq A(w) \vee v \Rightarrow |w| \leq \xi(|v|)$.

Proof (i) $\Rightarrow$ (ii) Follows from Theorem C.36.
(ii) $\Rightarrow$ (iii) Following Proposition C.46, define $\gamma_{ij}(r) := A_i(re_j)$ for all $i, j = 1, \ldots, n$ for which we have the representation $A_i(x) = \max_{j=1,\ldots,n} \gamma_{i,j}(x_j)$, for all $x \in \mathbb{R}^n_+$, and all $i = 1, \ldots, n$.

Assume that the assertion is not satisfied for some cycle:

$$\gamma_{i_1i_2} \circ \cdots \circ \gamma_{i_{k-1}i_k}(r) \geq r, \quad i_1 = i_k, , \quad \text{for some } r > 0.$$

Define the vector

$$\begin{aligned} s := re_{i_1} &+ e_{i_2}\gamma_{i_2i_3} \circ \cdots \circ \gamma_{i_{k-1}i_k}(r) \\ &+ e_{i_3}\gamma_{i_3i_4} \circ \cdots \circ \gamma_{i_{k-1}i_k}(r) + \ldots + e_{i_{k-1}}\gamma_{i_{k-1}i_k}(r). \end{aligned}$$

For any i_ν we have (recall that $i_k = i_1$)

$$A_{i_\nu}(s) = \max_{j=1,\ldots,n} \gamma_{i_\nu j}(s_j) \geq \gamma_{i_\nu i_{\nu+1}}(s_{i_{\nu+1}}) = \gamma_{i_\nu i_{\nu+1}} \circ \cdots \circ \gamma_{i_{k-1}i_k}(r).$$

The last expression equals s_{i_ν} if $\nu \in \{2, \ldots, k-1\}$ and is $\geq r = s_{i_1}$ if $\nu = 1$. This implies $A(s) \geq s$, which contradicts to (ii).
(iii) $\Rightarrow$ (iv) First let us show (ii). Assume that (ii) does not hold, i.e., there exists certain $s \in \mathbb{R}^n_+$ so that $s \neq 0$ and

$$A(s) = \left(\max_{j=1}^{n} \gamma_{1j}(s_j), \ldots, \max_{j=1}^{n} \gamma_{nj}(s_j)\right) \geq s. \tag{C.27}$$

Then there exists a sequence $(i_k)_{k=1}^n$:

$$\gamma_{ki_k}(s_{i_k}) \geq s_k.$$

Hence there exist $p \in \{1, \ldots, n\}$ and a sequence $(j_k)_{k=1}^{p+1}$, so that $j_{p+1} = j_1$ and for all $k = 1, \ldots, p$ it holds that

$$\gamma_{j_k j_{k+1}}(s_{j_{k+1}}) \geq s_{j_k}.$$

Due to these inequalities and since $\gamma_{ij} \in \mathcal{K}$ for all $i, j : i \neq j$, the following derivations hold

$$\begin{aligned} &\gamma_{j_1 j_2} \circ \gamma_{j_2 j_3} \circ \ldots \circ \gamma_{j_{p-1} j_p} \circ \gamma_{j_p j_{p+1}}(s_{j_{p+1}}) \\ &\geq \gamma_{j_1 j_2} \circ \gamma_{j_2 j_3} \circ \ldots \circ \gamma_{j_{p-1} j_p}(s_{j_p}) \\ &\vdots \\ &\geq \gamma_{j_1 j_2}(s_{j_2}) \\ &\geq s_{j_1}. \end{aligned}$$

Since $j_{p+1} = j_1$, we have that

$$\gamma_{j_1 j_2} \circ \gamma_{j_2 j_3} \circ \ldots \circ \gamma_{j_{p-2} j_{p-1}} \circ \gamma_{j_p j_1}(s_{j_1}) \geq s_{j_1}.$$

This inequality contradicts the assumption that all cycles of gains are contractions. Thus (ii) holds.

Now note that for any $x = (x_i)_{i=1}^n \in \mathbb{R}^n_+$ and for any $k \in \mathbb{N}$, it holds that

$$A^k(x) = \Big(\sup_{j_2, \ldots, j_{k+1} \in \{1, \ldots, n\}} \gamma_{i j_2} \circ \cdots \circ \gamma_{j_k j_{k+1}}(x_{j_{k+1}}) \Big)_{i=1}^n, \tag{C.28}$$

where for any $i = 1, \ldots, n$ we set $j_1 := i$.

The property that $A^k(x) \leq A^\times(x)$ for all $x \in \mathbb{R}^n_+$ and all $k \in \mathbb{N}$ follows from this formula and from the contractivity of all cycles.

(iv) $\Rightarrow$ (i) For any $x \in \mathbb{R}^n_+$ and all $k \in \mathbb{N}$ it holds that $A^k(x) \leq A^\times(x)$, and thus $O(x) \leq A^\times(x)$ and thus $O(x)$ is precompact. By Lemma C.32, it follows that $\lim_{k \to \infty} \phi(k, x) = 0$, and thus (C.2) is globally attractive. Now Theorem C.36 shows the claim.

(iv) $\Rightarrow$ (v) Pick any $w, v \in \mathbb{R}^n_+$ such that $w \leq A(w) \vee v$. It holds that

$$w \leq A(A(w) \vee v) \vee v = A^2(w) \vee A(v) \vee v,$$

and by induction, we obtain for each $s \in \mathbb{N}$ that

$$w \leq A^s(w) \vee \sup_{i=1}^{s-1} A^i(v).$$

As we assume that $A^k(v) \leq A^\times(v)$ for all $v \in \mathbb{R}^n_+$ and all $k \in \mathbb{N}$, we proceed to

$$w \leq A^s(w) \vee A^\times(v).$$

As (C.2) is UGAS by (i), it holds that $A^s(w) \to 0$ as $s \to \infty$. Thus, $w \le A^\times(v)$, and by monotonicity of the Euclidean norm with respect to the order, we have that $|w| \le |A^\times(v)|$.

As (ii) holds, $A(0) = 0$ by Lemma C.29. Now by Proposition C.47 there is $\xi \in \mathcal{K}_\infty$ such that $|w| \le \xi(|v|)$.

(v) $\Rightarrow$ (ii) Follows by Proposition C.19. □

Example C.49 One of the consequences of Theorem C.48 is that for max-preserving operators the items (ii) and (iii) of Proposition C.19 are equivalent. Nevertheless, the MBI property is still (and in contrast to the case of linear A) stronger than item (ii) ("max-form MBI property") and item (iii) (small-gain condition) of Proposition C.19, even combined with the UGAS property, as the following example shows.

Consider $A : \mathbb{R}^2_+ \to \mathbb{R}^2_+$, defined for any $(x_1, x_2) \in \mathbb{R}^2_+$ by:

$$A(x_1, x_2) := (\gamma(x_2), \gamma(x_1)), \quad \text{where } \gamma(r) = r(1 - e^{-r}), \quad r \ge 0. \tag{C.29}$$

Clearly, A is max-preserving. As $\gamma < \mathrm{id}$, also $\gamma \circ \gamma < \mathrm{id}$, and hence all cycles in A are contractions, and thus by Theorem C.48, the "max-form MBI property" and small-gain condition hold, and furthermore (C.29) is UGAS, which implies by Proposition C.19 that A satisfies the small-gain condition.

Assume that the MBI property for the operator $\mathrm{id} - A$ holds. Then there is $\xi \in \mathcal{K}_\infty$ such that for all $x_1 \ge 0$, it holds that

$$(\mathrm{id} - A)(x_1, x_1) = (x_1 e^{-x_1}, x_1 e^{-x_1}) \le (1, 1) \quad \Rightarrow \quad |(x_1, x_1)| \le \xi(|(1, 1)|).$$

However, the left part of the above implication is valid for all large enough x_1, which violates the inequality on the right-hand side of this implication. Hence $\mathrm{id} - A$ does not satisfy the MBI property.

Furthermore, by Proposition C.19, it follows that the strong small-gain condition for A does not hold. This shows that the strong small-gain condition is at least not weaker than the UGAS of an induced discrete-time system.

Now we would like to obtain some corollaries of Theorem C.48.

Proposition C.50 *Let $A \in \mathrm{MP}$ be continuous and satisfy any of the statements (i)–(v) in Theorem C.48. Then*

$$A^\times(x) = \sup_{i=0}^{\infty} A^i(x), \quad x \in \mathbb{R}^n_+, \tag{C.30}$$

and

$$A^\times = \mathrm{id} \vee (A^\times \circ A) = \mathrm{id} \vee (A \circ A^\times). \tag{C.31}$$

Proof By Theorem C.48, it holds that $A^\times(x) = \sup_{i=0}^{\infty} A^i(x)$ for all $x \in \mathbb{R}^n_+$. From this equality, the identity (C.31) easily follows. □

Under assumptions of Proposition C.50, and in view of (C.30), $A^\times$ is called the *strong transitive closure* of A, or *Kleene star*, see [2, Sect. 1.6.2], or just closure of A, see [16, Sect. 1.4]. It is of fundamental importance in max-algebra.

Proposition C.51 *Let $A \in \mathrm{MP}$ be continuous and satisfying $A(x) \not\geq x$ for all $x \in \mathbb{R}^n_+\backslash\{0\}$. Then*

$$A^\times\big(A(x)\big) = A \circ A^\times(x) \leq A^\times(x), \quad x \in \mathbb{R}^n_+, \tag{C.32}$$

and

$$A^\times\big(A(x)\big) = A \circ A^\times(x) < A^\times(x), \quad x \in \mathbb{R}^n_+\backslash\{0\}. \tag{C.33}$$

Proof Pick any $x \in \mathbb{R}^n_+$. By Theorem C.48, item (iv), $A^\times$ is well-defined on $\mathbb{R}^n_+$ and furthermore,

$$A \circ A^\times(x) = \sup_{k=1}^{n} A^k(x) \leq A^\times(x).$$

If $A \circ A^\times(x) = A^\times(x)$, then $A^\times(x) = 0$ by assumption of the proposition. As $A^\times(x) \geq x$, this implies that $x = 0$. □

Remark C.52 The estimate (C.32) shows that $\{A^\times(x) : x \in \mathbb{R}^n_+\}$ are all subeigenvectors of A, corresponding to the "eigenvalue" 1.

The set of nonstrict decay $\Psi(A)$ for an operator A can be represented in terms of the operator $A^\times$.

Proposition C.53 (Sets of nonstrict decay for MP-operators) *Let $A \in \mathrm{MP}$ be continuous and satisfy the small-gain condition (C.5). Then the operator $A^\times$ has the following properties:*

(i) The image of $A^\times$ is the set of nonstrict decay for A:

$$\Psi(A) = \{s \in \mathbb{R}^n_+ : A(s) \leq s\} = \mathrm{Im}A^\times.$$

This set is closed, contains $s = 0$, is cofinal and invariant with respect to A, i.e., $A(\mathrm{Im}A^\times) \subset \mathrm{Im}A^\times$.

(ii) $\mathrm{Im}A^\times$ *is* path-connected, *i.e., for any $s_1, s_2 \in \mathrm{Im}A^\times$ there is a continuous map $w : [0, 1] \to \mathrm{Im}A^\times$, such that $w(0) = s_1$ and $w(1) = s_2$.*

(iii) $A^\times \circ A^\times = A^\times$.

Proof (i) Proposition C.51 implies that $\mathrm{Im}A^\times \subset \{s \in \mathbb{R}^n_+ : A(s) \leq s\}$. Conversely, $A(s) \leq s$ implies $A^k(s) \leq s$ for all $k \geq 0$, and hence $A^\times(s) = s$ implying $s \in \mathrm{Im}A^\times$. Since A is continuous, $\mathrm{Im}A^\times$ is closed. As for each $s \in \mathbb{R}^n_+$ it holds that $s \leq A^\times(s) \in \mathrm{Im}A^\times$, the set $\mathrm{Im}A^\times$ is cofinal. As for any $s \in \mathrm{Im}A^\times$ it holds that $A(s) \leq s$, by monotonicity of A it holds that $A(A(s)) \leq A(s)$, thus $A(s) \in \Psi(A) = \mathrm{Im}A^\times$, which shows the invariance of $\mathrm{Im}A^\times$.

(ii) It suffices to prove that any point $s \in \mathrm{Im} A^\times$ can be connected with 0 by a continuous path. Hence, take any $s \in \mathbb{R}^n_+$ with $A(s) \le s$. Now define $s_\alpha := \alpha s + (1-\alpha)A(s)$ and observe that

$$\begin{aligned} A(s) = \alpha A(s) + (1-\alpha)A(s) &\le s_\alpha \\ = \alpha s + (1-\alpha)A(s) &\le \alpha s + (1-\alpha)s = s. \end{aligned}$$

Thus, $A(s) \le s_\alpha \le s$ and applying A once again and using its monotonicity yields $A(s_\alpha) \le A(s) \le s_\alpha$. This implies $s_\alpha \in \mathrm{Im} A^\times$ for all $\alpha \in [0, 1]$. We can repeat this process and thus construct a continuous (piecewise linear) path connecting s with $A^k(s)$ for any $k \ge 1$. Since $A^k(s) \to 0$ by assumption, it follows that s can be connected with 0 by a continuous path.
(iii) This follows immediately from the proof of (i).

As the first corollary of Proposition C.51, we obtain a construction of continuous *paths of nonstrict decay* for A.

Corollary C.54 (Paths of nonstrict decay for continuous MP-operators) *Let A be a continuous max-preserving operator, satisfying $A(x) \not\ge x$ for all $x \in \mathbb{R}^n_+\backslash\{0\}$. Pick any $a \in \mathrm{int}\,(\mathbb{R}^n_+)$. Then $\sigma : \mathbb{R}_+ \to \mathbb{R}^n_+$, defined by*

$$\sigma(t) = A^\times(at) \quad \forall t \ge 0, \tag{C.34}$$

is continuous and satisfies

$$A(\sigma(r)) \le \sigma(r) \quad \forall r > 0. \tag{C.35}$$

The estimate (C.33) indicates that the Kleene star $A^\times$ can be used to construct Lyapunov functions for discrete-time systems induced by $A \in \mathrm{MP}$. This is made precise by the following result

Proposition C.55 (Constructive converse Lyapunov theorem for MP-operators) *Let $A \in \mathrm{MP}$ be continuous and let any of the conditions (i)–(v) of Theorem C.48 hold. Pick any $a \in \mathrm{int}\,(\mathbb{R}^n_+)$ and consider the map $V : \mathbb{R}^n_+ \to \mathbb{R}_+$ defined by*

$$V(x) := a^T A^\times(x), \quad x \in \mathbb{R}^n_+. \tag{C.36}$$

Then V is continuous, monotone, and there are $\psi_1, \psi_2 \in \mathcal{K}_\infty$ such that

$$\psi_1(|x|) \le V(x) \le \psi_2(|x|), \quad x \in \mathbb{R}^n_+. \tag{C.37}$$

Furthermore, the following dissipative estimate is valid:

$$V(A(x)) < V(x), \quad x \in \mathbb{R}^n_+\backslash\{0\}. \tag{C.38}$$

Proof As $A \in \text{MP}$, A is monotone, and as it satisfies the condition $A(x) \not\geq x$ for all $x \in \mathbb{R}^n_+ \backslash \{0\}$, $A(0) = 0$ by Lemma C.29. Hence $A^\times(0) = 0$, and thus $V(0) = 0$. As A is continuous and monotone, then so are $A^\times$ and V.

Existence of ψ_2 as in (C.37) follows from item (ii) of Proposition C.47, and the existence of ψ_1 follows as $A^\times(x) \geq x$, and thus $V(x) \geq a^T x$ for all $x \in \mathbb{R}^n_+$.

The estimate (C.38) follows immediately from (C.33). □

C.9 Homogeneous Subadditive Operators

Finally, we characterize UGAS of monotone discrete-time systems governed by homogeneous subadditive operators, and derive rich information about the existence of paths of strict decay.

Definition C.56 An operator $A : \mathbb{R}^n_+ \to \mathbb{R}^n_+$ is called:

(i) *Homogeneous (of degree one)*, if $A(\lambda x) = \lambda A(x)$ for all $\lambda \in \mathbb{R}_+$ and $x \in \mathbb{R}^n_+$.
(ii) *Subadditive*, if $A(x^1 + x^2) \leq A(x^1) + A(x^2)$ for all $x^1, x^2 \in \mathbb{R}^n_+$.

As we deal in this book only with homogeneous operators of degree one, we will succinctly call them homogeneous operators.

All linear operators and max-preserving operators with linear gains are subadditive and homogeneous. The same properties are shared by "hybrids" of linear and max-linear operators, as, e.g.,

$$A(x_1, x_2)^T = (\gamma_{11}x_1 + \gamma_{12}x_2, \max\{\gamma_{21}x_1, \gamma_{22}x_2\})^T, \quad x_1, x_2 \geq 0,$$

for some $(\gamma_{ij}) \subset \mathbb{R}_+$.

Proposition C.57 *Let $A : \mathbb{R}^n_+ \to \mathbb{R}^n_+$ be monotone, subadditive, and homogeneous. Then A is globally Lipschitz continuous with a Lipschitz constant $|A(\mathbb{1})|$.*

Proof Using the notation for the positive part of vectors, we have for all $x, y \in \mathbb{R}^n_+$ that

$$\begin{aligned} A(x) - A(y) = A(x - y + y) - A(y) &\leq A((x-y)^+ + y) - A(y) \\ &\leq A((x-y)^+) + A(y) - A(y) = A((x-y)^+). \end{aligned}$$

Analogously, we obtain for all $x, y \in \mathbb{R}^n_+$ that

$$A(x) - A(y) \geq -A((y-x)^+).$$

Using homogeneity, we have for all $x, y \in \mathbb{R}^n_+$

$$
\begin{aligned}
|A(x) - A(y)| &\le \max\{|A((x-y)^+)|, |A((y-x)^+)|\} \\
&\le \max\left\{|(x-y)^+|\left|A\left(\frac{(x-y)^+}{|(x-y)^+|}\right)\right|, |(y-x)^+|\left|A\left(\frac{(y-x)^+}{|(y-x)^+|}\right)\right|\right\} \\
&\le \sup_{|z|=1} |A(z)||x-y| \\
&\le |A(\mathbb{1})||x-y|.
\end{aligned}
$$

We need one more important concept.

Definition C.58 Let $A \in \mathcal{M}(\mathbb{R}^n_+)$. A vector $z \in \operatorname{int}(\mathbb{R}_+)$ is called a *point of strict decay of A* if there is $\lambda \in (0, 1)$, such that

$$A(z) \le \lambda z. \tag{C.39}$$

The next lemma provides explicit formulas for the points of strict decay of homogeneous and subadditive operators that induce UGES systems.

Lemma C.59 *Let an operator $A \in \mathcal{M}(\mathbb{R}^n_+)$ be subadditive and homogeneous. Let also (C.2) be UGES with corresponding $a \in (0, 1)$ and $M > 0$ as in (C.4).*

Pick any $\lambda \in (a, 1)$, and define a map $z : \operatorname{int}(\mathbb{R}^n_+) \to \operatorname{int}(\mathbb{R}^n_+)$ for each $y \in \operatorname{int}(\mathbb{R}^n_+)$ by

$$z(y) := \sum_{k=0}^{\infty} \frac{A^k(y)}{\lambda^{k+1}}. \tag{C.40}$$

Then z is well-defined, continuous on $\mathbb{R}^n_+$, and

$$A(z(y)) \le \lambda z(y), \quad y \in \operatorname{int}(\mathbb{R}^n_+). \tag{C.41}$$

Proof **Step 1. z is well-defined.** By UGES property, we have

$$\sum_{k=0}^{\infty} \left|\frac{A^k(y)}{\lambda^{k+1}}\right| \le \frac{M}{\lambda} \sum_{k=0}^{\infty} \frac{a^k}{\lambda^k}|y| < \infty.$$

As $X = \mathbb{R}^n$ is a Banach space, absolute convergence implies convergence, see, e.g., [5, Lemma 1.22]. Since A is monotone, we have also $z(y) \ge \frac{1}{\lambda}y$, and thus $z(y) \in \operatorname{int}(\mathbb{R}^n_+)$ as $y \in \operatorname{int}(\mathbb{R}^n_+)$. Thus, $z : \operatorname{int}(\mathbb{R}^n_+) \to \operatorname{int}(\mathbb{R}^n_+)$ is well-defined.

Step 2. z is continuous. Take any $x \in \operatorname{int}(\mathbb{R}^n_+)$, any $\varepsilon > 0$, and the corresponding small enough $\delta > 0$ that will be specified later. For any $y \in \operatorname{int}(\mathbb{R}^n_+)$ such that $|x - y| \le \delta$ we have:

$$
\begin{aligned}
|z(x) - z(y)| &= \left| \sum_{k=0}^{\infty} \frac{A^k(x)}{\lambda^{k+1}} - \sum_{k=0}^{\infty} \frac{A^k(y)}{\lambda^{k+1}} \right| \\
&\leq \sum_{k=0}^{\infty} \frac{|A^k(x) - A^k(y)|}{\lambda^{k+1}} \\
&= \sum_{k=0}^{N} \frac{|A^k(x) - A^k(y)|}{\lambda^{k+1}} + \sum_{k=N+1}^{\infty} \frac{|A^k(x) - A^k(y)|}{\lambda^{k+1}}, \qquad \text{(C.42)}
\end{aligned}
$$

where N is any positive number that will be specified later.

As (C.2) is UGES, we have that

$$
\begin{aligned}
\sum_{k=N+1}^{\infty} \frac{|A^k(x) - A^k(y)|}{\lambda^{k+1}} &\leq \sum_{k=N+1}^{\infty} \frac{|A^k(x)| + |A^k(y)|}{\lambda^{k+1}} \\
&\leq (2|x| + \delta)\frac{M}{\lambda} \sum_{k=N+1}^{\infty} \frac{a^k}{\lambda^k}.
\end{aligned}
$$

As $a/\lambda \in (0, 1)$, for any $\varepsilon > 0$ we can find $N = N(\varepsilon) > 0$ large enough such that

$$
(2|x| + \delta)\frac{M}{\lambda} \sum_{k=N+1}^{\infty} \frac{a^k}{\lambda^k} < \frac{\varepsilon}{2}.
$$

As A is continuous, A^k is continuous for any $k \in \mathbb{N}$. Hence, we can choose $\delta > 0$ small enough such that

$$
\sum_{k=0}^{N} \frac{|A^k(x) - A^k(y)|}{\lambda^{k+1}} < \frac{\varepsilon}{2}.
$$

Substituting our findings in (C.42), we see that z is continuous.

Step 3. Strict decay. Now for any $y \in \operatorname{int}(\mathbb{R}^n_+)$ decompose $A(z(y))$ as

$$
\begin{aligned}
A(z(y)) &= (A - \lambda \mathrm{id} + \lambda \mathrm{id})\,(z(y)) \\
&= A\left(\sum_{k=0}^{\infty} \frac{A^k(y)}{\lambda^{k+1}}\right) - \sum_{k=0}^{\infty} \frac{A^k(y)}{\lambda^k} + \lambda z(y).
\end{aligned}
$$

By subadditivity and continuity of A, and then by homogeneity of A we obtain that

$$\begin{aligned}
A(z(y)) &\leq \sum_{k=0}^{\infty} A\left(\frac{A^k(y)}{\lambda^{k+1}}\right) - \sum_{k=0}^{\infty} \frac{A^k(y)}{\lambda^k} + \lambda z(y) \\
&= \sum_{k=0}^{\infty} \frac{A^{k+1}(y)}{\lambda^{k+1}} - \sum_{k=0}^{\infty} \frac{A^k(y)}{\lambda^k} + \lambda z(y) \\
&= -y + \lambda z(y) \leq \lambda z(y),
\end{aligned}$$

which implies the claim. □

Next we introduce the concept of a spectral radius motivated by Gelfand's formula for the spectral radius of a linear operator.

Definition C.60 For a continuous, subadditive, and homogeneous operator $A : \mathbb{R}^n_+ \to \mathbb{R}^n_+$ define the *spectral radius* $r(A)$ as

$$r(A) := \lim_{n\to\infty} \sup_{x\in\mathbb{R}^n_+,\ |x|=1} |A^n(x)|^{1/n}.$$

The following proposition (whose proof is left as Exercise C.17) empowers us with an alternative definition of the spectral radius and, at the same time, shows that $r(A)$ is well-defined for operators, which are continuous, subadditive, and homogeneous.

Proposition C.61 *Let A be a continuous, subadditive, and homogeneous operator. Then*

$$r(A) = \inf_{n\in\mathbb{N}} \sup_{x\in\mathbb{R}^n_+,\ |x|=1} |A^n(x)|^{1/n}. \tag{C.43}$$

Using the machinery developed in the previous sections, we can characterize the UGES property for discrete-time systems induced by homogeneous and subadditive operators.

Theorem C.62 *Let $A \in \mathcal{M}(\mathbb{R}^n_+)$ be homogeneous and subadditive. The following statements are equivalent:*

(i) $r(A) < 1$.
(ii) (C.2) is UGES.
(iii) (C.2) is UGAS.
(iv) There are $\lambda \in (0, 1)$ and $x^0 \in \operatorname{int}(\mathbb{R}^n_+)$ such that

$$A(x^0) \leq \lambda x^0. \tag{C.44}$$

(v) There is a path of strict decay with respect to A, with a linear decay rate ρ (see Definition C.38).
(vi) There is a path of strict decay with respect to A.

Proof (i) $\Leftrightarrow$ (ii) Follows from [12, Proposition 16 and its proof (in this part the subadditivity is used)].
(ii) $\Rightarrow$ (iii) Clear.
(iii) $\Rightarrow$ (ii) From the UGAS property and homogeneity of A, we have for a certain $\beta \in \mathcal{KL}$ and all $s \in \mathbb{R}^n_+$ that

$$|A^k(s)| = |s| \Big|A^k\Big(\frac{s}{|s|}\Big)\Big| \leq |s|\beta(1,k)$$

for all $0 \neq s \in \mathbb{R}^n_+$ and $k \in \mathbb{Z}_+$. As $\beta \in \mathcal{KL}$, there is $k_* \in \mathbb{N}$ such that $\beta(1, k_*) < 1$. For each $k \in \mathbb{Z}_+$, take $p, q \in \mathbb{Z}_+$ satisfying $k = pk_* + q$ with $q < k_*$. Then

$$|A^k(s)| \leq |A^q(s)|\beta(1,k_*)^p \leq |s|\beta(1,q)\beta(1,k_*)^p \leq |s|\beta(1,0)\beta(1,k_*)^{k/k_*}.$$

This implies that (ii) holds with $M = \beta(1, 0)$ and $a = \beta(1, k_*)^{1/k_*} < 1$.
(ii) $\Rightarrow$ (iv) Follows by Lemma C.59.
(iv) $\Rightarrow$ (v) Take $\lambda \in (0, 1)$ and $x^0 \in \operatorname{int}(\mathbb{R}^n_+)$ as in (iv), and define $\sigma(r) := rx^0$. As A is homogeneous, we have that $A(\sigma(r)) = rA(x^0) \leq \lambda r x^0 = \lambda\sigma(r)$, $r > 0$.
(v) $\Rightarrow$ (vi) Clear.
(vi) $\Rightarrow$ (iii) Follows by Proposition C.39. □

C.9.1 Computational Issues

Our primary application of the obtained characterizations for the condition $r(A) < 1$ is the small-gain analysis of networks performed in Sect. 3.5.1. In these results, we compute an ISS Lyapunov function for the whole network as a maximum of scaled Lyapunov functions of subsystems, and the scaling coefficients are inverses of the components of path of strict decay, corresponding to the gain operator of the network.

From this point of view, *a major simplification in the case of homogeneous and subadditive gain operators is that we do not need to have a path of strict decay in order to construct an ISS Lyapunov function for the network, but it is enough to have merely a point of strict decay* x^0. Moreover, Lemma C.59 provides us with an explicit expression for the points of strict decay.

For the practical computation of points of strict decay, the following variation of Lemma C.59 is of interest.

Lemma C.63 *Let an operator $A \in \mathcal{M}(\mathbb{R}^n_+)$ be monotone, subadditive, and homogeneous. Let also (C.2) be UGES with corresponding $a \in (0, 1)$ and $M > 0$.*

Pick any $\lambda \in (a, 1)$, and define the map $z_N : \operatorname{int}(\mathbb{R}^n_+) \to \operatorname{int}(\mathbb{R}^n_+)$ for each $y \in \operatorname{int}(\mathbb{R}^n_+)$ and $N \in \mathbb{N}$ by

$$z_N(y) := \sum_{k=0}^{N} \frac{A^k(y)}{\lambda^{k+1}}. \tag{C.45}$$

Then for each $y \in \text{int}(\mathbb{R}^n_+)$ *there is a finite* $N > 0$*, such that*

$$A(z_N(y)) \leq \lambda z_N(y). \tag{C.46}$$

Furthermore, for $y = \mathbb{1}$ *one can choose any* N *such that* $M(a/\lambda)^{N+1} \leq 1$.

Proof Arguing as in the proof of (C.59), we obtain

$$A(z_N(y)) \leq -y + \frac{A^{N+1}(y)}{\lambda^{N+1}} + \lambda z_N(y) \leq \lambda z_N(y) \tag{C.47}$$

provided that N is large enough.

Now choose $y := \mathbb{1}$. Then $|\frac{A^{N+1}(y)}{\lambda^{N+1}}| \leq M(a/\lambda)^{N+1}$. The last estimate in (C.47) holds if $M(a/\lambda)^{N+1} \leq 1$. From this relation we can compute N for the point of strict decay (C.46). □

The next proposition implies that the set of points of strict decay of the operators that we consider is open. This is good news, as the numerical computations of the points of strict decay are, as a rule, not exact. In particular, if the numerical errors are small enough, the computed point is still a point of strict decay.

Proposition C.64 *Let* $A : \mathbb{R}^n_+ \to \mathbb{R}^n_+$ *be a well-defined operator that is monotone, subadditive, and homogeneous. Assume that there are* $\lambda \in (0, 1)$ *and* $z \in \text{int}(\mathbb{R}^n_+)$ *such that*

$$A(z) \leq \lambda z. \tag{C.48}$$

Then for each $\varepsilon > 0$ *there is* $\delta > 0$*, such that*

$$A(y) \leq (\lambda + \varepsilon)y, \quad y \in B_\delta(z). \tag{C.49}$$

Proof As $z \in \text{int}(\mathbb{R}^n_+)$, for each $\varepsilon > 0$ there is $\delta > 0$ such that $B_\delta(0) \leq \varepsilon z$. This is equivalent to

$$-B_\delta(0) \leq \varepsilon z \Leftrightarrow z - B_\delta(z) \leq \varepsilon z \Leftrightarrow (1 - \varepsilon)z \leq B_\delta(z). \tag{C.50}$$

Choose now $\delta_2 \in (0, \delta)$ such that $A(B_{\delta_2}(0)) \subset B_\delta(0)$. This is possible, as $A(0) = 0$ and A is globally Lipschitz continuous due to Proposition C.57.

Now pick any $y \in B_{\delta_2}(z)$, i.e., $y = z + x$ for a certain $x \in B_{\delta_2}(0)$. Then

$$A(y) = A(z + x) \leq A(z + x^+) \leq A(z) + A(x^+) \leq \lambda z + \varepsilon z = (\lambda + \varepsilon)z \leq \frac{\lambda + \varepsilon}{1 - \varepsilon} y,$$

where we have used (C.50) for the last estimate in view of $y \in B_{\delta_2}(z) \subset B_\delta(z)$. The last estimate is the same as (C.49), after a redefinition of variables. □

Overall, if the system is UGES, then we have efficient and robust methods for the computation of the points of strict decay. To verify the UGES property, Theorem C.62 suggests that one can check the condition $r(A) < 1$. The next proposition shows that the spectral radius does not depend on the choice of the norm

Proposition C.65 *Let A be a continuous, subadditive, and homogeneous operator. Then $r(A)$ does not depend on the choice of the norm, i.e., for any norm $\|\cdot\|$ in $\mathbb{R}^n$ it holds that*

$$r(A) = \lim_{n\to\infty} \sup_{x\in\mathbb{R}^n_+,\ \|x\|=1} \|A^n(x)\|^{1/n}. \tag{C.51}$$

Proof As all the norms in finite-dimensional spaces are equivalent, there are $c, C > 0$, such that $c\|x\| \le |x| \le C\|x\|$ for all $x \in \mathbb{R}^n$. Using the homogeneity, we have

$$\begin{aligned} r(A) &= \lim_{n\to\infty} \sup_{x\in\mathbb{R}^n_+, x\neq 0} \frac{|A^n(x)|^{1/n}}{|x|^{1/n}} \\ &\le \lim_{n\to\infty} \sup_{x\in\mathbb{R}^n_+, x\neq 0} \frac{\|A^n(x)\|^{1/n}}{\|x\|^{1/n}} (C/c)^{1/n} \\ &= \lim_{n\to\infty} \sup_{x\in\mathbb{R}^n_+, x\neq 0} \frac{\|A^n(x)\|^{1/n}}{\|x\|^{1/n}} = \lim_{n\to\infty} \sup_{x\in\mathbb{R}^n_+,\ \|x\|=1} \|A^n(x)\|^{1/n}. \end{aligned}$$

The reverse inequality is obtained in the same way. □

In particular, consider the ℓ_∞-norm on $\mathbb{R}^n$: $\|x\|_\infty := \sup_{i=1}^n |x_i|$, $x \in \mathbb{R}^n$. If A is monotone, the expression (C.51) simplifies to

$$r(A) = \lim_{n\to\infty} \|A^n(\mathbb{1})\|_\infty^{1/n}. \tag{C.52}$$

The condition $r(A) < 1$ can be further simplified if A has a special structure, see Exercise C.16.

C.10 Summary of Properties of Nonlinear Monotone Operators and Induced Systems

Relations between the basic properties of nonlinear monotone operators are depicted in Fig. C.1.

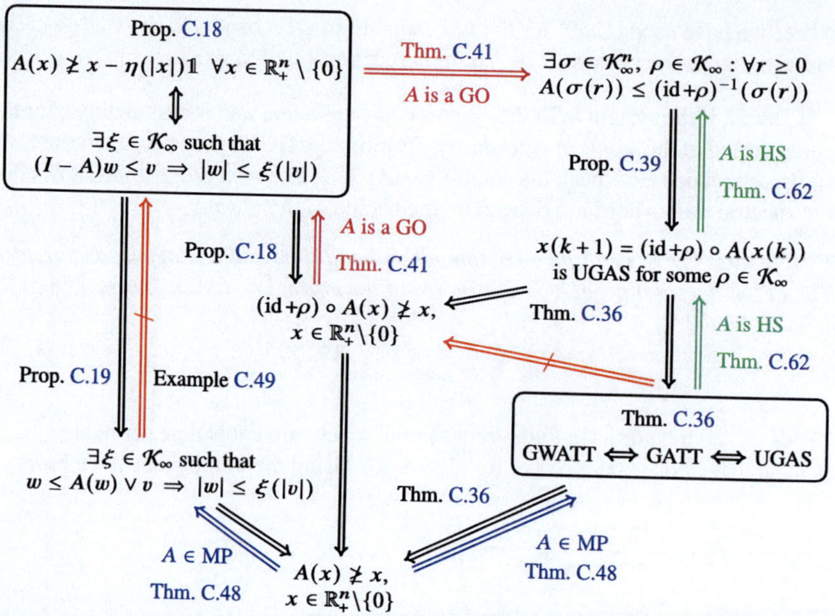

Fig. C.1 Main relations between stability properties of nonlinear monotone operators and induced systems. In this diagram: • Implications, represented by black arrows, are valid for all continuous monotone operators. • Blue arrows represent additional implications that are valid for continuous maximum preserving (MP) operators. In particular, for such operators, the conditions in the lower part of the figure are equivalent. • Magenta arrows represent additional implications that are valid for gain operators (GO). In particular, for such operators, the conditions in the upper part of the figure are equivalent. • Green arrows represent additional implications that are valid for homogeneous and subadditive operators (HS). In particular, for such operators, the conditions in the right-hand part of the figure are equivalent. • For linear systems, all the conditions depicted in this figure are equivalent

C.11 Concluding Remarks

The results of this chapter are based on the results in [3, 4, 6, 12, 13, 15, 16], [7, Sect. 5.2], [8, Sect. 2], [14, Chap. 2].

Proposition C.11 should be well-known, see, e.g., [14, Lemma 2.0.1] for some of the equivalences. Continuous-time finite-dimensional counterpart of some of the results in Proposition C.11 was shown in [18]. Proposition C.11 has been extended to a much more general setting of infinite-dimensional monotone discrete-time systems in ordered Banach spaces in [6], and Proposition C.11 follows from the main results in [6] as a special case.

Proposition C.18 has been shown in [12], and some of equivalences can be extended to monotone operators defined on infinite-dimensional ordered Banach spaces, see [12].

The results from Sect. C.4 are due to [4]. Proposition C.31 is due to [15, Proposition 4.1]. The notion of a path of strict decay was introduced (though in a slightly different manner) in [3], under the name Ω-path.

Theorem C.48 was shown in part in [14, Lemma 2.3.14, Theorem 2.2.8], and in part in [8, Proposition 2.7]. The cyclic small-gain condition in Theorem C.48 was inspired by Teel (even though he has not published his findings), as acknowledged both in [4, 8]. The idea to use the Kleene star operation in the context of small-gain conditions for max-preserving operators is due to [8]. In particular, Propositions C.46 and C.50 are due to [8, Propositions 2.6 and 2.7]. Corollary C.54 was motivated by [8, Remark 2.8]. Proposition C.55 is due to [16, Theorem 2].

The results in Sect. C.9 are due to [13], where the most part of the results has been extended also to the infinite-dimensional operators acting in ℓ_∞^+, except of the implication (vi) $\Rightarrow$ (ii) in Theorem C.62, as it (via several other lemmas) relies on a finite-dimensional topological fixed point theorem due to Knaster et al. [9]. The essential use of the KKM theorem in the finite-dimensional theory is one of the biggest challenges in extending the corresponding results to the infinite-dimensional setting.

Irreducibility of the operator A can be used to sharpen the stability criteria for the corresponding induced discrete-time system. We refer to [4, 3, 14, 15] for such results. Finally, in this chapter, we have considered the asymptotic stability of monotone systems without disturbances. Some results can also be extended to the case of ISS of monotone discrete-time systems with inputs, see [17].

C.12 Exercises

Exercise C.1 Prove Proposition C.45.

Exercise C.2 Let X be a real vector space. We say that $X^+ \subset X$ is a *positive cone* in X if the following conditions hold:

(i) $X^+ \cap (-X^+) = \{0\}$,
(ii) $\mathbb{R}_+ X^+ \subset X^+$,
(iii) $X^+ + X^+ \subset X^+$.

Show that if X^+ is a positive cone in X, then the binary relation "$\geq$" on X defined by $y \geq x \Leftrightarrow y - x \in X^+$ is a partial order relation.

Conversely, if X is an ordered vector space with an order $\geq$, then the set $X^+ := \{x \in X : x \geq 0\}$ is a positive cone in X.

Exercise C.3 Show that a linear operator $A : \mathbb{R}^n \to \mathbb{R}^n$ is monotone if and only if it is nonnegative.

Exercise C.4 Show that a linear operator $A : \mathbb{R}^n \to \mathbb{R}^n$

(i) is nonnegative iff the entries of the matrix of A in the canonical basis are all nonnegative,

(ii) is strictly positive, that is, $A(\mathbb{R}^n_+ \backslash \{0\}) \subset \text{int}\,(\mathbb{R}^n_+)$, iff the entries of the matrix of A in the canonical basis are all positive.

Exercise C.5 Let $A \in \mathbb{R}^{n\times n}$ satisfy $r(A) < 1$. Show that for any $\lambda \in (r(A), 1)$, and any $y \in \text{int}\,(\mathbb{R}^n_+)$, the vector $x := (\lambda \text{id} - T)^{-1}y$ satisfies the condition in Proposition C.11 (ix).

Exercise C.6 Let $A : \mathbb{R}^n_+ \to \mathbb{R}^n_+$ be a continuous map (not necessarily monotone). Show that if $\lim_{k\to\infty} A^k(x) = y$ for some $x, y \in \mathbb{R}^n_+$, then $A(y) = y$.

Exercise C.7 Let $A \in \mathbb{R}^{n\times n}$. Consider the system

$$x(k+1) = Ax(k) + u(k) \quad \text{for all } k \in \mathbb{Z}_+. \tag{C.53}$$

Denote $\mathcal{T} := (A^k)_{k\in\mathbb{Z}_+}$, and for each sequence $u : \mathbb{Z}_+ \to \mathbb{R}^n$ define the convolution $\mathcal{T} * u : \mathbb{Z}_+ \to \mathbb{R}^n$ by the formula

$$(\mathcal{T} * u)(k) = \sum_{j=0}^{k} A^{k-j}u(j) = \sum_{j=0}^{k} A^j u(k-j) \qquad \text{for all } k \in \mathbb{Z}_+.$$

Finally, let c_0 be the space of real sequences that converge to 0.

Show that if $r(A) < 1$ and $u \in c_0$, then $\mathcal{T} * u \in c_0$.

Exercise C.8 Fix real constants $\lambda \in (0, 1)$ and $\mu \ge 0$. Let $A : \mathbb{R}^2_+ \to \mathbb{R}^2_+$ be given by

$$A\begin{pmatrix} s_1 \\ s_2 \end{pmatrix} = \begin{pmatrix} \lambda s_1 + s_1^2 s_2 + \mu s_2 \\ \lambda s_2 \end{pmatrix}, \quad s_1, s_2 \in \mathbb{R}_+. \tag{C.54}$$

Show that

(i) A is continuous, monotone, $A(0) = 0$ and $A(x) \not\ge x$ for all $x \in \mathbb{R}^2_+ \backslash \{0\}$.
(ii) (C.54) is not UGAS.
(iii) $\Psi(A)$ is unbounded, but is not cofinal.

Exercise C.9 Show Lemma C.16.

Exercise C.10 Let $A : \mathbb{R}^n_+ \to \mathbb{R}^n_+$ be a monotone operator satisfying $A(x) \not\ge x$ for all $x \in \mathbb{R}^n_+ \backslash \{0\}$. Show that $A^k(x) \not\ge x$ for all $k \in \mathbb{N}$ and all $x \in \mathbb{R}^n_+ \backslash \{0\}$.

Exercise C.11 Let $A : \mathbb{R}^n_+ \to \mathbb{R}^n_+$ satisfy $A(x) \not\ge x$ for all $x \in \mathbb{R}^n_+ \backslash \{0\}$. Show that

$$\Big(\cup_{i=1}^n \Omega_i(A)\Big) \cup \{0\} = \mathbb{R}^n_+.$$

Exercise C.12 Let $A \in \text{MP}$, and all the gains γ_{ij} be linear. Assume that A induces a UGES discrete-time system. Show that for any $\varepsilon > 0$ and $x \in \text{int}\,(\mathbb{R}^n_+)$ the vector $y := \sup_{k\in\mathbb{Z}_+} (1+\varepsilon)^k A^k(x) \ge x$ is a vector of strict decay for A.

Exercise C.13 Let the cyclic small-gain condition (C.26) hold. Does it follow that each chain of gains $\gamma_{i_1 i_2} \circ \gamma_{i_2 i_3} \circ \ldots \circ \gamma_{i_n i_{n+1}}$ is a contraction, if the length of the chain exceeds some maximal length?

Exercise C.14 Give a simple and self-contained proof of the implication (v) ⇒ (iv) ⇒ (ii) in Theorem C.62.

Exercise C.15 Theorem C.62 gives several criteria for UGES of discrete-time systems (C.2) induced by homogeneous, subadditive, continuous, and monotone operators. These criteria extend some of the characterizations of UGES for linear systems, derived in Proposition C.11.

(i) Investigate, which further conditions from Proposition C.11 are equivalent to UGES for homogeneous, subadditive, and continuous operators.
(ii) Relate the results in Theorem C.62 with the results in [11, Theorem 2], where only max-linear operators have been considered. Can we recover some of results in [11, Theorem 2] from Theorem C.62?
(iii) Can we extend further results from [11] to the more general setting of continuous, homogeneous, and subadditive monotone operators?

Exercise C.16 Let $A = \Gamma_\otimes$ with all γ_{ij} as linear functions. Compute $r(A)$ and show that the property $r(A) < 1$ holds if and only if for all (j_i), such that $j_1 = j_{n+1}$ it holds that

$$\gamma_{j_1 j_2} \cdots \gamma_{j_n j_{n+1}} < 1. \tag{C.55}$$

This exercise shows that the cyclic small-gain condition from Theorem C.48 (iii) can be understood as a nonlinear counterpart of the condition $r(A) < 1$ for $A = \Gamma_\otimes$.

Exercise C.17 Show Proposition C.61.

References

1. Berman A, Plemmons RJ (1994) Nonnegative matrices in the mathematical sciences. Society for Industrial and Applied Mathematics
2. Butkovič P (2010) Max-linear systems: theory and algorithms. Springer Science & Business Media
3. Dashkovskiy S, Rüffer B, Wirth F (2010) Small gain theorems for large scale systems and construction of ISS Lyapunov functions. SIAM J Control Optim 48(6):4089–4118
4. Dashkovskiy S, Rüffer BS, Wirth FR (2007) An ISS small gain theorem for general networks. Math Control Signals Syst 19(2):93–122
5. Fabian M, Habala P, Hájek P, Montesinos V, Zizler V (2011) Banach space theory: the basis for linear and nonlinear analysis. Springer Science & Business Media
6. Glück J, Mironchenko A (2021) Stability criteria for positive linear discrete-time systems. Positivity 25(5):2029–2059

7. Karafyllis I, Jiang Z-P (2011) Stability and stabilization of nonlinear systems. Springer, London
8. Karafyllis I, Jiang Z-P (2011) A vector small-gain theorem for general non-linear control systems. IMA J Math Control Inf 28:309–344
9. Knaster B, Kuratowski C, Mazurkiewicz S (1929) Ein Beweis des Fixpunktsatzes für n-dimensionale Simplexe. Fundam Math 14(1):132–137
10. Krasa S, Yannelis NC (1994) An elementary proof of the Knaster–Kuratowski–Mazurkiewicz–Shapley theorem. Econ Theory 4(3):467–471
11. Lur Y-Y (2005) On the asymptotic stability of nonnegative matrices in max algebra. Linear Algebra Appl 407:149–161
12. Mironchenko A, Kawan C, Glück J (2021) Nonlinear small-gain theorems for input-to-state stability of infinite interconnections. Math Control Signals Syst 33:573–615
13. Mironchenko A, Noroozi N, Kawan C, Zamani M (2021) ISS small-gain criteria for infinite networks with linear gain functions. Syst Control Lett 157:105051
14. Rüffer B (2007) Monotone dynamical systems, graphs, and stability of large-scale interconnected systems. PhD thesis, Fachbereich 3 (Mathematik & Informatik) der Universität Bremen
15. Rüffer BS (2010) Monotone inequalities, dynamical systems, and paths in the positive orthant of Euclidean n-space. Positivity 14(2):257–283
16. Rüffer BS (2017) Nonlinear left and right eigenvectors for max-preserving maps. In: Positive systems. Lecture notes in control and information sciences, vol 471. Springer, Cham, pp 227–237
17. Rüffer BS, Sailer R (2014) Input-to-state stability for discrete-time monotone systems. In: Proceedings of 21st International Symposium on Mathematical Theory of Networks and Systems, pp 96–102
18. Stern RJ (1982) A note on positively invariant cones. Appl Math Optim 9:67–72
19. Teschl G (2012) Ordinary differential equations and dynamical systems. Am Math Soc

Index

A. Mironchenko, *Input-to-State Stability*, Communications and Control Engineering,
https://doi.org/10.1007/978-3-031-14674-9